Lecture Notes in Computer Science 5252

Commenced Publication in 1973
Founding and Former Series Editors:
Gerhard Goos, Juris Hartmanis, and Jan van Leeuwen

Jürgen Branke   Kalyanmoy Deb
Kaisa Miettinen   Roman Słowiński (Eds.)

Multiobjective Optimization

Interactive and Evolutionary Approaches

 Springer

Volume Editors

Jürgen Branke
University of Karlsruhe, Institute AIFB
76128 Karlsruhe, Germany
E-mail: branke@aifb.uni-karlsruhe.de

Kalyanmoy Deb
Indian Institute of Technology Kanpur, Department of Mechanical Engineering
Kanpur 208016, India
E-mail: deb@iitk.ac.in
and
Helsinki School of Economics, Department of Business Technology
P.O. Box 1210, 00101 Helsinki, Finland
E-mail: kalyanmoy.deb@hse.fi

Kaisa Miettinen
University of Jyväskylä, Department of Mathematical Information Technology
P.O. Box 35 (Agora), 40014 University of Jyväskylä, Finland
E-mail: kaisa.miettinen@jyu.fi

Roman Słowiński
Poznan University of Technology, Institute of Computing Science
60-965 Poznan, Poland
E-mail: roman.slowinski@cs.put.poznan.pl
and
Systems Research Institute, Polish Academy of Sciences
00-441 Warsaw, Poland

Library of Congress Control Number: 2008937576

CR Subject Classification (1998): F.1, F.2, G.1.6, G.2.1, G.1

LNCS Sublibrary: SL 1 – Theoretical Computer Science and General Issues

ISSN   0302-9743
ISBN   978-3-540-88907-6 Springer Berlin Heidelberg New York

Springer is a part of Springer Science+Business Media

springer.com

© Springer-Verlag Berlin Heidelberg 2008

Typesetting: Camera-ready by author, data conversion by Markus Richter, Heidelberg
Printed on acid-free paper   SPIN: 12542253   06/3180   5 4 3 2 1 0

Preface

Optimization is the task of finding one or more solutions which correspond to minimizing (or maximizing) one or more specified objectives and which satisfy all constraints (if any). A single-objective optimization problem involves a single objective function and usually results in a single solution, called an optimal solution. On the other hand, a multiobjective optimization task considers several conflicting objectives simultaneously. In such a case, there is usually no single optimal solution, but a set of alternatives with different trade-offs, called Pareto optimal solutions, or non-dominated solutions. Despite the existence of multiple Pareto optimal solutions, in practice, usually only one of these solutions is to be chosen. Thus, compared to single-objective optimization problems, in multiobjective optimization, there are at least two equally important tasks: an optimization task for finding Pareto optimal solutions (involving a computer-based procedure) and a decision-making task for choosing a single most preferred solution. The latter typically necessitates preference information from a decision maker (DM).

1 Modelling an Optimization Problem

Before any optimization can be done, the problem must first be modelled. As a matter of fact, to build an appropriate mathematical or computational model for an optimization problem is as important or as critical as the optimization task itself. Typically, most books devoted to optimization methods tacitly assume that the problem has been correctly specified. However, in practice, this is not necessarily always the case. Quantifying and discussing the modelling aspects largely depend on the actual context of the underlying problem and, thus, we do not consider modelling aspects in this book. However, we wish to highlight the following points.

First, building a suitable model (that is, the formulation of the optimization problem with specifying decision variables, objectives, constraints, and variable bounds) is an important task. Second, an optimization algorithm (single or multiobjective, alike) finds the optima of the model of the optimization problem specified and not of the true optimization problem. Due to these reasons, the optimal solutions found by an optimization algorithm must always be analyzed (through a post-optimality analysis) for their 'appropriateness' in the context of the problem. This aspect makes the optimization task iterative in the sense that if some discrepancies in the optimal solutions obtained are found in the post-optimality analysis, the optimization model may have to be modified and the optimization task must be performed again. For example, if the DM during the solution process of a multiobjective optimization problem learns that the interdependencies between the objectives do not correspond to his/her experience and understanding, one must get back to the modelling phase.

2 Why Use Multiple Objectives?

It is a common misconception in practice that most design or problem solving activities must be geared toward optimizing a single objective, for example, bringing maximum profit or causing the smallest cost, even though there may exist different conflicting goals for the optimization task. As a result, the different goals are often redefined to provide an equivalent cost or a profit value, thereby artificially reducing the number of apparently conflicting goals into a single objective. However, the correlation between objectives is usually rather complex and dependent on the alternatives available. Moreover, the different objectives are typically non-commensurable, so it is difficult to aggregate them into one synthetic objective. Let us consider the simple example of choosing a hotel for a night. If the alternatives are a one-star hotel for 70 euros, or a zero-star hotel for 20 euros, the user might prefer the one-star hotel. On the other hand, if the choice is between a five-star hotel for 300 euros, and a four-star hotel for 250 euros, the four-star hotel may be sufficient. That is, stars cannot be simply weighted with money. How much an extra star is valued depends on the alternatives. As a consequence, it may be very difficult to combine different objectives into a single goal function a priori, that is, before alternatives are known. It may be comparatively easier to choose among a given set of alternatives if appropriate decision support is available for the DM. Similarly, one cannot simply specify constraints on the objectives before alternatives are known, as the resulting feasible region may become empty, making the optimization problem impossible to solve.

It should be clear that multiobjective optimization consists of three phases: model building, optimization, and decision making (preference articulation). Converting a multiobjective optimization problem into a simplistic single-objective problem puts decision making before optimization, that is, before alternatives are known. As explained above, articulating preferences without a good knowledge of alternatives is difficult, and thus the resulting optimum may not correspond to the solution the user would have selected from the set of Pareto optimal solutions. Treating the problem as a true multiobjective problem means putting the preference articulation stage after optimization, or interlacing optimization and preference articulation. This will help the user gain a much better understanding of the problem and the available alternatives, thus leading to a more conscious and better choice. Furthermore, the resulting multiple Pareto optimal solutions can be analyzed to learn about interdependencies among decision variables, objectives, and constraints. Such knowledge about the interactions can be used to redefine the model of the optimization problem to get solutions that, on the one hand, correspond better to reality, and, on the other hand, satisfy better the DM's preferences.

3 Multiple Criteria Decision Making

The research field of considering decision problems with multiple conflicting objectives (or goals or criteria) is known as multiple criteria decision making (MCDM) or multiple criteria decision aiding (MCDA). It covers both discrete problems (with a finite set of alternatives, also called actions or solutions) and continuous problems (multiobjective optimization). Traditionally, in multiobjective optimization (also known as multicriteria optimization), mathematical programming techniques and decision making have been used in an intertwined manner, and the ultimate aim of solving a multiobjective optimization problem has been characterized as supporting the DM in finding the solution that best fits the DM's preferences. The alternating stages of decision making and optimization create typically an interactive procedure for finding the most preferred solution. The DM participates actively in this procedure, particularly in the decision-making stage. Decision making on alternatives discovered by optimization requires a more or less explicit model of DM's preferences, so as to find the most preferred solution among the alternatives currently considered, or to give indications for finding better solutions in the next optimization stage. Many interactive methods have been proposed to date, differing mainly in the way the DM is involved in the process, and in the type of preference model built on preference information elicited from the DM.

The origin of nonlinear multiobjective optimization goes back almost 60 years, when Kuhn and Tucker formulated optimality conditions. However, for example, the concept of Pareto optimality has a much earlier origin. More information about the history of the field can be found in Chap. 1 of this book. It is worth mentioning that biannual conferences on MCDM have been regularly organized since 1975 (first by active researchers in the field, then by a Special Interest Group formed by them and later by the International Society on Multiple Criteria Decision Making). In addition, in Europe a Working Group on Multiple Criteria Decision Aiding was established in 1975 within EURO (European Association of Operational Research Societies) and holds two meetings per year (it is presently in its 67th meeting). Furthermore, International Summer Schools on Multicriteria Decision Aid have been arranged since 1983. A significant number of monographs, journal articles, conference proceedings, and collections have been published during the years and the field is still active.

4 Evolutionary Multiobjective Optimization

In the 1960s, several researchers independently suggested adopting the principles of natural evolution, in particular Darwin's theory of the survival of the fittest, for optimization. These pioneers were Lawrence Fogel, John H. Holland, Ingo Rechenberg, and Hans-Paul Schwefel. One distinguishing feature of these so-called evolutionary algorithms (EAs) is that they work with a population of solutions. This is of particular advantage in the case of multiobjective optimization, as they can search for several Pareto optimal solutions simultaneously in one run, providing the DM with a set of alternatives to choose from.

Despite some early suggestions and studies, major research and application activities of EAs in multiobjective optimization, spurred by a unique suggestion by David E. Goldberg of a combined EA involving domination and niching, started only in the beginning of 1990s. But in the last 15 years, the field of evolutionary multiobjective optimization (EMO) has developed rapidly, with a regular, dedicated, biannual conference, commercial software, and more than 10 books on the topic. Although earlier studies focused on finding a representative set of solutions on the entire Pareto optimal set, EMO methodologies are also good candidates for finding only a part of the Pareto optimal set.

5 Genesis of This Book

Soon after initiating EMO activities, the leading researchers recognized the existence of the MCDM field and commonality in interests between the two fields. They realized the importance of exchanging ideas and engaging in collaborative studies. Since their first international conference in 2001 in Zurich, EMO conference organizers have always invited leading MCDM researchers to deliver keynote and invited lectures. The need for cross-fertilization was also realized by the MCDM community and they reciprocated. However, as each field tried to understand the other, the need for real collaborations became clear.

In the 2003 visit of Kalyanmoy Deb to the University of Karlsruhe to work on EMO topics with Jürgen Branke and Hartmut Schmeck, they came up with the idea of arranging a Dagstuhl seminar on multiobjective optimization along with two MCDM leading researchers, Kaisa Miettinen and Ralph E. Steuer.

The Dagstuhl seminar organized in November 2004 provided an ideal platform for bringing in the best minds from the two fields and exchanging the philosophies of each other's methodologies in solving multiobjective optimization problems. It became obvious that the fields did not yet know each other's approaches well enough. For example, some EMO researchers had developed ideas that have existed in the MCDM field for long and, on the other hand, the MCDM field welcomed the applicability of EMO approaches to problems where mathematical programming has difficulties.

The success of a multiobjective optimization application relies on the way the DM is allowed to interact with the optimization procedure. At the end of the 2004 Dagstuhl seminar, a general consensus clearly emerged that there is plenty of potential in combining ideas and approaches of MCDM and EMO fields and preparing hybrids of them. Examples of ideas that emerged were that more attention in the EMO field should be devoted to incorporating preference information into the methods and that EMO procedures can be used to parallelize the repetitive tasks often performed in an MCDM task. By sensing the opportunity of a collaborative effort, a second Dagstuhl seminar was organized in December 2006 and Roman Słowiński, who strongly advocated for inclusion of preference modelling into EMO procedures, was invited to the organizing team. The seminar brought together about 50 researchers from EMO and MCDM fields interested in bringing EMO and MCDM approaches closer to each other. We, the organizers, had a clear idea in mind. The presence of experts from both fields should be exploited so that the outcome could be written up in a single book for the benefit of both novices and experts from both fields.

6 Topics Covered

Before we discuss the topics covered in this book, we mention a few aspects of the MCDM field which we do not discuss here. Because of the large amount of research and publications produced in the MCDM field during the years, we have limited our review. We have mostly restricted our discussion to problems involving continuous problems, although some chapters include some extensions to discrete problems, as well. However, one has to mention that because the multiattribute or multiple criteria decision analysis methods have been developed for problems involving a discrete set of solution alternatives, they can directly be used for analyzing the final population of an EMO algorithm. In this way, there is a clear link between the two fields. Another topic not covered here is group decision making. This refers to situations where we have several DMs with different preferences. Instead, we assume that we have a single DM or a unanimous group of DMs involved.

We have divided the contents of this book into five parts. The first part is devoted to the basics of multiobjective optimization and introduces in three chapters the main methods and ideas developed in the field of nonlinear multiobjective optimization on the MCDM side (including both noninteractive and interactive approaches) and on the EMO side. This part lays a foundation for the rest of the book and should also allow newcomers to the field to get familiar with the topic. The second part introduces in four chapters recent developments in considering preference information or creating interactive methods. Approaches with both MCDM and EMO origin as well as their hybrids are included. The third part concentrates with Chap. 8 and 9 on visualization, both for individual solution candidates and the whole sets of Pareto optimal solutions. In Chap. 10-13 (Part Four), implementation issues including meta-modelling, parallel approaches, and software are of interest. In addition, various real-world applications are described in order to give some idea of the wide spectrum of disciplines and problems that can benefit from multiobjective optimization. Finally, in the last three chapters forming Part Five, some relevant topics including approximation quality in the EMO approaches and learning perspectives in decision making are studied. The last chapter points to some future challenges and encourages further research in the field. All 16 chapters matured during the 2006 Dagstuhl seminar. In particular, the last six chapters are outcomes of active working groups formed during the seminar.

7 Main Terminology and Notations Used

In order to avoid repeating basic concepts and problem formulations in each chapter, we present them here. We handle *multiobjective optimization problems* of the form

$$
\begin{aligned}
\text{minimize} \quad & \{f_1(\mathbf{x}), f_2(\mathbf{x}), \dots, f_k(\mathbf{x})\} \\
\text{subject to} \quad & \mathbf{x} \in S
\end{aligned}
\tag{1}
$$

involving k (≥ 2) conflicting *objective functions* $f_i : \mathbf{R}^n \to \mathbf{R}$ that we want to minimize simultaneously. The *decision (variable) vectors* $\mathbf{x} = (x_1, x_2, \dots, x_n)^T$ belong to the nonempty *feasible region* $S \subset \mathbf{R}^n$. In this general problem formulation we do not fix the types of constraints forming the feasible region. *Objective vectors* are images of decision vectors and consist of *objective (function) values* $\mathbf{z} = \mathbf{f}(\mathbf{x}) = (f_1(\mathbf{x}), f_2(\mathbf{x}), \dots, f_k(\mathbf{x}))^T$. Furthermore, the image of the feasible region in the objective space is called a *feasible objective region* $Z = \mathbf{f}(S)$.

In multiobjective optimization, objective vectors are regarded as optimal if none of their components can be improved without deterioration to at least one of the other components. More precisely, a decision vector $\mathbf{x}' \in S$ is called *Pareto optimal* if there does not exist another $\mathbf{x} \in S$ such that $f_i(\mathbf{x}) \leq f_i(\mathbf{x}')$ for all $i = 1, \dots, k$ and $f_j(\mathbf{x}) < f_j(\mathbf{x}')$ for at least one index j. The set of Pareto optimal decision vectors can be denoted by $P(S)$. Correspondingly, an objective vector is Pareto optimal if the corresponding decision vector is Pareto optimal and the set of Pareto optimal objective vectors can be denoted by $P(Z)$. The set of Pareto optimal solutions is a subset of the set of weakly Pareto optimal solutions. A decision vector $\mathbf{x}' \in S$ is *weakly Pareto optimal* if there does not exist another $\mathbf{x} \in S$ such that $f_i(\mathbf{x}) < f_i(\mathbf{x}')$ for all $i = 1, \ldots, k$. As above, here we can also denote two sets corresponding to decision and objective spaces by $WP(S)$ and $WP(Z)$, respectively.

The ranges of the Pareto optimal solutions in the feasible objective region provide valuable information about the problem considered if the objective functions are bounded over the feasible region. Lower bounds of the Pareto optimal set are available in the *ideal objective vector* $\mathbf{z}^\star \in \mathbf{R}^k$. Its components $z_i^\star$ are obtained by minimizing each of the objective functions individually subject to the feasible region. A vector strictly better than $\mathbf{z}^\star$ can be called a *utopian objective vector* $\mathbf{z}^{\star\star}$. In practice, we set $z_i^{\star\star} = z_i^\star - \varepsilon$ for $i = 1, \ldots, k$, where ε is some small positive scalar.

The upper bounds of the Pareto optimal set, that is, the components of a *nadir objective vector* $\mathbf{z}^{\mathrm{nad}}$, are usually difficult to obtain. Unfortunately, there exists no constructive way to obtain the exact nadir objective vector for nonlinear problems. It can be estimated using a payoff table but the estimate may be unreliable.

Because vectors cannot be ordered completely, all the Pareto optimal solutions can be regarded as equally desirable in the mathematical sense and we need a *decision maker* (DM) to identify the most preferred one among them. The DM is a person who can express preference information related to the conflicting objectives and we assume that less is preferred to more in each objective for her/him.

Besides a DM, we usually also need a so-called *analyst* to take part in the solution process. By an analyst we mean a person or a computer program responsible for the mathematical side of the solution process. The analyst may be, for example, responsible for selecting the appropriate method for optimization.

Acknowledgments

We would like to take this opportunity to thank all the participants of the 2004 and 2006 Dagstuhl seminars for their dedication and effort, without which this book would not have been possible. Andrzej Wierzbicki's suggestion to include a discussion on the importance of modelling issues in optimization in the preface is appreciated. We thank Springer for supporting our idea of this book. K. Deb and K. Miettinen acknowledge the support from the Academy of Finland (grant no. 118319) and the Foundation of the Helsinki School of Economics for completing this task.

The topics covered in this book are wide ranging; from presenting the basics of multiobjective optimization to advanced topics of incorporating diverse interactive features in multiobjective optimization and from practical real-world applications to software and visualization issues as well as various perspectives highlighting relevant research issues. With these contents, hopefully, the book remains useful to both beginners and current researchers including experts. Besides the coverage of the topics, this book will also remain a milestone achievement in the field of multiobjective optimization for another reason. This book is the first concrete approach in bringing two parallel fields of multiobjective optimization together. The 16 chapters of this book are contributed by 19 EMO and 22 MCDM researchers. Of the 16 chapters, six are written by a mix of EMO and MCDM researchers and all 16 chapters have been reviewed by at least one EMO and one MCDM researcher. We shall consider our efforts worthwhile if more such collaborative tasks are pursued in the coming years to develop hybrid ideas by sharing the strengths of different approaches.

June 2008

Jürgen Branke,
Kalyanmoy Deb,
Kaisa Miettinen,
Roman Słowiński

Most participants of the 2006 Dagstuhl seminar on "Practical Approaches to Multi-objective Optimization": 1 Eckart Zitzler, 2 Kalyanmoy Deb, 3 Kaisa Miettinen, 4 Joshua Knowles, 5 Carlos Fonseca, 6 Salvatore Greco, 7 Oliver Bandte, 8 Christian Igel, 9 Nirupam Chakraborti, 10 Silvia Poles, 11 Valerie Belton, 12 Jyrki Walle-nius, 13 Roman Słowiński, 14 Serpil Sayin, 15 Pekka Korhonen, 16 Lothar Thiele, 17 Włodzimierz Ogryczak, 18 Andrzej Osyczka, 19 Koji Shimoyama, 20 Daisuke Sasaki, 21 Johannes Jahn, 22 Günter Rudolph, 23 Jörg Fliege, 24 Matthias Ehrgott, 25 Petri Eskelinen, 26 Jerzy Błaszczyński, 27 Sanaz Mostaghim, 28 Pablo Funes, 29 Carlos Coello Coello, 30 Theodor Stewart, 31 José Figueira, 32 El-Ghazali Talbi, 33 Julian Molina, 34 Andrzej Wierzbicki, 35 Yaochu Jin, 36 Andrzej Jaszkiewicz, 37 Jürgen Branke, 38 Fransisco Ruiz, 39 Hirotaka Nakayama, 40 Tatsuya Okabe, 41 Alexander Lotov, 42 Hisao Ishibuchi

List of Contributors

Oliver Bandte
Icosystem Corporation, Cambridge,
MA 02138
oliver@icosystem.com

Valerie Belton
Department of Management Science,
University of
Strathclyde, 40 George Street,
Glasgow, UK, G1 1QE
val.belton@strath.ac.uk

Jürgen Branke
Institute AIFB
University of Karlsruhe
76128 Karlsruhe, Germany
branke@aifb.uni-karlsruhe.de

Heinrich Braun
SAP AG, Walldorf, Germany
heinrich.braun@sap.com

Nirupam Chakraborti
Indian Institute of Technology,
Kharagpur 721 302, India
nchakrab@iitkgp.ac.in

Carlos A. Coello Coello
CINVESTAV-IPN (Evolutionary
Computation Group), Depto. de
Computación
Av. IPN No 2508, Col. San Pedro
Zacatenco, D.F., 07360, Mexico
ccoello@cs.cinvestav.mx

Kalyanmoy Deb
Department of Mechanical Engineer-
ing, Indian Institute of Technology
Kanpur
Kanpur, PIN 208016, India
deb@iitk.ac.in
Department of Business Technology,
Helsinki School of Economics
PO Box 1210, 00101 Helsinki,
Finland
Kalyanmoy.Deb@hse.fi

Matthias Ehrgott
Department of Engineering Science,
The University of Auckland, Private
Bag 92019, Auckland 1142, New
Zealand
m.ehrgott@auckland.ac.nz

Petri Eskelinen
Helsinki School of Economics
P.O. Box 1210,
FI-00101 Helsinki, Finland
Petri.Eskelinen@hse.fi

José Rui Figueira
CEG-IST, Center for Management
Studies, Instituto
Superior Técnico, Technical Univer-
sity of Lisbon, Portugal
figueira@ist.utl.pt

Mathias Göbelt
SAP AG, Walldorf, Germany
mathias.goebelt@sap.com

Salvatore Greco
Faculty of Economics, University of
Catania, Corso Italia, 55,
95129 Catania, Italy
salgreco@unict.it

Hisao Ishibuchi
Department of Computer Science
and Intelligent Systems, Osaka
Prefecture University
Osaka 599-8531, Japan
hisaoi@cs.osakafu-u.ac.jp

Johannes Jahn
Department of Mathematics,
University of Erlangen-Nürnberg
Martensstrasse 3, 91058 Erlangen,
Germany
jahn@am.uni-erlangen.de

Andrzej Jaszkiewicz
Poznan University of Technology,
Institute of Computing Science
jaszkiewicz@cs.put.poznan.pl

Yaochu Jin
Honda Research Institute Europe,
63073 Offenbach, Germany
Yaochu.Jin@honda-ri.de

Joshua Knowles
School of Computer Science, University of Manchester, Oxford Road,
Manchester M13 9PL, UK
j.knowles@manchester.ac.uk

Pekka Korhonen
Helsinki School of Economics,
Department of Business
Technology, P.O. Box 1210, FI-00101
Helsinki, Finland

Alexander V. Lotov
Dorodnicyn Computing Centre of
Russian Academy of Sciences
Vavilova str. 40, Moscow 119333,
Russia
lotov08@ccas.ru

Benedetto Matarazzo
Faculty of Economics, University of
Catania, Corso Italia, 55,
95129 Catania, Italy
matarazz@unict.it

Kaisa Miettinen
Department of Mathematical
Information Technology,
P.O. Box 35 (Agora),
FI-40014 University of Jyväskylä,
Finland[1]
kaisa.miettinen@jyu.fi

Julián Molina
Department of Applied Economics
(Mathematics), University of
Málaga, Calle Ejido 6, E-29071
Málaga, Spain,
julian.molina@uma.es

Sanaz Mostaghim
Institute AIFB, University of
Karlsruhe
76128 Karlsruhe, Germany
mostaghim@aifb.uni-karlsruhe.de

Vincent Mousseau
LAMSADE, Université Paris-
Dauphine, Paris, France
mousseau@lamsade.dauphine.fr

---

[1] In 2007 also Helsinki
School of Economics, Helsinki, Finland

Hirotaka Nakayama
Konan University, Dept. of Information Science and Systems Engineering, 8-9-1 Okamoto, Higashinada, Kobe 658-8501, Japan
nakayama@konan-u.ac.jp

Wlodzimierz Ogryczak
Institute of Control & Computation Engineering, Faculty of Electronics & Information Technology, Warsaw University of Technology
ul. Nowowiejska 15/19, 00-665 Warsaw, Poland
w.ogryczak@ia.pw.edu.pl

Tatsuya Okabe
Honda Research Institute Japan Co., Ltd.
8-1 Honcho, Wako-City, Saitama, 351-0188, Japan
okabe@jp.honda-ri.com

Silvia Poles
ESTECO - Research Labs
Via Giambellino, 7 35129 Padova, Italy
silvia.poles@esteco.com

Günter Rudolph
Computational Intelligence Research Group, Chair of Algorithm Engineering (LS XI), Department of Computer Science, University of Dortmund
44227 Dortmund, Germany
Guenter.Rudolph@uni-dortmund.de

Francisco Ruiz
Department of Applied Economics (Mathematics), University of Málaga
Calle Ejido 6, E-29071 Málaga, Spain
rua@uma.es

Daisuke Sasaki
CFD Laboratory, Department of Engineering, University of Cambridge
Trumpington Street, Cambridge CB2 1PZ, UK
ds432@eng.cam.ac.uk

Koji Shimoyama
Institute of Fluid Science, Tohoku University
2-1-1 Katahira, Aoba-ku, Sendai, 980-8577, Japan
shimoyama@edge.ifs.tohoku.ac.jp

Roman Słowiński
Institute of Computing Science, Poznań University of Technology, 60-965 Poznań, and Systems Research Institute, Polish Academy of Sciences
00-441 Warsaw, Poland,
roman.slowinski@cs.put.poznan.pl

Danilo Di Stefano
Esteco Research Labs, 35129 Padova, Italy
danilo.distefano@esteco.com

Theodor Stewart
University of Cape Town, Rondebosch 7701, South Africa
theodor.stewart@uct.ac.za

El-Ghazali Talbi
Laboratoire d'Informatique Fondamentale de Lille Université des Sciences et Technologies de Lille
59655 - Villeneuve d'Ascq cedex, France
talbi@lifl.fr

Lothar Thiele
Computer Engineering and Networks Laboratory (TIK)
Department of Electrical Engineering and Information Technology
ETH Zurich, Switzerland
thiele@tik.ee.ethz.ch

Mariana Vassileva
Institute of Information Technologies, Bulgarian Academy of Sciences, Bulgaria
mvassileva@iinf.bas.bg

Rudolf Vetschera
Department of Business Administration, University of Vienna
Brünnerstrasse 72, 1210 Wien, Austria
rudolf.vetschera@univie.ac.at

Jyrki Wallenius
Helsinki School of Economics, Department of Business Technology, P.O. Box 1210, FI-00101 Helsinki, Finland
jyrki.wallenius@hse.fi

Andrzej P. Wierzbicki
21st Century COE
Program: Technology Creation Based on Knowledge Science, JAIST (Japan Advanced Institute of Science and Technology), Asahidai 1-1, Nomi, Ishikawa 923-1292, Japan and National Institute of Telecommunications, Szachowa Str. 1, 04-894 Warsaw, Poland,
andrzej@jaist.ac.jp

Eckart Zitzler
Computer Engineering and Networks Laboratory (TIK)
Department of Electrical Engineering and Information Technology
ETH Zurich, Switzerland
eckart.zitzler@tik.ee.ethz.ch

Table of Contents

1

Introduction to Multiobjective Optimization: Noninteractive Approaches

Kaisa Miettinen

Department of Mathematical Information Technology
P.O. Box 35 (Agora), FI-40014 University of Jyväskylä, Finland*
kaisa.miettinen@jyu.fi

Abstract. We give an introduction to nonlinear multiobjective optimization by covering some basic concepts as well as outlines of some methods. Because Pareto optimal solutions cannot be ordered completely, we need extra preference information coming from a decision maker to be able to select the most preferred solution for a problem involving multiple conflicting objectives. Multiobjective optimization methods are often classified according to the role of a decision maker in the solution process. In this chapter, we concentrate on noninteractive methods where the decision maker either is not involved or specifies preference information before or after the actual solution process. In other words, the decision maker is not assumed to devote too much time in the solution process.

1.1 Introduction

Many decision and planning problems involve multiple conflicting objectives that should be considered simultaneously (alternatively, we can talk about multiple conflicting criteria). Such problems are generally known as multiple criteria decision making (MCDM) problems. We can classify MCDM problems in many ways depending on the characteristics of the problem in question. For example, we talk about multiattribute decision analysis if we have a discrete, predefined set of alternatives to be considered. Here we study multiobjective optimization (also known as multiobjective mathematical programming) where the set of feasible solutions is not explicitly known in advance but it is restricted by constraint functions. Because of the aims and scope of this book, we concentrate on nonlinear multiobjective optimization (where at least one function in the problem formulation is nonlinear) and ignore approaches designed only for multiobjective linear programming (MOLP) problems (where all the functions are linear).

---

* In 2007 also Helsinki School of Economics, Helsinki, Finland
  Reviewed by: Nirupam Chakraborti, Indian Institute of Technology, India
  Hirotaka Nakayama, Konan University, Japan
  Roman Słowiński, Poznan University of Technology, Poland

J. Branke et al. (Eds.): Multiobjective Optimization, LNCS 5252, pp. 1–26, 2008.

In multiobjective optimization problems, it is characteristic that no unique solution exists but a set of mathematically equally good solutions can be identified. These solutions are known as nondominated, efficient, noninferior or Pareto optimal solutions (defined in Preface). In the MCDM literature, these terms are usually seen as synonyms. Multiobjective optimization problems have been intensively studied for several decades and the research is based on the theoretical background laid, for example, in (Edgeworth, 1881; Koopmans, 1951; Kuhn and Tucker, 1951; Pareto, 1896, 1906). As a matter of fact, many ideas and approaches have their foundation in the theory of mathematical programming. For example, while formulating optimality conditions of nonlinear programming, Kuhn and Tucker (1951) did also formulate them for multiobjective optimization problems.

Typically, in the MCDM literature, the idea of solving a multiobjective optimization problem is understood as helping a human decision maker (DM) in considering the multiple objectives simultaneously and in finding a Pareto optimal solution that pleases him/her the most. Thus, the solution process needs some involvement of the DM in the form of specifying preference information and the final solution is determined by his/her preferences in one way or the other. In other words, a more or less explicit preference model is built from preference information and this model is exploited in order to find solutions that better fit the DM's preferences. Here we assume that a single DM is involved. Group decision making with several DMs is discussed, e.g., in (Hwang and Lin, 1987; Fandel, 1990).

In general, the DM is a person who is assumed to know the problem considered and be able to provide preference information related to the objectives and/or different solutions in some form. Besides a DM, we usually also need an analyst when solving a multiobjective optimization problems. An *analyst* is a person or a computer program responsible for the mathematical modelling and computing sides of the solution process. The analyst is supposed to help the DM at various stages of the solution process, in particular, in eliciting preference information and in interpreting the information coming from the computations (see also Chapter 15).

We can list several desirable properties of multiobjective optimization methods. Among them are, for example, that the method should generate Pareto optimal solutions reliably, it should help the DM to get an overview of the set of Pareto optimal solutions, it should not require too much time from the DM, the information exchanged (given by the method and asked from the DM) should be understandable and not too demanding or complicated (cognitively or otherwise) and the method should support the DM in finding the most preferred solution as the final one so that the DM could be convinced of its relative goodness. The last-mentioned aim could be characterized as psychological convergence (differing from mathematical convergence which is emphasized in mathematical programming).

Surveys of methods developed for multiobjective optimization problems include (Chankong and Haimes, 1983; Hwang and Masud, 1979; Marler and

Arora, 2004; Miettinen, 1999; Sawaragi *et al.*, 1985; Steuer, 1986; Vincke, 1992). For example, in (Hwang and Masud, 1979; Miettinen, 1999), the methods are classified into the four following classes according to the role of the DM in the solution process. Sometimes, there is no DM and her/his preference information available and in those cases we must use so-called *no-preference methods*. Then, the task is to find some neutral compromise solution without any additional preference information. This means that instead of asking the DM for preference information, some assumptions are made about what a "reasonable" compromise could be like. In all the other classes, the DM is assumed to take part in the solution process.

In *a priori methods*, the DM first articulates preference information and one's aspirations and then the solution process tries to find a Pareto optimal solution satisfying them as well as possible. This is a straightforward approach but the difficulty is that the DM does not necessarily know the possibilities and limitations of the problem beforehand and may have too optimistic or pessimistic expectations. Alternatively, it is possible to use *a posteriori* methods, where a representation of the set of Pareto optimal solutions is first generated and then the DM is supposed to select the most preferred one among them. This approach gives the DM an overview of different solutions available but if there are more than two objectives in the problem, it may be difficult for the DM to analyze the large amount of information (because visualizing the solutions is no longer as straightforward as in a biobjective case) and, on the other hand, generating the set of Pareto optimal solutions may be computationally expensive. Typically, evolutionary multiobjective optimization algorithms (see Chapter 3) belong to this class but, when using them, it may happen that the real Pareto optimal set is not reached. This means that the solutions produced are nondominated in the current population but not necessarily actually Pareto optimal (if, e.g., the search is stopped too early).

In this chapter, we concentrate on the three classes of noninteractive methods where either no DM takes part in the solution process or (s)he expresses preference relations before or after the process. The fourth class devoted to *interactive methods* is the most extensive class of methods and it will be covered in Chapter 2. In interactive approaches, an iterative solution algorithm (which can be called a solution pattern) is formed and repeated (typically several times). After each iteration, some information is given to the DM and (s)he is asked to specify preference information (in the form that the method in question can utilize, e.g., by answering some questions). One can say that the analyst aims at determining the preference structure of the DM in an interactive way. What is noteworthy is that the DM can specify and adjust one's preferences between each iteration and at the same time learn about the interdependencies in the problem as well as about one's own preferences.

Methods in different classes have their strengths and weaknesses and for that reason different approaches are needed. Let us point out that the classification we use here is not complete or absolute. Overlapping and combinations of classes are possible and some methods can belong to more than one class depending on different interpretations. Other classifications are given, for example, by Cohon (1985); Rosenthal (1985).

The rest of this chapter is organized as follows. In Section 1.2, we augment the basic terminology and notation introduced in Preface. In other words, we discuss some more concepts of multiobjective optimization including optimality and elements of a solution process. After that we introduce two widely used basic methods, the weighting method and the ε-constraint method in Section 1.3. Sections 1.4–1.6 are devoted to some methods belonging to the three above-described classes, that is, no-preference methods, a posteriori methods and a priori methods, respectively. We also give references to further details. In Section 1.7, we summarize some properties of the methods described and, finally, we conclude with Section 1.8.

1.2 Some Concepts

1.2.1 Optimality

Continuous multiobjective optimization problems typically have an infinite number of Pareto optimal solutions (whereas combinatorial multiobjective optimization problems have a finite but possibly very large number of Pareto optimal solutions) and the Pareto optimal set (consisting of the Pareto optimal solutions) can be nonconvex and disconnected. Because the basic terminology and concepts of multiobjective optimization were defined in Preface, we do not repeat them here. However, it is important to note that the definitions of Pareto optimality and weak Pareto optimality (given in Preface) introduce *global Pareto optimality* and *global weak Pareto optimality*. Corresponding to nonlinear programming, we can also define local (weak) Pareto optimality in a small environment of the point considered. Let us emphasize that a locally Pareto optimal objective vector has no practical relevance (if it is not global) because it may be located in the interior of the feasible objective region (i.e., it is possible to improve all objective function values) whereas globally Pareto optimal solutions are always located on its boundary. Thus, it is important to use appropriate tools to get globally Pareto optimal solutions. We shall get back to this when we discuss scalarizing functions.

Naturally, any globally Pareto optimal solution is locally Pareto optimal. The converse is valid for convex problems, see, for example, (Miettinen, 1999). A multiobjective optimization problem can be defined to be convex if the feasible objective region is convex or if the feasible region is convex and the objective functions are quasiconvex with at least one strictly quasiconvex function.

Before we continue, it is important to briefly touch the existence of Pareto optimal solutions. It is shown in (Sawaragi *et al.*, 1985) that Pareto optimal solutions exist if we assume that the (nonempty) feasible region is compact and all the objective functions are lower semicontinuous. Alternatively, we can formulate the assumption in the form that the feasible objective region is nonempty and compact. We do not go into details of theoretical foundations here but assume in what follows that Pareto optimal solutions exist. Another important question besides the existence of Pareto optimal solutions is the stability of the Pareto optimal set with respect to perturbations of the feasible region, objective functions or domination structures of the DM. This topic is extensively discussed in (Sawaragi et al., 1985) and it is also touched in Chapter 9. Let us mention that sometimes, like by Steuer (1986), Pareto optimal decision vectors are referred to as efficient solutions and the term nondominated solution is used for Pareto optimal objective vectors.

If the problem is correctly specified, the final solution of a rational DM is always Pareto optimal. Thus, we can restrict our consideration to Pareto optimal solutions. For that reason, it is important that the multiobjective optimization method used can meet the following two needs: firstly, is must be able to cover, that is, find any Pareto optimal solution and, secondly, generate only Pareto optimal solutions (Sawaragi et al., 1985). However, weakly Pareto optimal solutions are often relevant from a technical point of view because they are sometimes easier to generate than Pareto optimal ones.

One more widely used optimality concepts is proper Pareto optimality. The properly Pareto optimal set is a subset of the Pareto optimal set which is a subset of the weakly Pareto optimal set. For an example of these three concepts of optimality and their relationships, see Figure 1.1. In the figure, the set of weakly Pareto optimal solutions is denoted by a bold line. The endpoints of the Pareto optimal set are denoted by circles and the endpoints of the properly Pareto optimal set by short lines (note that the sets can also be disconnected).

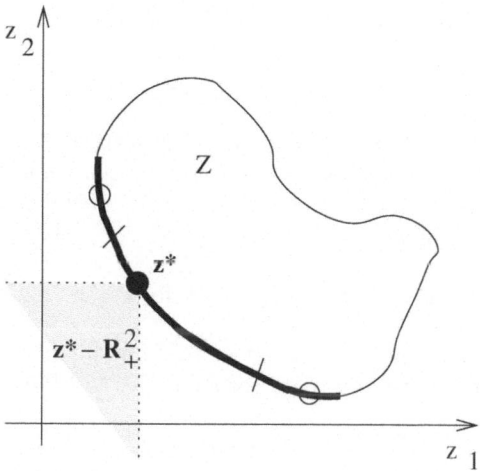

Fig. 1.1. Sets of properly, weakly and Pareto optimal solutions.

As a matter of fact, Pareto optimal solutions can be divided into improperly and properly Pareto optimal ones depending on whether unbounded trade-offs between objectives are allowed or not. Practically, a properly Pareto optimal solution with a very high trade-off does not essentially differ from a weakly Pareto optimal solution for a human DM. There are several definitions for proper Pareto optimality and they are not equivalent. The first definition was given by Kuhn and Tucker (1951) while they formulated optimality conditions for multiobjective optimization. Some of the definitions are collected, for example, in (Miettinen, 1999) and relationships between different definitions are analyzed in (Sawaragi *et al.*, 1985; Makarov and Rachkovski, 1999).

The idea of proper Pareto optimality is easily understandable in the definition of Geoffrion (1968): A decision vector $\mathbf{x}' \in S$ is *properly Pareto optimal (in the sense of Geoffrion)* if it is Pareto optimal and if there is some real number M such that for each f_i and each $\mathbf{x} \in S$ satisfying $f_i(\mathbf{x}) < f_i(\mathbf{x}')$ there exists at least one f_j such that $f_j(\mathbf{x}') < f_j(\mathbf{x})$ and

$$\frac{f_i(\mathbf{x}') - f_i(\mathbf{x})}{f_j(\mathbf{x}) - f_j(\mathbf{x}')} \leq M.$$

An objective vector is properly Pareto optimal if the corresponding decision vector is properly Pareto optimal. We can see from the definition that a solution is properly Pareto optimal if there is at least one pair of objectives for which a finite decrement in one objective is possible only at the expense of some reasonable increment in the other objective.

Let us point out that optimality can be defined in more general ways (than above) with the help of ordering cones (pointed convex cones) D defined in $\mathbf{R}^k$. The cone D can be used to induce a partial ordering in Z. In other words, for two objective vectors $\mathbf{z}$ and $\mathbf{z}'$ we can say that $\mathbf{z}'$ dominates $\mathbf{z}$ if

$$\mathbf{z} \in \mathbf{z}' + D \setminus \{\mathbf{0}\}.$$

Now we can say that a feasible decision vector is efficient and the corresponding objective vector is nondominated with respect to D if there exists no other feasible objective vector that dominates it. This definition is equivalent to Pareto optimality if we set

$$D = \mathbf{R}_+^k = \{\mathbf{z} \in \mathbf{R}^k \mid z_i \geq 0 \text{ for } i = 1, \ldots, k\},$$

that is, D is the nonnegative orthant of $\mathbf{R}^k$. For further details of ordering cones and different spaces we refer, for example, to (Jahn, 2004; Luc, 1989) and references therein.

As said, we can give an equivalent formulation to the definition of Pareto optimality (given in Preface) as follows: A feasible decision vector $\mathbf{x}^* \in S$ and the corresponding objective vector $\mathbf{z}^* = \mathbf{f}(\mathbf{x}^*) \in Z$ are Pareto optimal if

$$\mathbf{z}^* - \mathbf{R}_+^k \setminus \{\mathbf{0}\}) \cap Z = \emptyset.$$

For a visualization of this, see Figure 1.1, where a shifted cone at $\mathbf{z}^*$ is illustrated. This definition clearly shows why Pareto optimal objective vectors must be located on the boundary of the feasible objective region Z. After having introduced the definition of Pareto optimality in this form, we can give another definition for proper Pareto optimality. This definition (introduced by Wierzbicki (1986)) is both computationally usable and intuitive.

The above-defined vectors $\mathbf{x}^* \in S$ and $\mathbf{z}^* \in Z$ are ρ-*properly Pareto optimal* if

$$\mathbf{z}^* - \mathbf{R}_\rho^k \setminus \{\mathbf{0}\}) \cap Z = \emptyset,$$

where $\mathbf{R}_\rho^k$ is a slightly broader cone than $\mathbf{R}_+^k$. Now, trade-offs are bounded by ρ and $1/\rho$ and we have a relationship to M used in Geoffrion's definition as $M = 1 + 1/\rho$. For details, see, for example (Miettinen, 1999; Wierzbicki, 1986).

1.2.2 Solution Process and Some Elements in It

Mathematically, we cannot order Pareto optimal objective vectors because the objective space is only partially ordered. However, it is generally desirable to obtain one point as a final solution to be implemented and this solution should satisfy the preferences of the particular DM. Finding a solution to problem (1) defined in Preface is called a *solution process*. As mentioned earlier, it usually involves co-operation of the DM and an analyst. The analyst is supposed to know the specifics of the methods used and help the DM at various stages of the solution process. It is important to emphasize that the DM is not assumed to know MCDM or methods available but (s)he is supposed to be an expert in the problem domain, that is, understand the application considered. Sometimes, finding the set of Pareto optimal solutions is referred to as *vector optimization*. However, here by solving a multiobjective optimization problem we mean finding a feasible and Pareto optimal decision vector that satisfies the DM. Assuming such a solution exists, it is called a *final solution*.

The concepts of ideal and nadir objective vectors were defined in Preface for getting information about the ranges of the objective function values in the Pareto optimal set; provided the objective functions are bounded over the feasible region. As mentioned then, there is no constructive method for calculating the nadir objective vector for nonlinear problems. A *payoff table* (suggested by Benayoun et al. (1971)) is often used but it is not a reliable way as demonstrated, for example, by Korhonen et al. (1997); Weistroffer (1985). The payoff table has k objective vectors as its rows where objective function values are calculated at points optimizing each objective function individually. In other words, components of the ideal objective vector are located on the diagonal of the payoff table. An estimate of the nadir objective vector is obtained by finding the worst objective values in each column. This method gives accurate information only in the case of two objectives. Otherwise, it may be an over- or an underestimation (because of alternative optima, see, e.g., (Miettinen, 1999) for details). Let us mention that the nadir objective vector can also be estimated using evolutionary algorithms (Deb *et al.*, 2006).

Multiobjective optimization problems are usually solved by scalarization. *Scalarization* means that the problem involving multiple objectives is converted into an optimization problem with a single objective function or a family of such problems. Because this new problem has a real-valued objective function (that possibly depends on some parameters coming, e.g., from preference information), it can be solved using appropriate single objective optimizers. The real-valued objective function is often referred to as a *scalarizing function* and, as discussed earlier, it is justified to use such scalarizing functions that can be proven to generate Pareto optimal solutions. (However, sometimes it may be computationally easier to generate weakly Pareto optimal solutions.) Depending on whether a local or a global solver is used, we get either locally or globally Pareto optimal solutions (if the problem is not convex). As discussed earlier, locally Pareto optimal objective vectors are not of interest and, thus, we must pay attention that an appropriate solver is used. We must also keep in mind that when using numerical optimization methods, the solutions obtained are not necessarily optimal in practice (e.g., if the method used does not converge properly or if the global solver fails in finding the global optimum).

It is sometimes assumed that the DM makes decisions on the basis of an underlying function. This function representing the preferences of the DM is called a *value function* $v : \mathbf{R}^k \to \mathbf{R}$ (Keeney and Raiffa, 1976). In some methods, the value function is assumed to be known implicitly and it has been important in the development of solution methods and as a theoretical background. A utility function is often used as a synonym for a value function but we reserve that concept for stochastic problems which are not treated here. The value function is assumed to be non-increasing with the increase of objective values because we here assume that all objective functions are to be minimized, while the value function is to be maximized. This means that the preference of the DM will not decrease but will rather increase if the value of an objective function decreases, while all the other objective values remain unchanged (i.e., less is preferred to more). In this case, the solution maximizing v can be proven to be Pareto optimal. Regardless of the existence of a value function, it is usually assumed that less is preferred to more by the DM.

Instead of as a maximum of a value function, a final solution can be understood as a satisficing one. *Satisficing decision making* means that the DM does not intend to maximize any value function but tries to achieve certain aspirations (Sawaragi *et al.*, 1985). A Pareto optimal solution which satisfies all the aspirations of the DM is called a *satisficing solution*. In some rare cases, DMs may regard solutions satisficing even if they are not Pareto optimal. This may, for example, means that not all relevant objectives are explicitly expressed. However, here we assume DMs to be rational and concentrate on Pareto optimal solutions.

Not only value functions but, in general, any preference model of a DM may be explicit or implicit in multiobjective optimization methods. Examples of local preference models include aspiration levels and different distance measures. During solution processes, various kinds of information can be solicited from the DM. *Aspiration levels* $\bar{z}_i$ $(i = 1, \ldots, k)$ are such desirable or acceptable levels in the objective function values that are of special interest and importance to the DM. The vector $\bar{z} \in \mathbf{R}^k$ consisting of aspiration levels is called a *reference point*.

According to the definition of Pareto optimality, moving from one Pareto optimal solution to another necessitates trading off. This is one of the basic concepts in multiobjective optimization. A *trade-off* reflects the ratio of change in the values of the objective functions concerning the increment of one objective function that occurs when the value of some other objective function decreases. For details, see, e.g., (Chankong and Haimes, 1983; Miettinen, 1999) and Chapters 2 and 9.

As mentioned earlier, it is sometimes easier to generate weakly Pareto optimal solutions than Pareto optimal ones (because some scalarizing functions produce weakly Pareto optimal solutions). There are different ways to get solutions that can be proven to be Pareto optimal. Benson (1978) has suggested to check the Pareto optimality of the decision vector $\mathbf{x}^* \in S$ by solving the problem

$$
\begin{array}{ll}
\text{maximize} & \sum_{i=1}^{k} \varepsilon_i \\
\text{subject to} & f_i(\mathbf{x}) + \varepsilon_i = f_i(\mathbf{x}^*) \quad \text{for all} \ \ i = 1, \ldots, k, \\
& \varepsilon_i \geq 0 \quad \text{for all} \ \ i = 1, \ldots, k, \\
& \mathbf{x} \in S,
\end{array}
\tag{1.1}
$$

where both $\mathbf{x} \in \mathbf{R}^n$ and $\boldsymbol{\varepsilon} \in \mathbf{R}^k_+$ are variables. If the optimal objective function value of (1.1) is zero, then $\mathbf{x}^*$ can be proven to be Pareto optimal and if the optimal objective function value is finite and nonzero corresponding to a decision vector $\mathbf{x}'$, then $\mathbf{x}'$ is Pareto optimal. Note that the equality constraints in (1.1) can be replaced by inequalities $f_i(\mathbf{x}) + \varepsilon_i \leq f_i(\mathbf{x}^*)$. However, we must point out that problem (1.1) is computationally badly conditioned because it has only one feasible solution ($\varepsilon_i = 0$ for each i) if $\mathbf{x}^*$ is Pareto optimal and computational difficulties must be handled in practice, for example, using penalty functions. We shall introduce other ways to guarantee Pareto optimality in what follows in connection with some scalarizing functions.

Let us point out that in this chapter we do not concentrate on the theory behind multiobjective optimization, necessary and sufficient optimality conditions, duality results, etc. Instead, we refer, for example, to (Jahn, 2004; Luc, 1989; Miettinen, 1999; Sawaragi et al., 1985) and references therein.

In the following sections, we briefly describe some methods for solving multiobjective optimization problems. We introduce several philosophies and ways of approaching the problem. As mentioned in the introduction, we concentrate on the classes devoted to no-preference methods, a posteriori methods and a priori methods and remind that overlapping and combinations of classes are possible because no classification can fully cover the plethora of existing methods.

Methods in each class have their strengths and weaknesses and selecting a method to be used should be based on the desires and abilities of the DM as well as properties of the problem in question. Naturally, an analyst plays a crusial role when selecting a method because (s)he is supposed to know the properties of different methods available. Her/his recommendation should fit the needs and the psychological profile of the DM in question. In different methods, different types of information are given to the DM, the DM is assumed to specify preference information in different ways and different scalarizing functions are used. Besides the references given in each section, further details about the methods to be described, including proofs of theorems related to optimality, can be found in (Miettinen, 1999).

1.3 Basic Methods

Before we concentrate on the three classes of methods described in the introduction, we first discuss two well-known methods that can be called basic methods because they are so widely used. Actually, in many applications one can see them being used without necessarily recognizing them as multiobjective optimization methods. In other words, the difference between a modelling and an optimization phase are often blurred and these methods are used in order to convert the problem into a form where one objective function can be optimized with single objective solvers available. The reason for this may be that methods of single objective optimization are more widely known as those of multiobjective optimization. One can say that these two basic methods are the ones that first come to one's mind if there is a need to optimize multiple objectives simultaneously. Here we consider their strengths and weaknesses (which the users of these methods are not necessarily aware of) as well as show that many other (more advanced) approaches exist.

1.3.1 Weighting Method

In the weighting method (see, e.g., (Gass and Saaty, 1955; Zadeh, 1963)), we solve the problem

$$\text{minimize} \quad \sum_{i=1}^{k} w_i f_i(\mathbf{x}) \tag{1.2}$$
$$\text{subject to} \quad \mathbf{x} \in S,$$

where $w_i \geq 0$ for all $i = 1, \ldots, k$ and, typically, $\sum_{i=1}^{k} w_i = 1$. The solution of (1.2) can be proven to be weakly Pareto optimal and, furthermore, Pareto optimal if we have $w_i > 0$ for all $i = 1, \ldots, k$ or if the solution is unique (see, e.g., (Miettinen, 1999)).

The weighting method can be used as an a posteriori method so that different weights are used to generate different Pareto optimal solutions and then the DM is asked to select the most satisfactory one. Alternatively, the DM can be asked to specify the weights in which case the method is used as an a priori method.

As mentioned earlier, it is important in multiobjective optimization that Pareto optimal solutions are generated and that any Pareto optimal solution can be found. In this respect, the weighting method has a serious shortcoming. It can be proven that any Pareto optimal solution can be found by altering the weights only if the problem is convex. Thus, it may happen that some Pareto optimal solutions of nonconvex problems cannot be found no matter how the weights are selected. (Conditions under which the whole Pareto optimal set can be generated by the weighting method with positive weights are presented in (Censor, 1977).) Even though linear problems are not considered here, we should point out that despite MOLP problems being convex, the weighting method may not behave as expected even when solving them. This is because, when altering the weights, the method may jump from one vertex to another leaving intermediate solutions undetected. This is explained by the fact that linear solvers typically produce vertex solutions.

Unfortunately, people who use the weighting method do not necessarily know that that it does not work correctly for nonconvex problems. This is a serious and important aspect because it is not always easy to check the convexity in real applications if the problem is based, for example, on some simulation model or solving some systems like systems of partial differential equations. If the method is used in nonconvex problems for generating a representation of the Pareto optimal set, the DM gets a completely misleading impression about the feasible solutions available when some parts of the Pareto optimal set remain uncovered.

It is advisable to normalize the objectives with some scaling so that different magnitudes do not confuse the method. Systematic ways of perturbing the weights to obtain different Pareto optimal solutions are suggested, e.g., in (Chankong and Haimes, 1983). However, as illustrated by Das and Dennis (1997), an evenly distributed set of weights does not necessarily produce an evenly distributed representation of the Pareto optimal set, even if the problem is convex.

On the other hand, if the method is used as an a priori method, the DM is expected to be able to represent her/his preferences in the form of weights. This may be possible if we assume that the DM has a linear value function (which then corresponds to the objective function in problem (1.2)). However, in general, the role of the weights may be greatly misleading. They are often said to reflect the relative importance of the objective functions but, for example, Roy and Mousseau (1996) show that it is not at all clear what underlies this notion. Moreover, the relative importance of objective functions is usually understood globally, for the entire decision problem, while many practical applications show that the importance typically varies for different objective function values, that is, the concept is meaningful only locally. (For more discussion on ordering objective functions by importance, see, e.g., (Podinovski, 1994).)

One more reason why the DM may not get satisfactory solutions with the weighting method is that if some of the objective functions correlate with each other, then changing the weights may not produce expected solutions at all but, instead, seemingly bad weights may result with satisfactory solutions and vice versa (see, e.g., (Steuer, 1986)). This is also shown in (Tanner, 1991) with an example originally formulated by P. Korhonen. With this example of choosing a spouse (where three candidates are evaluated with five criteria) it is clearly demonstrated how weights representing the preferences of the DM (i.e., giving the clearly biggest weight to the most important criterion) result with a spouse who is the worst in the criterion that the DM regarded as the most important one. (In this case, the undesired outcome may be explained by the compensatory character of the weighting method.)

In particular for MOLP problems, weights that produce a certain Pareto optimal solution are not necessarily unique and, thus, dramatically different weights may produce similar solutions. On the other hand, it is also possible that a small change in the weights may cause big differences in objective values. In all, we can say that it is not necessarily easy for the DM (or the analyst) to control the solution process with weights because weights behave in an indirect way. Then, the solution process may become an interactive one where the DM tries to guess such weights that would produce a satisfactory solution and this is not at all desirable because the DM can not be properly supported and (s)he is likely to get frustrated. Instead, in such cases it is advisable to use real interactive methods where the DM can better control the solution process with more intuitive preference information. For further details, see Chapter 2.

1.3.2 ε-Constraint Method

In the ε-constraint method, one of the objective functions is selected to be optimized, the others are converted into constraints and the problem gets the form

$$\begin{array}{ll} \text{minimize} & f_\ell(\mathbf{x}) \\ \text{subject to} & f_j(\mathbf{x}) \leq \varepsilon_j \ \text{ for all } \ j = 1, \ldots, k, \ \ j \neq \ell, \\ & \mathbf{x} \in S, \end{array} \quad (1.3)$$

where $\ell \in \{1, \ldots, k\}$ and ε_j are upper bounds for the objectives $(j \neq \ell)$. The method has been introduced in (Haimes *et al.*, 1971) and widely discussed in (Chankong and Haimes, 1983).

As far as optimality is concerned, the solution of problem (1.3) can be proven to always be weakly Pareto optimal. On the other hand, $\mathbf{x}^* \in S$ can be proven to be Pareto optimal if and only if it solves (1.3) for every $\ell = 1, \ldots, k$, where $\varepsilon_j = f_j(\mathbf{x}^*)$ for $j = 1, \ldots, k, \ j \neq \ell$. In addition, a unique solution of

(1.3) can be proven to be Pareto optimal for any upper bounds. In other words, to ensure Pareto optimality we must either solve k different problems (and solving many problems for each Pareto optimal solution increases computational cost) or obtain a unique solution (which is not necessarily easy to verify). However, a positive fact is that finding any Pareto optimal solution does not necessitate convexity (as was the case with the weighting method). In other words, this method works for both convex and nonconvex problems.

In practice, it may be difficult to specify the upper bounds so that the resulting problem (1.3) has solutions, that is, the feasible region will not become empty. This difficulty is emphasized when the number of objective functions increases. Systematic ways of perturbing the upper bounds to obtain different Pareto optimal solutions are suggested in (Chankong and Haimes, 1983). In this way, the method can be used as an a posteriori method. Information about the ranges of objective functions in the Pareto optimal set is useful in perturbing the upper bounds. On the other hand, it is possible to use the method in an a priori way and ask the DM to specify the function to be optimized and the upper bounds. Specifying upper bounds can be expected to be easier for the DM than, for example, weights because objective function values are understandable as such for the DM. However, the drawback here is that if there is a promising solution really close to the bound but on the infeasible side, it will never be found. In other words, the bounds are a very stiff way of specifying preference information.

In what follows, we discuss three method classes described in the introduction and outline some methods belonging to each of them. Again, proofs of theorems related to optimality as well as further details about the methods can be found in (Miettinen, 1999).

1.4 No-Preference Methods

In no-preference methods, the opinions of the DM are not taken into consideration in the solution process. Thus, the problem is solved using some relatively simple method and the idea is to find some compromise solution typically 'in the middle' of the Pareto optimal set because there is no preference information available to direct the solution process otherwise. These methods are suitable for situations where there is no DM available or (s)he has no special expectations of the solution. They can also be used to produce a starting point for interactive methods.

One can question the name of no-preference methods because there may still exist an underlying preference model (e.g., the acceptance of a global criterion by a DM, like the one in the method to be described in the next subsection, can be seen as a preference model). However, we use the term of no-preference method in order to emphasize the fact that no explicit preferences from the DM are available and and, thus, they cannot be used. These methods can also be referred to as methods of neutral preferences.

1.4.1 Method of Global Criterion

In the method of global criterion or compromise programming (Yu, 1973; Zeleny, 1973), the distance between some desirable reference point in the objective space and the feasible objective region is minimized. The analyst selects the reference point used and a natural choice is to set it as the ideal objective vector. We can use, for example, the L_p-metric or the Chebyshev metric (also known as the L_∞-metric) to measure the distance to the ideal objective vector $\mathbf{z}^\star$ or the utopian objective vector $\mathbf{z}^{\star\star}$ (see definitions in Preface) and then we need to solve the problem

$$\begin{array}{ll} \text{minimize} & \left(\sum_{i=1}^{k} |f_i(\mathbf{x}) - z_i^\star|^p\right)^{1/p} \\ \text{subject to} & \mathbf{x} \in S, \end{array} \tag{1.4}$$

(where the exponent $1/p$ can be dropped) or

$$\begin{array}{ll} \text{minimize} & \max_{i=1,\dots,k} \left[\, |f_i(\mathbf{x}) - z_i^\star| \,\right] \\ \text{subject to} & \mathbf{x} \in S, \end{array} \tag{1.5}$$

respectively. Note that if we here know the real ideal objective vector, we can ignore the absolute value signs because the difference is always positive (according to the definition of the ideal objective vector).

It is demonstrated, for example, in (Miettinen, 1999) that the choice of the distance metric affects the solution obtained. We can prove that the solution of (1.4) is Pareto optimal and the solution of (1.5) is weakly Pareto optimal. Furthermore, the latter can be proven to be Pareto optimal if it is unique.

Let us point out that if the objective functions have different magnitudes, the method works properly only if we scale the objective functions to a uniform, dimensionless scale. This means, for example, that we divide each absolute value term involving f_i by the corresponding range of f_i in the Pareto optimal set characterized by nadir and utopian objective vectors (defined in Preface), that is, by $z_i^{\text{nad}} - z_i^{\star\star}$ (for each i). As the utopian objective vector dominates all Pareto optimal solutions, we use the utopian and not the ideal objective values in order to avoid dividing by zero in all occasions. (Connections of this method to utility or value functions are discussed in (Ballestero and Romero, 1991).)

1.4.2 Neutral Compromise Solution

Another simple way of generating a solution without the involvement of the DM is suggested in (Wierzbicki, 1999) and referred to as a *neutral compromise solution*. The idea is to project a point located 'somewhere in the middle' of the ranges of objective values in the Pareto optimal set to become feasible. Components of such a point can be obtained as the average of the ideal (or utopian) and nadir values of each objective function. We can get a neutral compromise solution by solving the problem

$$\text{minimize} \quad \max_{i=1,\dots,k} \left[\frac{f_i(\mathbf{x}) - ((z_i^\star + z_i^{\text{nad}})/2)}{z_i^{\text{nad}} - z_i^{\star\star}} \right]$$
$$\text{subject to} \quad \mathbf{x} \in S. \tag{1.6}$$

As can be seen, this problem uses the utopian and the nadir objective vectors or other reliable approximations about the ranges of the objective functions in the Pareto optimal set for scaling purposes (in the denominator), as mentioned above. The solution is weakly Pareto optimal. We shall later return to scalarizing functions of this type later and discuss how Pareto optimality can be guaranteed. Naturally, the average in the numinator can be taken between components of utopian and nadir objective vectors, instead of the ideal and nadir ones.

1.5 A Posteriori Methods

In what follows, we assume that we have a DM available to take part in the solution process. A posteriori methods can be called *methods for generating Pareto optimal solutions*. Because there usually are infinitely many Pareto optimal solutions, the idea is to generate a representation of the Pareto optimal set and present it to the DM who selects the most satisfactory solution of them as the final one. The idea is that once the DM has seen an overview of different Pareto optimal solutions, it is easier to select the most preferred one. The inconveniences here are that the generation process is usually computationally expensive and sometimes in part, at least, difficult. On the other hand, it may be hard for the DM to make a choice from a large set of alternatives. An important question related to this is how to represent and display the alternatives to the DM in an illustrative way (Miettinen, 2003, 1999). Plotting the objective vectors on a plane is a natural way of displaying them only in the case of two objectives. In that case, the Pareto optimal set can be generated parametrically (see, e.g., (Benson, 1979; Gass and Saaty, 1955)). The problem becomes more complicated with more objectives. For visualizing sets of Pareto optimal solutions, see Chapter 8. Furthermore, visualization and approximation of Pareto optimal sets are discussed in Chapter 9. It is also possible to use so-called box-indices to represent Pareto optimal solutions to be compared by using a rough enough scale in order to let the DM easily recognize the main characteristics of the solutions at a glance (Miettinen *et al.*, 2008).

Remember that the weighting method and the ε-constraint method can be used as a posteriori methods. Next we outline some other methods in this class.

1.5.1 Method of Weighted Metrics

In the method of weighted metrics, we generalize the idea of the method of global criterion where the distance between some reference point and the feasible objective region is minimized. The difference is that we can produce different solutions by weighting the metrics. The weighted approach is also sometimes called compromise programming (Zeleny, 1973).

Again, the solution obtained depends greatly on the distance measure used. For $1 \leq p < \infty$, we have a problem

$$\text{minimize} \quad \left(\sum_{i=1}^{k} w_i\big(f_i(\mathbf{x}) - z_i^\star\big)^p\right)^{1/p} \tag{1.7}$$
$$\text{subject to} \quad \mathbf{x} \in S.$$

The exponent $1/p$ can be dropped. Alternatively, we can use a *weighted Chebyshev problem*

$$\text{minimize} \quad \max_{i=1,\ldots,k} \left[w_i(f_i(\mathbf{x}) - z_i^\star) \right] \tag{1.8}$$
$$\text{subject to} \quad \mathbf{x} \in S.$$

Note that we have here ignored the absolute values assuming we know the global ideal (or utopian) objective vector. As far as optimality is concerned, we can prove that the solution of (1.7) is Pareto optimal if either the solution is unique or all the weights are positive. Furthermore, the solution of (1.8) is weakly Pareto optimal for positive weights. Finally, (1.8) has at least one Pareto optimal solution. On the other hand, convexity of the problem is needed in order to be able to prove that every Pareto optimal solution can be found by (1.7) by altering the weights. However, any Pareto optimal solution can be found by (1.8) assuming that the utopian objective vector $\mathbf{z}^{\star\star}$ is used as a reference point.

The objective function in (1.8) is nondifferentiable and, thus single objective optimizers using gradient information cannot be used to solve it. But if all the functions in the problem considered are differentiable, we can use an equivalent differentiable variant of (1.8) by introducing one more variable and new constraints of the form

$$\text{minimize} \quad \alpha$$
$$\text{subject to} \quad \alpha \geq w_i(f_i(\mathbf{x}) - z_i^\star) \text{ for all } i = 1, \ldots, k, \tag{1.9}$$
$$\mathbf{x} \in S,$$

where both $\mathbf{x} \in \mathbf{R}^n$ and $\alpha \in \mathbf{R}$ are variables. With this formulation, single objective solvers assuming differentiability can be used.

Because problem (1.8) with $\mathbf{z}^{\star\star}$ seems a promising approach (as it can find any Pareto optimal solution), it would be nice to be able to avoid weakly Pareto optimal solutions. This can be done by giving a slight slope to the contours of the scalarizing function used (see, e.g., (Steuer, 1986)). In other words, we can formulate a so-called *augmented Chebyshev problem* in the form

$$\text{minimize} \quad \max_{i=1,\ldots,k} \left[w_i(f_i(\mathbf{x}) - z_i^{\star\star}) \right] + \rho \sum_{i=1}^{k} (f_i(\mathbf{x}) - z_i^{\star\star}) \tag{1.10}$$
$$\text{subject to} \quad \mathbf{x} \in S,$$

where ρ is a sufficiently small positive scalar. Strictly speaking, (1.10) generates properly Pareto optimal solutions and any properly Pareto optimal solution can be found (Kaliszewski, 1994). In other words, we are not actually able to find any Pareto optimal solution but only such solutions having a finite trade-off. However, when solving real-life problems, it is very likely that the DM is not interested in improperly Pareto optimal solutions after all. Here ρ corresponds to the bound for desirable or acceptable trade-offs (see definition of ρ-proper Pareto optimality in Section 1.2.1). Let us mention that an augmented version of the differentiable problem formulation (1.9) is obtained by adding the augmentation term (i.e., the term multiplied by ρ) to the objective function α.

Alternatively, it is possible to generate provably Pareto optimal solutions by solving two problems in a row. In other words, problem (1.8) is first solved and then another optimization problem is solved in the set of optimal solutions to (1.8). To be more specific, let $\mathbf{x}^*$ be the solution of the first problem (1.8). Then the second problem is the following

$$
\begin{aligned}
&\text{minimize} \quad \sum_{i=1}^{k}\left(f_i(\mathbf{x}) - z_i^{\star\star}\right) \\
&\text{subject to} \quad \max_{i=1,\ldots,k}\left[w_i(f_i(\mathbf{x}) - z_i^{\star\star})\right] \leq \max_{i=1,\ldots,k}\left[w_i(f_i(\mathbf{x}^*) - z_i^{\star\star})\right], \\
&\qquad\qquad \mathbf{x} \in S.
\end{aligned}
$$

One should mention that the resulting problem may be computationally badly conditioned if the problem has only one feasible solution. With this so-called lexicographic approach it is possible to reach any Pareto optimal solution. Unfortunately, the computational cost increases because two optimization problems must be solved for each Pareto optimal solution (Miettinen $et\ al.$, 2006).

1.5.2 Achievement Scalarizing Function Approach

Scalarizing functions of a special type are called $achievement\ (scalarizing)$ $functions$. They have been introduced, for example, in (Wierzbicki, 1982, 1986). These functions are based on an arbitrary reference point $\bar{\mathbf{z}} \in \mathbf{R}^k$ and the idea is to project the reference point consisting of desirable aspiration levels onto the set of Pareto optimal solutions. Different Pareto optimal solutions can be produced with different reference points. The difference to the previous method (i.e., method of weighted metrics) is that no distance metric is used and the reference point does not have to be fixed as the ideal or utopian objective vector. Because of these characteristics, Pareto optimal solutions are obtained no matter how the reference point is selected in the objective space.

Achievement functions can be formulated in different ways. As an example we can mention the problem

$$
\begin{aligned}
&\text{minimize} \quad \max_{i=1,\ldots,k}\left[w_i(f_i(\mathbf{x}) - \bar{z}_i)\right] + \rho\sum_{i=1}^{k}(f_i(\mathbf{x}) - \bar{z}_i) \\
&\text{subject to} \quad \mathbf{x} \in S,
\end{aligned} \tag{1.11}
$$

where $\mathbf{w}$ is a fixed normalizing factor, for example, $w_i = 1/(z_i^{\text{nad}} - z_i^{\star\star})$ for all i and $\rho > 0$ is an augmentation multiplier as in (1.10). And corresponding to (1.10), we can prove that solutions of this problem are properly Pareto optimal and any properly Pareto optimal solution can be found. To be more specific, the solutions obtained are ρ-properly Pareto optimal (as defined in Section 1.2). If the augmentation term is dropped, the solutions can be proven to be weakly Pareto optimal. Pareto optimality can also be guaranteed and proven if the lexicographic approach described above is used. Let us point out that problem (1.6) uses an achievement scalarizing function where the reference point is fixed. The problem could be augmented as in (1.11).

Note that when compared to the method of weighted metrics, we do not use absolute value signs here in any case. No matter which achievement function formulation is used, the idea is the same: if the reference point is feasible, or actually to be more exact, $\bar{z} \in Z + \mathbf{R}_+^k$, then the minimization of the achievement function subject to the feasible region allocates slack between the reference point and Pareto optimal solutions producing a Pareto optimal solution. In other words, in this case the reference point is a Pareto optimal solution for the problem in question or it is dominated by some Pareto optimal solution. On the other hand, if the reference point is infeasible, that is, $\bar{z} \notin Z + \mathbf{R}_+^k$, then the minimization produces a solution that minimizes the distance between $\bar{z}$ and Z. In both cases, we can say that we project the reference point on the Pareto optimal set. Discussion on how the projection direction can be varied in the achievement function can be found in (Luque *et al.*, 2009).

As mentioned before, achievement functions can be formulated in many ways and they can be based on so-called reservation levels, besides aspiration levels. For more details about them, we refer, for example, to (Wierzbicki, 1982, 1986, 1999, 2000) and Chapter 2.

1.5.3 Approximation Methods

During the years, many methods have been developed for approximating the set of Pareto optimal solutions in the MCDM literature. Here we do not go into their details. A survey of such methods is given in (Ruzika and Wiecek, 2005). Other approximation algorithms (not included there) are introduced in (Lotov *et al.*, 2004). For more information about approximation methods we also refer to Chapter 9.

1.6 A Priori Methods

In a priori methods, the DM must specify her/his preference information (for example, in the form of aspirations or opinions) before the solution process. If the solution obtained is satisfactory, the DM does not have to invest too much time in the solution process. However, unfortunately, the DM does not necessarily know beforehand what it is possible to attain in the problem and how realistic her/his expectations are. In this case, the DM may be disappointed at the solution obtained and may be willing to change one's preference information. This easily leads to a desire of using an interactive approach (see Chapter 2). As already mentioned, the basic methods introduced earlier can be used as a priori methods. It is also possible to use the achievement scalarizing function approach as an a priori method where the DM specifies the reference point and the Pareto optimal solution closest to it is generated. Here we briefly describe three other methods.

1.6.1 Value Function Method

The value function method (Keeney and Raiffa, 1976) was already mentioned in Section 1.2.2. It is an excellent method if the DM happens to know an explicit mathematical formulation for the value function and if that function can capture and represent all her/his preferences. Then the problem to be solved is

$$\text{maximize} \quad v(\mathbf{f}(\mathbf{x}))$$
$$\text{subject to} \quad \mathbf{x} \in S.$$

Because the value function provides a complete ordering in the objective space, the best Pareto optimal solution is found in this way. Unfortunately, it may be difficult, if not impossible, to get that mathematical expression of v. For example, in (deNeufville and McCord, 1984), the inability to encode the DM's underlying value function reliably is demonstrated by experiments. On the other hand, the value function can be difficult to optimize because of its possible complicated nature. Finally, even if it were possible for the DM to express her/his preferences globally as a value function, the resulting preference structure may be too simple since value functions cannot represent intransitivity or incomparability. In other words, the DM's preferences must satisfy certain conditions (like consistent preferences) so that a value function can be defined on them. For more discussion see, for example, (Miettinen, 1999).

1.6.2 Lexicographic Ordering

In lexicographic ordering (Fishburn, 1974), the DM must arrange the objective functions according to their absolute importance. This means that a more important objective is infinitely more important than a less important objective. After the ordering, the most important objective function is minimized subject to the original constraints. If this problem has a unique solution, it is the final one and the solution process stops. Otherwise, the second most important objective function is minimized. Now, a new constraint is introduced to guarantee that the most important objective function preserves its optimal value. If this problem has a unique solution, the solution process stops. Otherwise, the process goes on as above. (Let us add that computationally it is not trivial to check the uniqueness of solutions. Then the next problem must be solved just to be sure. However, if the next problem has a unique solution, the problem is computationally badly conditioned, as discussed earlier.)

The solution of lexicographic ordering can be proven to be Pareto optimal. The method is quite simple and one can claim that people often make decisions successively. However, the DM may have difficulties in specifying an absolute order of importance. Besides, the method is very rough and it is very likely that the process stops before less important objective functions are taken into consideration. This means that all the objectives that were regarded as relevant while formulating the problem are not taken into account at all, which is questionable.

The notion of absolute importance is discussed in (Roy and Mousseau, 1996). Note that lexicographic ordering does not allow a small increment of an important objective function to be traded off with a great decrement of a less important objective. Yet, the DM might find this kind of trading off appealing. If this is the case, lexicographic ordering is not likely to produce a satisficing solution.

1.6.3 Goal Programming

Goal programming is one of the first methods expressly created for multiobjective optimization (Charnes *et al.*, 1955; Charnes and Cooper, 1961). It has been originally developed for MOLP problems (Ignizio, 1985).

In goal programming, the DM is asked to specify aspiration levels $\bar{z}_i$ ($i = 1, \ldots, k$) for the objective functions. Then, deviations from these aspiration levels are minimized. An objective function jointly with an aspiration level is referred to as a *goal*. For minimization problems, goals are of the form $f_i(\mathbf{x}) \leq \bar{z}_i$ and the aspiration levels are assumed to be selected so that they are not achievable simultaneously. After the goals have been formed, the *deviations* $\delta_i = \max[0, f_i(\mathbf{x}) - \bar{z}_i]$ of the objective function values are minimized.

The method has several variants. In the *weighted goal programming approach* (Charnes and Cooper, 1977), the weighted sum of the deviations is minimized. This means that in addition to the aspiration levels, the DM must specify positive weights. Then we solve a problem

$$
\begin{aligned}
&\text{minimize} && \sum_{i=1}^{k} w_i \delta_i \\
&\text{subject to} && f_i(\mathbf{x}) - \delta_i \leq \bar{z}_i \ \text{ for all } \ i = 1, \ldots, k, \\
& && \delta_i \geq 0 \ \text{ for all } \ i = 1, \ldots, k, \\
& && \mathbf{x} \in S,
\end{aligned}
\tag{1.12}
$$

where $\mathbf{x} \in \mathbf{R}^n$ and δ_i ($i = 1, \ldots, k$) are the variables.

On the other hand, in the *lexicographic goal programming approach*, the DM must specify a lexicographic order for the goals in addition to the aspiration levels. After the lexicographic ordering, the problem with the deviations as objective functions is solved lexicographically subject to the constraints of (1.12) as explained in Section 1.6.2. It is also possible to use a combination of the weighted and the lexicographic approaches. In this case, several

objective functions may belong to the same class of importance in the lexicographic order. In each priority class, a weighted sum of the deviations is minimized. Let us also mention a so-called *min-max goal programming approach*
(Flavell, 1976) where the maximum of deviations is minimized and meta-goal
programming (Rodríguez Uría *et al.*, 2002), where different variants of goal
programming are incorporated.

Let us next discuss optimality. The solution of a goal programming problem can be proven to be Pareto optimal if either the aspiration levels form a
Pareto optimal reference point or all the variables δ_i have positive values at
the optimum. In other words, if the aspiration levels form a feasible point, the
solution is equal to that reference point which is not necessarily Pareto optimal. We can say that the basic formulation of goal programming presented
here works only if the aspiration levels are overoptimistic enough. Pareto optimality of the solutions obtained is discussed, for example, in (Jones *et al.*,
1998).

Goal programming is a very widely used and popular solution method.
Goal-setting is an understandable and easy way of making decisions. The
specification of the weights or the lexicographic ordering may be more difficult (the weights have no direct physical meaning). For further details, see
(Romero, 1991). Let us point out that goal programming is related to the
achievement scalarizing function approach (see Section 1.5.2) because they
both are based on reference points. The advantage of the latter is that it is
able to produce Pareto optimal solutions independently of how the reference
point is selected.

Let us finally add that goal programming has been used in a variety of
further developments and modifications. Among others, goal programming
is related to some fuzzy multiobjective optimization methods where fuzzy
sets are used to express degrees of satisfaction from the attainment of goals
and from satisfaction of soft constraints (Rommelfanger and Slowinski, 1998).
Some more applications of goal programming will be discussed in further
chapters of this book.

1.7 Summary

In this section we summarize some of the properties of the nine methods
discussed so far. We provide a collection of different properties in Figure 1.2.
We pay attention to the class the method can be regarded to belong to as well
as properties of solutions obtained. We also briefly comment the format of
preference information used. In some connections, we use the notation (X) to
indicate that the statement or property is true under assumptions mentioned
when describing the method.

	weighting method	e–constraint method	method of global criterion	neutral compromise solution	method of weighted metrics	achievement scalarizing function	value function method	lexicographic ordering	goal programming
no–preference method			×	×					
a priori method	×	×				×	×	×	×
a posteriori method	×	×			×	×			
can find any Pareto optimal solution		×			(×)	×			×
solution always Pareto optimal	(×)	(×)	(×)	(×)	(×)	(×)	×	×	
type of preference information									
weights	×								(×)
bounds		×							
reference point						×			×
value function							×		
lexicographic order								×	(×)

Fig. 1.2. Summary of some properties of the methods described.

1.8 Conclusions

The aim of this chapter has been to briefly describe some basics of MCDM methods. For this, we have concentrated on some noninteractive methods developed for multiobjective optimization. A large variety of methods exists and it is impossible to cover all of them. In this chapter, we have concentrated on methods where the DM either specifies no preferences or specifies them after or before the solution process. The methods can be combined, hybridized and further developed in many ways, for example, with evolutionary algorithms. Other chapters of this book will discuss possibilities of such developments more.

None of the methods can be claimed to be superior to the others in every aspect. When selecting a solution method, the specific features of the problem to be solved must be taken into consideration. In addition, the opinions and abilities of the DM are important. The theoretical properties of the methods can rather easily be compared but, in addition, practical applicability also plays an important role in the selection of an appropriate method. One can say that selecting a multiobjective optimization method is a problem with multiple objectives itself! Some methods may suit some problems and some DMs better than others. A decision tree is provided in (Miettinen, 1999) for easing the selection. Specific methods for different areas of application that take into account the characteristics of the problems may also be useful.

Acknowledgements

I would like to give my most sincere thanks to Professors Alexander Lotov, Francisco Ruiz and Andrzej Wierzbicki for their valuable comments that improved this chapter. This work was partly supported by the Foundation of the Helsinki School of Economics.

References

Ballestero, E., Romero, C.: A theorem connecting utility function optimization and compromise programming. Operations Research Letters 10(7), 421–427 (1991)

Benayoun, R., de Montgolfier, J., Tergny, J., Laritchev, O.: Programming with multiple objective functions: Step method (STEM). Mathematical Programming 1(3), 366–375 (1971)

Benson, H.P.: Existence of efficient solutions for vector maximization problems. Journal of Optimization Theory and Application 26(4), 569–580 (1978)

Benson, H.P.: Vector maximization with two objective functions. Journal of Optimization Theory and Applications 28(3), 253–257 (1979)

Censor, Y.: Pareto optimality in multiobjective problems. Applied Mathematics and Optimization 4(1), 41–59 (1977)

Chankong, V., Haimes, Y.Y.: Multiobjective Decision Making: Theory and Methodology. Elsevier Science Publishing, New York (1983)

Charnes, A., Cooper, W.W.: Management Models and Industrial Applications of Linear Programming, vol. 1. Wiley, New York (1961)

Charnes, A., Cooper, W.W.: Goal programming and multiple objective optimization; part 1. European Journal of Operational Research 1(1), 39–54 (1977)

Charnes, A., Cooper, W.W., Ferguson, R.O.: Optimal estimation of executive compensation by linear programming. Management Science 1(2), 138–151 (1955)

Cohon, J.L.: Multicriteria programming: Brief review and application. In: Gero, J.S. (ed.) Design Optimization, pp. 163–191. Academic Press, London (1985)

Das, I., Dennis, J.E.: A closer look at drawbacks of minimizing weighted sums of objectives for Pareto set generation in multicriteria optimization problems. Structural Optimization 14(1), 63–69 (1997)

Deb, K., Chaudhuri, S., Miettinen, K.: Towards estimating nadir objective vector using evolutionary approaches. In: Keijzer, M., et al. (eds.) Proceedings of the 8th Annual Genetic and Evolutionary Computation Conference (GECCO-2006), Seattle, vol. 1, pp. 643–650. ACM Press, New York (2006) deNeufville, R., McCord, M.: Unreliable measurement of utility: Significant problems for decision analysis. In: Brans, J.P. (ed.) Operational Research '84, pp. 464–476. Elsevier, Amsterdam (1984)

Edgeworth, F.Y.: Mathematical Psychics: An Essay on the Application of Mathematics to the Moral Sciences. C. Kegan Paul & Co., London (1881), University Microfilms International (Out-of-Print Books on Demand) (1987)

Fandel, G.: Group decision making: Methodology and applications. In: Bana e Costa, C. (ed.) Readings in Multiple Criteria Decision Aid, pp. 569–605. Berlin (1990)

Fishburn, P.C.: Lexicographic orders, utilities and decision rules: A survey. Management Science 20(11), 1442–1471 (1974)

Flavell, R.B.: A new goal programming formulation. Omega 4(6), 731–732 (1976)

Gass, S., Saaty, T.: The computational algorithm for the parametric objective function. Naval Research Logistics Quarterly 2, 39–45 (1955)

Geoffrion, A.M.: Proper efficiency and the theory of vector maximization. Journal of Mathematical Analysis and Applications 22(3), 618–630 (1968)

Haimes, Y.Y., Lasdon, L.S., Wismer, D.A.: On a bicriterion formulation of the problems of integrated system identification and system optimization. IEEE Transactions on Systems, Man, and Cybernetics 1, 296–297 (1971)

Hwang, C.-L., Lin, M.-J.: Group Decision Making under Multiple Criteria: Methods and Applications. Springer, New York (1987)

Hwang, C.L., Masud, A.S.M.: Multiple Objective Decision Making – Methods and Applications: A State-of-the-Art Survey. Springer, Berlin (1979)

Ignizio, J.P.: Introduction to Linear Goal Programming. Sage Publications, Beverly Hills (1985)

Jahn, J.: Vector Optimization. Springer, Berlin (2004)

Jones, D.F., Tamiz, M., Mirrazavi, S.K.: Intelligent solution and analysis of goal programmes: the GPSYS system. Decision Support Systems 23(4), 329–332 (1998)

Kaliszewski, I.: Quantitative Pareto Analysis by Cone Separation Technique. Kluwer, Dordrecht (1994)

Keeney, R.L., Raiffa, H.: Decisions with Multiple Objectives: Preferences and Value Tradeoffs. Wiley, Chichester (1976)

Koopmans, T.: Analysis and production as an efficient combination of activities. In: Koopmans, T. (ed.) Activity Analysis of Production and Allocation: Proceedings of a Conference, pp. 33–97. Wiley, New York (1951), Yale University Press, London (1971)

Korhonen, P., Salo, S., Steuer, R.E.: A heuristic for estimating nadir criterion values in multiple objective linear programming. Operations Research 45(5), 751–757 (1997)

Kuhn, H., Tucker, A.: Nonlinear programming. In: Neyman, J. (ed.) Proceedings of the Second Berkeley Symposium on Mathematical Statistics and Probability, pp. 481–492. University of California Press, Berkeley (1951)

Lotov, A.V., Bushenkov, V.A., Kamenev, G.K.: Interactive Decision Maps. Approximation and Visualization of Pareto Frontier. Kluwer Academic Publishers, Boston (2004)

Luc, D.T.: Theory of Vector Optimization. Springer, Berlin (1989)

Luque, M., Miettinen, K., Eskelinen, P., Ruiz, F.: Incorporating preference information in interactive reference point methods for multiobjective optimization. Omega 37(2), 450–462 (2009)

Makarov, E.K., Rachkovski, N.N.: Unified representation of proper efficiency by means of dilating cones. Journal of Optimization Theory and Applications 101(1), 141–165 (1999)

Marler, R., Arora, J.: Survey of multi-objective optimization methods for engineering. Structural and Multidisciplinary Optimization 26(6), 369–395 (2004)

Miettinen, K.: Nonlinear Multiobjective Optimization. Kluwer Academic Publishers, Boston (1999)

Miettinen, K.: Graphical illustration of Pareto optimal solutions. In: Tanino, T., Tanaka, T., Inuiguchi, M. (eds.) Multi-Objective Programming and Goal Programming: Theory and Applications, pp. 197–202. Springer, Berlin (2003)

Miettinen, K., Mäkelä, M.M., Kaario, K.: Experiments with classification-based scalarizing functions in interactive multiobjective optimization. European Journal of Operational Research 175(2), 931–947 (2006)

Miettinen, K., Molina, J., González, M., Hernández-Díaz, A., Caballero, R.: Using box indices in supporting comparison in multiobjective optimization. European Journal of Operational Research, to appear (2008), doi:10.1016/j.ejor.2008.05.103

Pareto, V.: Cours d'Economie Politique. Rouge, Lausanne (1896)

Pareto, V.: Manuale di Economia Politica. Piccola Biblioteca Scientifica, Milan (1906), Translated into English by Schwier, A.S., Manual of Political Economy, MacMillan, London (1971)

Podinovski, V.V.: Criteria importance theory. Mathematical Social Sciences 27(3), 237–252 (1994)

Rodríguez Uría, M., Caballero, R., Ruiz, F., Romero, C.: Meta-goal programming. European Journal of Operational Research 136(2), 422–429 (2002)

Romero, C.: Handbook of Critical Issues in Goal Programming. Pergamon Press, Oxford (1991)

Rommelfanger, H., Slowinski, R.: Fuzzy linear programming with single or multiple objective functions. In: Slowinski, R. (ed.) Fuzzy Sets in Decision Analysis, Operations Research and Statistics, pp. 179–213. Kluwer Academic Publishers, Boston (1998)

Rosenthal, R.E.: Principles of Multiobjective Optimization. Decision Sciences 16(2), 133–152 (1985)

Roy, B., Mousseau, V.: A theoretical framework for analysing the notion of relative importance of criteria. Journal of Multi-Criteria Decision Analysis 5(2), 145–159 (1996)

Ruzika, S., Wiecek, M.M.: Approximation methods in multiobjective programming. Journal of Optimization Theory and Applications 126(3), 473–501 (2005)

Sawaragi, Y., Nakayama, H., Tanino, T.: Theory of Multiobjective Optimization. Academic Press, Orlando (1985)

Steuer, R.E.: Multiple Criteria Optimization: Theory, Computation, and Application. Wiley, New York (1986)

Tanner, L.: Selecting a text-processing system as a qualitative multiple criteria problem. European Journal of Operational Research 50(2), 179–187 (1991)

Vincke, P.: Multicriteria Decision-Aid. Wiley, Chichester (1992)

Weistroffer, H.R.: Careful usage of pessimistic values is needed in multiple objectives optimization. Operations Research Letters 4(1), 23–25 (1985)

Wierzbicki, A.P.: A mathematical basis for satisficing decision making. Mathematical Modelling 3, 391–405 (1982)

Wierzbicki, A.P.: On the completeness and constructiveness of parametric characterizations to vector optimization problems. OR Spectrum 8(2), 73–87 (1986)

Wierzbicki, A.P.: Reference point approaches. In: Gal, T., Stewart, T.J., Hanne, T.
(eds.) Multicriteria Decision Making: Advances in MCDM Models, Algorithms,
Theory, and Applications, pp. 9-1–9-39. Kluwer, Boston (1999)

Wierzbicki, A.P.: Reference point methodology. In: Wierzbicki, A.P., Makowski, M.,
Wessels, J. (eds.) Model-Based Decision Support Methodology with Environmental Applications, pp. 71–89. Kluwer Academic Publishers, Dordrecht (2000)

Yu, P.L.: A class of solutions for group decision problems. Management Science 19(8), 936–946 (1973)

Zadeh, L.: Optimality and non-scalar-valued performance criteria. IEEE Transactions on Automatic Control 8, 59–60 (1963)

Zeleny, M.: Compromise programming. In: Cochrane, J.L., Zeleny, M. (eds.) Multiple
Criteria Decision Making, pp. 262–301. University of South Carolina, Columbia,
SC (1973)

2

Introduction to Multiobjective Optimization: Interactive Approaches

Kaisa Miettinen[1], Francisco Ruiz[2], and Andrzej P. Wierzbicki[3]

[1] Department of Mathematical Information Technology, P.O. Box 35 (Agora), FI-40014 University of Jyväskylä, Finland, kaisa.miettinen@jyu.fi[*]
[2] Department of Applied Economics (Mathematics), University of Málaga, Calle Ejido 0, E-29071 Málaga, Spain, rua@uma.es
[3] 21st Century COE Program: Technology Creation Based on Knowledge Science, JAIST (Japan Advanced Institute of Science and Technology), Asahidai 1-1, Nomi, Ishikawa 923-1292, Japan and National Institute of Telecommunications, Szachowa Str. 1, 04-894 Warsaw, Poland, andrzej@jaist.ac.jp

Abstract. We give an overview of interactive methods developed for solving nonlinear multiobjective optimization problems. In interactive methods, a decision maker plays an important part and the idea is to support her/him in the search for the most preferred solution. In interactive methods, steps of an iterative solution algorithm are repeated and the decision maker progressively provides preference information so that the most preferred solution can be found. We identify three types of specifying preference information in interactive methods and give some examples of methods representing each type. The types are methods based on trade-off information, reference points and classification of objective functions.

2.1 Introduction

Solving multiobjective optimization problems typically means helping a human decision maker (DM) in finding the most preferred solution as the final one. By the *most preferred solution* we refer to a Pareto optimal solution which the DM is convinced to be her/his best option. Naturally, finding the most preferred solution necessitates the participation of the DM who is supposed to have insight into the problem and be able to specify preference information related to the objectives considered and different solution alternatives, as discussed in Chapter 1. There we presented four classes for multiobjective optimization methods according to the role of the DM in the solution process.

---

[*] In 2007 also Helsinki School of Economics, Helsinki, Finland
Reviewed by: Andrzej Jaszkiewicz, Poznan University of Technology, Poland
Wlodzimierz Ogryczak, Warsaw University of Technology, Poland
Roman Słowiński, Poznan University of Technology, Poland

J. Branke et al. (Eds.): Multiobjective Optimization, LNCS 5252, pp. 27–57, 2008.
© Springer-Verlag Berlin Heidelberg 2008

This chapter is a direct continuation to Chapter 1 and here we concentrate on the fourth class, that is, interactive methods.

As introduced in Chapter 1, in interactive methods, an iterative solution algorithm (which can be called a solution pattern) is formed, its steps are repeated and the DM specifies preference information progressively during the solution process. In other words, the phases of preference elicitation (decision phase) and solution generation (optimization stage) alternate until the DM has found the most preferred solution (or some stopping criterion is satisfied, or there is no satisfactory solution for the current problem setting). After every iteration, some information is given to the DM and (s)he is asked to answer some questions concerning a critical evaluation of the proposed solutions or to provide some other type of information to express her/his preferences. This information is used to construct a more or less explicit model of the DM's local preferences and new solutions (which are supposed to better fit the DM's preferences) are generated based on this model. In this way, the DM directs the solution process and only a part of the Pareto optimal solutions has to be generated and evaluated. Furthermore, the DM can specify and correct her/his preferences and selections during the solution process.

In brief, the main steps of a general interactive method are the following: (1) initialize (e.g., calculate ideal and nadir values and showing them to the DM), (2) generate a Pareto optimal starting point (some neutral compromise solution or solution given by the DM), (3) ask for preference information from the DM (e.g., aspiration levels or number of new solutions to be generated), (4) generate new Pareto optimal solution(s) according to the preferences and show it/them and possibly some other information about the problem to the DM. If several solutions were generated, ask the DM to select the best solution so far, and (6) stop, if the DM wants to. Otherwise, go to step (3).

Because of the structure of an iterative approach, the DM does not need to have any global preference structure and (s)he can learn (see Chapter 15) during the solution process. This is a very important benefit of interactive methods because getting to know the problem, its possibilities and limitations is often very valuable for the DM. To summarize, we can say that interactive methods overcome weaknesses of a priori and a posteriori methods because the DM does not need a global preference structure and only such Pareto optimal solutions are generated that are interesting to the DM. The latter means savings in computational cost and, in addition, avoids the need to compare many Pareto optimal solutions simultaneously.

If the final aim is to choose and implement a solution, then the goal of applying a multiobjective optimization method is to find a single, most pre-ferred, final solution. However, in some occasions it may be preferable that instead of one, we find several solutions. This may be particularly true in case of robustness considerations when some aspects of uncertainty, imprecision or inconsistency in the data or in the model are to be taken into account (but typically, eventually, one of them will still have to be chosen). In what follows, as the goal of using an interactive solution process we consider finding a single most preferred solution.

A large variety of interactive methods has been developed during the years. We can say that none of them is generally superior to all the others and some methods may suit different DMs and problems better than the others. The most important assumption underlying the successful application of interactive methods is that the DM must be available and willing to actively participate in the solution process and direct it according to her/his preferences.

Interactive methods differ from each other by both the style of interaction and technical elements. The former includes the form in which information is given to the DM and the form and type of preference information the DM specifies. On the other hand, the latter includes the type of final solution obtained (i.e., whether it is weakly, properly or Pareto optimal or none of these), the kind of problems handled (i.e., mathematical assumptions set on the problem), the mathematical convergence of the method (if any) and what kind of a scalarizing function is used. It is always important that the DM finds the method worthwhile and acceptable and is able to use it properly, in other words, the DM must find the style of specifying preference information understandable and preferences easy and intuitive to provide in the style selected. We can often identify two phases in the solution process: a learning phase when the DM learns about the problem and feasible solutions in it and a decision phase when the most preferred solution is found in the region identified in the first phase. Naturally, the two phases can also be used iteratively if so desired.

In fact, solving a multiobjective optimization problem interactively is a constructive process where, while learning, the DM is building a conviction of what is possible (i.e., what kind of solutions are available) and confronting this knowledge with her/his preferences that also evolve. In this sense, one should generally speak about a psychological convergence in interactive methods, rather than about a mathematical one.

Here we identify three types of specifying preference information in interactive methods and discuss the main characteristics of each type as well as give some examples of methods. The types are methods based on trade-off information, reference point approaches and classification-based methods. However, it is important to point out that other interactive methods do also exist. For example, it is possible to generate a small sample of Pareto optimal solutions using different weights in the weighted Chebyshev problem (1.8) or (1.10) introduced in Chapter 1 when minimizing the distance to the utopian objective vector. Then we can ask the DM to select the most preferred one of them and the next sample of Pareto optimal solutions is generated so that it concentrates on the neighbourhood of the selected one. See (Steuer, 1986, 1989) for details of this so-called Tchebycheff method.

Different interactive methods are described, for example, in the monographs (Chankong and Haimes, 1983; Hwang and Masud, 1979; Miettinen, 1999; Sawaragi et al., 1985; Steuer, 1986; Vincke, 1992). Furthermore, methods with applications to large-scale systems and industry are presented in (Haimes *et al.*, 1990; Statnikov, 1999; Tabucanon, 1988). Let us also mention examples of reviews of methods including (Buchanan, 1986; Stewart, 1992) and collections of interactive methods like (Korhonen, 2005; Shin and Ravindran, 1991; Vanderpooten and Vincke, 1989).

2.2 Trade-off Based Methods

2.2.1 Different Trade-off Concepts

Several definitions of trade-offs are available in the MCDM literature. Intuitively speaking, a trade-off is an exchange, that is, a loss in one aspect of the problem, in order to gain additional benefit in another aspect. In our multiobjective optimization language, a trade-off represents giving up in one of the objectives, which allows the improvement of another objective. More precisely, how much must we give up of a certain objective in order to improve another one to a certain quantity. Here, one important distinction must be made. A trade-off can measure, attending just to the structure of the problem, the change in one objective in relation to the change in another one, when moving from a feasible solution to another one. This is what we call an *objective trade-off*. On the other hand, a trade-off can also measure how much the DM considers desirable to sacrifice in the value of some objective function in order to improve another objective to a certain quantity. Then we talk about a *subjective trade-off*. As it will be seen, both concepts may be used within an interactive scheme in order to move from a Pareto optimal solution to another. For objective trade-offs, let us define the following concepts:

Definition 1. *Let us consider two feasible solutions* $\mathbf{x}^1$ *and* $\mathbf{x}^2$, *and the corresponding objective vectors* $\mathbf{f}(\mathbf{x}^1)$ *and* $\mathbf{f}(\mathbf{x}^2)$. *Then, the* ratio of change *between* f_i *and* f_j *is denoted by* $T_{ij}(\mathbf{x}^1, \mathbf{x}^2)$, *where*

$$T_{ij}(\mathbf{x}^1, \mathbf{x}^2) = \frac{f_i(\mathbf{x}^1) - f_i(\mathbf{x}^2)}{f_j(\mathbf{x}^1) - f_j(\mathbf{x}^2)}.$$

$T_{ij}(\mathbf{x}^1, \mathbf{x}^2)$ *is said to be a* partial trade-off *involving* f_i *and* f_j *between* $\mathbf{x}^1$ *and* $\mathbf{x}^2$ *if* $f_l(\mathbf{x}^1) = f_l(\mathbf{x}^2)$ *for all* $l = 1, \ldots, k$, $l \neq i, j$. *If, on the other hand, there exists an index* $l \in \{1, \ldots, k\} \setminus \{i, j\}$ *such that* $f_l(\mathbf{x}^1) \neq f_l(\mathbf{x}^2)$, *then* $T_{ij}(\mathbf{x}^1, \mathbf{x}^2)$ *is called the* total trade-off *involving* f_i *and* f_j *between* $\mathbf{x}^1$ *and* $\mathbf{x}^2$.

When moving from one Pareto optimal solution to another, there is at least a pair of objective functions such that one of them is improved and the other one gets worse. These trade-off concepts help the DM to study the effect of changing the current solution. For continuously differentiable problems, the finite increments quotient represented by $T_{ij}(\mathbf{x}^1, \mathbf{x}^2)$ can be changed by an infinitesimal change trend when moving from a certain Pareto optimal solution $\mathbf{x}^0$ along a feasible direction $\mathbf{d}$. This yields the following concept.

Definition 2. *Given a feasible solution* $\mathbf{x}^0$ *and a feasible direction* $\mathbf{d}$ *emanating from* $\mathbf{x}^0$ *(i.e., there exists* $\alpha_0 > 0$ *so that* $\mathbf{x}^0 + \alpha\mathbf{d} \in S$ *for* $0 \leq \alpha \leq \alpha_0$*), we define the* total trade-off rate *at* $\mathbf{x}^0$*, involving* f_i *and* f_j*, along the direction* $\mathbf{d}$ *as*

$$t_{ij}(\mathbf{x}^0, \mathbf{d}) = \lim_{\alpha \to 0} T_{ij}(\mathbf{x}^0 + \alpha\mathbf{d}, \mathbf{x}^0).$$

If $\mathbf{d}$ *is a feasible direction with the property that there exists* $\bar{\alpha} > 0$ *such that* $f_l(\mathbf{x}^0 + \alpha\mathbf{d}) = f_l(\mathbf{x}^0)$ *for all* $l \notin \{i, j\}$ *and for all* $0 \leq \alpha \leq \bar{\alpha}$*, then we shall call the corresponding* $t_{ij}(\mathbf{x}^0, \mathbf{d})$ *the* partial trade-off rate.

The following result is straightforward:

Proposition 1. *Let us assume that all the objective functions* f_i *are continuously differentiable. Then,*

$$t_{ij}(\mathbf{x}^0, \mathbf{d}) = \frac{\nabla f_i(\mathbf{x}^0)^T \mathbf{d}}{\nabla f_j(\mathbf{x}^0)^T \mathbf{d}}.$$

It must be pointed out that the expression given in Definition 2 for the trade-off rate makes it necessary for direction $\mathbf{d}$ to be feasible. Nevertheless, the characterization given in Proposition 1 for the continuously differentiable case can be extended to (non-feasible) tangent directions.

Now, let us proceed to the definition of subjective trade-off concepts. The term subjective means that the DM's preferences are somehow taken into account. That is, subjective trade-offs are desirable trade-offs for the DM. This idea often implies the existence of an underlying (implicit) value function $v(z_1, \ldots, z_k)$ which defines the DM's subjective preferences among the feasible solutions of the problem. If the objective functions are to be minimized, then v is assumed to be strictly decreasing with respect to each one of its variables. Very frequently, the concavity of v is also assumed. If two alternatives are equally desired for the DM, this means that they lie on the same indifference curve (i.e., an isoquant of the value function), $v(z_1, \ldots, z_k) = v^0$.

This yields the following definition:

Definition 3. *Given two solutions* $\mathbf{x}^1$ *and* $\mathbf{x}^2$*, if* $\mathbf{f}(\mathbf{x}^1)$ *and* $\mathbf{f}(\mathbf{x}^2)$ *lie on the same indifference curve, the corresponding trade-off* $T_{ij}(\mathbf{x}^1, \mathbf{x}^2)$ *whether total or partial, is usually known as the* indifference trade-off *involving* f_i *and* f_j *between* $\mathbf{x}^1$ *and* $\mathbf{x}^2$.

Let us assume that all functions f_i are continuously differentiable, and suppose we are studying the indifference trade-offs between f_i and f_j at a fixed point $\mathbf{x}^0$, with objective vector $\mathbf{z}^0 = \mathbf{f}(\mathbf{x}^0)$, which lies on the indifference curve $v(z_1, \ldots, z_k) = v^0$. If $\partial v(\mathbf{z}^0)/\partial z_i \neq 0$, we can express, z_i as an implicit function of the remaining objectives (including z_j):

$$z_i = z_i(z_1, \ldots, z_{i-1}, z_{i+1}, \ldots, z_k). \tag{2.1}$$

This expression allows us to obtain the trade-off rate between two functions when moving along an indifference curve:

Definition 4. *Given a solution* $\mathbf{x}^0$ *and the corresponding* $\mathbf{z}^0$, *the* indifference trade-off rate *or* marginal rate of substitution *(MRS) between* f_i *and* f_j *at* $\mathbf{x}^0$ *is defined as follows:*

$$m_{ij}(\mathbf{x}^0) = \frac{\partial v(\mathbf{z}^0)}{\partial z_j} \bigg/ \frac{\partial v(\mathbf{z}^0)}{\partial z_i} .$$

Let us observe that, if the chain rule is applied to expression (2.1), then

$$m_{ij}(\mathbf{x}^0) = - \frac{\partial z_i(z_1, \ldots, z_{i-1}, z_{i+1}, \ldots, z_k)}{\partial z_j} \bigg|_{\mathbf{z} = \mathbf{z}^0}. \tag{2.2}$$

Therefore, the marginal rate of substitution between f_i and f_j at $\mathbf{x}^0$ represents the amount of decrement in the value of the objective function f_i that compensates an infinitesimal increment in the value of the objective f_j, while the values of all the other objectives remain unaltered. Alternatively, it can be viewed as the (absolute value of the) slope of the indifference curve v^0 at $\mathbf{f}(\mathbf{x}^0)$, if z_i and z_j are represented on the axes. This, given that v is strictly decreasing, implies that

$$N_v(\mathbf{x}^0) = (-m_{i1}(\mathbf{x}^0), \ldots, -m_{ii-1}(\mathbf{x}^0), -1, -m_{ii+1}(\mathbf{x}^0), \ldots, -m_{ik}(\mathbf{x}^0)) \tag{2.3}$$

is a normal vector to the indifference curve at $\mathbf{z}^0$ (see Figure 2.1).

2.2.2 Obtaining Objective Trade-offs over the Pareto Optimal Set

When solving a multiobjective optimization problem using an interactive method, it can be important and useful to know the objective trade-offs when moving from a Pareto optimal solution to another one. This knowledge can allow the DM to decide whether to search for more preferred Pareto optimal solutions in certain directions. A key issue for many trade-off based interactive methods is to obtain partial trade-off rates for a Pareto optimal solution. The ε-constraint problem plays a key role in this task.

Given the multiobjective problem, a vector $\varepsilon \in R^{k-1}$, and an objective function f_i to be optimized, let us consider problem (1.3) defined in Chapter 1. We can denote this problem by $P_i(\varepsilon)$. If the feasible set of $P_i(\varepsilon)$ is nonempty and $\mathbf{x}^0$ is an optimal solution, then let us denote by λ_{ij}^* the optimal Karush-Kuhn-Tucker (KKT) multipliers associated with the f_j constraints. Chankong and Haimes (1983) prove (under certain regularity and second order conditions) that if all the optimal KKT multipliers are strictly positive, then $-\lambda_{ij}^*$ is a partial trade-off rate between objectives f_i and f_j, along a direction $\mathbf{d}_j$:

$$- \lambda_{ij}^* = \frac{\partial f_i(\mathbf{x}^0)}{\partial z_j} = t_{ij}(\mathbf{x}^0, \mathbf{d}_j), \qquad j \neq i, \tag{2.4}$$

where

$$\mathbf{d}_j = \frac{\partial \mathbf{x}(\varepsilon^0)}{\partial \varepsilon_j} \tag{2.5}$$

is an efficient direction, that is, a direction that is tangent to the Pareto optimal set (i.e., frontier) at $\mathbf{x}^0$. Therefore, graphically, $-\lambda_{ij}^*$ can be viewed as the slope of the Pareto optimal frontier at $\mathbf{z}^0$, if functions f_i and f_j are represented on the axes. For objective functions to be minimized, a vector

$$N^*(\mathbf{x}^0) = (-\lambda_{i1}^*, \ldots, -\lambda_{ii-1}^*, -1, -\lambda_{ii+1}^*, \ldots, -\lambda_{ik}^*) \tag{2.6}$$

can be interpreted as a normal vector to the Pareto optimal frontier at $\mathbf{z}^0$ (see Figure 2.1). In fact, this expression matches the traditional definition of a normal vector when the efficient frontier is differentiable.

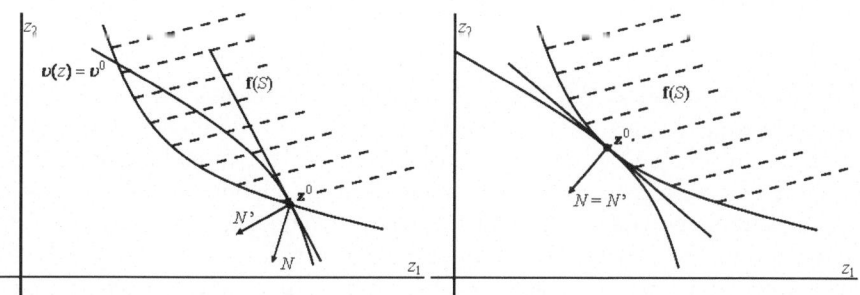

Fig. 2.1. On the left, N, given by the optimal KKT multipliers, is the normal vector to the Pareto optimal frontier at $\mathbf{z}^0$, and N', given by the MRS, is the normal vector to the indifference curve at $\mathbf{z}^0$. On the right, the convergence condition given in (2.7) holds.

If the strict positivity conditions on the multipliers is removed, a more general result is also proved in (Chankong and Haimes, 1983). Namely, if all the optimal multipliers are strictly positive, then $-\lambda_{kj}^*$ is a partial trade-off rate between objectives f_k and f_j, along direction $\mathbf{d}_j$, as defined in (2.5). If, on the other hand, there are optimal multipliers equal to zero, then $-\lambda_{ij}^*$ is a total trade-off rate between f_i and f_j, along direction $\mathbf{d}_j$, as defined in (2.5).

Let us consider a Pareto optimal solution $\mathbf{x}^0$ and the objective vector $\mathbf{z}^0 = \mathbf{f}(\mathbf{x}^0)$. If the Pareto optimal set is connected as well as of full dimensionality $(k-1)$ and continuously differentiable, there is an alternative way of obtaining a normal vector to the Pareto optimal frontier at $\mathbf{z}^0$. It does not require the second order sufficiency conditions to be satisfied and works by means of solving problem (1.9) formulated in Chapter 1 with

$$w_i = \frac{1}{z_i^0 - z_i^{\star\star}}, \quad i = 1, \ldots, k,$$

where $\mathbf{z}^{\star\star}$ is the utopian objective vector. If λ_{ij}^* are the optimal KKT multipliers of this problem associated with the function constraints, then the vector

$$N = (-w_1\lambda^*_{i1}, \ldots, -w_k\lambda^*_{ik})$$

is a normal vector to the Pareto optimal frontier at $\mathbf{z}^0$ (Yang and Li, 2002).

2.2.3 The Use of Trade-offs within Interactive Methods

Among the different trade-off based interactive methods reported in the literature there are two most commonly used schemes:

- to determine at each iteration objective trade-offs, which are shown to the DM, who must give some kind of answer about the desirability of such trade-offs, or
- to ask the DM to provide subjective trade-offs, which are used to find a Pareto optimal solution with a better value of the DM's (implicit) value function.

The Zionts-Wallenius (Z-W) method (Zionts and Wallenius, 1976) belongs to the first group. In this method, the DM is shown several objective trade-offs at each iteration, and (s)he is expected to say whether (s)he likes, dislikes or is indifferent with respect to each trade-off. More elaborated information is required from the DM in the ISWT method (Chankong and Haimes, 1983), where several objective trade-offs are shown at each iteration to the DM who must rank each one of them in a scale from -10 to 10, depending on its desirability (or from -2 to 2, as suggested by Tarvainen (1984)).

In the second group, there are three important methods. The Geoffrion-Dyer-Feinberg (GDF) method (Geoffrion et al., 1972) uses a Frank–Wolfe algorithm in order to perform a line search using the subjective trade-off information given by the DM to determine the search direction. In the SPOT method (Sakawa, 1982), the subjective trade-offs given by the DM are also used to determine a search direction, but a proxy function is used to calculate the optimal step length. Finally, the GRIST method(Yang, 1999) uses the normal vector in order to project the direction given by the subjective trade-offs onto the tangent plane to the Pareto optimal frontier.

All these five methods will be briefly described in the following section. In most of them, the relation between the objective and the subjective trade-offs is very important in order to determine the final solution. Namely, at a final solution, that is, a Pareto optimal solution that maximizes the DM's value function, the indifference curve of the value function must be tangent to the Pareto optimal frontier. This implies that if the indifference trade-off rate is defined like in (2.2) and the objective partial trade-off rate is defined as in (2.4), then in a final solution the relations

$$m_{ij}(\mathbf{x}^0) = \lambda^*_{ij} \quad j = 1, \ldots, k \quad j \neq i \tag{2.7}$$

must hold, which in turn implies that the normal vectors (2.3) and (2.6) must coincide (see Figure 2.1). Note that, in this case, the i-th component of both the vectors is equal to 1, and that is why they must exactly coincide. If the normal vector to the Pareto optimal frontier is defined as in (2.6), then the equality given in expression (2.7) or the equality between the normal vectors should be replaced by a proportionality condition.

2.2.4 Some Trade-off Based Methods

In this section, we will briefly give the basic ideas of the previously mentioned methods. For further details, the reader may follow the references given. See also (Miettinen, 1999) for other trade-off based interactive techniques not mentioned here.

The Z-W method was originally proposed in (Zionts and Wallenius, 1976), and it is based on piecewise linearizations of the problem and the use of the properties of Pareto optimal solutions of linear problems. The assumptions of this method are the following:

- An implicit value function v exists, and it is assumed to be concave.
- The objective functions and the feasible region are convex.

Although not strictly necessary from the theoretical point of view, the authors mention that the additive separability of the objective functions is convenient for practical reasons. Another version of the algorithm exists for a class of pseudoconcave value functions (Zionts and Wallenius, 1983). As the method is based on piecewise linear approximations of the functions, we will briefly describe it for MOLP problems representing one of these piecewise approximations. The idea of the method is the following: A Pareto optimal solution is found using the weighting method (see Section 1.3.1 in Chapter 1). Then adjacent Pareto optimal vertices to the current solution are identified and the corresponding trade-offs are shown to the DM, who is asked to say whether (s)he prefers each of them to the current solution or not. Making use of this information, the weights are actualized, and a new solution is found.

The interactive surrogate worth trade-off method (ISWT) is an interactive version of the original surrogate worth trade-off method (Haimes and Hall, 1974). The basic idea lies on the concept of surrogate worth, which is a valuation by the DM of the desirability of the trade-offs obtained at a Pareto optimal solution. The interactive method was first reported in (Chankong and Haimes, 1978), and both versions are also described in (Chankong and Haimes, 1983). The basic assumptions of this method are the following:

- An implicit value function v exists, and it is assumed to be continuously differentiable and monotonically decreasing.
- All the functions are twice continuously differentiable.
- The feasible region S is compact.
- Optimal KKT multipliers provide partial trade-offs.

This method proceeds as follows. A Pareto optimal solution is determined using the ε-constraint problem (see Section 1.3.2 in Chapter 1). The objective trade-offs at the current solution are obtained and shown to the DM, who is asked to assess their desirability using a scale from -10 to 10. This information is used to actualize the vector of upper bounds, and to produce a new solution.

The basic idea of the interactive Geoffrion, Dyer and Feinberg (GDF) algorithm (Geoffrion *et al.*, 1972) is the following. The existence of an implicit value function v is assumed, which the DM wishes to maximize over the feasible region. The Franke-Wolfe algorithm is applied to solve the intermediate problems formed. The assumptions of the GDF method are:

- The feasible region S is compact and convex.
- Objective functions are continuously differentiable and convex.
- An implicit value function v exists, and is assumed to be continuously differentiable, monotonically decreasing and concave.

In the GDF method, given the current solution, the DM is asked to provide marginal rates of substitution, which are used to determine an ascent direction for the value function. Then, the optimal step-length is approximated using an evaluation scheme, and the next iteration is generated.

The sequential proxy optimization technique (SPOT) is an interactive algorithm developed by Sakawa (1982). The basic idea of this method is to assume the existence of an implicit value function v of the DM, that has to be maximized over the feasible region. This maximization is done using a feasible direction scheme. In order to determine the optimal step-length, a proxy function is used to simulate locally the behavior of the (unknown) value function (the author proposes several options for this function). The ε-constraint problem is used to determine Pareto optimal solutions and to obtain trade-off information at the current solution, and the DM is asked to provide marginal rates of substitution. The assumptions of this method are the following:

- The implicit value function v exists, and is continuously differentiable, strictly decreasing and concave.
- All objective functions f_i are convex and twice continuously differentiable.
- The feasible region S is compact and convex.
- Optimal KKT multipliers provide partial trade-offs.

The idea of the iterations is the following. The ε-constraint problem is used to determine a Pareto optimal solution. The DM is asked to give the MRSs, which are used to find a search direction. The optimal step-length is approximated using the proxy function, and the vector of bounds is updated so as to find the next iteration.

Finally, we introduce the gradient based interactive step trade-off method (GRIST) proposed by Yang (1999). This interactive technique has been designed to deal with general (non necessarily convex) differentiable problems with a differentiable, connected and full dimensional $(k-1)$ Pareto optimal set. The main idea consists of a projection of the vector determined by the marginal rates of substitution given by the DM onto a tangent plane to the Pareto optimal frontier at the current iteration. This projection is proved to be an increasing direction of the DM's underlying value function. Then a reference point is obtained following this direction, which is in turn projected onto the Pareto optimal frontier to generate the next iteration. The assumptions of this method are the following ones:

- The implicit value function v exists, and it is continuously differentiable and strictly decreasing.
- All objective functions are continuously differentiable.
- The feasible region S is compact.
- The Pareto optimal set is differentiable, connected and of full dimensionality $(k - 1)$.
- All the solutions generated are regular.

In GRIST, given the current solution, the DM is asked to give the MRSs. The corresponding vector is projected onto the tangent plane to the Pareto optimal frontier, and a step-length is approximated by an evaluation scheme. The point obtained is projected onto the Pareto optimal set, and this yields a new solution.

2.2.5 Summary

Finally, let us point out some of the most outstanding features of the methods described.

Convergence. The mathematical convergence of all the methods to the optimum of the implicit value function can be proved, given that the value function satisfies for each method the assumptions mentioned in Section 2.2.4. (Moreover, for MOLP problems, the Z-W method can be proved to converge in a finite number of iterations.)

Information. The preference information required from the DM can be regarded as not very hard for the Z-W method (the DM has to say whether the trade-offs proposed are desirable or not), hard for the SPOT, GDF and GRIST methods (the DM has to give MRSs at the current iteration) and very hard for the ISWT method (where a surrogate worth for each trade-off has to be given). Besides, in some methods the step-length has to be estimated evaluating different solutions.

Consistency. In all the methods the consistency of the responses of the DM is vital for the real convergence of the method. In the special case of the SPOT method, there are hard consistency tests for choosing the proxy functions and their parameters. Although all the methods (except Z-W) allow to revisit solutions and to go back in the process, they usually do not perform well with inconsistent answers.

Type of problem. The Z-W method handles convex problems for which a piecewise linearization can be carried out, although it is mainly used in practice for linear problems. The GDF method is designed for convex problems. The SPOT and ISWT methods do not assume convexity, but second order sufficient conditions must be satisfied at the iterations. Finally, the GRIST method assumes that the Pareto optimal set is differentiable, connected and full dimensional. These conditions may be very hard to be assured a priori.

Pareto optimality. The Pareto optimality of the final solution is not guaranteed in the GDF method. All the other methods assure that the final solution is Pareto optimal (although only extreme Pareto optimal solutions are obtained with the Z-W method).

2.3 Reference Point Approaches

2.3.1 Fundamental Assumptions of Reference Point Approaches

Reference point approaches have a long history and multiple practical applications (Wierzbicki, 1977, 1980, 1999; Wierzbicki *et al.*, 2000). However, we shall limit the description here to their fundamental philosophy, a short indication of their basic features and of some contemporary, new developments related to this class of approaches. During over 30 years of development of reference point approaches, including their diverse applications, several methodological postulates describing desirable features of the decision process supported by these approaches have been clarified, most of them expressing lessons learned from the practice of decision making. These postulates are:

1) Separation of preferential and substantive models. This indicates the conviction that in a good decision support system, we should carefully distinguish between the subjective part of knowledge represented in this system, concerning the preferences of the user, thus called a *preferential model* (including, but understood more broadly than a preference model of the DM) of the decision situation, and the objective part, representing in this system some selected knowledge about pertinent aspects of the decision situation – obviously selected never fully objectively, but formulated with objectivity as a goal – called a *substantive model* (sometimes *core model*) of the decision situation. For example, objective trade-offs, defined in Section 2.2 are part of the substantive model, while subjective trade-offs belong to the preferential model. Typically, a substantive model has the following general form:

$$\mathbf{y} = \mathbf{F}(\mathbf{x}, \mathbf{v}, \mathbf{a}); \ \mathbf{x} \in \mathbf{S}, \qquad (2.8)$$

where

- $\mathbf{y}$ is a vector of outcomes (outputs) y_j, used for measuring the consequences of implementation of decisions;
- $\mathbf{x}$ is a vector of decisions (controls, inputs to the decision making process), which are controlled by the user;
- $\mathbf{v}$ is a vector of external impacts (external inputs, perturbations), which are not controlled by the user;
- $\mathbf{a}$ is a vector of model parameters;

- **F** is a vector of functions (including such that are conventionally called objectives and constraints), describing the relations between decisions **x**, impacts **v**, parameters **a**, and outcomes **y**;
- S is the set of feasible decisions.

The form of (2.8) is only slightly more complex but essentially equivalent to what in the other parts of this book is written simply as $\mathbf{z} = \mathbf{f}(\mathbf{x})$, where **z** denotes objectives selected between outcomes **y**. This short notation is a great oversimplification of the actual complexity of models involving multiple objectives or criteria. However, even the form of (2.8) is misleading by its compactness, since it hides the actual complexity of the underlying knowledge representation: a large model today may have several millions of variables and constraints, even when the number of decision and outcome variables is usually much smaller (Makowski, 2005).

Additionally, the substantive model includes constraint specification (symbolically denoted by $\mathbf{x} \in \mathbf{S}$) that might have the form of feasible bounds on selected model outcomes (or be just a list of considered decision options in a discrete case). While the reference point approach is typically described for the continuous case (with a nonempty interior of S, thus an infinite number of solutions in this set), it is as well (or even better) applicable to the discrete case, with a finite number of decision options. The reason for this is that the reference point approach is specifically designed to be effective for nonconvex problems (which is typical for the discrete case).

The actual issue of the separation of preferential and substantive models is that the substantive model should not represent the preferences of the DM, except in one aspect: *the number of decision outcomes in this model should be large enough* for using them in a separate representation of a preferential structure $P(\mathbf{x}, \mathbf{y})$ of the user, needed for selecting a manageable subset of solutions (decisions) that correspond best to user's preferences. The separate representation of preferential structure (the structure of preferential information) can have several degrees of specificity, while the reference point approaches assume that this specification should be as general as possible, since a more detailed specification violates the sovereign right of a DM to change her/his mind.

The most general specification contains a selection of outcomes y_j that are chosen by the DM to measure the quality of decisions (or solutions), called objectives (values of objective functions) or criteria (quality measures, quality indicators) and denoted by $z_j, j = 1, \ldots, k$. This specification is accompanied by defining a partial order in the space of objectives – simply asking the DM which objectives should be maximized and which minimized (while another option, stabilizing some objectives around given reference levels, is also possible in reference point approaches (Wierzbicki *et al.*, 2000)). Here we consider the simplest case when all the objectives are to be minimized.

The second level of specificity in reference point approaches is assumed to consist of specification of reference points – generally, desired objective function values. These reference points might be interval-type, double, including aspiration levels, denoted here by z_j^a (objective function values that the DM would like to achieve) and reservation levels z_j^r (objective values that should be achieved according to the DM). Specification of reference levels is treated as an alternative to trade-off or weighting coefficient information that leads usually to linear representation of preferences and unbalanced decisions as discussed below, although some reference point approaches (Nakayama, 1995) combine reference levels with trade-off information.

The detailed specification of preferences might include full or gradual identification of value functions, see Section 2.2 on trade-off methods or (Keeney and Raiffa, 1976; Keeney, 1992). This is avoided in *reference point approaches that stress learning instead of value identification.* According to the reference point philosophy, the DM should learn during the interaction with a decision support system (DSS), hence her/his preferences might change in the decision making process and (s)he has full, sovereign right or even necessity to be inconsistent.

2) Nonlinearity of preferences. According to a general conviction that human preferences have essentially a nonlinear character, including a preference for balanced solutions. Any linear approximation of preferences (e.g., by a weighted sum distorts them, favoring unbalanced solutions. This is in opposition to the methods taught usually as "the basic approaches" to MCDM. These methods consist of determining (by diverse approaches, between which the AHP (Saaty, 1982) is one of the most often used) weighting coefficients and solving the weighted problem (1.2) discussed in Chapter 1. Such a linear aggregation might be sometimes necessary, but it has several limitations as discussed in Chapter 1. The most serious ones are the following:

- The weighted sum tends to promote decisions with unbalanced objectives, as illustrated by the Korhonen paradox mentioned in Chapter 1. In order to accommodate the natural human preference for balanced solutions, a nonlinear aggregation is necessary.
- The weighted sum is based on a tacit (unstated) assumption that a trade-off analysis is applicable to all objective functions: a worsening of the value of one objective function might be compensated by the improvement of the value of another one. While often encountered in economic applications, this compensatory character of objectives is usually not encountered in interdisciplinary applications.

Educated that weighting methods are basic, the legislators in Poland introduced a public tender law. This law requires that any institution preparing a tender using public money should publish beforehand all objectives of ranking the offers and all weighting coefficients used to aggregate the objectives. This legal innovation backfired: while the law was intended to make public tenders more transparent and accountable, the practical outcome was opposite because of effects similar to the Korhonen paradox. Organizers of the tenders soon discovered that they are forced either to select the offer that is the cheapest and worst in quality or the best in quality but most expensive one. In order to counteract, they either limited the solution space drastically by diverse side constraints (which is difficult but consistent with the spirit of the law) or added additional poorly defined objectives such as the degree of satisfaction (which is simple and legal but fully inconsistent with the spirit of the law, since it makes the tender less transparent and opens a hidden door for graft).

To summarize, a linear weighted sum aggregation is simple but too simplistic in representing typical human preferences that are often nonlinear. Using this simplistic approach may result in adverse and unforeseen side-effects.

3) Holistic perception of objectives. The third basic assumption of reference point approaches is that the DM selects her/his decision using a holistic assessment of the decision situation. In order to help her/him in such a holistic evaluation, a DSS should compute and inform the DM about relevant ranges of objective function values. Such ranges can be defined in diverse ways, while the two basic ones are the following:

- Total ranges of objective functions involve the definition of the lower z_j^{lo} and the upper bound z_j^{up}, over all feasible decisions $\mathbf{x} \in \mathbf{S}$ $(j = 1, \ldots, k)$.
- Pareto optimal ranges of objectives are counted only over Pareto optimal solutions. The lower bound is the utopian or ideal objective vector $z_j^\star$ and is typically equal to z_j^{lo}. The upper bound is the nadir objective vector z_j^{nad} (as discussed in Preface and Chapter 1).

Generally, $z_j^{nad} \leq z_j^{up}$ and the nadir objective vector is easy to determine only in the case of biobjective problems (Ehrgott and Tenfelde-Podehl, 2000) (for continuous models; for discrete models the determination of a nadir point is somewhat simpler). No matter which ranges of objectives we use, it is often useful to assume that all objective functions or quality indicators and their values $f_j(\mathbf{x})$ for decision vectors $\mathbf{x} \in S$ are scaled down to a relative scale by the transformation:

$$z_j^{rel} = f_j^{rel}(\mathbf{x}) = (f_j(\mathbf{x}) - z_j^{lo})/(z_j^{up} - z_j^{lo}) \times 100\%.$$

4) Reference points as tools of holistic learning. Another basic assumption of reference point approaches is that reference (aspiration, reservation) levels and points are treated not as a fixed expression of preferences but as a tool of adaptive, holistic learning about the decision situation as described by the substantive model. Thus, even if the convergence of reference point approaches to a solution most preferred by the DM can be proved (Wierzbicki, 1999), this aspect is never stressed. More important aspects relate to other properties of these approaches. Even if the reference points might be determined in some objective fashion, independently of the preferences of the DM, we stress again a diversity of such objective determinations, thus making possible comparisons of resulting optimal solutions.

5) Achievement functions as ad hoc approximations of value. Given the partial information about preferences (the partial order in the objective space) and their assumed nonlinearity, and the information about the positioning of reference points inside known objective function ranges, the simplest ad hoc approximation of a nonlinear value function consistent with this information and promoting balanced solutions can be proposed. Such an ad hoc approximation takes the form of achievement functions discussed later; see (2.9)–(2.10). (A simple example of them was also introduced in Chapter 1 as problem (1.11). Note that that problem was formulated so that it was to be minimized but here a different variant is described where the achievement function is maximized.)

Achievement functions are determined essentially by max-min terms that favour solutions with balanced deviations from reference points and express the Rawlsian principle of justice (concentrating the attention on worst off members of society or on issues worst provided for (Rawls, 1971)). These terms are slightly corrected by regularizing terms, resulting in the Pareto optimality of the solutions that maximize achievement functions. See also (Lewandowski and Wierzbicki, 1989) for diverse applications where the partial order in the objective space (called also the dominance relation) is not assumed a priori but defined interactively with the DM.

6) Sovereignty of the DM. It can be shown (Wierzbicki, 1986) that achievement functions have the property of full controllability. This means that any Pareto optimal solution can be selected by the DM by modifying reference points and maximizing the achievement function. This provides for the full sovereignty of the DM. Thus, a DSS based on a reference point approach behaves analogously to a perfect analytic section staff in a business organization (Wierzbicki, 1983). The CEO (boss) can outline her/his preferences to the staff and specify the reference points. The perfect staff will tell the boss that her/his aspirations or even reservations are not attainable, if this is the case; but the staff computes in this case also the Pareto optimal solution that comes closest to the aspirations or reservations. If, however, the aspirations are attainable and not Pareto optimal (a better decision might be found), the perfect staff will present to the boss the decision that results in the aspirations and also a Pareto optimal solution corresponding to a uniform improvement of all objectives over the aspirations. In a special case when the aspirations or reservations are Pareto optimal, the perfect staff responds with the decision that results precisely in attaining these aspirations (reservations) – and does not argue that another decision is better, even if such a decision might result from a trade-off analysis performed by the staff. (Only a computerized DSS, not a human staff, can behave in such a perfect fashion.)

7) Final aims: intuition support versus rational objectivity. To summarize the fundamental assumptions and philosophy of reference point approaches, the basic aim when supporting an individual, subjective DM, is to enhance her/his power of intuition (Wierzbicki, 1997) by enabling holistic learning about the decision situation as modelled by the substantive model. The same applies, actually, when using reference point approaches for supporting negotiations and group decision making (Makowski, 2005).

2.3.2 Basic Features of Reference Point Approaches

The large disparity between the opposite ends of the spectrum of preference elicitation (full value function identification versus a weighted sum approach) indicates the need of a middle-ground approach, simple enough and easily adaptable but not too simplistic. An interactive decision making process in a DSS using a reference point approach consists typically of the following steps:

- The decision maker (DM) specifies reference points (e.g., aspiration and reservation levels for all objective functions). To help her/him in starting the process, the DSS can compute a *neutral solution,* a response to reference levels situated in the middle of objective function ranges (see problem (1.6) in Chapter 1);
- The DSS responds by maximizing an achievement function, a relatively simple but nonlinear aggregation of objective functions interpreted as an ad hoc and adaptable approximation of the value function of the DM based on the information contained in the estimates of the ranges of objective functions and in the positioning of aspiration and reservation levels inside these ranges;
- The DM is free to modify the reference points as (s)he will. (S)he is supposed to use this freedom to learn about the decision situation and to explore interesting parts of the Pareto optimal set;
- Diverse methods can be used to help the DM in this exploration (we comment on them later), but the essential point is that they should not limit the freedom of the DM.

In order to formulate an achievement function, we first count achievements for each individual objective function by transforming it (piece-wise linearly) (for objective functions to be minimized) as

$$\sigma_j(z_j, z_j^a, z_j^r) = \begin{cases} 1 + \alpha(z_j^a - z_j)/(z_j^a - z_j^{lo}), & if \ z_j^{lo} \le z_j \le z_j^a \\ (z_j^r - z_j)/(z_j^r - z_j^a), & if \ z_j^a < z_j \le z_j^r, \\ \beta(z_j^r - z_j)/(z_j^{up} - z_j^r), & if \ z_j^r < z_j \le z_j^{up} \end{cases} \quad (2.9)$$

The coefficients α and β are typically selected to assure the concavity of this function (Wierzbicki *et al.,* 2000); but the concavity is needed only for problems with a continuous (nonempty interior) set of solutions, for an easy transformation to a linear programming problem. The value $\sigma_j = \sigma_j(z_j, z_j^a, z_j^r)$ of this achievement function (where $z_j = f_j(\mathbf{x})$ for a given decision vector $\mathbf{x} \in S$) signifies the satisfaction level with the quality indicator or objective j for this decision vector. If we assign the values of satisfaction from -1 to 0 for $z_j^r < z_j \le z_j^{up}$, values from 0 to 1 for $z_j^a < z_j \le z_j^r$, values from 1 to 2 for $z_j^{lo} \leq z_j \leq z_j^a$, then we can just set $\alpha = \beta = 1$. After this transformation of all objective function values, we might use then the following form of the overall achievement function to be maximized[1]:

$$\sigma(\mathbf{z}, \mathbf{z}^a, \mathbf{z}^r) = \min_{j=1,\ldots,k} \sigma_j(\mathbf{z_j}, \mathbf{z_j^a}, \mathbf{z_j^r}) + \rho \sum_{j=1,\ldots,k} \sigma_j(\mathbf{z_j}, \mathbf{z_j^a}, \mathbf{z_j^r}), \qquad (2.10)$$

where $\mathbf{z} = \mathbf{f}(\mathbf{x})$ is the objective vector and $\mathbf{z}^a = (\mathbf{z_1^a}, \ldots, \mathbf{z_k^a})$ and $\mathbf{z}^r = (\mathbf{z_1^r}, \ldots, \mathbf{z_k^r})$ the vectors of aspiration and reservation levels, respectively. Furthermore, $\rho > 0$ is a small regularizing coefficient (as discussed in Chapter 1.

There are many possible forms of achievement functions besides (2.9)–(2.10), as shown in (Wierzbicki *et al.*, 2000). All of them, however, have an important property of partial order approximation: their level sets approximate closely the positive cone defining the partial order (Wierzbicki, 1986). As indicated above, the achievement function has also a very important theoretical property of controllability, not possessed by value functions nor by weighted sums: for sufficiently small values of ρ, given any point $\mathbf{z}^*$ in the set of (properly) Pareto optimal objective vectors, we can always choose such reference levels that the maximum of the achievement function (2.10) is attained precisely at this point. In fact, it suffices to set aspiration levels equal to the components of $\mathbf{z}^*$. Conversely, if $\rho > 0$, all maxima of the achievement function (2.10) correspond to Pareto optimal solutions (because of the monotonicity of this function with respect to the partial order in the objective space.) Thus, the behaviour of achievement functions corresponds in this respect to value functions and weighted sums. However, let us emphasize that this is not the case in the distance norm used in goal programming (see Section 1.6.3 in Chapter 1), since the norm is not monotone when passing zero. As noted above, precisely the controllability property results in a fully sovereign control of the DSS by the user.

Alternatively, as shown in (Ogryczak, 2006), we can assume $\rho = 0$ and use the nucleolar minimax approach. In this approach, we consider first the minimal, worst individual objective-wise achievement computed as in (2.9)–(2.10) with $\rho = 0$. If two (or more) solutions have the same achievement value, we order them according to the second worst individual objective-wise achievement and so on.

There are many modifications, variants and extensions (Wierzbicki *et al.*, 2000) or approaches related to the basic reference point approach, mostly designed for helping the search phase in the Pareto optimal set. For example,

- the Tchebycheff method (Steuer, 1986) was developed independently but actually is equivalent to using weighting coefficients implied by reference levels;

---

[1] Even if in this book objectives are supposed to be typically minimized, achievements are here maximized.

- Pareto Race (Korhonen and Laakso, 1986) is a visual method based on reference points distributed along a direction in the objective space, or the REF-LEX method for nonlinear problems (Miettinen and Kirilov, 2005);
- the satisficing trade-off method, or the NIMBUS method, both described in a later section, or the 'light beam search' method (Jaszkiewicz and Słowiński, 1999), or several other approaches were motivated by the reference point approach.

In this section, we have presented some of the basic assumptions and philosophy of reference point approaches, stressing their unique concentration on the sovereignty of the subjective DM. Next we concentrate on classification-based methods.

2.4 Classification-Based Methods

2.4.1 Introduction to Classification of Objective Functions

According to the definition of Pareto optimality, moving from one Pareto optimal solution to another implies trading off. In other words, it is possible to move to another Pareto optimal solution and improve some objective function value(s) only by allowing some other objective function value(s) to get worse. This idea is used as such in classification-based methods. By classification-based methods we mean methods where the DM indicates her/his preferences by classifying objective functions. The idea is to tell which objective functions should improve and which ones could impair from their current values. In other words, the DM is shown the current Pareto optimal solution and asked what kind of changes in the objective function values would lead to a more preferred solution. It has been shown by Larichev (1992) that the classification of objective functions is a cognitively valid way of expressing preference information for a DM.

We can say that classification is a very intuitive way for the DM to direct the solution process in order to find the most preferred solution because no artificial concepts are used. Instead, the DM deals with objective function values that are as such meaningful and understandable for her/him. The DM can express hopes about improved solutions and directly see and compare how well the hopes could be attained when the next solution is generated.

To be more specific, when classifying objective functions (at the current Pareto optimal solution) the DM indicates which function values should improve, which one are acceptable as such and which ones are allowed to impair. In addition, desirable amounts of improvement or allowed amounts of impairments may be asked from the DM. There exist several classification-based interactive multiobjective optimization methods. They differ from each other, for example, in the number of classes available, the preference information asked from the DM and how this information is used to generate new Pareto optimal solutions.

Let us point out that closely related to classification is the idea of expressing preference information as a reference point (Miettinen and Mäkelä, 2002; Miettinen *et al.*, 2006). The difference is that while classification assumes that some objective function must be allowed to get worse, a reference point can be selected more freely. Naturally, it is not possible to improve all objective function values of a Pareto optimal solution, but the DM can express preferences without paying attention to this and then see what kind of solutions are feasible. However, when using classification, the DM can be more in control and select functions to be improved and specify amounts of relaxation for the others.

As far as stopping criteria are concerned, classification-based methods share the philosophy of reference point based methods (discussed in the previous section) so that the DM's satisfaction is the most important stopping criterion. This means that the search process continues as long as the DM wants to and the mathematical convergence is not essential (as in trade-off based methods) but rather the psychological convergence is emphasized (discussed in the introduction). This is justified by the fact that DMs typically want to feel being in control and do not necessarily want the method to tell them when they have found their most preferred solutions. After all, the most important task of interactive methods is to support the DM in decision making.

In what follows, we briefly describe the step method, the satisficing trade-off method and the NIMBUS method. Before that, we introduce some common notation.

Throughout this section, we denote the current Pareto optimal solution by $\mathbf{z}^h = \mathbf{f}(\mathbf{x}^h)$. When the DM classifies the objective functions at the current solution, we can say that (s)he assigns each of them into some class and the number of classes available varies in different methods. In general, we have the following classes for functions f_i $(i = 1, \ldots, k)$

- $I^<$ whose values should be improved (i.e., decrease) from the current level,
- $I^\leq$ whose values should improve till some desired aspiration level $\hat{z}_i < z_i^h$,
- $I^=$ whose values are acceptable in the current solution,
- $I^\geq$ whose values can be impaired (i.e., increase) till some upper bound $\varepsilon_i > z_i^h$ and,
- $I^\diamond$ whose values are temporarily allowed to change freely.

The aspiration levels and the upper bounds corresponding to the classification are elicited from the DM, if they are needed. According to the definition of Pareto optimality, a classification is feasible only if $I^< \cup I^\leq \neq \emptyset$ and $I^\geq \cup I^\diamond \neq \emptyset$ and the DM has to classify all the objective functions, that is, $I^< \cup I^\leq \cup I^= \cup I^\geq \cup I^\diamond = \{1, \ldots, k\}$.

2.4.2 Step Method

The step method (STEM) (Benayoun $et\ al.$, 1971) uses only two classes. STEM is one of the first interactive methods introduced for multiobjective optimization and it was originally developed for MOLP problems. However, here we describe variants for nonlinear problems according to Eschenauer $et\ al.$ (1990); Sawaragi $et\ al.$ (1985); Vanderpooten and Vincke (1989).

In STEM, the DM is assumed to classify the objective functions at the current solution $\mathbf{z}^h$ into those that have acceptable values $I^\geq$ and those whose values are too high, that is, functions that have unacceptable values $I^<$. Then the DM is supposed to give up a little in the value(s) of some acceptable objective function(s) in order to improve the values of some unacceptable objective functions. In other words, the DM is asked to specify upper bounds $\varepsilon_i^h > z_i^h$ for the functions in $I^\geq$. All the objective functions must be classified and, thus, $I^< \cup I^\geq = \{1, \ldots, k\}$.

It is assumed that the objective functions are bounded over S because distances are measured to the (global) ideal objective vector. STEM uses the weighted Chebyshev problem 1.8 introduced in Chapter 1 to generate new solutions. The weights are used to make the scales of the objective functions similar. The first problem to be solved is

$$\text{minimize} \quad \max_{i=1,\ldots,k} \left[\frac{e_i}{\sum_{j=1}^k e_j} (f_i(\mathbf{x}) - z_i^\star) \right] \tag{2.11}$$
$$\text{subject to} \quad \mathbf{x} \in S,$$

where $e_i = \frac{1}{z_i^\star} \frac{z_i^{\text{nad}} - z_i^\star}{z_i^{\text{nad}}}$ as suggested by Eschenauer $et\ al.$ (1990). Alternatively, we can set $e_i = \frac{z_i^{\text{nad}} - z_i^\star}{\max\left[|z_i^{\text{nad}}|, |z_i^\star| \right]}$ as suggested by Vanderpooten and Vincke (1989). (Naturally we assume that the denominators are not equal to zero.) It can be proved that the solution of (2.11) is weakly Pareto optimal. The solution obtained is the starting point for the method and the DM is asked to classify the objective functions at this point.

Then the feasible region is restricted according to the information given by the DM. The weights of the relaxed objective functions are set equal to zero, that is $e_i - 0$ for $i \in I^\geq$. Then a new distance minimization problem

$$\text{minimize} \quad \max_{i=1,\ldots,k} \left[\frac{e_i}{\sum_{j=1}^k e_j} (f_i(\mathbf{x}) - z_i^\star) \right]$$
$$\text{subject to} \quad f_i(\mathbf{x}) \leq \varepsilon_i^h \ \text{ for all } \ i \in I^\geq,$$
$$f_i(\mathbf{x}) \leq f_i(\mathbf{x}^h) \ \text{ for all } \ i \in I^<,$$
$$\mathbf{x} \in S$$

is solved.

The DM can classify the objective functions at the solution obtained and the procedure continues until the DM does not want to change the current solution. In STEM, the idea is to move from one weakly Pareto optimal solution to another. Pareto optimality of the solutions could be guaranteed, for example, by using augmentation terms as discussed in Chapter 1 (see also (Miettinen and Mäkelä, 2006)). The idea of classification is quite simple for the DM. However, it may be difficult to estimate how much the other functions should be relaxed in order to potentiate the desired amounts of improvement in the others. The next method aims at resolving this kind of a difficulty.

2.4.3 Satisficing Trade-off Method

The satisficing trade-off method (STOM) (Nakayama, 1995; Nakayama and Sawaragi, 1984) is based on ideas very similar to those in reference point approaches. As its name suggests, it concentrates on finding a satisficing solution (see Chapter 1).

The DM is asked to classify the objective functions at $\mathbf{z}^h$ into three classes. The classes are the objective functions whose values should be improved $I^{\leq}$, the functions whose values can be relaxed $I^{\geq}$ and the functions whose values are acceptable as such $I^{=}$. The DM is supposed to specify desirable aspiration levels for functions in $I^{\leq}$. Here, $I^{\leq} \cup I^{\geq} \cup I^{=} = \{1, \ldots, k\}$.

Because of so-called *automatic trade-off*, the DM only has to specify desirable levels for functions in $I^{\leq}$ and the upper bounds for functions in $I^{\geq}$ are derived from trade-off rate information. The idea is to decrease the burden set on the DM so that the amount of information to be specified is reduced. Functions are assumed to be twice continuously differentiable. Under some special assumptions, trade-off information can be obtained from the KKT multipliers related to the scalarizing function used (corresponding to Section 2.2).

By putting together information about desirable function values, upper bounds deduced using automatic trade-off and current acceptable function values, we get a reference point $\bar{\mathbf{z}}^h$. Different scalarizing functions can be used in STOM but in general, the idea is to minimize the distance to the utopian objective vector $\mathbf{z}^{\star\star}$. We can, for example, solve the problem

$$\text{minimize} \quad \max_{i=1,\ldots,k} \left[\frac{f_i(\mathbf{x}) - z_i^{\star\star}}{\bar{z}_i^h - z_i^{\star\star}} \right] + \rho \sum_{i=1}^{k} \frac{f_i(\mathbf{x})}{\bar{z}_i^h - z_i^{\star\star}} \quad (2.12)$$
$$\text{subject to} \quad \mathbf{x} \in S,$$

where we must have $\bar{z}_i > z_i^{\star\star}$ for all $i = 1, \ldots, k$. It can be proved that the solutions obtained are properly Pareto optimal . Furthermore, it can be proved that the solution $\mathbf{x}^*$ is satisficing (see Chapter 1) if the reference point is feasible (Sawaragi *et al.*, 1985). This means that $f_i(\mathbf{x}^*) \leq \bar{z}_i^h$ for all i. Let us point out that if some objective function f_i is not bounded from below in S, then some small scalar value can be selected as $z_i^{\star\star}$. Other weighting coefficients can also be used instead of $1/(\bar{z}_i^h - z_i^{\star\star})$ (Nakayama, 1995).

The solution process can be started, for example, by asking the DM to specify a reference point and solving problem (2.12). Then, at this point, the

DM is asked to classify the objective functions and specify desirable aspiration levels for the functions to be improved. The solution process continues until the DM does not want to improve or relax any objective function value.

In particular, if the problem has many objective functions, the DM my appreciate the fact that (s)he does not have to specify upper bound values. Naturally, the DM may modify the calculated values if they are not agreeable. It is important to note that STOM can be used even if the assumptions enabling automatic trade-off are not valid. In this case, the DM has to specify both aspiration levels and upper bounds.

2.4.4 NIMBUS Method

The NIMBUS method is described in (Miettinen, 1999; Miettinen and Mäkelä, 1995, 1999, 2000; Miettinen *et al*, 1998) and it is based on the classification of the objective functions into up to five classes. It has been developed for demanding nonlinear multiobjective optimization. In NIMBUS, the DM can classify objective functions at $\mathbf{z}^h$ into any of the five classes introduced at the beginning of this section, that is, functions to be decreased $I^<$ and to be decreased till an aspiration level $I^\leq$, functions that are satisfactory at the moment $I^=$, functions that can increase till an upper bound $I^\geq$ and functions that can change freely $I^\diamond$ and $I^< \cup I^\leq \cup I^= \cup I^\geq \cup I^\diamond = \{1, \ldots, k\}$. We assume that we have the ideal objective vector and a good approximation of the nadir objective vector available.

The difference between the classes $I^<$ and $I^\leq$ is that the functions in $I^<$ are to be minimized as far as possible but the functions in $I^\leq$ only till the aspiration level (specified by the DM). There are several different variants of NIMBUS but here we concentrate on the so-called synchronous method (Miettinen and Mäkelä, 2006). After the DM has made the classification, we form a scalarizing function and solve the problem

$$
\begin{aligned}
\text{minimize} \quad & \max_{\substack{i \in I^< \\ j \in I^\leq}} \left[\frac{f_i(\mathbf{x}) - z_i^\star}{z_i^{\text{nad}} - z_i^{\star\star}}, \frac{f_j(\mathbf{x}) - \hat{z}_j}{z_j^{\text{nad}} - z_j^{\star\star}} \right] + \rho \sum_{i=1}^{k} \frac{f_i(\mathbf{x})}{z_i^{\text{nad}} - z_i^{\star\star}} \\
\text{subject to} \quad & f_i(\mathbf{x}) \leq f_i(\mathbf{x}^h) \quad \text{for all} \quad i \in I^< \cup I^\leq \cup I^=, \\
& f_i(\mathbf{x}) \leq \varepsilon_i \quad \text{for all} \quad i \in I^\geq, \\
& \mathbf{x} \in S,
\end{aligned}
\tag{2.13}
$$

where $\rho > 0$ is a relatively small scalar. The weighting coefficients $1/(z_j^{\text{nad}} - z_j^{\star\star})$ (scaling the objective function values) have proven to facilitate capturing the preferences of the DM well. They also increase computational efficiency (Miettinen *et al.*, 2006). By solving problem (2.13) we get a provably (properly) Pareto optimal solution that satisfies the classification as well as possible.

In the synchronous NIMBUS method, the DM can ask for up to four different Pareto optimal solutions be generated based on the classification once expressed. This means that solutions are produced that take the classification information into account in slightly different ways. In practice, we form a reference point $\bar{z}$ based on the classification information specified as follows: $\bar{z}_i = z_i^\star$ for $i \in I^<$, $\bar{z}_i = \hat{z}_i$ for $i \in I^{\leq}$, $\bar{z}_i = z_i^h$ for $i \in I^=$, $\bar{z}_i = \varepsilon_i$ for $i \in I^{\geq}$ and $\bar{z}_i = z_i^{\text{nad}}$ for $i \in I^\circ$. (This, once again, demonstrates the close relationship between classification and reference points.) Then we can use reference point based scalarizing functions to generate new solutions. In the synchronous NIMBUS method, the scalarizing functions used are those coming from STOM (problem (2.12) in Section 2.4.3), reference point method (problem (1.11) defined in Chapter 1) and GUESS (Buchanan, 1997). See (Miettinen and Mäkelä, 2002) for details on how they were selected. Let us point out that all the solutions produced are guaranteed to be properly Pareto optimal.

Further details and the synchronous NIMBUS algorithm are described in (Miettinen and Mäkelä, 2006). The main steps are the following: Once the DM has classified the objective functions at the current solution $\mathbf{z}^h$ and specified aspiration levels and upper bounds, if needed, (s)he is asked how many new solutions (s)he wants to see and compare. As many solutions are generated (as described above) and the DM can select any of them as the final solution or as the starting point of a new classification. It is also possible to select any of the solutions generated so far as a starting point for a new classification. The DM can also control the search process by asking for a desired number of intermediate solutions to be generated between any two interesting solutions found so far. In this case, steps of equal length are taken in the decision space and corresponding objective vectors are used as reference points to get Pareto optimal solutions that the DM can compare.

The starting point can be, for example, a neutral compromise solution (see problem (1.6) in Chapter 1) or any point specified by the DM (which has been projected to the Pareto optimal set). In NIMBUS, the DM expresses iteratively her/his desires. Unlike some other methods based on classification, the success of the solution process does not depend entirely on how well the DM manages in specifying the classification and the appropriate parameter values. It is important to note that the classification is not irreversible. Thus, no irrevocable damage is caused in NIMBUS if the solution obtained is not what was expected. The DM is free to go back or explore intermediate points. (S)he can easily get to know the problem and its possibilities by specifying, for example, loose upper bounds and examining intermediate solutions.

The method has been implemented as a WWW-NIMBUS® system operating on the Internet (Miettinen and Mäkelä, 2000, 2006). Via the Internet, the computing can be centralized to one server computer and the WWW is a way of distributing the graphical user interface to the computers of each individual user and the user always has the latest version of the method available. The most important aspect of WWW-NIMBUS® is that it is easily accessible and available to any academic Internet user at http://nimbus.it.jyu.fi/. For a discussion on how to design user interfaces for a software implementing a classification-based interactive method, see (Miettinen and Kaario, 2003). (When the first version of WWW-NIMBUS® was implemented in 1995 it was a pioneering interactive optimization system on the Internet.) Another implementation IND-NIMBUS$^®$ for MS-Windows and Linux operating systems also exists (Miettinen, 2006) (see Chapter 12). Many successful applications, for example, in the fields of optimal control, optimal shape design and process design have shown the usefulness of the method (Hakanen *et al.*, 2005, 2007; Hämäläinen *et al.*, 2003; Heikkola *et al.*, 2006; Miettinen *et al.*, 1998).

2.4.5 Other Classification-Based Methods

Let us briefly mention some more classification-based methods. Among them are the interactive reference direction algorithm for convex nonlinear integer problems (Vassilev *et al.*, 2001) which uses three classes $I^\leq$, $I^\geq$ and $I^=$ and the reference direction approach for nonlinear problems (Narula *et al.*, 1994) using the same three classes and generating several solutions in the reference direction (pointing from the current solution towards the reference point). Furthermore, the interactive decision making approach NIDMA (Kaliszewski and Michalowski, 1999) asks for both a classification and maximal acceptable global trade-offs from the DM.

A method where NIMBUS (see Subsection 2.4.4) is hybridized with the feasible goals method (Lotov *et al.*, 2004) is described in (Miettinen *et al.*, 2003). Because the feasible goals method produces visual interactive displays of the variety of feasible objective vectors, the hybrids introduced help the DM in getting understanding of what kinds of solutions are feasible, which helps when specifying classifications for NIMBUS. (There are also classification-based methods developed for MOLP problems which we do not touch here.)

2.5 Discussion

Due to the large variety of interactive methods available in the literature, it is a hard task to choose the most appropriate method for each decision situation. Here, a "decision situation" must be understood in a wide sense: a DM, with a given attitude (due to many possible facts) facing (a part of) a decision problem. This issue will be discussed in detail in Chapter 15. In order to accommodate the variety of methods in a single decision system, some authors have already proposed the creation of open architectures or combined systems (Gardiner and Steuer, 1994; Kaliszewski, 2004; Luque *et al.*, 2007b). Some of such integrated systems have already been implemented. For example, MKO and PROMOIN are described in Chapter 12. It is also worth pointing out that some relations among the different types of information (like weights, trade-offs, reference points etc.) that the interactive methods may require from the DM, are investigated in (Luque *et al.*, 2007a).

One direction for developing new, improved, methods is hybridizing advantages of different methods in order to overcome their weaknesses. For example, hybridizing a posteriori and interactive methods has a lot of potential. The

DM can, for example, first get a rough impression of the possibilities of the problem and then can interactively locate the most preferred solution. One approach in this direction was already mentioned with NIMBUS (Lotov *et al.*, 2004). Another idea is to combine reference points with an algorithm that generates an approximation of the Pareto optimal set (Klamroth and Miettinen, 2008). This means that only those parts of the Pareto optimal set are approximated more accurately that the DM is interested in and (s)he can conveniently control which parts to study by using a reference point. Steuer *et al.* (1993) provide an example of hybridizing ideas of two interactive methods by combining ideas of the Tchebycheff method and reference point approaches.

When dealing with human DMs, behavioural issues cannot be ignored. For example, some points of view in this respect are collected in (Korhonen and Wallenius, 1996). Let us also mention an interesting practical observation mentioned by (Buchanan, 1997). Namely, DMs seem to be easily satisfied if there is a small difference between their hopes and the solution obtained. Somehow they feel a need to be satisfied when they have almost achieved what they wanted for even if they still were in the early steps of the learning process. In this case they may stop iterating 'too early.' Naturally, the DM is allowed to stop the solution process if the solution really is satisfactory but the coincidence of setting the desires near an attainable solution may unnecessarily increase the DM's satisfaction (see also Chapter 15).

2.6 Conclusions

We have characterized some basic properties of interactive methods developed for multiobjective optimization and considered three types of methods based on trade-offs, reference points and classification. An important requirement for using interactive methods is that the DM must have time and interest in taking part in the iterative solution process. On the other hand, the major advantage of these methods is that they give the DM a unique possibility to learn about the problem considered. In this way, the DM is much better able to justify why the final solution is the most preferred one.

As has been stressed, a large variety of methods exists and none of them can be claimed to be superior to the others in every aspect. When selecting a solution method, the opinions of the DM are important because (s)he must feel comfortable with the way (s)he is expected to provide preference information. In addition, the specific features of the problem to be solved must be taken into consideration. One can say that selecting a multiobjective optimization method is a problem with multiple objectives itself.

When dealing with interactive methods, the importance of user-friendliness is emphasized. This is also a topic for future research. Methods must be even better able to correspond to the characteristics of the DM. If the aspirations of the DM change during the solution process, the algorithm must be able to cope with this situation.

DMs want to feel in control of the solution process and, consequently, they must understand what is happening. Thus, the preferences expressed must be reflected in the Pareto optimal solutions generated. But if the DM needs support in identifying the most preferred region of the Pareto optimal set, this should be available, as well. Thus, the aim is to have methods that support learning so that guidance is given whenever necessary. The DM can be supported by using visual illustrations (see, e.g. Chapters 8 and 9) and further development of such tools is essential. In particular, when designing DSSs for DMs, user interfaces play a central role. Special-purpose methods for different areas of application that take into account the characteristics of the problems are also important.

Acknowledgements

The work of K. Miettinen was partly supported by the Foundation of the Helsinki School of Economics. The work of F. Ruiz was partly supported by the Spanish Ministry of Education and Science.

References

Benayoun, R., de Montgolfier, J., Tergny, J., Laritchev, O.: Programming with multiple objective functions: Step method (STEM). Mathematical Programming 1, 366–375 (1971)

Buchanan, J.T.: Multiple objective mathematical programming: A review. New Zealand Operational Research 14, 1–27 (1986)

Buchanan, J.T.: A naïve approach for solving MCDM problems: The GUESS method. Journal of the Operational Research Society 48, 202–206 (1997)

Chankong, V., Haimes, Y.Y.: The interactive surrogate worth trade-off (ISWT) method for multiobjective decision making. In: Zionts, S. (ed.) Multiple Criteria Problem Solving, pp. 42–67. Springer, Berlin (1978)

Chankong, V., Haimes, Y.Y.: Multiobjective Decision Making. Theory and Methodology. North-Holland, New York (1983)

Ehrgott, M., Tenfelde-Podehl, D.: Nadir values: Computation and use in compromise programming. Technical report, Universität Kaiserslautern Fachbereich Mathematik (2000)

Eschenauer, H.A., Osyczka, A., Schäfer, E.: Interactive multicriteria optimization in design process. In: Eschenauer, H., Koski, J., Osyczka, A. (eds.) Multicriteria Design Optimization Procedures and Applications, pp. 71–114. Springer, Berlin (1990)

Gardiner, L., Steuer, R.E.: Unified interactive multiobjective programming. European Journal of Operational Research 74, 391–406 (1994)

Geoffrion, A.M., Dyer, J.S., Feinberg, A.: An interactive approach for multi-criterion optimization, with an application to the operation of an academic department. Management Science 19, 357–368 (1972)

Haimes, Y.Y., Hall, W.A.: Multiobjectives in water resources systems analysis: the surrogate worth trade off method. Water Resources Research 10, 615–624 (1974)

Haimes, Y.Y., Tarvainen, K., Shima, T., Thadathil, J.: Hierarchical Multiobjective Analysis of Large-Scale Systems. Hemisphere Publishing Corporation, New York (1990)

Hakanen, J., Miettinen, K., Mäkelä, M., Manninen, J.: On interactive multiobjective optimization with NIMBUS in chemical process design. Journal of Multi-Criteria Decision Analysis 13, 125–134 (2005)

Hakanen, J., Kawajiri, Y., Miettinen, K., Biegler, L.: Interactive multi-objective optimization for simulated moving bed processes. Control and Cybernetics 36, 283–302 (2007)

Hämäläinen, J., Miettinen, K., Tarvainen, P., Toivanen, J.: Interactive solution approach to a multiobjective optimization problem in paper machine headbox design. Journal of Optimization Theory and Applications 116, 265–281 (2003)

Heikkola, E., Miettinen, K., Nieminen, P.: Multiobjective optimization of an ultrasonic transducer using NIMBUS. Ultrasonics 44, 368–380 (2006)

Hwang, C.L., Masud, A.S.M.: Multiple Objective Decision Making – Methods and Applications: A State-of-the-Art Survey. Springer, Berlin (1979)

Jaszkiewicz, A., Słowiński, R.: The 'light beam search' approach - an overview of methodology and applications. European Journal of Operational Research 113, 300–314 (1999)

Kaliszewski, I.: Out of the mist–towards decision-maker-friendly multiple criteria decision support. European Journal of Operational Research 158, 293–307 (2004)

Kaliszewski, I., Michalowski, W.: Searching for psychologically stable solutions of multiple criteria decision problems. European Journal of Operational Research 118, 549–562 (1999)

Keeney, R.: Value Focused Thinking, a Path to Creative Decision Making. Harvard University Press, Harvard (1992)

Keeney, R., Raiffa, H.: Decisions with Multiple Objectives: Preferences and Value Tradeoffs. Wiley, New York (1976)

Klamroth, K., Miettinen, K.: Integrating approximation and interactive decision making in multicriteria optimization. Operations Research 56, 222–234 (2008)

Korhonen, P.: Interactive methods. In: Figueira, J., Greco, S., Ehrgott, M. (eds.) Multiple Criteria Decision Analysis. State of the Art Surveys, pp. 641–665. Springer, New York (2005)

Korhonen, P., Laakso, J.: A visual interactive method for solving the multiple criteria problem. European Journal of Operational Research 24, 277–287 (1986)

Korhonen, P., Wallenius, J.: Behavioural issues in MCDM: Neglected research questions. Journal of Multi-Criteria Decision Analysis 5, 178–182 (1996)

Larichev, O.: Cognitive validity in design of decision aiding techniques. Journal of Multi-Criteria Decision Analysis 1, 127–138 (1992)

Lewandowski, A., Wierzbicki, A.P.: Aspiration Based Decision Support Systems. Theory, Software and Applications. Springer, Berlin (1989)

Lotov, A.V., Bushenkov, V.A., Kamenev, G.K.: Interactive Decision Maps. Approximation and Visualization of Pareto Frontier. Kluwer Academic Publishers, Boston (2004)

Luque, M., Caballero, R., Molina, J., Ruiz, F.: Equivalent information for multiobjective interactive procedures. Management Science 53, 125–134 (2007a)

Luque, M., Ruiz, F., Miettinen, K.: GLIDE – general formulation for interactive multiobjective optimization. Technical Report W-432, Helsinki School of Economics, Helsinki (2007b)

Makowski, M.: Model-based decision making support for problems with conflicting goals. In: Proceedings of the 2nd International Symposium on System and Human Science, Lawrence Livermore National Laboratory, Livermore (2005)

Miettinen, K.: Nonlinear Multiobjective Optimization. Kluwer Academic Publishers, Boston (1999)

Miettinen, K.: IND-NIMBUS for demanding interactive multiobjective optimization. In: Trzaskalik, T. (ed.) Multiple Criteria Decision Making '05, pp. 137–150. The Karol Adamiecki University of Economics, Katowice (2006)

Miettinen, K., Kaario, K.: Comparing graphic and symbolic classification in interactive multiobjective optimization. Journal of Multi-Criteria Decision Analysis 12, 321–335 (2003)

Miettinen, K., Kirilov, L.: Interactive reference direction approach using implicit parametrization for nonlinear multiobjective optimization. Journal of Multi-Criteria Decision Analysis 13, 115–123 (2005)

Miettinen, K., Mäkelä, M.M.: Interactive bundle-based method for nondifferentiable multiobjective optimization: NIMBUS. Optimization 34, 231–246 (1995)

Miettinen, K., Mäkelä, M.M.: Comparative evaluation of some interactive reference point-based methods for multi-objective optimisation. Journal of the Operational Research Society 50, 949–959 (1999)

Miettinen, K., Mäkelä, M.M.: Interactive multiobjective optimization system WWW-NIMBUS on the Internet. Computers & Operations Research 27, 709–723 (2000)

Miettinen, K., Mäkelä, M.M.: On scalarizing functions in multiobjective optimization. OR Spectrum 24, 193–213 (2002)

Miettinen, K., Mäkelä, M.M.: Synchronous approach in interactive multiobjective optimization. European Journal of Operational Research 170, 909–922 (2006)

Miettinen, K., Mäkelä, M.M., Männikkö, T.: Optimal control of continuous casting by nondifferentiable multiobjective optimization. Computational Optimization and Applications 11, 177–194 (1998)

Miettinen, K., Lotov, A.V., Kamenev, G.K., Berezkin, V.E.: Integration of two multiobjective optimization methods for nonlinear problems. Optimization Methods and Software 18, 63–80 (2003)

Miettinen, K., Mäkelä, M.M., Kaario, K.: Experiments with classification-based scalarizing functions in interactive multiobjective optimization. European Journal of Operational Research 175, 931–947 (2006)

Nakayama, H.: Aspiration level approach to interactive multi-objective programming and its applications. In: Pardalos, P.M., Siskos, Y., Zopounidis, C. (eds.) Advances in Multicriteria Analysis, pp. 147–174. Kluwer Academic Publishers, Dordrecht (1995)

Nakayama, H., Sawaragi, Y.: Satisficing trade-off method for multiobjective programming. In: Grauer, M., Wierzbicki, A.P. (eds.) Interactive Decision Analysis, pp. 113–122. Springer, Heidelberg (1984)

Narula, S.C., Kirilov, L., Vassilev, V.: Reference direction approach for solving multiple objective nonlinear programming problems. IEEE Transactions on Systems, Man, and Cybernetics 24, 804–806 (1994)

Ogryczak, W.: On multicriteria optimization with fair aggregation of individual achievements. In: CSM'06: 20th Workshop on Methodologies and Tools for Complex System Modeling and Integrated Policy Assessment, IIASA, Laxenburg, Austria (2006), http://www.iiasa.ac.at/~marek/ftppub/Pubs/csm06/ogryczak_pap.pdf

Rawls, J.: A Theory of Justice. Belknap Press, Cambridge (1971)

Saaty, T.: Decision Making for Leaders: the Analytical Hierarchy Process for Decisions in a Complex World. Lifetime Learning Publications, Belmont (1982)

Sakawa, M.: Interactive multiobjective decision making by the sequential proxy optimization technique. European Journal of Operational Research 9, 386–396 (1982)

Sawaragi, Y., Nakayama, H., Tanino, T.: Theory of Multiobjective Optimization. Academic Press, Orlando (1985)

Shin, W.S., Ravindran, A.: Interactive multiple objective optimization: Survey I – continuous case. Computers & Operations Research 18, 97–114 (1991)

Statnikov, R.B.: Multicriteria Design: Optimization and Identification. Kluwer Academic Publishers, Dordrecht (1999)

Steuer, R.E.: Multiple Criteria Optimization: Theory, Computation, and Applications. Wiley, Chichester (1986)

Steuer, R.E.: The Tchebycheff procedure of interactive multiple objective programming. In: Karpak, B., Zionts, S. (eds.) Multiple Criteria Decision Making and Risk Analysis Using Microcomputers, pp. 235–249. Springer, Berlin (1989)

Steuer, R.E., Silverman, J., Whisman, A.W.: A combined Tchebycheff/aspiration criterion vector interactive multiobjective programming procedure. Management Science 39, 1255–1260 (1993)

Stewart, T.J.: A critical survey on the status of multiple criteria decision making theory and practice. Omega 20, 569–586 (1992)

Tabucanon, M.T.: Multiple Criteria Decision Making in Industry. Elsevier Science Publishers, Amsterdam (1988)

Tarvainen, K.: On the implementation of the interactive surrogate worth tradeoff (ISWT) method. In: Grauer, M., Wierzbicki, A.P. (eds.) Interactive Decision Analysis, pp. 154–161. Springer, Berlin (1984)

Vanderpooten, D., Vincke, P.: Description and analysis of some representative interactive multicriteria procedures. Mathematical and Computer Modelling 12, 1221–1238 (1989)

Vassilev, V.S., Narula, S.C., Gouljashki, V.G.: An interactive reference direction algorithm for solving multi-objective convex nonlinear integer programming problems. International Transactions in Operational Research 8, 367–380 (2001)

Vincke, P.: Multicriteria Decision-Aid. Wiley, Chichester (1992)

Wierzbicki, A.P.: Basic properties of scalarizing functionals for multiobjective optimization. Mathematische Operationsforschung und Statistik – Optimization 8, 55–60 (1977)

Wierzbicki, A.P.: The use of reference objectives in multiobjective optimization. In: Fandel, G., Gal, T. (eds.) Multiple Criteria Decision Making, Theory and Applications, pp. 468–486. Springer, Berlin (1980)

Wierzbicki, A.P.: A mathematical basis for satisficing decision making. Mathematical Modeling 3, 391–405 (1983)

Wierzbicki, A.P.: On the completeness and constructiveness of parametric characterizations to vector optimization problems. OR Spectrum 8, 73–87 (1986)

Wierzbicki, A.P.: On the role of intuition in decision making and some ways of multicriteria aid of intuition. Journal of Multi-Criteria Decision Analysis 6, 65–78 (1997)

Wierzbicki, A.P.: Reference point approaches. In: Gal, T., Stewart, T.J., Hanne, T. (eds.) Multicriteria Decision Making: Advances in MCDM Models, Algorithms, Theory, and Applications, pp. 9-1–9-39, Kluwer, Dordrecht (1999)

Wierzbicki, A.P., Makowski, M., Wessels, J. (eds.): Decision Support Methodology with Environmental Applications. Kluwer Academic Publishers, Dordrecht (2000)

Yang, J.B.: Gradient projection and local region search for multiobjective optimisation. European Journal of Operational Research 112, 432–459 (1999)

Yang, J.B., Li, D.: Normal vector identification and interactive tradeoff analysis using minimax formulation in multiobjective optimisation. IEEE Transactions on Systems, Man and Cybernetics 32, 305–319 (2002)

Zionts, S., Wallenius, J.: An interactive programming method for solving the multiple criteria problem. Management Science 22, 652–663 (1976)

Zionts, S., Wallenius, J.: An interactive multiple objective linear programming method for a class of underlying utility functions. Management Science 29, 519–529 (1983)

3

Introduction to Evolutionary Multiobjective Optimization

Kalyanmoy Deb

[1] Department of Mechanical Engineering, Indian Institute of Technology Kanpur, Kanpur, PIN 208016, India
deb@iitk.ac.in
http://www.iitk.ac.in/kangal/deb.htm

[2] Department of Business Technology, Helsinki School of Economics, PO Box 1210, 00101 Helsinki, Finland
Kalyanmoy.Deb@hse.fi

Abstract. In its current state, evolutionary multiobjective optimization (EMO) is an established field of research and application with more than 150 PhD theses, more than ten dedicated texts and edited books, commercial softwares and numerous freely downloadable codes, a biannual conference series running successfully since 2001, special sessions and workshops held at all major evolutionary computing conferences, and full-time researchers from universities and industries from all around the globe. In this chapter, we provide a brief introduction to EMO principles, illustrate some EMO algorithms with simulated results, and outline the current research and application potential of EMO. For solving multiobjective optimization problems, EMO procedures attempt to find a set of well-distributed Pareto-optimal points, so that an idea of the extent and shape of the Pareto-optimal front can be obtained. Although this task was the early motivation of EMO research, EMO principles are now being found to be useful in various other problem solving tasks, enabling one to treat problems naturally as they are. One of the major current research thrusts is to combine EMO procedures with other multiple criterion decision making (MCDM) () tools so as to develop hybrid and interactive multiobjective optimization algorithms for finding a set of trade-off optimal solutions and then choose a preferred solution for implementation. This chapter provides the background of EMO principles and their potential to launch such collaborative studies with MCDM researchers in the coming years.

3.1 Introduction

In a short span of about fourteen years since the suggestion of the first set of successful algorithms, evolutionary multiobjective optimization (EMO) has

---

Reviewed by: Matthias Ehrgott, The University of Auckland, New Zealand
Christian Igel, Ruhr-Universität Bochum, Germany

J. Branke et al. (Eds.): Multiobjective Optimization, LNCS 5252, pp. 59–96, 2008.

now become a popular and useful field of research and application. In a recent survey announced during the World Congress on Computational Intelligence (WCCI) in Vancouver 2006, EMO has been judged as one of the three fastest growing fields of research and application among all computational intelligence topics. Evolutionary optimization (EO) algorithms use a population based approach in which more than one solution participates in an iteration and evolves a new population of solutions in each iteration. The reasons for their popularity are many. Some of them are: (i) EOs do not require any derivative information (ii) EOs are relatively simple to implement and (iii) EOs are flexible and have a wide-spread applicability. For solving single-objective optimization problems or in other tasks focusing on finding a single optimal solution, the use of a population of solutions in each iteration may at first seem like an overkill (but they help provide an implicit parallel search ability, thereby making EOs computationally efficient (Holland, 1975; Goldberg, 1989)), in solving multiobjective optimization problems an EO procedure is a perfect match (Deb, 2001). The multiobjective optimization problems, by nature, give rise to a set of Pareto-optimal solutions which need a further processing to arrive at a single preferred solution. To achieve the first task, it becomes quite a natural proposition to use an EO, because the use of population in an iteration helps an EO to simultaneously find multiple non-dominated solutions, which portrays a trade-off among objectives, in a single simulation run.

In this chapter, we begin with a brief description of an evolutionary optimization procedure for single-objective optimization. Thereafter, we describe the principles of evolutionary multiobjective optimization and sketch a brief history of how the field has evolved over the past one-and-a-half decades. To generate interest in the minds of the readers, we also provide a description of a few well-known EMO procedures with some representative simulation studies. Finally, we discuss the achievements of EMO research and its current focus. It is clear from these discussions that EMO is not only being found to be useful in solving multiobjective optimization problems, it is also helping to solve other kinds of optimization problems in a better manner than they are traditionally solved. As a by-product, EMO-based solutions are helping to reveal important hidden knowledge about a problem – a matter which is difficult to achieve otherwise.

However, much of the research focus in EMO had been to find a set of near Pareto-optimal solutions and not much studies have been made yet to execute the remaining half of a multiobjective optimization task – selection of a single preferred Pareto-optimal solution for implementation. Although a few studies have just scratched the surface in this direction, this book remains as a major step towards achieving possible EMO and multi-criterion decision-making hybrids. This chapter provides a sketch of the EMO background, some representative EMO procedures, and EMO's achievements and potential, so that readers can get attracted to the EMO literature and engage in collaborative research and application using EMO and MCDM.

3.2 Evolutionary Optimization (EO): A Brief Introduction

Evolutionary optimization principles are different from classical optimization methodologies in the following main ways (Goldberg, 1989):

- An EO procedure does not usually use gradient information in its search process. Thus, EO methodologies are direct search procedures, allowing them to be applied to a wide variety of optimization problems. Due to this reason, EO methodologies may not be competitive with specific gradient-based optimization procedures in solving more structured and well-behaved optimization problems such as convex programming, linear or quadratic programming.
- An EO procedure uses more than one solution (a *population* approach) in an iteration, unlike in most classical optimization algorithms which updates one solution in each iteration (a *point* approach). The use of a population has a number of advantages: (i) it provides an EO with a parallel processing power achieving a computationally quick overall search, (ii) it allows an EO to find multiple optimal solutions, thereby facilitating the solution of multi-modal and multiobjective optimization problems, and (iii) it provides an EO with the ability to normalize decision variables (as well as objective and constraint functions) within an evolving population using the population-best minimum and maximum values. However, the flip side of working with a population of solutions is the computational cost and memory associated with executing each iteration.
- An EO procedure uses stochastic operators, unlike deterministic operators used in most classical optimization methods. The operators tend to achieve a desired effect by using biased probability distributions towards desirable outcomes, as opposed to using predetermined and fixed transition rules. This allows an EO algorithm to negotiate multiple optima and other complexities better and provide them with a global perspective in their search.

An EO begins its search with a population of solutions usually created at random within a specified lower and upper bound on each variable. If bounds are not supplied in an optimization problem, suitable values can be assumed only for the initialization purpose. Thereafter, the EO procedure enters into an iterative operation of updating the current population to create a new population by the use of four main operators: selection, crossover, mutation and elite-preservation. The operation stops when one or more pre-specified termination criteria are met. Thus, an EO takes the following simple structure in which the terms shown in bold (but not underlined) are changeable by the user.

The initialization procedure is already described above. If in a problem the knowledge of some good solutions is available, it is wise to use such information in creating the initial population P_0. Elsewhere (Deb *et al.*, 2003a),

---

Algorithm 1 An Evolutionary Optimization Procedure:

$t = 0$;
$Initialization(P_t)$;
do
$Evaluation(P_t)$;
$P'_t = Selection(P_t)$;
$P''_t = Variation(P'_t)$;
$P_{t+1} = Elitism(P_t, P''_t)$;
$t = t + 1$; **od**;
while
$(Termination(P_t, P_{t+1}))$; **od**

---

it is highlighted that for solving complex real-world optimization problems, such a customized initialization is useful and also helpful in achieving a faster search.

The evaluation of a population means computation of each population member for its objective function value, constraint values and determining if the solution is feasible. Since this requires evaluation of multiple functions, this procedure also requires a *relative* preference order (or sorting) of solutions (say from best to worst) in the population. Often, such an ordering can be established by creating a real-valued *fitness* function derived from objective and constraint values. For multiobjective optimization problems, one of the ways to achieve the ordering is to sort the population based on a domination principle Goldberg (1989). It is interesting to note that since an ordering of best-to-worst is enough for the evaluation purpose, EO procedures allow handling of different problem types: dynamically changing problems, problems which are not mathematically expressed, problems which are procedure-based, and others.

After the population members are evaluated, the selection operator chooses above-average (in other words, better) solutions with a larger probability to fill an intermediate mating pool. For this purpose, several stochastic selection operators exist in the EO literature. In its simplest form (called the *tournament* selection (Deb, 1999a)), two solutions can be picked at random from the evaluated population and the better of the two (in terms of its evaluated order) can be picked.

The 'variation' operator is a collection of a number of operators (such as crossover, mutation etc.) which are used to generate a modified population. The purpose of the is to pick two or more solutions (parents) randomly from the mating pool and create one or more solutions by exchanging information among the parent solutions. The crossover operator is applied with a crossover probability ($p_c \in [0, 1]$), indicating the proportion of population members participating in the crossover operation. The remaining $(1 - p_c)$ proportion of the population is simply copied to the modified (child) population. In the context of real-parameter optimization having n real-valued variables and involving a crossover with two parent solutions, each variable may be crossed at a time. A probability distribution which depends on the difference between the two parent variable values is often used to create two new numbers as child values around the two parent values. One way to set the probability distribution is that if the two parent values are quite different, the created child values are also set to different from their parent values, thereby allowing a broader search to take place. This process is usually helpful during the early generations of an EO when population members are quite different from each other. However, after some generations when population members tend to get close to each other to converge to an interesting region in the search space due to the interactions of EO operators, the same crossover operator must then focus the search by creating child values closer to parent values. This varying action of the EO crossover operator without any intervention from the user provides a *self-adaptive* feature to a real-parameter EO algorithm. There exists a number of probability distributions for achieving such a self-adaptive crossover operation (Deb, 2001; Herrera *et al.*, 1998). A unimodal probability distribution with its mode at the parent variable value is used to design the simulated binary crossover (SBX) (SBX|seesimulated binary crossover (SBX)) and is popularly used in the EO literature. Besides the variable-wise recombination operators, vector-wise recombination operators also suggested to propagate the correlation among variables of parent solutions to the created child solutions (Deb *et al.*, 2002a; Storn and Price, 1997).

Each child solution, created by the crossover operator, is then perturbed in its vicinity by a mutation operator (Goldberg, 1989). Every variable is mutated with a mutation probability p_m, usually set as $1/n$ (n is the number of variables), so that on an average one variable gets mutated per solution. In the context of real-parameter optimization, a simple Gaussian probability distribution with a predefined variance can be used with its mean at the child variable value (Deb, 2001). This operator allows an EO to search locally around a solution and is independent on the location of other solutions in the population.

The elitism operator combines the old population with the newly created population and chooses to keep better solutions from the combined population. Such an operation makes sure that an algorithm has a monotonically non-degrading performance. Rudolph (1994) proved an of a specific EO but having elitism and mutation as two essential operators.

Finally, the user of an EO needs to choose . Often, a predetermined number of generations is used as a termination criterion. For goal attainment problems, an EO can be terminated as soon as a solution with a predefined goal or a target solution is found. In many studies (Goldberg, 1989; Michalewicz, 1992; Gen and Cheng, 1997; Bäck *et al.*, 1997), a termination criterion based on the statistics of the current population vis-a-vis that of the previous population to determine the rate of convergence is used. In other more recent studies, theoretical optimality conditions (such as the extent of satisfaction of Karush-Kuhn-Tucker (KKT) conditions) are used to determine the termination of a real-parameter EO algorithm (Deb *et al.*, 2007). Although EOs are heuristic based, the use of such theoretical optimality concepts in an EO can also be used to test their converging abilities towards local optimal solutions.

Thus, overall an EO procedure is a population-based stochastic search procedure which iteratively emphasizes its better population members, uses them to recombine and perturb locally in the hope of creating new and better populations until a predefined termination criterion is met. The use of a population helps to achieve an *implicit parallelism* (Goldberg, 1989; Holland, 1975; Vose *et al.*, 2003) in an EO's search mechanism (an inherent parallel search in different regions of the search space), a matter which makes an EO computationally attractive for solving difficult problems. In the context of certain Boolean functions, a computational time saving to find the optimum varying polynomial to the population size is proven (Jansen and Wegener, 2001). On one hand, the EO procedure is flexible, thereby allowing a user to choose suitable operators and problem-specific information to suit a specific problem. On the other hand, the flexibility comes with an onus on the part of a user to choose appropriate and tangible operators so as to create an efficient and consistent search (Radcliffe, 1991). However, the benefits of having a flexible optimization procedure, over their more rigid and specific optimization algorithms, provide hope in solving difficult real-world optimization problems involving non-differentiable objectives and constraints, non-linearities, discreteness, multiple optima, large problem sizes, uncertainties in computation of objectives and constraints, uncertainties in decision variables, mixed type of variables, and others. A wiser approach to solving optimization problems of the real world would be to first understand the niche of both EO and classical methodologies and then adopt hybrid procedures employing the better of the two as the search progresses over varying degrees of search-space complexity from start to finish.

3.2.1 EO Terminologies

Evolutionary optimization literature uses somewhat different terminologies than those used in classical optimization. The following brief descriptions may be beneficial for the readers from other optimization fields:

Children: New solutions (or decision variable vectors) created by a combined effect of crossover and mutation operators.

Crossover: An operator in which two or more parent solutions are used to create (through recombination) one or more child solutions.

Crossover probability: The probability of performing a crossover operation. This means, on average, the proportion of population members participating in crossover operation in a generation.

Differential evolution (DE): A particular evolutionary optimization procedure which is usually applied to solve real-parameter optimization problems. A good reference on DE is available from Price *et al.* (2005).

Distribution index: A non-negative parameter (η_c) used for implementing the real-parameter simulated binary crossover (SBX). For achieving a mutation operation in the real-valued search space, a polynomial probability distribution with its mode at the parent variable value and with monotonically reducing probability for creating child values away from parent was suggested (Deb, 2001). This operator involves a non-negative parameter (η_m), which must also be set the user. In both the above recombination and mutation operations, a large distribution index value means a probability distribution with a small variance. For single-objective EO, $\eta_c = 2$ is commonly used and for multiobjective EAs, $\eta_c \in [5, 10]$ is used, whereas $\eta_m \approx 20$ is usually used in both cases. A study (Deb and Agrawal, 1999) has shown that with η_m, a perturbation to the order of $O(\eta_m^{-1})$ takes place to the mutated variable from its parent solution to the mutated solution.

Elitism: An operator which preserves the better of parent and child solutions (or populations) so that a previously found better solution is never deleted.

Evolutionary algorithm: A generic name given to an algorithm which applies Darwinian survival-of-the-fittest evolutionary principles along with genetically motivated recombination and mutation principles in a stochastic manner usually to a population of solutions to iteratively create a new and hopefully better population of solutions in the context of a stationary or a dynamic fitness landscape.

Evolutionary optimization (EO): An EA which is designed to solve an optimization problem.

Evolutionary programming (EP): An EA originally applied to a set of finite-state machines to evolve better learning machines (Fogel *et al.*, 1966). Now, EPs are used for various optimization tasks including real-parameter optimization and are somewhat similar to evolution strategies. EP mainly depends on its selection and mutation operators.

Evolution strategy (ES): An EA which is mostly applied to real-valued decision variables and is mainly driven by a selection and a mutation operator, originally suggested by P. Bienert, I. Rechenberg and H.-P. Schwefel of the Technical University of Berlin. Early applications were experimental based (Rechenberg, 1965; Schwefel, 1968) and some texts (Rechenberg, 1973; Schwefel, 1995) provide details about ES procedures.

Fitness: A fitness or a fitness landscape is a function derived from objective function(s), constraint(s) and other problem descriptions which is used in the selection (or reproduction) operator of an EA. A solution is usually called better than the other, if its fitness function value is better.

Generation: An iteration of an EA.

Generational EA: An EA in which a complete set of child solutions (equal to the population size N) is first created by EA operators (usually, selection, crossover and mutation) before comparing with parent solutions to decide which solutions qualify as members of the new population. Thus, in one iteration, a complete population of N members replaces the old population.

Genetic algorithm (GA): An early version of an EA, originally proposed by Holland (1962, 1975), which uses three main operators – selection, crossover and mutation – on a population of solutions at every generation. In binary-coded GAs, solutions are represented in a string of binary digits (bits). In real-parameter GAs, solutions are represented as a vector of real-parameter decision variables. Other representations can also be used to suit the handling of a problem.

Genetic programming (GP): An EA which works on computer programs (usually C or Lisp codes) or on graphs, trees, etc., representing a solution methodology to mainly evolve optimal learning strategies (Koza, 1992).

Individual: An EA population member representing a solution to the problem at hand.

Mating pool: An intermediate population (usually created by the selection operator) used for creating new solutions by crossover and mutation operators.

Mutation: An EA operator which is applied to a single solution to create a new perturbed solution. A fundamental difference with a crossover operator is that mutation is applied to a single solution, whereas crossover is applied to more than one solution.

Mutation probability: The probability of performing a mutation operation. This refers to, on average, the proportion of decision variables participating in a mutation operation to a solution.

Niching: A niching is an operator by which selection pressure of population members are controlled so as to not allow a single solution to take over the population. Thus, niching helps to maintain a diverse population. A number of niching techniques exist, but the *sharing* approach (Goldberg and Richardson, 1987) is popularly used to find a set of multiple optimal solutions in a multi-modal optimization problem.

Offspring: Same as Children, defined above.

Parent: A solution used during crossover operation to create a child solution.

Particle swarm optimization (PSO): An EA which updates each population member by using a weighted sum of two concepts borrowed from the movement of natural swarm of birds (or fishes etc.) – inclination to move towards its own individual best position and towards the swarm's best position in the decision variable space. A good source for more information is (Kennedy *et al.*, 2001).

Population: A set of solutions used in one generation of an EA. The number of solutions in a population is called 'population size'.

Recombination: Same as crossover, defined above.

Reproduction: An EA operator which mimics Darwin's survival of the fittest principle by making duplicate copies of above-average solutions in the population at the expense of deleting below-average solutions. Initial EA studies used a *proportionate* reproduction procedure in which multiple copies of a population member are assigned to the mating pool proportionate to the individual's fitness. Thus, this operator is used for maximization problems and for fitness values which are non-negative. Current studies use tournament selection which compares two population members based on their fitness values and sends the better solution to the mating pool. This operator does not have any limitation on fitness function.

Selection: Same as Reproduction, defined above.

Selection pressure: The extent of emphasis given to above-average solution in a selection operator. No real quantifiable definition exists, but loosely it is considered as the number of copies allocated to the population-best solution by the selection operator Goldberg *et al.* (1993).

Sharing strategy: It is niching operation in which each population member's fitness is divided by a niche-count (in some sense, a niche-count is an estimate of number of other population members around an individual) and a shared fitness is computed. A proportionate selection procedure is then used with the shared fitness values to create the mating pool.

Solution: An EA population member, same as an 'individual'.

Steady-state EA: An EA in which only one new population member is added to the population in one generation.

String: In a binary-coded GA, a population member, made of a collection of bits, is called a string.

3.3 Evolutionary Multiobjective Optimization (EMO)

A multiobjective optimization problem involves a number of objective functions which are to be either minimized or maximized. As in a single-objective optimization problem, the multiobjective optimization problem may contain a number of constraints which any feasible solution (including all optimal solutions) must satisfy. Since objectives can be either minimized or maximized, we state the multiobjective optimization problem in its general form:

$$\left.\begin{array}{ll} \text{Minimize/Maximize } f_m(\mathbf{x}), & m = 1, 2, \ldots, M; \\ \text{subject to } g_j(\mathbf{x}) \geq 0, & j = 1, 2, \ldots, J; \\ h_k(\mathbf{x}) = 0, & k = 1, 2, \ldots, K; \\ x_i^{(L)} \leq x_i \leq x_i^{(U)}, & i = 1, 2, \ldots, n. \end{array}\right\} \quad (3.1)$$

A solution $\mathbf{x} \in \mathbf{R}^n$ is a vector of n decision variables: $\mathbf{x} = (x_1, x_2, \ldots, x_n)^T$. The solutions satisfying the constraints and variable bounds constitute a *feasible decision variable space* $S \subset \mathbf{R}^n$. One of the striking differences between single-objective and multiobjective optimization is that in multiobjective optimization the objective functions constitute a multi-dimensional space, in addition to the usual decision variable space. This additional M-dimensional space is called the *objective space*, $Z \subset \mathbf{R}^M$. For each solution $\mathbf{x}$ in the decision variable space, there exists a point $\mathbf{z} \in \mathbf{R}^M$) in the objective space, denoted by $\mathbf{f}(\mathbf{x}) = \mathbf{z} = (z_1, z_2, \ldots, z_M)^T$. To make the descriptions clear, we refer a 'solution' as a variable vector and a 'point' as the corresponding objective vector.

The optimal solutions in multiobjective optimization can be defined from a mathematical concept of *partial ordering*. In the parlance of multiobjective optimization, the term *domination* is used for this purpose. In this section, we restrict ourselves to discuss unconstrained (without any equality, inequality or bound constraints) optimization problems. The domination between two solutions is defined as follows (Deb, 2001; Miettinen, 1999):

Definition 1. *A solution $\mathbf{x}^{(1)}$ is said to dominate the other solution $\mathbf{x}^{(2)}$, if both the following conditions are true:*

1. *The solution $\mathbf{x}^{(1)}$ is no worse than $\mathbf{x}^{(2)}$ in all objectives. Thus, the solutions are compared based on their objective function values (or location of the corresponding points ($\mathbf{z}^{(1)}$ and $\mathbf{z}^{(2)}$) on the objective space).*
2. *The solution $\mathbf{x}^{(1)}$ is strictly better than $\mathbf{x}^{(2)}$ in at least one objective.*

Although this definition is defined between two solution vectors here, the domination is determined with their objective vectors and the above definition is identical to the one outlined in Section 7 of the preface. For a given set of solutions (or corresponding points on the objective space, for example, those shown in Figure 3.1(a)), a pair-wise comparison can be made using the above definition and whether one point dominates the other can be established. All points which are not dominated by any other member of the set are called the non-dominated points of class one, or simply the non-dominated points. For the set of six solutions shown in the figure, they are points 3, 5, and 6. One property of any two such points is that a gain in an objective from one point to the other happens only due to a sacrifice in at least one other objective. This *trade-off* property between the non-dominated points makes the practitioners interested in finding a wide variety of them before making a final choice. These points make up a front when viewed them together on the objective space; hence the non-dominated points are often visualized to represent a *non-domination front*. The computational effort needed to select the points of the non-domination front from a set of N points is $O(N \log N)$ for 2 and 3 objectives, and $O(N \log^{M-2} N)$ for $M > 3$ objectives (Kung *et al.*, 1975).

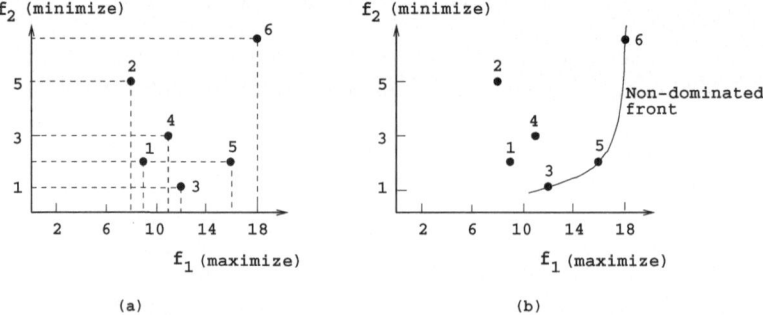

Fig. 3.1. A set of points and the first non-domination front are shown.

With the above concept, now it is easier to define the *Pareto-optimal solutions* in a multiobjective optimization problem. If the given set of points for the above task contain all points in the search space (assuming a countable number), the points lying on the non-domination front, by definition, do not get dominated by any other point in the objective space, hence are Pareto-optimal points (together they constitute the Pareto-optimal front) and the corresponding pre-images (decision variable vectors) are called Pareto-optimal solutions. However, more mathematically elegant definitions of Pareto-optimality (including the ones for continuous search space problems) exist in the multiobjective literature and are discussed in Chapters 1 and 2.

Similar to local optimal solutions in single objective optimization, local Pareto-optimal solutions are also defined in multiobjective optimization (Deb, 2001; Miettinen, 1999):

Definition 2. *If for every member* $\mathbf{x}$ *in a set* $\underline{P}$ *there exists no solution* $\mathbf{y}$ *(in the neighborhood of* $\mathbf{x}$ *such that* $\|\mathbf{y} - \mathbf{x}\|_\infty \leq \epsilon$, *where* ϵ *is a small positive scalar) dominating any member of the set* $\underline{P}$, *then solutions belonging to the set* $\underline{P}$ *constitute a local Pareto-optimal set.*

3.3.1 EMO Principles

In the context of multiobjective optimization, the extremist principle of finding the optimum solution cannot be applied to one objective alone, when the rest of the objectives are also important. Different solutions may produce trade-offs (conflicting outcomes among objectives) among different objectives. A solution that is extreme (in a better sense) with respect to one objective requires a compromise in other objectives. This prohibits one to choose a solution which is optimal with respect to only one objective. This clearly suggests two ideal goals of multiobjective optimization:

1. Find a set of solutions which lie on the Pareto-optimal front, and
2. Find a set of solutions which are diverse enough to represent the entire range of the Pareto-optimal front.

Evolutionary multiobjective optimization (EMO) algorithms attempt to follow both the above principles similar to the other a posteriori MCDM methods (refer to Chapter 1).

Although one fundamental difference between single and multiple objective optimization lies in the cardinality in the optimal set, from a practical standpoint a user needs only one solution, no matter whether the associated optimization problem is single or multiobjective. The user is now in a dilemma. Since a number of solutions are optimal, the obvious question arises: Which of these optimal solutions must one choose? This is not an easy question to answer. It involves higher-level information which is often non-technical, qualitative and experience-driven. However, if a set of many trade-off solutions are already worked out or available, one can evaluate the pros and cons of each of these solutions based on all such non-technical and qualitative, yet still important, considerations and compare them to make a choice. Thus, in a multiobjective optimization, ideally the effort must be made in finding the set of trade-off optimal solutions by considering all objectives to be important. After a set of such trade-off solutions are found, a user can then use higher-level qualitative considerations to make a choice. Since an EMO procedure deals with a population of solutions in every iteration, it makes them intuitive to be applied in multiobjective optimization to find a set of non-dominated solutions. Like other a posteriori MCDM methodologies, an EMO based procedure works with the following principle in handling multiobjective optimization problems:

Step 1 Find multiple non-dominated points as close to the Pareto-optimal front as possible, with a wide trade-off among objectives.

Step 2 Choose one of the obtained points using higher-level information.

Figure 3.2 shows schematically the principles, followed in an EMO procedure. Since EMO procedures are heuristic based, they may not guarantee in finding Pareto-optimal points, as a theoretically provable optimization method would do for tractable (for example, linear or convex) problems. But EMO procedures have essential operators to constantly improve the evolving non-dominated points (from the point of view of convergence and diversity discussed above) similar to the way most natural and artificial evolving systems continuously improve their solutions. To this effect, a recent simulation study (Deb *et al.*, 2007) has demonstrated that a particular EMO procedure, starting from random non-optimal solutions, can progress towards theoretical Karush-Kuhn-Tucker (KKT) points with iterations in real-valued multiobjec-

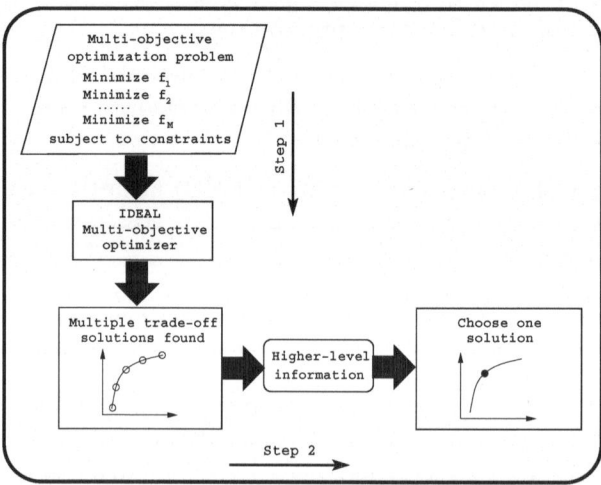

Fig. 3.2. Schematic of a two-step multiobjective optimization procedure.

tive optimization problems. The main difference and advantage of using an EMO compared to a posteriori MCDM procedures is that multiple trade-off solutions can be found in a single simulation run, as most a posteriori MCDM methodologies would require multiple applications.

In Step 1 of the EMO-based multiobjective optimization (the task shown vertically downwards in Figure 3.2), multiple trade-off, non-dominated points are found. Thereafter, in Step 2 (the task shown horizontally, towards the right), higher-level information is used to choose one of the obtained trade-off points. This dual task allows an interesting feature, if applied for solving single-objective optimization problems. It is easy to realize that a single-objective optimization is a degenerate case of multiobjective optimization, as shown in details in another study (Deb and Tiwari, 2008). In the case of single-objective optimization having only one globally optimal solution, Step 1 will ideally find only one solution, thereby not requiring us to proceed to Step 2. However, in the case of single-objective optimization having multiple global optima, both steps are necessary to first find all or multiple global optima, and then to choose one solution from them by using a higher-level information about the problem. Thus, although seems ideal for multiobjective optimization, the framework suggested in Figure 3.2 can be ideally thought as a generic principle for both single and multiple objective optimization.

3.3.2 A Posteriori MCDM Methods and EMO

In the 'a posteriori' MCDM approaches (also known as 'generating MCDM methods'), the task of finding multiple Pareto-optimal solutions is achieved by executing many independent single-objective optimizations, each time finding a single Pareto-optimal solution (see Miettinen (1999) and Chapter 1 of this book). A parametric scalarizing approach (such as the weighted-sum approach, ε-constraint approach, and others) can be used to convert multiple objectives into a parametric single-objective objective function. By simply varying the parameters (weight vector or ε-vector) and optimizing the scalarized function, different Pareto-optimal solutions can be found. In contrast, in an EMO, multiple Pareto-optimal solutions are attempted to be found in a single simulation by emphasizing multiple non-dominated and isolated solutions. We discuss a little later some EMO algorithms describing how such dual emphasis is provided, but now discuss qualitatively the difference between a posteriori MCDM and EMO approaches.

Consider Figure 3.3, in which we sketch how multiple independent parametric single-objective optimizations may find different Pareto-optimal solutions. The Pareto-optimal front corresponds to global optimal solutions of several scalarized objectives. However, during the course of an optimization task, algorithms must overcome a number of difficulties, such as infeasible regions, local optimal solutions, flat regions of objective functions, isolation of optimum, etc., to converge to the global optimal solution. Moreover, due to practical limitations, an optimization task must also be completed in a

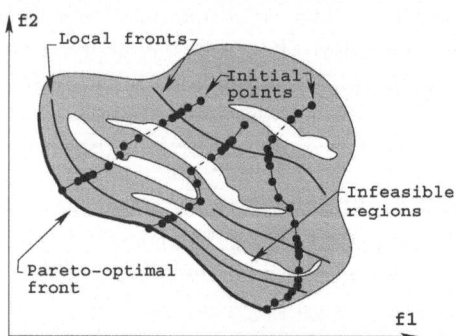

Fig. 3.3. Posteriori MCDM methodology employs independent single-objective optimizations.

reasonable computational time. This requires an algorithm to strike a good balance between the extent of these tasks its search operators must do to overcome the above-mentioned difficulties reliably and quickly. When multiple simulations are to performed to find a set of Pareto-optimal solutions, the above balancing act must have to performed in every single simulation. Since simulations are performed independently, no information about the success or failure of previous simulations is used to speed up the process. In difficult multiobjective optimization problems, such memory-less a posteriori methods may demand a large overall computational overhead to get a set of Pareto-optimal solutions. Moreover, even though the convergence can be achieved in some problems, independent simulations can never guarantee finding a good distribution among obtained points.

EMO, as mentioned earlier, constitutes an inherent parallel search. When a population member overcomes certain difficulties and make a progress towards the Pareto-optimal front, its variable values and their combination reflect this fact. When a recombination takes place between this solution and other population members, such valuable information of variable value combinations gets shared through variable exchanges and blending, thereby making the overall task of finding multiple trade-off solutions a parallelly processed task.

3.4 History of EMO and Non-elitist Methodologies

During the early years, EA researchers have realized the need of solving multiobjective optimization problems in practice and mainly resorted to using weighted-sum approaches to convert multiple objectives into a single goal (Rosenberg, 1967; Fogel et al., 1966).

However, the first implementation of a real multiobjective evolutionary algorithm (vector-evaluated GA or VEGA) was suggested by David Schaffer in the year 1984 (Schaffer, 1984). Schaffer modified the simple three-operator genetic algorithm (with selection, crossover, and mutation) by performing independent selection cycles according to each objective. The selection method is repeated for each individual objective to fill up a portion of the mating pool. Then the entire population is thoroughly shuffled to apply crossover and mutation operators. This is performed to achieve the mating of individuals of different subpopulation groups. The algorithm worked efficiently for some generations but in some cases suffered from its bias towards some individuals or regions (mostly individual objective champions). This does not fulfill the second goal of EMO, discussed earlier.

Ironically, no significant study was performed for almost a decade after the pioneering work of Schaffer, until a revolutionary 10-line sketch of a new non-dominated sorting procedure suggested by David E. Goldberg in his seminal book on GAs (Goldberg, 1989). Since an EA needs a fitness function for reproduction, the trick was to find a single metric from a number of objective functions. Goldberg's suggestion was to use the concept of *domination* to assign more copies to non-dominated individuals in a population. Since diversity is the other concern, he also suggested the use of a *niching* strategy (Goldberg and Richardson, 1987) among solutions of a non-dominated class. Getting this clue, at least three independent groups of researchers developed different versions of multiobjective evolutionary algorithms during 1993-94. Basically, these algorithms differ in the way a fitness assignment scheme is introduced to each individual. We discuss them briefly in the following paragraphs.

Fonseca and Fleming (1993) suggested a multiobjective GA (MOGA), in which all non-dominated population members are assigned a rank one. Other individuals are ranked by calculating how many solutions (say k) dominated a particular solution. That solution is then assigned a rank $(k+1)$. The selection procedure then chooses lower rank solutions to form the mating pool. Since the fitness of a population member is the same as its rank, many population members will have an identical fitness. MOGA applies a niching technique on solutions having identical fitness to maintain a diverse population. But instead of performing niching on the parameter values, they suggested niching on objective function values. The ranking of individuals according to their non-dominance in the population is an important aspect of the work.

Horn *et al.* (1994) used *Pareto domination tournaments* in their niched-Pareto GA (NPGA). In this method, a *comparison set* comprising of a specific number (t_{dom}) of individuals is picked at random from the population at the beginning of each selection process. Two random individuals are picked from the population for selecting a winner according to the following procedure. Both individuals are compared with the members of the comparison set for domination. If one of them is non-dominated and the other is dominated, then the non-dominated point is selected. On the other hand, if both are either non-dominated or dominated, a *niche-count* is calculated by simply counting the number of points in the entire population within a certain distance (σ_{share}) in the variable space from an individual. The individual with least niche-count is selected. The proposers of this algorithm has reported that the outcome of the algorithm depends on the chosen value of t_{dom}. Nevertheless, the concept of niche formation among the non-dominated points using the tournament selection is an important aspect of the work.

Srinivas and Deb (1994) developed a non-dominated sorting GA (NSGA) which differs from MOGA in two ways: fitness assignment and the way niching is performed. After the population members belonging to the first non-domination class are identified, they are assigned a dummy fitness value equal to N (population size). A sharing strategy is then used on parameter values (instead of objective function values) to find the niche-count for each individual of the best class. Here, the parameter niche-count of an individual is estimated by calculating *sharing function* values (Goldberg and Richardson, 1987). Interested readers may refer to the original study or (Deb, 2001). For each individual, a shared fitness is then found by dividing the assigned fitness N by the niche-count. Thereafter, the second class of non-dominated solutions (obtained by temporarily discounting solutions of first non-domination class and then finding new non-dominated points) is assigned a dummy fitness value smaller than the least shared fitness of solutions of the previous non-domination class. This process is continued till all solutions are assigned a fitness value. This fitness assignment procedure ensured two matters: (i) a dominated solution is assigned a smaller shared fitness value than any solution which dominated it and (ii) in each non-domination class an adequate diversity is ensured. On a number of test problems and real-world optimization problems, NSGA has been found to provide a wide-spread Pareto-optimal or near Pareto-optimal solutions. However, one difficulty of NSGA is to choose an appropriate niching parameter, which directly affects the maximum distance between two neighboring solutions obtained by NSGA. Although most studies used a fixed value of the niching parameter, an adaptive niche-sizing strategy has been suggested elsewhere (Fonseca and Fleming, 1993).

3.5 Elitist Methodologies

The above-mentioned non-elitist EMO methodologies gave a good head-start to the research and application of EMO, but suffered from the fact that they did not use an elite-preservation mechanism in their procedures. It is mentioned before that an addition of elitism in an EO provides a monotonically non-degrading performance. The second generation EMO algorithms implemented an elite-preserving operator in different ways and gave birth to elitist EMO procedures, such as NSGA-II (Deb *et al.*, 2002b), Strength Pareto EA (SPEA) (Zitzler and Thiele, 1999), Pareto-archived ES (PAES) (Knowles and Corne, 2000), and others. Since these EMO algorithms are state-of-the-art procedures, we describe one of these algorithms in detail and briefly discuss two other procedures commonly used in EMO studies.

3.5.1 Elitist Non-dominated Sorting GA or NSGA-II

The NSGA-II procedure (Deb *et al.*, 2002b) is one of the popularly used EMO procedures which attempt to find multiple Pareto-optimal solutions in a multiobjective optimization problem and has the following three features:

1. It uses an elitist principle,
2. it uses an explicit diversity preserving mechanism, and
3. it emphasizes non-dominated solutions.

At any generation t, the offspring population (say, Q_t) is first created by using the parent population (say, P_t) and the usual genetic operators. Thereafter, the two populations are combined together to form a new population (say, R_t) of size $2N$. Then, the population R_t classified into different non-domination classes. Thereafter, the new population is filled by points of different non-domination fronts, one at a time. The filling starts with the first non-domination front (of class one) and continues with points of the second non-domination front, and so on. Since the overall population size of R_t is $2N$, not all fronts can be accommodated in N slots available for the new population. All fronts which could not be accommodated are deleted. When the last allowed front is being considered, there may exist more points in the front than the remaining slots in the new population. This scenario is illustrated in Figure 3.4. Instead of arbitrarily discarding some members from the last front, the points which will make the diversity of the selected points the highest are chosen.

The crowded-sorting of the points of the last front which could not be accommodated fully is achieved in the descending order of their *crowding!distance values* and points from the top of the ordered list are chosen. The crowding distance d_i of point i is a measure of the objective space around i which is not occupied by any other solution in the population. Here, we simply calculate this quantity d_i by estimating the perimeter of the cuboid

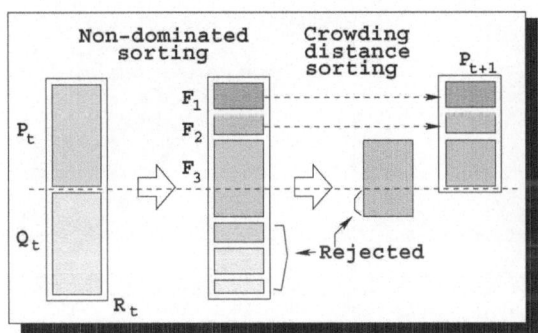

Fig. 3.4. Schematic of the NSGA-II procedure.

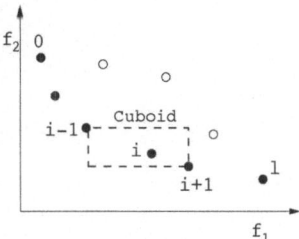

Fig. 3.5. The crowding distance calculation.

(Figure 3.5) formed by using the nearest neighbors in the objective space as the vertices (we call this the *crowding distance*).

Sample Simulation Results

Here, we show simulation results of NSGA-II on two test problems. The first problem (ZDT2) is two-objective, 30-variable problem with a concave Pareto-optimal front:

$$\text{ZDT}: \begin{cases} \text{Minimize } f_1(\mathbf{x}) = x_1, \\ \text{Minimize } f_2(\mathbf{x}) = g(\mathbf{x}) \left[1 - (f_1(\mathbf{x})/g(\mathbf{x}))^2\right], \\ \text{where } \quad g(\mathbf{x}) = 1 + \frac{9}{29} \sum_{i=2}^{30} x_i \\ 0 \le x_1 \le 1, \\ -1 \le x_i \le 1, \quad i = 2, 3, \dots, 30. \end{cases} \tag{3.2}$$

The second problem (KUR), with three variables, has a disconnected Pareto-optimal front:

$$\text{KUR}: \begin{cases} \text{Minimize } f_1(\mathbf{x}) = \sum_{i=1}^{2} \left[-10 \exp(-0.2\sqrt{x_i^2 + x_{i+1}^2})\right], \\ \text{Minimize } f_2(\mathbf{x}) = \sum_{i=1}^{3} \left[|x_i|^{0.8} + 5 \sin(x_i^3)\right], \\ -5 \le x_i \le 5, \quad i = 1, 2, 3. \end{cases} \tag{3.3}$$

NSGA-II is run with a population size of 100 and for 250 generations. The variables are used as real numbers and an SBX recombination operator with $p_c = 0.9$ and distribution index of $\eta_c = 10$ and a polynomial mutation operator (Deb, 2001) with $p_m = 1/n$ (n is the number of variables) and distribution index of $\eta_m = 20$ are used. Figures 3.6 and 3.7 show that NSGA-II converges on the Pareto-optimal front and maintains a good spread of solutions on both test problems.

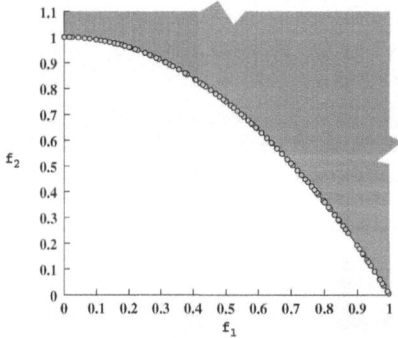

Fig. 3.6. NSGA-II on ZDT2.

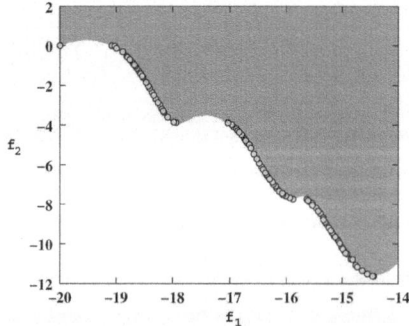

Fig. 3.7. NSGA-II on KUR.

3.5.2 Strength Pareto EA (SPEA) and SPEA2

Zitzler and Thiele (1998) suggested an elitist multi-criterion EA with the concept of non-domination in their strength Pareto EA (SPEA). They suggested maintaining an external population in every generation storing all non-dominated solutions discovered so far, beginning from the initial population. This external population participates in genetic operations. In each generation, a combined population with the external and the current population is first constructed. All non-dominated solutions in the combined population are assigned a fitness based on the number of solutions they dominate and all dominated solutions are assigned a fitness equal to one more than the sum of fitness of solutions which dominate it. This assignment of fitness makes sure that the search is directed towards the non-dominated solutions and simultaneously diversity among dominated and non-dominated solutions are maintained. Diversity is maintained by performing a clustering procedure to maintain a fixed size archive. On knapsack problems, they have reported better results than other methods used in that study.

In their subsequent improved version (SPEA2) (Zitzler *et al.*, 2001b), three changes have been made. First, the archive size is always kept fixed by adding dominated solutions from the EA population, if needed. Second, the fitness assignment procedure for the dominated solutions is slightly different and a density information is used to resolve ties between solutions having identical fitness values. Third, a modified clustering algorithm is used from the k-th nearest neighbor distance estimates for each cluster and special attention is made to preserve the boundary elements.

3.5.3 Pareto Archived ES (PAES) and Pareto Envelope based Selection Algorithms (PESA and PESA2)

Knowles and Corne (2000) suggested Pareto-archived ES (PAES) with one parent and one child. The child is compared with the parent. If the child dominates the parent, the child is accepted as the next parent and the iteration continues. On the other hand, if the parent dominates the child, the child is discarded and a new mutated solution (a new child) is found. However, if the child and the parent do not dominate each other, the choice between child or a parent is resolved by using a crowding procedure. To maintain diversity, an archive of non-dominated solutions found so far is maintained. The child is compared with the archive to check if it dominates any member of the archive. If yes, the child is accepted as the new parent and the dominated solution is eliminated from the archive. If the child does not dominate any member of the archive, both parent and child are checked for their proximity (in terms of Euclidean distance in the objective space) to the archive members. If the child resides in the least crowded area in the objective space compared to other archive members, it is accepted as a parent and a copy is added to the archive. Later, they suggested a multi-parent PAES with similar principles as above, but applied with a multi-parent evolution strategy framework (Schwefel, 1995). In their subsequent version, called the Pareto Envelope based Selection Algorithm (PESA) (Corne et al., 2000), they combined good aspects of SPEA and PAES. Like SPEA, PESA carries two populations (a smaller EA population and a larger archive population). Non-domination and the PAES crowding concept is used to update the archive with the newly created child solutions.

In an extended version of PESA (Corne et al., 2001), instead of applying the selection procedure on population members, hyperboxes in the objective space are selected based on the number of solutions residing in the hyperboxes. After hyperboxes are selected, a random solution from the chosen hyperboxes is kept. This region-based selection procedure has shown to perform better than the individual-based selection procedure of PESA. In some sense, the PESA2 selection scheme is similar in concept to ϵ-dominance (Laumanns et al., 2002) in which predefined ϵ values determine the hyperbox dimensions. Other ϵ-dominance based EMO procedures (Deb et al., 2003b) have shown computationally faster and better distributed solutions than NSGA-II or SPEA2.

There also exist other competent EMOs, such as multiobjective messy GA (MOMGA) (Veldhuizen and Lamont, 2000), multiobjective micro-GA (Coello and Toscano, 2000), neighborhood constraint GA (Loughlin and Ranjithan, 1997), ARMOGA (Sasaki et al., 2001), and others. Besides, there exists other EA based methodologies, such as particle swarm EMO (Coello and Lechuga, 2002; Mostaghim and Teich, 2003), ant-based EMO (McMullen, 2001; Gravel et al., 2002), and differential evolution based EMO (Babu and Jehan, 2003).

3.6 Applications of EMO

Since the early development of EMO algorithms in 1993, they have been applied to many real-world and interesting optimization problems. Descriptions of some of these studies can be found in books (Deb, 2001; Coello et al., 2002; Osyczka, 2002), dedicated conference proceedings (Zitzler et al., 2001a; Fonseca et al., 2003; Coello et al., 2005; Obayashi et al., 2007), and domain-specific books, journals and proceedings. In this section, we describe one case study which clearly demonstrates the EMO philosophy which we described in Section 3.3.1.

3.6.1 Spacecraft Trajectory Design

Coverstone-Carroll et al. (2000) proposed a multiobjective optimization technique using the original non-dominated sorting algorithm (NSGA) (Srinivas and Deb, 1994) to find multiple trade-off solutions in a spacecraft trajectory optimization problem. To evaluate a solution (trajectory), the SEPTOP (Solar

Electric Propulsion Trajectory Optimization) software (Sauer, 1973) is called for, and the delivered payload mass and the total time of flight are calculated. The multiobjective optimization problem has eight decision variables controlling the trajectory, three objective functions: (i) maximize the delivered payload at destination, (ii) maximize the negative of the time of flight, and (iii) maximize the total number of heliocentric revolutions in the trajectory, and three constraints limiting the SEPTOP convergence error and minimum and maximum bounds on heliocentric revolutions.

On the Earth–Mars rendezvous mission, the study found interesting trade-off solutions (Coverstone-Carroll *et al.*, 2000). Using a population of size 150, the NSGA was run for 30 generations. The obtained non-dominated solutions are shown in Figure 3.8 for two of the three objectives and some selected solutions are shown in Figure 3.9. It is clear that there exist short-time flights with smaller delivered payloads (solution marked 44) and long-time flights with larger delivered payloads (solution marked 36). Solution 44 can deliver a mass of 685.28 kg and requires about 1.12 years. On other hand, an intermediate solution 72 can deliver almost 862 kg with a travel time of about 3 years. In these figures, each continuous part of a trajectory represents a *thrusting* arc and each dashed part of a trajectory represents a *coasting* arc. It is interesting to note that only a small improvement in delivered mass occurs in the solutions between 73 and 72 with a sacrifice in flight time of about an year.

The multiplicity in trade-off solutions, as depicted in Figure 3.9, is what we envisaged in discovering in a multiobjective optimization problem by using a posteriori procedure, such as an EMO algorithm. This aspect was also discussed in Figure 3.2. Once such a set of solutions with a good trade-off among objectives is obtained, one can analyze them for choosing a particular solution. For example, in this problem context, it makes sense to not choose a

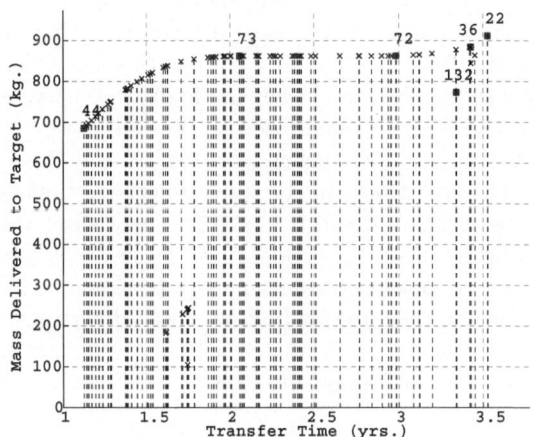

Fig. 3.8. Obtained non-dominated solutions using NSGA.

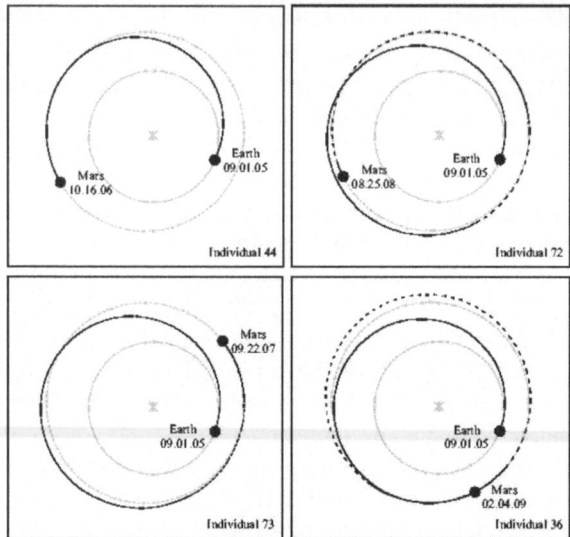

Fig. 3.9. Four trade-off trajectories.

solution between points 73 and 72 due to poor trade-off between the objectives in this range. On the other hand, choosing a solution within points 44 and 73 is worthwhile, but which particular solution to choose depends on other mission related issues. But by first finding a wide range of possible solutions and revealing the shape of front, EMO can help narrow down the choices and allow a decision maker to make a better decision. Without the knowledge of such a wide variety of trade-off solutions, a proper decision-making may be a difficult task. Although one can choose a scalarized objective (such as the ϵ-constraint method with a particular ϵ vector) and find the resulting optimal solution, the decision-maker will always wonder what solution would have been derived if a different ϵ vector was chosen. For example, if $\epsilon_1 = 2.5$ years is chosen and mass delivered to the target is maximized, a solution in between points 73 and 72 will be found. As discussed earlier, this part of the Pareto-optimal front does not provide the best trade-offs between objectives that this problem can offer. A lack of knowledge of good trade-off regions before a decision is made may allow the decision maker to settle for a solution which, although optimal, may not be a good compromised solution. The EMO procedure allows a flexible and a pragmatic procedure for finding a well-diversified set of solutions simultaneously so as to enable picking a particular region for further analysis or a particular solution for implementation.

3.7 Constraint Handling in EMO

The constraint handling method modifies the binary tournament selection, where two solutions are picked from the population and the better solution is chosen. In the presence of constraints, each solution can be either feasible or infeasible. Thus, there may be at most three situations: (i) both solutions are feasible, (ii) one is feasible and other is not, and (iii) both are infeasible. We consider each case by simply redefining the domination principle as follows (we call it the *constrained-domination* condition for any two solutions $\mathbf{x}^{(i)}$ and $\mathbf{x}^{(j)}$):

Definition 3. *A solution $\mathbf{x}^{(i)}$ is said to 'constrained-dominate' a solution $\mathbf{x}^{(j)}$ (or $\mathbf{x}^{(i)} \preceq_c \mathbf{x}^{(j)}$), if any of the following conditions are true:*

1. *Solution $\mathbf{x}^{(i)}$ is feasible and solution $\mathbf{x}^{(j)}$ is not.*
2. *Solutions $\mathbf{x}^{(i)}$ and $\mathbf{x}^{(j)}$ are both infeasible, but solution $\mathbf{x}^{(i)}$ has a smaller constraint violation, which can be computed by adding the normalized violation of all constraints:*

$$
\mathrm{CV}(\mathbf{x}) = \sum_{j=1}^{J} \langle \bar{g}_j(\mathbf{x}) \rangle + \sum_{k=1}^{K} \mathrm{abs}(\bar{h}_k(\mathbf{x})),
$$

where $\langle \alpha \rangle$ is $-\alpha$, if $\alpha < 0$ and is zero, otherwise. The normalization is achieved with the population minimum ($\langle g_j \rangle_{\min}$) and maximum ($\langle g_j \rangle_{\max}$) constraint violations: $\bar{g}_j(\mathbf{x}) = (\langle g_j(\mathbf{x}) \rangle - \langle g_j \rangle_{\min})/(\langle g_j \rangle_{\max} - \langle g_j \rangle_{\min})$.
3. *Solutions $\mathbf{x}^{(i)}$ and $\mathbf{x}^{(j)}$ are feasible and solution $\mathbf{x}^{(i)}$ dominates solution $\mathbf{x}^{(j)}$ in the usual sense (Definition 1).*

The above change in the definition requires a minimal change in the NSGA-II procedure described earlier. Figure 3.10 shows the non-domination fronts on a six-membered population due to the introduction of two constraints (the minimization problem is described as CONSTR elsewhere (Deb, 2001)). In the absence of the constraints, the non-domination fronts (shown by dashed lines) would have been ((1,3,5), (2,6), (4)), but in their presence, the new fronts are ((4,5), (6), (2), (1), (3)). The first non-domination front consists of the "best" (that is, non-dominated and feasible) points from the population and any feasible point lies on a better non-domination front than an infeasible point.

3.8 Test Problems with Known Optima

After the initial three non-elitist EMO procedures came out around 1993-94, a plethora of EMO algorithms was suggested starting around 2000. Since EMO algorithms are heuristic based, it is essential to compare them against each other and against true optimal solutions, so as to test the algorithms to investigate if they are adequate enough to handle different vagaries of problem

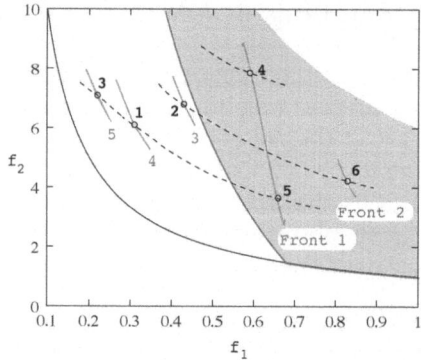

Fig. 3.10. Non-constrained-domination fronts.

difficulties real-world problems would offer. Since the purpose of EMO algorithms is to find a set of well-converged and well-spread solutions on or as close to the Pareto-optimal front as possible, it was necessary to develop numerical test problems for which the location and extent of the Pareto-optimal front are exactly known and test the performance of EMO algorithms on such test problems. The availability of test problems has not only boosted the research in EMO but has also brought researchers closer together in understanding how a population-based EMO solves multiobjective optimization problems.

Veldhuizen and Lamont (1998) collected a set of test problems from various literature including classical multiobjective optimization books. In 1999, Deb (1999b) devised a simple procedure of designing two-objective optimization problems in which different kinds of difficulties commonly found and studied in single-objective EO problems were introduced in a multiobjective optimization problem. A later study (Zitzler *et al.*, 2000) used Deb's original idea and suggested six test problems (now popularly known as Zitzler-Deb-Thiele or ZDT) which tested a multiobjective optimization for the following complexities:

1. Effect of increasing the number of decision variables (all six ZDT problems),
2. effect of convexity and non-convexity of the Pareto-optimal front (ZDT1 versus ZDT2),
3. effect of discontinuities and disconnectedness in the Pareto-optimal front (ZDT1 or ZDT2 versus ZDT3),
4. effect of multiple local Pareto-optimal fronts in arriving at the global Pareto-optimal front (ZDT1 versus ZDT4),
5. effect of *isolation* (presence of potentially bad solutions in the neighborhood of Pareto-optimal solutions) and *deception* (presence of a wider and

a specific type of basin of attraction for the local Pareto-optimal front) (ZDT1 versus ZDT5), and

6. effect of non-uniform density of points (on the objective space) across the Pareto-optimal front (ZDT2 versus ZDT6).

ZDT problems served the EMO research for quite many years since their suggestion in 2000. To introduce further complexities, Okabe *et al.* (2004) and Huband *et al.* (2005) have suggested test problems with strong correlations among variables. An extension of ZDT test problems to include correlations among variables was suggested in the original study Deb (1999b) and specific test problems are suggested recently (Deb *et al.*, 2006a; Igel *et al.*, 2007).

A set of nine test problems scalable to the number of objectives and variables were suggested in 2002 (Deb *et al.*, 2002c, 2005). Some of these problems, (DTLZ1 to DTLZ7 minimization problems) use a bottom-up strategy in which the functional shape of the Pareto-optimal front is first assumed and then an objective space is constructed with the desired Pareto-optimal front as the lower boundary. Thereafter, a mapping between the decision space and the objective space is constructed to introduce different kinds of complexities which can cause an EMO procedure sustained difficulties in converging and finding a widely distributed set of solutions. The remaining two test problems (DTLZ8 and DTLZ9) use a principle of starting with a hyperbox as the objective space and then introduces a number of constraints which eliminate portions of the hyperbox to help generate a Pareto-optimal front. These test problems have served EMO researchers in various ways, particularly in developing better algorithms and in testing the performance of their algorithms in solving various multiobjective optimization problems (Deb *et al.*, 2006a; Igel *et al.*, 2007).

In addition, there also exist a number of constrained test problems with non-linear objective functions and constraints. These so-called CTP problems (Deb, 2001; Deb *et al.*, 2001) introduce different complexities: disconnected feasible objective space, long narrow feasible objective space to reach to Pareto-optimal points, and others. The purpose of these test problems was to simulate different difficulties which real-world problems may have. The problems provide a tunable degree of such difficulties and this way, if an algorithm is capable of negotiating them well in test problems, the algorithm is also expected to perform well in a real-world scenario having similar difficulties.

3.9 Performance Measures and Evaluation of EMO Methodologies

When algorithms are developed and test problems with known Pareto-optimal fronts are available, an important task is to have one or more performance measures with which the EMO algorithms can be evaluated. Thus, a major focus of EMO research has been spent in developing different performance measures and is a major area of EMO research. Chapter 14 presents a detail account of this topic. Since the focus in an EMO task is multi-faceted – convergence to the Pareto-optimal front and diversity of solutions along the entire front, it is also expected that one performance measure to evaluate EMO algorithms will be unsatisfactory. In the early years of EMO research, three different sets of performance measures were used:

1. Metrics evaluating convergence to the known Pareto-optimal front (such as error ratio, distance from reference set, etc.),
2. Metrics evaluating spread of solutions on the known Pareto-optimal front (such as spread, spacing, etc.), and
3. Metrics evaluating certain combinations of convergence and spread of solutions (such as hypervolume, coverage, R-metrics, etc.).

Some of these metrics are described in texts (Coello *et al.*, 2002; Deb, 2001). A detailed study (Knowles and Corne, 2002) comparing most existing performance metrics based on out-performance relations has recommended the use of the S-metric (or the hypervolume metric) and R-metrics suggested by Hansen and Jaskiewicz (1998). A recent study has argued that a single unary performance measure or any finite combination of them (for example, any of the first two metrics described above in the enumerated list or both together) cannot adequately determine whether one set is better than another (Zitzler *et al.*, 2003). That study also concluded that binary performance metrics (indicating usually two different values when a set of solutions A is compared against B and B is compared against A), such as epsilon-indicator, binary hypervolume indicator, utility indicators R1 to R3, etc., are better measures for multiobjective optimization. The flip side is that the chosen binary metric must be computed $K(K-1)$ times when comparing K different sets to make a fair comparison, thereby making the use of binary metrics computationally expensive in practice. Importantly, these performance measures have allowed researchers to use them directly as fitness measures within indicator based EAs (IBEAs) (Zitzler and Künzli, 2004). In addition, attainment indicators of Fonseca and Fleming (1996); Fonseca *et al.* (2005) provide further information about location and inter-dependencies among obtained solutions.

3.10 Other Current EMO Research and Practices

With the initial fast development of efficient EMO procedures, availability of free and commercial softwares (some of which are described in the Appendix), and applications to a wide variety of problems, EMO research and application is in its peak at the current time. It is difficult to cover every aspect of current research in a single paper. Here, we outline four main broad areas of research and application.

3.10.1 Hybrid EMO Procedures

The search operators used in EMO are heuristic based. Thus these methodologies are not guaranteed to find Pareto-optimal solutions with a finite number of solution evaluations in an arbitrary problem. However, as discussed
in this chapter, EMO methodologies provide adequate emphasis to currently
non-dominated and isolated solutions so that population members progress
towards the Pareto-optimal front iteratively. To make the overall procedure
faster and to perform the task with a more theoretical basis, EMO methodologies can be combined with mathematical optimization techniques having local
convergence properties. A simple-minded approach would be to start the process with an EMO and the solutions obtained from EMO can be improved by
optimizing a composite objective derived from multiple objectives to ensure a
good spread by using a local search technique. Another approach would be to
use a local search technique as a mutation-like operator in an EMO so that
all population members are at least guaranteed local optimal solutions. To
save computational time, the local search based mutation can be performed
after a few generations. In single-objective EA research, hybridization of EAs
is common for ensuring an optima, it is time that more studies on developing
hybrid EMO are pursued to ensure finding true Pareto-optimal solutions.

3.10.2 EMO and Decision-Making

This book is designed to cover this topic in detail. It will suffice to point out in
this chapter that finding a set of Pareto-optimal solutions by using an EMO
fulfills only one aspect of multiobjective optimization, as choosing a particular
solution for an implementation is the remaining decision-making task which
is also equally important. In the view of the author, the decision-making task
can be considered from two main aspects:

1. **Generic consideration:** There are some aspects which most practical
   users would like to use in narrowing down their choice. For example, in the
   presence of uncertainties in decision variables and/or problem parameters,
   the users are usually interested in finding *robust* solutions which demonstrate a insensitiveness in objective variation due to a perturbation in
   decision variable values or parameters. In the presence of such variations,
   no one is interested in Pareto-optimal but sensitive solutions. Practitioners
   are interested in sacrificing a global optimal solution to achieve a robust
   solution which when mapped to the objective space lie on a relatively flat
   part or which lie away from the constraint boundaries. In such scenarios
   and in the presence of multiple objectives, instead of finding the global
   Pareto-optimal front, the user may be interested in finding the *robust front*
   which may be partially or totally different from the Pareto-optimal front.
   A couple of definitions for a robust front are discussed in Chapter 16.
   A recent study has developed a modified EMO procedure for finding the
   robust front in a problem (Deb and Gupta, 2005).

In addition, instead of finding the entire Pareto-optimal front, the users may be interested in finding some specific solutions on the Pareto-optimal front, such as *knee* points (requiring a large sacrifice in at least one objective to achieve a small gain in another thereby making it discouraging to move out from a knee point (Branke *et al.*, 2004)), Pareto-optimal points depicting certain pre-specified relationship between objectives, Pareto-optimal points having multiplicity (say, at least two or more solutions in the decision variable space mapping to identical objective values), Pareto-optimal solutions which do not lie close to variable boundaries, Pareto-optimal points having certain mathematical properties, such as all Lagrange multipliers having more or less identical magnitude – a condition often desired to have an equal importance to all constraints, and others. These considerations are motivated from the fundamental and practical aspects of optimization and may be applied to most multiobjective problem solving tasks, without any consent of a decision-maker.

2. **Subjective consideration:** In this category, any problem-specific information can be used to narrow down the choices and the process may even lead to a single preferred solution at the end. Most decision-making procedures use some preference information (utility functions, reference points (Wierzbicki, 1980), reference directions (Korhonen and Laakso, 1986), marginal rate of return and a host of other considerations (Miettinen, 1999)) to select a subset of Pareto-optimal solutions. This book is dedicated to the discussion of many such multicriteria decision analysis (MCDA) tools and collaborative suggestions of using EMO with such MCDA tools. Some hybrid EMO and MCDA algorithms are also suggested in the recent past (Deb *et al.*, 2006b; Deb and Kumar, 2007b,a; Thiele *et al.*, 2007; Luque *et al.*, 2009).

3.10.3 Multi-objectivization

Interestingly, the act of finding multiple trade-off solutions using an EMO procedure has found its application outside the realm of solving multiobjective optimization problems per se. The concept of finding near-optimal trade-off solutions is applied to solve other kinds of optimization problems as well. For example, the EMO concept is used to solve constrained single-objective optimization problems by converting the task into a two-objective optimization task of additionally minimizing an aggregate constraint violation (Coello, 2000). This eliminates the need to specify a penalty parameter while using a penalty based constraint handling procedure. If viewed this way, the usual penalty function approach used in classical optimization studies is a special weighted-sum approach to the bi-objective optimization problem of minimizing the objective function and minimizing the constraint violation, for which the weight vector is a function of penalty parameter. A well-known difficulty in genetic programming studies, called the 'bloating', arises due to the continual increase in size of genetic programs with iteration. The reduction of bloating by minimizing the size of programs as an additional objective helped find high-performing solutions with a smaller size of the code (Bleuler *et al.*, 2001). Minimizing the intra-cluster distance and maximizing inter-cluster distance simultaneously in a bi-objective formulation of a clustering problem is found to yield better solutions than the usual single-objective minimization of the ratio of the intra-cluster distance to the inter-cluster distance (Handl and Knowles, 2007). An EMO is used to solve minimum spanning tree problem better than a single-objective EA (Neumann and Wegener, 2005). A recent edited book (Knowles *et al.*, 2008) describes many such interesting applications in which EMO methodologies have helped solve problems which are otherwise (or traditionally) not treated as multiobjective optimization problems.

3.10.4 Applications

EMO methodologies including other a posteriori multiobjective optimization methods must be applied to more interesting real-world problems to demonstrate the utility of finding multiple trade-off solutions. Although some recent studies are finding that EMO procedures are not computationally efficient to find multiple and widely distributed sets of solutions on problems having a large number of objectives (say more than five objectives) (Deb and Saxena, 2006; Corne and Knowles, 2007), EMO procedures are still applicable in very large problems if the attention is changed to finding only a preferred region on the Pareto-optimal front, instead of the complete front. Some such preference based EMO studies (Deb *et al.*, 2006b; Deb and Kumar, 2007a; Branke and Deb, 2004) are applied to 10 or more objectives. In certain many-objective problems, the Pareto-optimal front can be low-dimensional mainly due to the presence of redundant objectives and EMO procedures can again be effective in solving such problems (Deb and Saxena, 2006; Brockhoff and Zitzler, 2007). In addition, the use of reliability based EMO (Deb *et al.*, 2007) and robust EMO (Deb and Gupta, 2005) procedures are ready to be applied to real-world multiobjective design optimization problems. Application studies are also of interest from the point of demonstrating how an EMO procedure and a subsequent MCDA approach can be combined in an iterative manner together to solve a multicriteria decision making problem. Such efforts may lead to development of GUI-based softwares and approaches for solving the task and will demand addressing other important issues such as visualization of multidimensional data, parallel implementation of EMO and MCDA procedures, meta-modeling approaches, and others.

Besides solving real-world multiobjective optimization problems, EMO procedures are also found to be useful for a knowledge discovery task related to a better understanding of a problem. After a set of trade-off solutions are found by an EMO, these solutions can be compared against each other to unveil interesting principles which are common to all these solutions. These common properties among high-performing solutions will provide useful insights about what makes a solution optimal in a particular problem. Such useful information mined from the obtained EMO trade-off solutions have been discovered in many real-world engineering design problems in the recent past and is termed as the task of 'innovization' (Deb and Srinivasan, 2006).

3.11 Conclusions

This chapter has provided a brief introduction to the fast-growing field of multiobjective optimization based on evolutionary algorithms. First, the principles of single-objective evolutionary optimization (EO) techniques have been discussed so that readers can visualize the differences between evolutionary optimization and classical optimization methods. Although the main difference seems to be in the population approach, EO methodologies do not use any derivative information and they possess a parallel search ability through their operators which makes them computationally efficient procedures.

The EMO principle of handling multiobjective optimization problems is to first attempt to find a set of Pareto-optimal solutions and then choose a preferred solution. Since an EO uses a population of solutions in each iteration, EO procedures are potentially viable techniques to capture a number of trade-off near-optimal solutions in a single simulation run. Thus, EMO procedures work in achieving two goals: (i) convergence to as close to the Pareto-optimal front as possible and (ii) maintenance of a well-distributed set of trade-off solutions. This chapter has described a number of popular EMO methodologies, presented some simulation studies on test problems, and discussed how EMO principles can be useful in solving real-world multiobjective optimization problems through a case study of spacecraft trajectory optimization.

Since early EMO research concentrated on finding a set of well-converged and well-distributed set of near-optimal trade-off solutions, EMO researchers concentrated on developing better and computationally faster algorithms by developing scalable test problems and adequate performance metrics to evaluate EMO algorithms.

Finally, this chapter has discussed the potential of EMO and its current research activities. Interestingly, in addition to EMO, these applications can also be achieved with a posteriori MCDM techniques. Besides their routine applications in solving multiobjective optimization problems, EMO and a posteriori MCDM methodologies are capable of solving other types of optimization problems, such as single-objective constrained optimization, clustering problems etc. in a better manner than they are usually solved. EMO and a posteriori MCDM methodologies are capable of unveiling important hidden knowledge about what makes a solution optimal in a problem. EMO techniques are increasingly being found to have tremendous potential to be used in conjunction with interactive multiple criterion decision making tasks in not only finding a set of optimal solutions but also to aid in selecting a preferred solution at the end.

Before closing the chapter, we provide some useful information about the EMO research and literature in the Appendix.

Acknowledgement

The author acknowledges the support of the Academy of Finland and Foundation of Helsinki School of Economics (Grant # 118319).

Appendix: EMO Repository

Here, we outline some dedicated literature in the area of multiobjective optimization. Further references can be found from http://www.lania.mx/~ccoello/EMOO/.

Some Relevant Books in Print

- A. Abraham, L. C. Jain and R. Goldberg. *Evolutionary Multiobjective Optimization: Theoretical Advances and Applications*, London: Springer-Verlag, 2005.
  This is a collection of the latest state-of-the-art theoretical research, design challenges and applications in the field of EMO.
- C. A. C. Coello, D. A. VanVeldhuizen, and G. Lamont. *Evolutionary Algorithms for Solving Multi-Objective Problems*. Boston, MA: Kluwer Academic Publishers, 2002.
  A good reference book with a good citation of most EMO studies. A revised version is in print.
- Y. Collette and P. Siarry. *Multiobjective Optimization: Principles and Case Studies*, Berlin: Springer, 2004.
  This book describes multiobjective optimization methods including EMO and decision-making, and a number of engineering case studies.
- K. Deb. *Multi-objective optimization using evolutionary algorithms*. Chichester, UK: Wiley, 2001. (Third edition, with exercise problems)
  A comprehensive text-book introducing the EMO field and describing major EMO methodologies and salient research directions.
- N. Nedjah and L. de Macedo Mourelle (Eds.). *Real-World Multi-Objective System Engineering*, New York: Nova Science Publishers, 2005.
  This edited book discusses recent developments and application of multiobjective optimization including EMO.
- A. Osyczka. *Evolutionary algorithms for single and multicriteria design optimisation*, Heidelberg: Physica-Verlag, 2002.
  A book describing single and multiobjective EAs with many engineering applications.
- M. Sakawa. *Genetic Algorithms and Fuzzy Multiobjective Optimization*, Norwell, MA: Kluwer, 2002.
  This book discusses EMO for 0-1 programming, integer programming, nonconvex programming, and job-shop scheduling problems under multiobjectiveness and fuzziness.

- K. C. Tan and E. F. Khor and T. H. Lee. *Multiobjective Evolutionary Algorithms and Applications*, London, UK: Springer-Verlag, 2005.
  A book on various methods of preference-based EMO and application case studies covering areas such as control and scheduling.

Some Review Papers

- C. A. C. Coello. Evolutionary Multi-Objective Optimization: A Critical Review. In R. Sarker, M. Mohammadian and X. Yao (Eds), *Evolutionary Optimization*, pp. 117–146, Kluwer Academic Publishers, New York, 2002.
- C. Dimopoulos. A Review of Evolutionary Multiobjective Optimization Applications in the Area of Production Research. *2004 Congress on Evolutionary Computation (CEC'2004)*, IEEE Service Center, Vol. 2, pp. 1487–1494, 2004.
- S. Huband, P. Hingston, L. Barone and L. While. A Review of Multiobjective Test Problems and a Scalable Test Problem Toolkit, *IEEE Transactions on Evolutionary Computation*, Vol. 10, No. 5, pp. 477–506, 2006.

Dedicated Conference Proceedings

- S. Obayashi, K. Deb, C. Poloni, and T. Hiroyasu (Eds.)., *Evolutionary Multi-Criterion Optimization (EMO-07) Conference Proceedings*, Also available as LNCS 4403. Berlin, Germany: Springer, 2007.
- C. A. Coello Coello and A. H. Aguirre and E. Zitzler (Eds.), *Evolutionary Multi-Criterion Optimization (EMO-05) Conference Proceedings*, Also available as LNCS 3410. Berlin, Germany: Springer, 2005.
- Fonseca, C., Fleming, F., Zitzler, E., Deb, K., and Thiele, L. (Eds.), *Evolutionary Multi-Criterion Optimization (EMO-03) Conference Proceedings*. Also available as LNCS 2632. Heidelberg: Springer, 2003.
- Zitzler, E., Deb, K., Thiele, L., Coello, C. A. C. and Corne, D. (Eds.), *Evolutionary Multi-Criterion Optimization (EMO-01) Conference Proceedings*, Also available as LNCS 1993. Heidelberg: Springer, 2001.

Some Public-Domain Source Codes

- NIMBUS: http://www.mit.jyu.fi/MCDM/soft.html
- NSGA-II in C: http://www.iitk.ac.in/kangal/soft.htm
- PISA: http://www.tik.ee.ethz.ch/sop/pisa/
- Shark in C++: http://shark-project.sourceforge.net
- SPEA2 in C++: http://www.tik.ee.ethz.ch/~zitzler

Some Commercial Codes Implementing EMO

- GEATbx in Matlab (http://www.geatbx.com/)
- iSIGHT and FIPER from Engineous (http://www.engineous.com/)
- MAX from CENAERO (http://www.cenaero.be/)
- modeFRONTIER from Esteco (http://www.esteco.com/)

References

Babu, B.V., Jehan, M.M.L.: Differential Evolution for Multi-Objective Optimization. In: Proceedings of the 2003 Congress on Evolutionary Computation (CEC'2003), Canberra, Australia, December 2003, vol. 4, pp. 2696–2703. IEEE Computer Society Press, Los Alamitos (2003)

Bäck, T., Fogel, D., Michalewicz, Z.: Handbook of Evolutionary Computation. Oxford University Press, Oxford (1997)

Bleuler, S., Brack, M., Zitzler, E.: Multiobjective genetic programming: Reducing bloat using spea2. In: Proceedings of the 2001 Congress on Evolutionary Computation, pp. 536–543 (2001)

Branke, J., Deb, K.: Integrating user preferences into evolutionary multi-objective optimization. In: Jin, Y. (ed.) Knowledge Incorporation in Evolutionary Computation, pp. 461–477. Springer, Heidelberg (2004)

Branke, J., Deb, K., Dierolf, H., Osswald, M.: Finding knees in multi-objective optimization. In: Yao, X., Burke, E.K., Lozano, J.A., Smith, J., Merelo-Guervós, J.J., Bullinaria, J.A., Rowe, J.E., Tiňo, P., Kabán, A., Schwefel, H.-P. (eds.) PPSN 2004. LNCS, vol. 3242, pp. 722–731. Springer, Heidelberg (2004)

Brockhoff, D., Zitzler, E.: Dimensionality Reduction in Multiobjective Optimization: The Minimum Objective Subset Problem. In: Waldmann, K.H., Stocker, U.M. (eds.) Operations Research Proceedings 2006, Saarbücken, Germany, pp. 423–429. Springer, Heidelberg (2007)

Coello Coello, C.A.: Treating objectives as constraints for single objective optimization. Engineering Optimization 32(3), 275–308 (2000)

Coello Coello, C.A., Lechuga, M.S.: MOPSO: A Proposal for Multiple Objective Particle Swarm Optimization. In: Congress on Evolutionary Computation (CEC'2002), May 2002, vol. 2, pp. 1051–1056. IEEE Service Center, Piscataway (2002)

Coello Coello, C.A., Toscano, G.: A micro-genetic algorithm for multi-objective optimization. Technical Report Lania-RI-2000-06, Laboratoria Nacional de Informatica Avanzada, Xalapa, Veracruz, Mexico (2000)

Coello, C.A.C., VanVeldhuizen, D.A., Lamont, G.: Evolutionary Algorithms for Solving Multi-Objective Problems. Kluwer Academic Publishers, Boston (2002)

Coello Coello, C.A., Hernández Aguirre, A., Zitzler, E. (eds.): EMO 2005. LNCS, vol. 3410. Springer, Heidelberg (2005)

Corne, D.W., Knowles, J.D.: Techniques for highly multiobjective optimisation: some nondominated points are better than others. In: GECCO'07: Proceedings of the 9th annual conference on Genetic and evolutionary computation, pp. 773–780. ACM Press, New York (2007)

Corne, D.W., Knowles, J.D., Oates, M.: The Pareto envelope-based selection algorithm for multiobjective optimization. In: Deb, K., Rudolph, G., Lutton, E., Merelo, J.J., Schoenauer, M., Schwefel, H.-P., Yao, X. (eds.) PPSN 2000. LNCS, vol. 1917, pp. 839–848. Springer, Heidelberg (2000)

Corne, D.W., Jerram, N.R., Knowles, J.D., Oates, M.J.: PESA-II: Region-based selection in evolutionary multiobjective optimization. In: Proceedings of the Genetic and Evolutionary Computation Conference (GECCO-2001), pp. 283–290. Morgan Kaufmann, San Francisco (2001)

Coverstone-Carroll, V., Hartmann, J.W., Mason, W.J.: Optimal multi-objective low-thurst spacecraft trajectories. Computer Methods in Applied Mechanics and Engineering 186(2–4), 387–402 (2000)

Deb, K.: An introduction to genetic algorithms. Sādhanā 24(4), 293–315 (1999a)

Deb, K.: Multi-objective genetic algorithms: Problem difficulties and construction of test problems. Evolutionary Computation Journal 7(3), 205–230 (1999b)

Deb, K.: Multi-objective optimization using evolutionary algorithms. Wiley, Chichester (2001)

Deb, K., Agrawal, S.: A niched-penalty approach for constraint handling in genetic algorithms. In: Proceedings of the International Conference on Artificial Neural Networks and Genetic Algorithms (ICANNGA-99), pp. 235–243. Springer, Heidelberg (1999)

Deb, K., Gupta, H.: Searching for robust pareto-optimal solutions in multi-objective optimization. In: Coello Coello, C.A., Hernández Aguirre, A., Zitzler, E. (eds.) EMO 2005. LNCS, vol. 3410, pp. 150–164. Springer, Heidelberg (2005)

Deb, K., Kumar, A.: Interactive evolutionary multi-objective optimization and decision-making using reference direction method. In: Proceedings of the Genetic and Evolutionary Computation Conference (GECCO-2007), pp. 781–788. ACM, New York (2007a)

Deb, K., Kumar, A.: Light beam search based multi-objective optimization using evolutionary algorithms. In: Proceedings of the Congress on Evolutionary Computation (CEC-07), pp. 2125–2132 (2007b)

Deb, K., Saxena, D.: Searching for pareto-optimal solutions through dimensionality reduction for certain large-dimensional multi-objective optimization problems. In: Proceedings of the World Congress on Computational Intelligence (WCCI-2006), pp. 3352–3360 (2006)

Deb, K., Srinivasan, A.: Innovization: Innovating design principles through optimization. In: Proceedings of the Genetic and Evolutionary Computation Conference (GECCO-2006), pp. 1629–1636. ACM, New York (2006)

Deb, K., Tiwari, S.: Omni-optimizer: A generic evolutionary algorithm for global optimization. European Journal of Operational Research (EJOR) 185(3), 1062–1087 (2008)

Deb, K., Pratap, A., Meyarivan, T.: Constrained test problems for multi-objective evolutionary optimization. In: Zitzler, E., Deb, K., Thiele, L., Coello Coello, C.A., Corne, D.W. (eds.) EMO 2001. LNCS, vol. 1993, pp. 284–298. Springer, Heidelberg (2001)

Deb, K., Anand, A., Joshi, D.: A computationally efficient evolutionary algorithm for real-parameter optimization. Evolutionary Computation Journal 10(4), 371–395 (2002a)

Deb, K., Agrawal, S., Pratap, A., Meyarivan, T.: A fast and elitist multi-objective genetic algorithm: NSGA-II. IEEE Transactions on Evolutionary Computation 6(2), 182–197 (2002b)

Deb, K., Thiele, L., Laumanns, M., Zitzler, E.: Scalable multi-objective optimization test problems. In: Proceedings of the Congress on Evolutionary Computation (CEC-2002), pp. 825–830 (2002c)

Deb, K., Reddy, A.R., Singh, G.: Optimal scheduling of casting sequence using genetic algorithms. Journal of Materials and Manufacturing Processes 18(3), 409–432 (2003a)

Deb, K., Mohan, R.S., Mishra, S.K.: Towards a quick computation of well-spread pareto-optimal solutions. In: Fonseca, C.M., Fleming, P.J., Zitzler, E., Deb, K., Thiele, L. (eds.) EMO 2003. LNCS, vol. 2632, pp. 222–236. Springer, Heidelberg (2003b)

Deb, K., Thiele, L., Laumanns, M., Zitzler, E.: Scalable test problems for evolutionary multi-objective optimization. In: Abraham, A., Jain, L., Goldberg, R. (eds.) Evolutionary Multiobjective Optimization, pp. 105–145. Springer, London (2005)

Deb, K., Sinha, A., Kukkonen, S.: Multi-objective test problems, linkages and evolutionary methodologies. In: Proceedings of the Genetic and Evolutionary Computation Conference (GECCO-2006), pp. 1141–1148. ACM, New York (2006a)

Deb, K., Sundar, J., Uday, N., Chaudhuri, S.: Reference point based multi-objective optimization using evolutionary algorithms. International Journal of Computational Intelligence Research (IJCIR) 2(6), 273–286 (2006b)

Deb, K., Padmanabhan, D., Gupta, S., Mall, A.K.: Reliability-based multi-objective optimization using evolutionary algorithms. In: Obayashi, S., Deb, K., Poloni, C., Hiroyasu, T., Murata, T. (eds.) EMO 2007. LNCS, vol. 4403, pp. 66–80. Springer, Heidelberg (2007)

Deb, K., Tiwari, R., Dixit, M., Dutta, J.: Finding trade-off solutions close to KKT points using evolutionary multi-objective optimisation. In: Proceedings of the Congress on Evolutionary Computation (CEC-2007), Singapore, pp. 2109–2116 (2007)

Fogel, L.J., Owens, A.J., Walsh, M.J.: Artificial Intelligence Through Simulated Evolution. Wiley, New York (1966)

Fonseca, C.M., Fleming, P.J., Zitzler, E., Deb, K., Thiele, L. (eds.): EMO 2003. LNCS, vol. 2632. Springer, Heidelberg (2003)

Fonseca, C.M., Fleming, P.J.: Genetic algorithms for multiobjective optimization: Formulation, discussion, and generalization. In: Proceedings of the Fifth International Conference on Genetic Algorithms, pp. 416–423 (1993)

Fonseca, C.M., Fleming, P.J.: On the performance assessment and comparison of stochastic multiobjective optimizers. In: Ebeling, W., Rechenberg, I., Voigt, H.-M., Schwefel, H.-P. (eds.) PPSN 1996. LNCS, vol. 1141, pp. 584–593. Springer, Heidelberg (1996)

Fonseca, C.M., da Fonseca, V.G., Paquete, L.: Exploring the performance of stochastic multiobjective optimisers with the second-order attainment function. In: Coello Coello, C.A., Hernández Aguirre, A., Zitzler, E. (eds.) EMO 2005. LNCS, vol. 3410, pp. 250–264. Springer, Heidelberg (2005)

Gen, M., Cheng, R.: Genetic Algorithms and Engineering Design. Wiley, Chichester (1997)

Goldberg, D.E.: Genetic Algorithms for Search, Optimization, and Machine Learning. Addison-Wesley, Reading (1989)

Goldberg, D.E., Richardson, J.: Genetic algorithms with sharing for multimodal function optimization. In: Proceedings of the First International Conference on Genetic Algorithms and Their Applications, pp. 41–49 (1987)

Goldberg, D.E., Deb, K., Thierens, D.: Toward a better understanding of mixing in genetic algorithms. Journal of the Society of Instruments and Control Engineers (SICE) 32(1), 10–16 (1993)

Gravel, M., Price, W.L., Gagné, C.: Scheduling continuous casting of aluminum using a multiple objective ant colony optimization metaheuristic. European Journal of Operational Research 143(1), 218–229 (2002)

Handl, J., Knowles, J.D.: An evolutionary approach to multiobjective clustering. IEEE Transactions on Evolutionary Computation 11(1), 56–76 (2007)

Hansen, M.P., Jaskiewicz, A.: Evaluating the quality of approximations to the non-dominated set. Technical Report IMM-REP-1998-7, Institute of Mathematical Modelling, Technical University of Denmark, Lyngby (1998)

Herrera, F., Lozano, M., Verdegay, J.L.: Tackling real-coded genetic algorithms: Operators and tools for behavioural analysis. Artificial Intelligence Review 12(4), 265–319 (1998)

Holland, J.H.: Concerning efficient adaptive systems. In: Yovits, M.C., Jacobi, G.T., Goldstein, G.B. (eds.) Self-Organizing Systems, pp. 215–230. Spartan Press, New York (1962)

Holland, J.H.: Adaptation in Natural and Artificial Systems. MIT Press, Ann Arbor (1975)

Horn, J., Nafploitis, N., Goldberg, D.E.: A niched Pareto genetic algorithm for multi-objective optimization. In: Proceedings of the First IEEE Conference on Evolutionary Computation, pp. 82–87 (1994)

Huband, S., Barone, L., While, L., Hingston, P.: A scalable multi-objective test problem toolkit. In: Coello Coello, C.A., Hernández Aguirre, A., Zitzler, E. (eds.) EMO 2005. LNCS, vol. 3410, pp. 280–295. Springer, Heidelberg (2005)

Igel, C., Hansen, N., Roth, S.: Covariance matrix adaptation for multi-objective optimization evolutionary computation. Evolutionary Computation Journal 15(1), 1–28 (2007)

Jansen, T., Wegener, I.: On the utility of populations. In: Proceedings of the Genetic and Evolutionary Computation Conference (GECCO 2001), pp. 375–382. Morgan Kaufmann, San Francisco (2001)

Kennedy, J., Eberhart, R.C., Shi, Y.: Swarm intelligence. Morgan Kaufmann, San Francisco (2001)

Knowles, J., Corne, D., Deb, K.: Multiobjective Problem Solving from Nature. Springer, Heidelberg (2008)

Knowles, J.D., Corne, D.W.: Approximating the non-dominated front using the Pareto archived evolution strategy. Evolutionary Computation Journal 8(2), 149–172 (2000)

Knowles, J.D., Corne, D.W.: On metrics for comparing nondominated sets. In: Congress on Evolutionary Computation (CEC-2002), pp. 711–716. IEEE Press, Piscataway (2002)

Korhonen, P., Laakso, J.: A visual interactive method for solving the multiple criteria problem. European Journal of Operational Reseaech 24, 277–287 (1986)

Koza, J.R.: Genetic Programming: On the Programming of Computers by Means of Natural Selection. MIT Press, Cambridge (1992)

Kung, H.T., Luccio, F., Preparata, F.P.: On finding the maxima of a set of vectors. Journal of the Association for Computing Machinery 22(4), 469–476 (1975)

Laumanns, M., Thiele, L., Deb, K., Zitzler, E.: Combining convergence and diversity in evolutionary multi-objective optimization. Evolutionary Computation 10(3), 263–282 (2002)

Loughlin, D.H., Ranjithan, S.: The neighborhood constraint method: A multiobjective optimization technique. In: Proceedings of the Seventh International Conference on Genetic Algorithms, pp. 666–673 (1997)

Luque, M., Miettinen, K., Eskelinen, P., Ruiz, F.: Incorporating preference information in interactive reference point methods for multiobjective optimization. Omega 37(2), 450–462 (2009)

McMullen, P.R.: An ant colony optimization approach to addessing a JIT sequencing problem with multiple objectives. Artificial Intelligence in Engineering 15, 309–317 (2001)

Michalewicz, Z.: Genetic Algorithms + Data Structures = Evolution Programs. Springer, Berlin (1992)

Miettinen, K.: Nonlinear Multiobjective Optimization. Kluwer, Boston (1999)

Mostaghim, S., Teich, J.: Strategies for Finding Good Local Guides in Multiobjective Particle Swarm Optimization (MOPSO). In: 2003 IEEE Swarm Intelligence Symposium Proceedings, Indianapolis, Indiana, USA, April 2003, pp. 26–33. IEEE Computer Society Press, Los Alamitos (2003)

Neumann, F., Wegener, I.: Minimum spanning trees made easier via multi-objective optimization. In: GECCO '05: Proceedings of the 2005 conference on Genetic and evolutionary computation, pp. 763–769. ACM Press, New York (2005)

Obayashi, S., Deb, K., Poloni, C., Hiroyasu, T., Murata, T. (eds.): EMO 2007. LNCS, vol. 4403. Springer, Heidelberg (2007)

Okabe, T., Jin, Y., Olhofer, M., Sendhoff, B.: On test functions for evolutionary multi-objective optimization. In: Yao, X., Burke, E.K., Lozano, J.A., Smith, J., Merelo-Guervós, J.J., Bullinaria, J.A., Rowe, J.E., Tiňo, P., Kabán, A., Schwefel, H.-P. (eds.) PPSN 2004. LNCS, vol. 3242, pp. 792–802. Springer, Heidelberg (2004)

Osyczka, A.: Evolutionary algorithms for single and multicriteria design optimization. Physica-Verlag, Heidelberg (2002)

Price, K.V., Storn, R., Lampinen, J.: Differential Evolution: A Practical Approach to Global Optimization. Springer-Verlag, Berlin (2005)

Radcliffe, N.J.: Forma analysis and random respectful recombination. In: Proceedings of the Fourth International Conference on Genetic Algorithms, pp. 222–229 (1991)

Rechenberg, I.: Cybernetic solution path of an experimental problem. Royal Aircraft Establishment, Library Translation Number 1122, Farnborough, UK (1965)

Rechenberg, I.: Evolutionsstrategie: Optimierung Technischer Systeme nach Prinzipien der Biologischen Evolution. Frommann-Holzboog Verlag, Stuttgart (1973)

Rosenberg, R.S.: Simulation of Genetic Populations with Biochemical Properties. Ph.D. thesis, Ann Arbor, MI, University of Michigan (1967)

Rudolph, G.: Convergence analysis of canonical genetic algorithms. IEEE Transactions on Neural Network 5(1), 96–101 (1994)

Sasaki, D., Morikawa, M., Obayashi, S., Nakahashi, K.: Aerodynamic shape optimization of supersonic wings by adaptive range multiobjective genetic algorithms. In: Zitzler, E., Deb, K., Thiele, L., Coello Coello, C.A., Corne, D.W. (eds.) EMO 2001. LNCS, vol. 1993, pp. 639–652. Springer, Heidelberg (2001)

Sauer, C.G.: Optimization of multiple target electric propulsion trajectories. In: AIAA 11th Aerospace Science Meeting, Paper Number 73-205 (1973)

Schaffer, J.D.: Some Experiments in Machine Learning Using Vector Evaluated Genetic Algorithms. Ph.D. thesis, Vanderbilt University, Nashville, TN (1984)

Schwefel, H.-P.: Projekt MHD-Staustrahlrohr: Experimentelle optimierung einer zweiphasendüse, teil I. Technical Report 11.034/68, 35, AEG Forschungsinstitut, Berlin (1968)

Schwefel, H.-P.: Evolution and Optimum Seeking. Wiley, New York (1995)

Srinivas, N., Deb, K.: Multi-objective function optimization using non-dominated sorting genetic algorithms. Evolutionary Computation Journal 2(3), 221–248 (1994)

Storn, R., Price, K.: Differential evolution – A fast and efficient heuristic for global optimization over continuous spaces. Journal of Global Optimization 11, 341–359 (1997)

Thiele, L., Miettinen, K., Korhonen, P., Molina, J.: A preference-based interactive evolutionary algorithm for multiobjective optimization. Technical Report W-412, Helsingin School of Economics, Helsingin Kauppakorkeakoulu, Finland (2007)

Veldhuizen, D.V., Lamont, G.B.: Multiobjective evolutionary algorithm research: A history and analysis. Technical Report TR-98-03, Department of Electrical and Computer Engineering, Air Force Institute of Technology, Dayton, OH (1998)

Veldhuizen, D.V., Lamont, G.B.: Multiobjective evolutionary algorithms: Analyzing the state-of-the-art. Evolutionary Computation Journal 8(2), 125–148 (2000)

Vose, M.D., Wright, A.H., Rowe, J.E.: Implicit parallelism. In: Cantú-Paz, E., Foster, J.A., Deb, K., Davis, L., Roy, R., O'Reilly, U.-M., Beyer, H.-G., Kendall, G., Wilson, S.W., Harman, M., Wegener, J., Dasgupta, D., Potter, M.A., Schultz, A., Dowsland, K.A., Jonoska, N., Miller, J., Standish, R.K. (eds.) GECCO 2003. LNCS, vol. 2723, Springer, Heidelberg (2003)

Wierzbicki, A.P.: The use of reference objectives in multiobjective optimization. In: Fandel, G., Gal, T. (eds.) Multiple Criteria Decision Making Theory and Applications, pp. 468–486. Springer, Berlin (1980)

Zitzler, E., Künzli, S.: Indicator-Based Selection in Multiobjective Search. In: Yao, X., Burke, E.K., Lozano, J.A., Smith, J., Merelo-Guervós, J.J., Bullinaria, J.A., Rowe, J.E., Tiňo, P., Kabán, A., Schwefel, H.-P. (eds.) PPSN 2004. LNCS, vol. 3242, pp. 832–842. Springer, Heidelberg (2004)

Zitzler, E., Thiele, L.: Multiobjective optimization using evolutionary algorithms - A comparative case study. In: Eiben, A.E., Bäck, T., Schoenauer, M., Schwefel, H.-P. (eds.) PPSN 1998. LNCS, vol. 1498, pp. 292–301. Springer, Heidelberg (1998)

Zitzler, E., Thiele, L.: Multiobjective evolutionary algorithms: A comparative case study and the strength pareto approach. IEEE Transactions on Evolutionary Computation 3(4), 257–271 (1999)

Zitzler, E., Deb, K., Thiele, L.: Comparison of multiobjective evolutionary algorithms: Empirical results. Evolutionary Computation Journal 8(2), 125–148 (2000)

Zitzler, E., Deb, K., Thiele, L., Coello Coello, C.A., Corne, D.W. (eds.): EMO 2001. LNCS, vol. 1993. Springer, Heidelberg (2001a)

Zitzler, E., Laumanns, M., Thiele, L.: SPEA2: Improving the strength pareto evolutionary algorithm for multiobjective optimization. In: Giannakoglou, K.C., Tsahalis, D.T., Périaux, J., Papailiou, K.D., Fogarty, T. (eds.) Evolutionary Methods for Design Optimization and Control with Applications to Industrial Problems, pp. 95–100. International Center for Numerical Methods in Engineering (Cmine), Athens, Greece (2001b)

Zitzler, E., Thiele, L., Laumanns, M., Fonseca, C.M., Fonseca, V.G.: Performance assessment of multiobjective optimizers: An analysis and review. IEEE Transactions on Evolutionary Computation 7(2), 117–132 (2003)

4

Interactive Multiobjective Optimization Using a Set of Additive Value Functions

José Rui Figueira[1], Salvatore Greco[2], Vincent Mousseau[3], and Roman Słowiński[4,5]

[1] CEG-IST, Center for Management Studies, Instituto Superior Técnico, Technical University of Lisbon, Portugal, figueira@ist.utl.pt
[2] Faculty of Economics, University of Catania, Corso Italia, 55, 95129 Catania, Italy, salgreco@unict.it
[3] LAMSADE, Université Paris-Dauphine, 75775 Paris, France, mousseau@lamsade.dauphine.fr
[4] Institute of Computing Science, Poznań University of Technology, 60-965 Poznań, Poland, roman.slowinski@cs.put.poznan.pl
[5] Systems Research Institute, Polish Academy of Sciences, 01-447 Warsaw, Poland

Abstract. In this chapter, we present a new interactive procedure for multiobjective optimization, which is based on the use of a set of value functions as a preference model built by an ordinal regression method. The procedure is composed of two alternating stages. In the first stage, a representative sample of solutions from the Pareto optimal set (or from its approximation) is generated. In the second stage, the Decision Maker (DM) is asked to make pairwise comparisons of some solutions from the generated sample. Besides pairwise comparisons, the DM may compare selected pairs from the viewpoint of the intensity of preference, both comprehensively and with respect to a single criterion. This preference information is used to build a preference model composed of all general additive value functions compatible with the obtained information. The set of compatible value functions is then applied on the whole Pareto optimal set, which results in possible and necessary rankings of Pareto optimal solutions. These rankings are used to select a new sample of solutions, which is presented to the DM, and the procedure cycles until a satisfactory solution is selected from the sample or the DM comes to conclusion that there is no satisfactory solution for the current problem setting. Construction of the set of compatible value functions is done using ordinal regression methods called UTA^{GMS} and GRIP. These two methods generalize UTA-like methods and they are competitive to AHP and MACBETH methods. The interactive procedure will be illustrated through an example.

Reviewed by: Jerzy Błaszczyński, Poznań University of Technology, Poland
Daisuke Sasaki, University of Cambridge, UK
Kalyanmoy Deb, Indian Institute of Technology Kanpur, India

J. Branke et al. (Eds.): Multiobjective Optimization, LNCS 5252, pp. 97–119, 2008.

4.1 Introduction

Over the last decade, research on MultiObjective Optimization (MOO) has been mainly devoted to generation of exact Pareto optimal solutions, or of an approximation of the Pareto optimal set (also called Pareto Frontier – PF), for problems with both combinatorial and multiple criteria structure. Only little attention has been paid to the inclusion of Decision Maker's (DM's) preferences in the generation process. MOO has been thus considered merely from the point of view of mathematical programming, while limited work is devoted to the point of view of decision aiding (see, however, Chapter 6 and Chapter 7, where preference information is used in evolutionary multiobjective optimization). There is no doubt, that the research about the inclusion of preferences within MOO is not sufficient, and thus the link between MOO and decision aiding should be strengthened. With this aim, in this chapter we propose to use the ordinal regression paradigm as a theoretically sound foundation for handling preference information in an interactive process of solving MOO problems.

In the following, we assume that the interactive procedure explores the PF of an MOO problem, however, it could be as well an approximation of this set.

The ordinal regression paradigm has been originally applied to multiple criteria decision aiding in the UTA method (Jacquet-Lagrèze and Siskos, 1982). This paradigm assumes construction of a criteria aggregation model compatible with preference information elicited from the DM. In the context of MOO, this information has the form of holistic judgments on a reference subset of the PF. The criteria aggregation model built in this way, is a DM's preference model. It is applied on the whole PF to show how the PF solutions compare between them using this model. The ordinal regression paradigm, gives a new sense to the interaction with the DM. The preference information is collected in a very easy way and concerns a small subset of PF solutions playing the role of a training sample. Elicitation of holistic pairwise comparisons of some solutions from the training sample, as well as comparisons of the intensity of preferences between some selected pairs of solutions, require from the DM a relatively small cognitive effort. The ordinal regression paradigm is also appropriate for designing an interactive process of solving a MOO problem, as a constructive learning process. This allows a DM to learn progressively about his/her preferences and make revisions of his/her judgments in successive iterations.

Designing an interactive process in a constructive learning perspective is based on the hypothesis that beyond the model definition, one of the prominent roles of the interactive process is to build a conviction in the mind of the DM on how solutions compare between them. Elaborating such a conviction is grounded on two aspects: (1) the preexisting elements, such as the DM's value system, past experience related to the decision problem; and (2) the elements of information presented to the DM in the dialogue stage, showing how the preference information from the previous iterations induces comparisons of PF solutions. In order to be more specific about the nature of the constructive learning of preferences, it is important to say that there is a clear feedback in the process. On one hand, the preference information provided by the DM contributes to the construction of a preference model and, on the other hand, the use of the preference model shapes the DM's preferences or, at least, makes the DM's conviction evolve.

An interactive MOO procedure using the ordinal regression has been proposed in (Jacquet-Lagrèze et al., 1987). The ordinal regression implemented in this procedure is the same as in the UTA method, thus the preference model being used is a single additive value function with piecewise-linear components.

The interactive procedure proposed in this chapter is also based on ordinal regression, however, it is quite different from the previous proposal because it is using a preference model being a set of value functions, as considered in UTA^{GMS} and GRIP methods (Greco et al., 2003, 2008; Figueira et al., 2008). The value functions have a general additive form and they are compatible with preference information composed of pairwise comparisons of some solutions, and comparisons of the intensities of preference between pairs of selected solutions. UTA^{GMS} and GRIP methods extend the original UTA method in several ways: (1) all additive value functions compatible with the preference information are taken into account, while UTA is using only one such a function; (2) the marginal value functions are general monotone non-decreasing functions, and not only piecewise-linear ones, as in UTA. Moreover, the DM can provide a preference information which can be less demanding than in UTA (partial preorder of some solutions instead of a complete preorder), on one hand, and richer than in UTA (comparison of intensities of preference between selected pairs of solutions, e.g., the preference of a over b is stronger than the one of c over b), on the other hand. Lastly, these methods provide as results necessary rankings that express statements that hold for all compatible value functions, and possible rankings that express statements which hold for at least one compatible value function, respectively. The two extensions of the UTA method appear to be very useful for organizing an interactive search of the most satisfactory solution of a MOO problem – the interaction with the DM is organized such that the preference information is provided incrementally, with the possibility of checking the impact of particular pieces of information on the preference structure of the PF.

The chapter is organized as follows. Section 4.2 is devoted to presentation of the general scheme of the constructive learning interactive procedure. Section 4.3 provides a brief reminder on learning of one compatible additive piecewise-linear value function for multiple criteria ranking problems using the UTA method. In Section 4.4, the GRIP method is presented, which is currently the most general of all UTA-like methods. GRIP is also competitive to the current main methods in the field of multiple criteria decision aiding. In particular, it is competitive to the AHP method (Saaty, 1980), which requires pairwise comparisons of solutions and criteria, and yields a priority ranking of solutions. Then, GRIP is competitive to the MACBETH method (Bana e Costa and Vansnick, 1994), which also takes into account a preference order of solutions and intensity of preference for pairs of solutions. The preference information used in GRIP does not need, however, to be complete: the DM is asked to provide comparisons of only those pairs of selected solutions on particular criteria for which his/her judgment is sufficiently certain. This is an important advantage when comparing GRIP to methods which, instead, require comparison of all possible pairs of solutions on all the considered criteria. Section 4.5 presents an application of the proposed interactive procedure for MOO; the possible pieces of preference information that can be considered in an interactive protocol are the following: ordinal pairwise comparisons of selected PF solutions, and ordinal comparisons of intensities of preference between selected pairs of PF solutions. In the last Section, some conclusions and further research directions are provided.

4.2 Application of an Ordinal Regression Method within a Multiobjective Interactive Procedure

In the following, we assume that the Pareto optimal set of a MOO problem is generated prior to an interactive exploration of this set. Instead of the whole and exact Pareto optimal set of a MOO problem, one can also consider a proper representation of this set, or its approximation. In any case, an interactive exploration of this set should lead the DM to a conviction that either there is no satisfactory solution to the considered problem, or there is at least one such a solution. We will focus our attention on the interactive exploration, and the proposed interactive procedure will be valid for any finite set of solutions to be explored. Let us denote this set by A. Note that such set A can be computed using a MOO or EMO algorithm (see Chapters 1 and 3).

In the course of the interactive procedure, the preference information provided by the DM concerns a small subset of A, called reference or training sample, and denoted by A^R. The preference information is transformed by an ordinal regression method into a DM's preference model. We propose to use at this stage the GRIP method, thus the preference model is a set of general additive value functions compatible with the preference information. A compatible value function compares the solutions from the reference sample in the same way as the DM. The obtained preference model is then applied on the whole set A, which results in necessary and possible rankings of solutions. These rankings are used to select a new sample of reference solutions, which is presented to the DM, and the procedure cycles until a satisfactory solution is selected from the sample or the DM comes to conclusion that there is no satisfactory solution for the current problem setting.

The proposed interactive procedure is composed of the following steps:

***Step* 1.** Select a representative reference sample A^R of solutions from set A.

***Step* 2.** Present the sample A^R to the DM.

***Step* 3.** If the DM is satisfied with at least one solution from the sample, then this is the most preferred solution and the procedure stops. The procedure also stops in this step if, after several iterations, the DM concludes that there is no satisfactory solution for the current problem setting. Otherwise continue.

***Step* 4.** Ask the DM to provide information about his/her preferences on set A^R in the following terms:

– pairwise comparison of some solutions from A^R,
– comparison of intensities of comprehensive preferences between some pairs of solutions from A^R,
– comparison of intensities of preferences on single criteria between some pairs of solutions from A^R.

***Step* 5.** Use the GRIP method to build a set of general additive value functions compatible with the preference information obtained from the DM in *Step* 4.

***Step* 6.** Apply the set of compatible value functions built in *Step* 5 on the whole set A, and present the necessary and possible rankings (see sub-section 4.4.2) resulting from this application to the DM.

***Step* 7.** Taking into account the necessary and possible rankings on set A, let the DM select a new reference sample of solutions $A^R \subseteq A$, and go to *Step* 2.

In *Step* 4, the information provided by the DM may lead to a set of constraints which define an empty polyhedron of the compatible value functions. In this case, the DM is informed what items of his/her preference information make the polyhedron empty, so as to enable revision in the next round. This point is explained in detail in (Greco *et al.*, 2008; Figueira *et al.*, 2008). Moreover, information provided by the DM in *Step* 4 cannot be considered as irreversible. Indeed, the DM can come back to one of previous iterations and continue from this point. This feature is concordant with the spirit of a learning oriented conception of multiobjective interactive optimization, i.e. it confirms the idea that the interactive procedure permits the DM to learn about his/her preferences and about the "shape" of the Pareto optimal set (see Chapter 15).

Notice that the proposed approach allows to elicit incrementally preference information by the DM. Remark that in *Step* 7, the "new" reference sample A^R is not necessarily different from the previously considered, however, the preference information elicited by the DM in the next iteration is richer than previously, due to the learning effect. This permits to build and refine progressively the preference model: in fact, each new item of information provided in *Step* 4 reduces the set of compatible value functions and defines the DM's preferences more and more precisely.

Let us also observe that information obtained from the DM in *Step* 4 and information given to the DM in *Step* 6 is composed of very simple and easy to understand statements: preference comparisons in *Step* 4, and necessary and possible rankings in *Step* 6 (i.e., a necessary ranking that holds for all compatible value functions, and a possible ranking that holds for at least one compatible value function, see sub-section 4.4.2). Thus, the nature of information exchanged with the DM during the interaction is purely ordinal. Indeed, monotonically increasing transformations of evaluation scales of considered criteria have no influence on the final result.

Finally, observe that a very important characteristic of our method from the point of view of learning is that the DM can observe the impact of information provided in *Step* 4 in terms of necessary and possible rankings of solutions from set *A*.

4.3 The Ordinal Regression Method for Learning One Compatible Additive Piecewise-Linear Value Function

The preference information may be either direct or indirect, depending whether it specifies directly values of some parameters used in the preference model (e.g. trade-off weights, aspiration levels, discrimination thresholds, etc.) or, whether it specifies some examples of holistic judgments from which compatible values of the preference model parameters are induced. Eliciting direct preference information from the DM can be counterproductive in real-world decision making situations because of a high cognitive effort required. Consequently, asking directly the DM to provide values for the parameters seems to make the DM uncomfortable. Eliciting indirect preference is less demanding of the cognitive effort. Indirect preference information is mainly used in the ordinal regression paradigm. According to this paradigm, a holistic preference information on a subset of some reference or training solutions is known first and then a preference model compatible with the information is built and applied to the whole set of solutions in order to rank them.

The ordinal regression paradigm emphasizes the discovery of intentions as an interpretation of actions rather than as a priori position, which was called by March the posterior rationality (March, 1978). It has been known for at least fifty years in the field of multidimensional analysis. It is also concordant with the induction principle used in machine learning. This paradigm has been applied within the two main Multiple Criteria Decision Aiding (MCDA) approaches mentioned above: those using a value function as preference model (Srinivasan and Shocker, 1973; Pekelman and Sen, 1974; Jacquet-Lagrèze and Siskos, 1982; Siskos *et al.*, 2005), and those using an outranking relation as preference model (Kiss *et al.*, 1994; Mousseau and Slowinski, 1998). This paradigm has also been used since mid nineties' in MCDA methods involving a new, third family of preference models - a set of dominance decision rules induced from rough approximations of holistic preference relations (Greco *et al.*, 1999, 2001, 2005; Słowiński *et al.*, 2005).

Recently, the ordinal regression paradigm has been revisited with the aim of considering the whole set of value functions compatible with the preference information provided by the DM, instead of a single compatible value function used in UTA-like methods (Jacquet-Lagrèze and Siskos, 1982; Siskos *et al.*, 2005). This extension has been implemented in a method called UTA^{GMS} (Greco *et al.*, 2003, 2008), further generalized in another method called GRIP (Figueira *et al.*, 2008). UTA^{GMS} and GRIP are not revealing to the DM one compatible value function, but they are using the whole set of compatible (general, not piecewise-linear only) additive value functions to set up a necessary weak preference relation and a possible weak preference relation in the whole set of considered solutions.

4.3.1 Concepts: Definitions and Notation

We are considering a multiple criteria decision problem where a finite set of solutions $A = \{x, \ldots, y, \ldots w, \ldots\}$ is evaluated on a family $F = \{g_1, g_2, \ldots, g_n\}$ of n criteria. Let $I = \{1, 2, \ldots, n\}$ denote the set of criteria indices. We assume, without loss of generality, that the greater $g_i(x)$, the better solution x on criterion g_i, for all $i \in I$, $x \in A$. A DM is willing to rank the solutions of A from the best to the worst, according to his/her preferences. The ranking can be complete or partial, depending on the preference information provided by the DM and on the way of exploiting this information. The family of criteria F is supposed to satisfy consistency conditions, i.e. completeness (all relevant criteria are considered), monotonicity (the better the evaluation of a solution on the considered criteria, the more it is preferable to another), and non-redundancy (no superfluous criteria are considered), see (Roy and Bouyssou, 1993).

Such a decision-making problem statement is called *multiple criteria ranking problem*. It is known that the only information coming out from the formulation of this problem is the dominance ranking. Let us recall that in the dominance ranking, solution $x \in A$ is preferred to solution $y \in A$ if and only if $g_i(x) \geq g_i(y)$ for all $i \in I$, with at least one strict inequality. Moreover, x is indifferent to y if and only if $g_i(x) = g_i(y)$ for all $i \in I$. Hence, for any pair of solutions $x, y \in A$, one of the four situations may arise in the dominance ranking: x is preferred to y, y is preferred to x, x is indifferent to y, or x is incomparable to y. Usually, the dominance ranking is very poor, i.e. the most frequent situation is: x incomparable to y.

In order to enrich the dominance ranking, the DM has to provide preference information which is used to construct an aggregation model making the solutions more comparable. Such an aggregation model is called preference model. It induces a preference structure on set A, whose proper exploitation permits to work out a ranking proposed to the DM.

In what follows, the evaluation of each solution $x \in A$ on each criterion $g_i \in F$ will be denoted either by $g_i(x)$ or x_i.

Let G_i denote the value set (scale) of criterion g_i, $i \in I$. Consequently,

$$G = \prod_{i=1}^{n} G_i$$

represents the evaluation space, and $x \in G$ denotes a profile of a solution in such a space. We consider a weak preference relation $\succsim$ on A which means, for each pair of solutions $x, y \in A$,

$$x \succsim y \Leftrightarrow \text{“}x \text{ is at least as good as } y\text{”}.$$

This weak preference relation can be decomposed into its asymmetric and symmetric parts, as follows,

1) $x \succ y \equiv [x \succsim y \text{ and } not(y \succsim x)] \Leftrightarrow$ "x is preferred to y", and
2) $x \sim y \equiv [x \succsim y \text{ and } y \succsim x] \Leftrightarrow$ "x is indifferent to y".

From a pragmatic point of view, it is reasonable to assume that $G_i \subseteq \mathbb{R}$, for $i = 1, \ldots, n$. More specifically, we will assume that the evaluation scale on each criterion g_i is bounded, such that $G_i = [\alpha_i, \beta_i]$, where $\alpha_i, \beta_i, \alpha_i < \beta_i$ are the worst and the best (finite) evaluations, respectively. Thus, $g_i : A \rightarrow G_i$, $i \in I$. Therefore, each solution $x \in A$ is associated with an evaluation solution denoted by $\underline{g}(x) = (x_1, x_2, \ldots, x_n) \in G$.

4.3.2 The UTA Method for a Multiple Criteria Ranking Problem

In this sub-section, we recall the principle of the ordinal regression *via* linear programming, as proposed in the original UTA method, see (Jacquet-Lagrèze and Siskos, 1982).

Preference Information

The preference information is given in the form of a complete preorder on a subset of reference solutions $A^R \subseteq A$ (where $|A^R| = p$), called *reference preorder*. The reference solutions are usually those contained in set A for which the DM is able to express holistic preferences. Let $A^R = \{a, b, c, \ldots\}$ be the set of reference solutions.

An Additive Model

The additive value function is defined on A such that for each $\underline{g}(x) \in G$,

$$U(\underline{g}(x)) = \sum_{i=1}^{n} u_i(g_i(x_i)), \qquad (4.1)$$

where, u_i are non-decreasing marginal value functions, $u_i : G_i \to \mathbb{R}$, $i \in I$. For the sake of simplicity, we shall write (4.1) as follows,

$$U(x) = \sum_{i=1}^{n} u_i(x_i). \qquad (4.2)$$

In the UTA method, the marginal value functions u_i are assumed to be piecewise-linear functions. The ranges $[\alpha_i, \beta_i]$ are divided into $\gamma_i \geq 1$ equal sub-intervals,

$$[x_i^0, x_i^1], [x_i^1, x_i^2], \dots, [x_i^{\gamma_i - 1}, x_i^{\gamma_i}]$$

where,

$$x_i^j = \alpha_i + \frac{j}{\gamma_i}(\beta_i - \alpha_i), \quad j = 0, \dots, \gamma_i, \text{ and } i \in I.$$

The marginal value of a solution $x \in A$ is obtained by linear interpolation,

$$u_i(x) = u_i(x_i^j) + \frac{x_i - x_i^j}{x_i^{j+1} - x_i^j}(u_i(x_i^{j+1}) - u_i(x_i^j)), \quad \text{for } x_i \in [x_i^j, x_i^{j+1}]. \quad (4.3)$$

The piecewise-linear additive model is completely defined by the marginal values at the breakpoints, i.e. $u_i(x_i^0) = u_i(\alpha_i), u_i(x_i^1), u_i(x_i^2), \dots, u_i(x_i^{\gamma_i}) = u_i(\beta_i)$.

In what follows, the principle of the UTA method is described as it was recently presented by Siskos *et al.* (2005).

Therefore, a value function $U(x) = \sum_{i=1}^{n} u_i(x_i)$ is compatible if it satisfies the following set of constraints

$$
\left.
\begin{aligned}
U(a) > U(b) &\Leftrightarrow a \succ b \\
U(a) = U(b) &\Leftrightarrow a \sim b
\end{aligned}
\right\} \quad \forall \ a, b \in A^R
$$
$$u_i(x_i^{j+1}) - u_i(x_i^j) \geq 0, \ i - 1, \dots, n, \ j = 1, \dots, \gamma_i \quad 1 \qquad (4.4)$$
$$u_i(\alpha_i) = 0, \ i = 1, \dots, n$$
$$\sum_{i=1}^{n} u_i(\beta_i) = 1$$

Checking for Compatible Value Functions through Linear Programming

To verify if a compatible value function $U(x) = \sum_{i=1}^{n} u_i(x_i)$ restoring the reference preorder $\succsim$ on A^R exists, one can solve the following linear programming problem, where $u_i(x_i^j), i = 1, \dots, n, \ j = 1, \dots, \gamma_i$, are unknown, and $\sigma^+(a), \sigma^-(a) \ (a \in A^R)$ are auxiliary variables:

$$Min \to F = \sum_{a \in A^R} (\sigma^+(a) + \sigma^-(a))$$

s.t.

$$\left.\begin{array}{l} U(a) + \sigma^+(a) - \sigma^-(a) \geq \\ \qquad U(b) + \sigma^+(b) - \sigma^-(b) + \varepsilon \;\Leftrightarrow\; a \succ b \\ U(a) + \sigma^+(a) - \sigma^-(a) = \\ \qquad U(b) + \sigma^+(b) - \sigma^-(b) \;\Leftrightarrow\; a \sim b \end{array}\right\} \forall a, b \in A^R \quad (4.5)$$

$$u_i(x_i^{j+1}) - u_i(x_i^j) \geq 0, \; i = 1, ..., n, \; j = 1, ..., \gamma_i - 1$$
$$u_i(\alpha_i) = 0, \; i = 1, ..., n$$
$$\sum_{i=1}^n u_i(\beta_i) = 1$$
$$\sigma^+(a), \; \sigma^-(a) \geq 0, \; \forall a \in A^R$$

where, ε is an arbitrarily small positive value so that $U(a) + \sigma^+(a) - \sigma^-(a) > U(b) + \sigma^+(b) - \sigma^-(b)$ in case of $a \succ b$.

If the optimal value of the objective function of program (4.5) is equal to zero ($F^* = 0$), then there exists at least one value function $U(x) = \sum_{i=1}^n u_i(x_i)$ satisfying (4.4), i.e. compatible with the reference preorder on A^R. In other words, this means that the corresponding polyhedron (4.4) of feasible solutions for $u_i(x_i^j)$, $i = 1, ..., n$, $j = 1, ..., \gamma_i$, is not empty.

Let us remark that the transition from the preorder $\succsim$ to the marginal value function exploits the ordinal character of the criterion scale G_i. Note, however, that the scale of the marginal value function is a conjoint interval scale. More precisely, for the considered additive value function $U(x) = \sum_{i=1}^n u_i(x_i)$, the admissible transformations on the marginal value functions $u_i(x_i)$ have the form $u_i^*(x_i) = k \times u_i(x_i) + h_i$, $h_i \in \mathbb{R}$, $i = 1, \ldots, n$, $k > 0$, such that for all $[x_1, ..., x_n], [y_1, ..., y_n] \in G$

$$\sum_{i=1}^n u_i(x_i) \geq \sum_{i=1}^n u_i(y_i) \;\Leftrightarrow\; \sum_{i=1}^n u_i^*(x_i) \geq \sum_{i=1}^n u_i^*(y_i).$$

An alternative way of representing the same preference model is:

$$U(x) = \sum_{i=1}^n w_i \hat{u}_i(x), \qquad (4.6)$$

where $\hat{u}(\alpha_i) = 0$, $\hat{u}(\beta_i) = 1$, $w_i \geq 0$ $i = 1, 2, \ldots, n$ and $\sum_{i=1}^n w_i = 1$. Note that the correspondence between (4.6) and (4.2) is such that $w_i = u_i(\beta_i)$, $\forall i \in G$. Due to the cardinal character of the marginal value function scale, the parameters w_i can be interpreted as tradeoff weights among marginal value functions $\hat{u}_i(x)$. We will use, however, the preference model (4.2) with normalization constraints bounding $U(x)$ to the interval $[0, 1]$.

When the optimal value of the objective function of the program (4.5) is greater than zero ($F^* > 0$), then there is no value function $U(x) = \sum_{i=1}^n u_i(x_i)$ compatible with the reference preorder on A^R. In such a case, three possible moves can be considered:

- increasing the number of linear pieces γ_i for one or several marginal value function u_i could make it possible to find an additive value function compatible with the reference preorder on A^R;
- revising the reference preorder on A^R could lead to find an additive value function compatible with the new preorder;
- searching over the relaxed domain $F \leq F^* + \eta$ could lead to an additive value function giving a preorder on A^R sufficiently close to the reference preorder (in the sense of Kendall's τ).

4.4 The Ordinal Regression Method for Learning the Whole Set of Compatible Value Functions

Recently, two new methods, UTA^{GMS} (Greco et al., 2008) and GRIP (Figueira et al., 2008), have generalized the ordinal regression approach of the UTA method in several aspects:

- taking into account all additive value functions (4.1) compatible with the preference information, while UTA is using only one such function,
- considering marginal value functions of (4.1) as general non-decreasing functions, and not piecewise-linear, as in UTA,
- asking the DM for a ranking of reference solutions which is not necessarily complete (just pairwise comparisons),
- taking into account additional preference information about intensity of preference, expressed both comprehensively and with respect to a single criterion,
- avoiding the use of the exogenous, and not neutral for the result, parameter ε in the modeling of strict preference between solutions.

UTA^{GMS} produces two rankings on the set of solutions A, such that for any pair of solutions $a, b \in A$:

- in the *necessary* ranking, a is ranked at least as good as b if and only if, $U(a) \geq U(b)$ for *all* value functions compatible with the preference information,
- in the *possible* ranking, a is ranked at least as good as b if and only if, $U(a) \geq U(b)$ for *at least one* value function compatible with the preference information.

GRIP produces four more necessary and possible rankings on the set of solutions $A \times A$ as it can bee seen in sub-section 4.4.2.

The necessary ranking can be considered as robust with respect to the preference information. Such robustness of the necessary ranking refers to the fact that any pair of solutions compares in the same way whatever the additive value function compatible with the preference information. Indeed, when no preference information is given, the necessary ranking boils down to the weak dominance relation (i.e., a is necessarily at least as good as b, if $g_i(a) \geq g_i(b)$ for all $g_i \in F$), and the possible ranking is a complete relation. Every new pairwise comparison of reference solutions, for which the dominance relation does not hold, is enriching the necessary ranking and it is impoverishing the possible ranking, so that they converge with the growth of the preference information.

Moreover, such an approach has another feature which is very appealing in the context of MOO. It stems from the fact that it gives space for inter-activity with the DM. Presentation of the necessary ranking, resulting from a preference information provided by the DM, is a good support for generating reactions from part of the DM. Namely, (s)he could wish to enrich the ranking or to contradict a part of it. Such a reaction can be integrated in the preference information considered in the next calculation stage.

The idea of considering the whole set of compatible value functions was originally introduced in UTA^{GMS}. GRIP (Generalized Regression with Intensities of Preference) can be seen as an extension of UTA^{GMS} permitting to take into account additional preference information in form of comparisons of intensities of preference between some pairs of reference solutions. For solutions $x, y, w, z \in A$, these comparisons are expressed in two possible ways (not exclusive): (i) comprehensively, on all criteria, like "x is preferred to y at least as much as w is preferred to z"; and, (ii) partially, on each criterion, like "x is preferred to y at least as much as w is preferred to z, on criterion $g_i \in F$ ". Although UTA^{GMS} was historically the first method among the two, as GRIP incorporates and extends UTA^{GMS}, in the following we shall present only GRIP.

4.4.1 The Preference Information Provided by the Decision Maker

The DM is expected to provide the following preference information in the dialogue stage of the procedure:

- A partial preorder $\succsim$ on A^R whose meaning is: for some $x, y \in A^R$

$$x \succsim y \Leftrightarrow \text{"}x \text{ is at least as good as } y\text{"}.$$

  Moreover, $\succ$ (preference) is the asymmetric part of $\succsim$, and $\sim$ (indifference) is its symmetric part.
- A partial preorder $\succsim^*$ on $A^R \times A^R$, whose meaning is: for some $x, y, w, z \in A^R$,

$$(x, y) \succsim^* (w, z) \Leftrightarrow \text{"}x \text{ is preferred to } y \text{ at least as much as } w \text{ is preferred to } z\text{"}.$$

  Also in this case, $\succ^*$ is the asymmetric part of $\succsim^*$, and $\sim^*$ is its symmetric part.
- A partial preorder $\succsim_i^*$ on $A^R \times A^R$, whose meaning is: for some $x, y, w, z \in A^R$, $(x, y) \succsim_i^* (w, z) \Leftrightarrow \text{"}x \text{ is preferred to } y \text{ at least as much as } w \text{ is preferred to } z\text{"}$ on criterion g_i, $i \in I$.

In the following, we also consider the weak preference relation $\succsim_i$ being a complete preorder whose meaning is: for all $x, y \in A$,

$$x \succsim_i y \quad \Leftrightarrow \quad \text{``}x \text{ is at least as good as } y\text{'' on criterion } g_i, \quad i \in I.$$

Weak preference relations $\succsim_i$, $i \in I$, are not provided by the DM, but they are obtained directly from the evaluation of solutions x and y on criteria g_i, i.e., $x \succsim_i y \Leftrightarrow g_i(x) \geq g_i(y)$, $i \in I$.

4.4.2 Necessary and Possible Binary Relations in Set A and in Set $A \times A$

When there exists at least one value function compatible with the preference information provided by the DM, the method produces the following rankings:

- a necessary ranking $\succsim^N$, for all pairs of solutions $(x, y) \in A \times A$;
- a possible ranking $\succsim^P$, for all pairs of solutions $(x, y) \in A \times A$;
- a necessary ranking $\succsim^{*^N}$, with respect to the comprehensive intensities of preferences for all $((x, y), (w, z)) \in A \times A \times A \times A$;
- a possible ranking $\succsim^{*^P}$, with respect to the comprehensive intensities of preferences for all $((x, y), (w, z)) \in A \times A \times A \times A$;
- a necessary ranking $\succsim_i^{*^N}$, with respect to the partial intensities of preferences for all $((x, y), (w, z)) \in A \times A \times A \times A$ and for all criteria g_i, $i \in I$;
- a possible ranking $\succsim_i^{*^P}$, with respect to the partial intensities of preferences for all $((x, y), (w, z)) \in A \times A \times A \times A$ and for all criteria g_i, $i \in I$.

4.4.3 Linear Programming Constraints

In this sub-section, we present a set of constraints that interprets the preference information in terms of conditions on the compatible value functions.

To be compatible with the provided preference information, the value function $U : A \to [0, 1]$ should satisfy the following constraints corresponding to DM's preference information:

a) $U(w) > U(z)$ if $w \succ z$
b) $U(w) = U(z)$ if $w \sim z$
c) $U(w) - U(z) > U(x) - U(y)$ if $(w, z) \succ^* (x, y)$
d) $U(w) - U(z) = U(x) - U(y)$ if $(w, z) \sim^* (x, y)$
e) $u_i(w) \geq u_i(z)$ if $w \succsim_i z$, $i \in I$
f) $u_i(w) - u_i(z) > u_i(x) - u_i(y)$ if $(w, z) \succ_i^* (x, y)$, $i \in I$
g) $u_i(w) - u_i(z) = u_i(x) - u_i(y)$ if $(w, z) \sim_i^* (x, y)$, $i \in I$

Let us remark that within UTA-like methods, constraint $a)$ is written as $U(w) \geq U(z) + \varepsilon$, where $\varepsilon > 0$ is a threshold exogenously introduced. Analogously, constraints $c)$ and $f)$ should be written as,

$$U(w) - U(z) \geq U(x) - U(y) + \varepsilon$$

and

$$u_i(w) - u_i(z) \geq u_i(x) - u_i(y) + \varepsilon.$$

However, we would like to avoid the use of any exogenous parameter and, therefore, instead of setting an arbitrary value of ε, we consider it as an auxiliary variable, and we test the feasibility of constraints $a)$, $c)$, and $f)$ (see sub-section 4.4.4). This permits to take into account all possible value functions, even those having a very small preference threshold ε. This is also safer from the viewpoint of "objectivity" of the used methodology. In fact, the value of ε is not meaningful in itself and it is useful only because it permits to discriminate preference from indifference.

Moreover, the following normalization constraints should also be taken into account:

$h)$ $u_i(x_i^*) = 0$, where x_i^* is such that $x_i^* = \min\{g_i(x) \ : \ x \in A\}$;

$i)$ $\sum_{i \in I} u_i(y_i^*) = 1$, where y_i^* is such that $y_i^* = \max\{g_i(x) \ : \ x \in A\}$.

4.4.4 Computational Issues

In order to conclude the truth or falsity of binary relations $\succsim^N$, $\succsim^P$, $\succsim^{*N}$, $\succsim^{*P}$, $\succsim_i^{*N}$ and $\succsim_i^{*P}$, we have to take into account that, for all $x, y, w, z \in A$ and $i \in I$:

1) $x \succsim^N y \Leftrightarrow \inf \left\{ U(x) - U(y) \right\} \geq 0$,

2) $x \succsim^P y \Leftrightarrow \inf \left\{ U(y) - U(x) \right\} \leq 0$,

3) $(x, y) \succsim^{*N} (w, z) \Leftrightarrow \inf \left\{ \left(U(x) - U(y) \right) - \left(U(w) - U(z) \right) \right\} \geq 0$,

4) $(x, y) \succsim^{*P} (w, z) \Leftrightarrow \inf \left\{ \left(U(w) - U(z) \right) - \left(U(x) - U(y) \right) \right\} \leq 0$,

5) $(x, y) \succsim_i^{*N} (w, z) \Leftrightarrow \inf \left\{ \left(u_i(x_i) - u_i(y_i) \right) - \left(u_i(w_i) - u_i(z_i) \right) \right\} \geq 0$,

6) $(x, y) \succsim_i^{*P} (w, z) \Leftrightarrow \inf \left\{ \left(u_i(w_i) - u_i(z_i) \right) - \left(u_i(x_i) - u_i(y_i) \right) \right\} \leq 0$,

with the infimum calculated on the set of value functions satisfying constraints from $a)$ to $i)$. Let us remark, however, that the linear programming is not able to handle strict inequalities such as the above $a)$, $c)$, and $f)$. Moreover, linear programming permits to calculate the minimum or the maximum of an objective function and not an infimum. Nevertheless, reformulating properly the above properties 1) to 6), a result presented in (Marichal and Roubens, 2000) permits to use linear programming for testing the truth of binary relations, $\succsim^N$, $\succsim^P$, $\succsim^{*N}$, $\succsim^{*P}$, $\succsim_i^{*N}$ and $\succsim_i^{*P}$.

In order to use such a result, constraints $a)$, $c)$ and $f)$ have to be reformulated as follows:

a') $U(x) \geq U(y) + \varepsilon$ if $x \succ y$;
c') $U(x) - U(y) \geq U(w) - U(z) + \varepsilon$ if $(x, y) \succ^* (w, z)$;
f') $u_i(x) - u_i(y) \geq u_i(w) - u_i(z) + \varepsilon$ if $(x, y) \succ_i^* (w, z)$.

Notice that constraints a), c) and f) are equivalent to a'), c'), and f') whenever $\varepsilon > 0$.

After properties 1) − 6) have to be reformulated such that the search of the infimum is replaced by the calculation of the maximum value of ε on the set of value functions satisfying constraints from a) to i), with constraints a), c), and f) transformed to a'), c'), and f'), plus constraints specific for each point:

1') $x \succsim^P y \Leftrightarrow \varepsilon^* > 0$,
 where $\varepsilon^* = \max \varepsilon$, subject to the constraints a'), b), c'), d), e), f'), plus the constraint $U(x) \geq U(y)$;
2') $x \succsim^N y \Leftrightarrow \varepsilon^* \leq 0$,
 where $\varepsilon^* = \max \varepsilon$, subject to the constraints a'), b), c'), d), e), f'), plus the constraint $U(y) \geq U(x) + \varepsilon$;
3') $(x, y) \succsim^{*P} (w, z) \Leftrightarrow \varepsilon^* > 0$,
 where $\varepsilon^* = \max \varepsilon$, subject to the constraints a'), b), c'), d), e), f'), plus the constraint $\big((U(x) - U(y)) - (U(w) - U(z))\big) \geq 0$;
4') $(x, y) \succsim^{*N} (w, z) \Leftrightarrow \varepsilon^* \leq 0$,
 where $\varepsilon^* = \max \varepsilon$, subject to the constraints a'), b), c'), d), e), f'), plus the constraint $\big((U(w) - U(z)) - (U(x) - U(y))\big) \geq \varepsilon$;
5') $(x, y) \succsim_i^{*P} (w, z) \Leftrightarrow \varepsilon^* > 0$,
 where $\varepsilon^* = \max \varepsilon$, subject to the constraints a'), b), c'), d), e), f'), plus the constraint $\big(u_i(x_i) - u_i(y_i)\big) - \big(u_i(w_i) - u_i(z_i)\big) \geq 0$;
6') $(x, y) \succsim_i^{*N} (w, z) \Leftrightarrow \varepsilon^* \leq 0$,
 where $\varepsilon^* = \max \varepsilon$, subject to the constraints a'), b), c'), d), e), f'), plus the constraint $\big((u_i(w_i) - u_i(z_i)) - (u_i(x_i) - u_i(y_i))\big) \geq \varepsilon$.

4.4.5 Comparison of GRIP with the Analytical Hierarchy Process

In AHP (Saaty, 1980, 2005), criteria should be compared pairwise with respect to their importance. Actions (solutions) are also compared pairwise on particular criteria with respect to intensity of preference. The following nine point scale of preference is used: 1 - equal importance, 3 - moderate importance, 5 - strong importance, 7 - very strong or demonstrated importance, and 9 - extreme importance. 2, 4, 6 and 8 are intermediate values between the two adjacent judgements. The intensity of importance of criterion g_i over criterion g_j is the inverse of the intensity of importance of g_j over g_i. Analogously, the intensity of preference of action x over action y is the inverse of the intensity of preference of y over x. The above scale is a ratio scale. Therefore, the intensity of importance is read as the ratio of weights w_i and w_j corresponding to criteria g_i and g_j, and the intensity of preference is read as the ratio of the attractiveness of x and the attractiveness of y, with respect to the considered criterion g_i. In terms of value functions, the intensity of preference can be interpreted as the ratio $\frac{u_i(g_i(x))}{u_i(g_i(y))}$. Thus, the problem is how to obtain values of w_i and w_j from ratio $\frac{w_i}{w_j}$, and values of $u_i(g_i(x))$ and $u_i(g_i(y))$ from ratio $\frac{u_i(g_i(x))}{u_i(g_i(y))}$.

In AHP, it is proposed that these values are supplied by principal eigenvectors of matrices composed of the ratios $\frac{w_i}{w_j}$ and $\frac{u_i(g_i(x))}{u_i(g_i(y))}$. The marginal value functions $u_i(g_i(x))$ are then aggregated by means of a weighted-sum using the weights w_i.

Comparing AHP with GRIP, we can say that with respect to a single criterion, the type of questions addressed to the DM is the same: express the intensity of preference in qualitative-ordinal terms (equal, moderate, strong, very strong, extreme). However, differently from GRIP, this intensity of preference is translated into quantitative terms (the scale from 1 to 9) in a quite arbitrary way. In GRIP, instead, the marginal value functions are just a numerical representation of the original qualitative-ordinal information, and no intermediate transformation into quantitative terms is exogenously imposed.

Other differences between AHP and GRIP are related to the following aspects.

1) In GRIP, the value functions $u_i(g_i(x))$ depend mainly on holistic judgements, i.e. comprehensive preferences involving jointly all the criteria, while this is not the case in AHP.
2) In AHP, the weights w_i of criteria g_i are calculated on the basis of pairwise comparisons of criteria with respect to their importance; in GRIP, this is not the case, because the value functions $u_i(g_i(x))$ are expressed on the same scale and thus they can be summed up without any further weighting.
3) In AHP, all non-ordered pairs of actions must be compared from the viewpoint of the intensity of preference with respect to each particular criterion. Therefore, if m is the number of actions, and n the number of criteria, then the DM has to answer $n \times \frac{m \times (m-1)}{2}$ questions. Moreover, the DM has to answer questions relative to $\frac{n \times (n-1)}{2}$ pairwise comparisons of considered criteria with respect to their importance. This is not the case in GRIP, which accepts partial information about preferences in terms of pairwise comparison of some reference actions. Finally, in GRIP there is no question about comparison of relative importance of criteria.

As far as point 2) is concerned, observe that the weights w_i used in AHP represent tradeoffs between evaluations on different criteria. For this reason it is doubtful if they could be inferred from answers to questions concerning comparison of importance. Therefore, AHP has a problem with meaningfulness of its output with respect to its input, and this is not the case of GRIP.

4.4.6 Comparison of GRIP with MACBETH

MACBETH (Measuring Attractiveness by a Categorical Based Evaluation TecHnique) (Bana e Costa and Vansnick, 1994; Bana e Costa *et al.*, 2005) is a method for multiple criteria decision analysis that appeared in the early nineties. This approach requires from the DM qualitative judgements about differences of value to quantify the relative attractiveness of actions (solutions) or criteria.

When using MACBETH, the DM is asked to provide preference information composed of a strict order of all actions from A, and a qualitative judgement of the difference of attractiveness between all two non-indifferent actions. Seven semantic categories of the difference of attractiveness are considered: null, very weak, weak, moderate, strong, very strong, and extreme. The difference of attractiveness reflects the intensity of preferences.

The main idea of MACBETH is to build an interval scale from the preference information provided by the DM. It is, however, necessary that the above categories correspond to disjoint intervals (represented in terms of the real numbers). The bounds for such intervals are not arbitrarily fixed *a priori*, but they are calculated so as to be compatible with the numerical values of all particular actions from A, and to ensure compatibility between these values (see Bana e Costa et al. 2005). Linear programming models are used for these calculations. In case of inconsistent judgments, MACBETH provides the DM with information in order to eliminate such inconsistency.

When comparing MACBETH with GRIP, the following aspects should be considered:

- both deal with qualitative judgements;
- both need a set of comparisons of actions or pairs of actions to work out a numerical representation of preferences, however, MACBETH depends on the specification of two characteristic levels on the original scale, "neutral" and "good", to obtain the numerical representation of preferences, while GRIP does not need this information;
- GRIP adopts the "disaggregation-aggregation" approach and, therefore, it considers mainly holistic judgements relative to comparisons involving jointly all the criteria, which is not the case of MACBETH;
- GRIP is more general than MACBETH since it can take into account the same kind of qualitative judgments as MACBETH (the difference of attractiveness between pairs of actions) and, moreover, the intensity of preferences of the type "x is preferred to y at least as much as z is preferred to w".

As for the last item, it should be noticed that the intensity of preference considered in MACBETH and the intensity coming from comparisons of the type "x is preferred to y at least as strongly as w is preferred to z" (i.e., the quaternary relation $\succsim^*$) are substantially the same. In fact, the intensities of preference are equivalence classes of the preorder generated by $\succsim^*$. This means that all the pairs (x, y) and (w, z), such that x is preferred to y with the same intensity as w is preferred to z, belong to the same semantic category of difference of attractiveness considered in MACBETH. To be more precise, the structure of intensity of preference considered in MACBETH is a particular case of the structure of intensity of preference represented by $\succsim^*$ in GRIP. Still more precisely, GRIP has the same structure of intensity as MACBETH when $\succsim^*$ is a complete preorder. When this does not occur, MACBETH cannot be used while GRIP can naturally deal with this situation.

Comparison of GRIP and MACBETH could be summarized in the following points:

1. GRIP is using preference information relative to: 1) comprehensive preference on a subset of reference actions with respect to all criteria, 2) marginal intensity of preference on some single criteria, and 3) comprehensive intensity of preference with respect to all criteria, while MACBETH requires preference information on all pairs of actions with respect to each one of the considered criteria.
2. Information about marginal intensity of preference is of the same nature in GRIP and MACBETH (equivalence classes of relation $\succsim_i^*$ correspond to qualitative judgements of MACBETH), but in GRIP it may not be complete.
3. GRIP is a "disaggregation-aggregation" approach while MACBETH makes use of the "aggregation" approach and, therefore, it needs weights to aggregate evaluations on the criteria.
4. GRIP works with all compatible value functions, while MACBETH builds a single interval scale for each criterion, even if many such scales would be compatible with preference information.
5. Distinguishing necessary and possible consequences of using all value functions compatible with preference information, GRIP includes a kind of robustness analysis instead of using a single "best-fit" value function.
6. The necessary and possible preference relations considered in GRIP have several properties of general interest for MCDA.

4.5 An Illustrative Example

In this section, we illustrate how our approach can support the DM to specify his/her preferences on a set of Pareto optimal solutions. In this didactic example, we shall imagine an interaction with a fictitious DM so as to exemplify and illustrate the type of interaction proposed in our method.

We consider a MOO problem that involves five objectives that are to be maximized. Let us consider a subset A of the Pareto Frontier of a MOO problem consisting of 20 solutions (see Table 4.1). Note that this set A can be computed using a MOO or EMO algorithm (see Chapters 2 and 3). Let us suppose that the reference sample A^R of solutions from set A is the following:

Table 4.1. The whole set of Pareto optimal solutions for the example MOO problem

s_1	=	$(14.5, 147, 4, 1014, 5.25)$
s_2	=	$(13.25, 199.125, 4, 1014, 4)$
s_3	=	$(15.75, 164.375, 16.5, 838.25, 5.25)$
s_4	=	$(12, 181.75, 16.5, 838.25, 4)$
s_5	=	$(12, 164.375, 54, 838.25, 4)$
s_6	=	$(13.25, 199.125, 29, 662.5, 5.25)$
s_7	=	$(13.25, 147, 41.5, 662.5, 5.25)$
s_8	=	$(17, 216.5, 16.5, 486.75, 1.5)$
s_9	=	$(17, 147, 41.5, 486.75, 5.25)$
s_{10}	=	$(15.75, 216.5, 41.5, 662.5, 1.5)$
s_{11}	=	$(15.75, 164.375, 41.5, 311, 6.5)$
s_{12}	=	$(13.25, 181.75, 41.5, 311, 4)$
s_{13}	=	$(12, 199.125, 41.5, 311, 2.75)$
s_{14}	=	$(17, 147, 16.5, 662.5, 5.25)$
s_{15}	=	$(15.75, 199.125, 16.5, 311, 6.5)$
s_{16}	=	$(13.25, 164.375, 54, 311, 4)$
s_{17}	=	$(17, 181.75, 16.5, 486.75, 5.25)$
s_{18}	=	$(14.5, 164.375, 41.5, 838.25, 4)$
s_{19}	=	$(15.75, 181.75, 41.5, 135.25, 5.25)$
s_{20}	=	$(15.75, 181.75, 41.5, 311, 2.75)$

$A^R = \{s_1, s_2, s_4, s_5, s_8, s_{10}\}$. For the sake of simplicity, we shall consider the set A^R constant across iterations (although the interaction scheme permits A^R to evolve during the process). For the same reason, we will suppose that the DM expresses preference information only in terms of pairwise comparisons of solutions from A^R (intensity of preference will not be expressed in the preference information).

The DM does not see any satisfactory solution in the reference sample A^R (s_1, s_2, s_4 and s_5 have too weak evaluations on the first criterion, while s_8 and s_{10} have the worst evaluation in A on the last criterion), and wishes to find a satisfactory solution in A. Obviously, solutions in A are not comparable unless preference information is expressed by the DM. In this perspective, he/she provides a first comparison: $s_1 \succ s_2$.

Considering the provided preference information, we can compute the necessary and possible rankings on set A (computation of this example were performed using the GNU-UTA software package (Chakhar and Mousseau, 2007); note that the UTA^{GMS} and GRIP methods are also implemented in the Decision Deck software platform (Consortium, 2008)). The DM decided to consider the necessary ranking only, as it has more readable graphical representation than the possible ranking at the stage of relatively poor preference information. The partial preorder of the necessary ranking is depicted in Fig. 4.1 and shows the comparisons that hold for all additive value functions compatible with the information provided by the DM (i.e., $s_1 \succ s_2$). It should be observed that the computed partial preorder contains the preference information provided by the DM (dashed arrow), but also additional comparisons that result from the initial information (continuous arrows); for instance, $s_3 \succ^N s_4$ holds because $U(s_3) \geq U(s_4)$ for each compatible value function (this gives $s_3 \succsim^N s_4$) and $U(s_3) > U(s_4)$ for at least one value function (this gives $not(s_4 \succsim^N s_3)$).

Analyzing this first result, the DM observes that the necessary ranking is still very poor which makes it difficult to discriminate among the solutions in A. He/she reacts by stating that s_4 is preferred to s_5. Considering this new piece of preference information, the necessary ranking is computed again

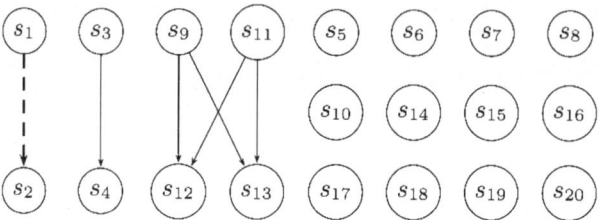

Fig. 4.1. Necessary partial ranking at the first iteration

and shown in Fig. 4.2. At this second iteration, it should be observed that the resulting necessary ranking has been enriched as compared to the first iteration (bold arrows), narrowing the set of "best choices", i.e., solutions that are not preferred by any other solution in the necessary ranking: $\{s_1, s_3, s_6, s_8, s_{10}, s_{14}, s_{15}, s_{17}, s_{18}, s_{19}, s_{20}\}$.

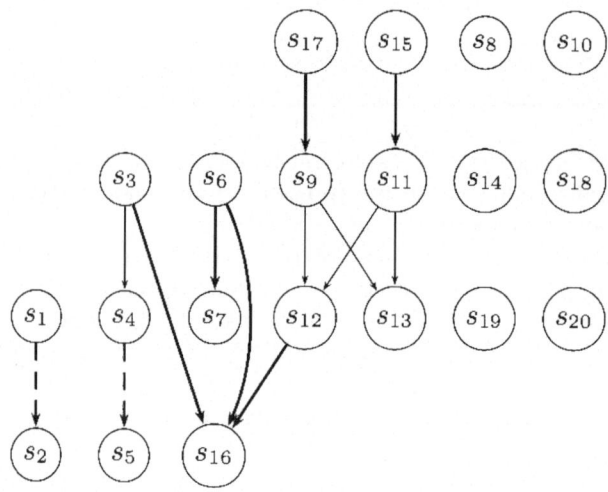

Fig. 4.2. Necessary partial ranking at the second iteration

The DM believes that this necessary ranking is still insufficiently decisive and adds a new comparison: s_8 is preferred to s_{10}. Once again, the necessary ranking is computed and shown in Fig. 4.3.

At this stage, the set of possible "best choices" has been narrowed down to a limited number of solutions, among which s_{14} and s_{17} are judged satisfactory by the DM. In fact, these two solutions have a very good performance on the first criterion without any "dramatic" evaluation on the other criteria.

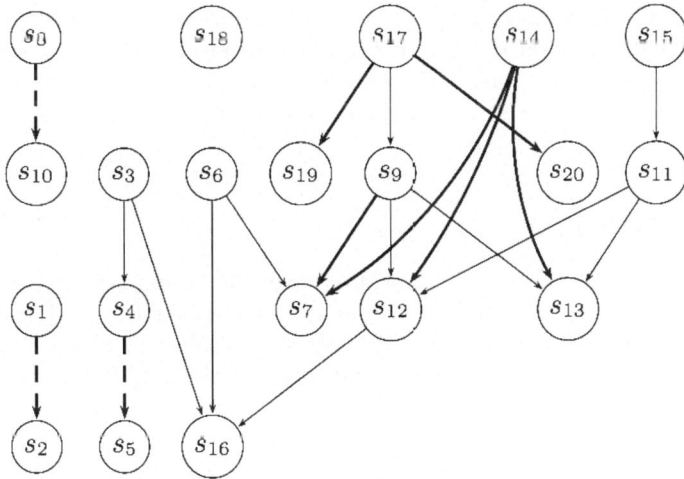

Fig. 4.3. Necessary partial ranking at the third iteration

The current example stops at this step, but the DM could then decide to provide further preference information to enrich the necessary ranking. He/she could also compute new Pareto optimal solutions "close" to s_{14} and s_{17} to zoom investigations in this area. In this example we have shown that the proposed interactive process supports the DM in choosing most satisfactory solutions, without imposing any strong cognitive effort, as the only information required is a holistic preference information.

4.6 Conclusions and Further Research Directions

In this chapter, we introduced a new interactive procedure for multiobjective optimization. It consists in an interactive exploration of a Pareto optimal set, or its approximation, generated prior to the exploration using a MOO or EMO algorithm. The procedure represents a constructive learning approach, because on one hand, the preference information provided by the DM contributes to the construction of a preference model and, on the other hand, the use of the preference model shapes the DM's preferences or, at least, makes DM's convictions evolve.

Contrary to many existing MCDA methods, the proposed procedure does not require any excessive cognitive effort from the DM because the preference information is of a holistic nature and, moreover, it can be partial. Due to distinguishing necessary and possible consequences of using all value functions compatible with the preference information, the procedure is also robust, comparing to methods using a single "best-fit" value function. This is a feature of uttermost importance in MOO.

An almost immediate extension of the procedure could consist in admitting preference information in form of a sorting of selected Pareto optimal solutions into some pre-defined and preference ordered classes. Providing such an information could be easier for some DMs than making the pairwise comparisons.

Acknowledgements

The first and the third authors acknowledge the support from Luso-French PESSOA bilateral cooperation. The fourth author wishes to acknowledge financial support from the Polish Ministry of Science and Higher Education. All authors acknowledge, moreover, the support of the COST Action IC0602 "Algorithmic Decision Theory".

References

Bana e Costa, C.A., Vansnick, J.C.: MACBETH: An interactive path towards the construction of cardinal value functions. International Transactions in Operational Research 1(4), 387–500 (1994)

Bana e Costa, C.A., De Corte, J.M., Vansnick, J.C.: On the mathematical foundation of MACBETH. In: Figueira, J., Greco, S., Ehrgott, M. (eds.) Multiple Criteria Decision Analysis: State of the Art Surveys, pp. 409–443. Springer Science + Business Media Inc., New York (2005)

Chakhar, S., Mousseau, V.: GNU-UTA: a GNU implementation of UTA methods (2007), http://www.lamsade.dauphine.fr/~mousseau/GNU-UTA

Consortium, D.D.: Decision deck: an open-source software platform for mcda methods (2006-2008), www.decision-deck.org

Figueira, J., Greco, S., Słowiński, R.: Building a set of additive value functions representing a reference preorder and intensities of preference: GRIP method. European Journal of Operational Research, to appear

Greco, S., Matarazzo, B., Słowiński, R.: The use of rough sets and fuzzy sets in MCDM. In: Gal, T., Hanne, T., Stewart, T. (eds.) Multicriteria Decision Making: Advances in MCDM Models, Algorithms, Theory and Applications, pp. 1–14. Kluwer Academic Publishers, Dordrecht (1999)

Greco, S., Matarazzo, B., Slowinski, R.: Rough sets theory for multicriteria decision analysis. European Journal of Operational Research 129, 1–47 (2001)

Greco, S., Mousseau, V., Słowiński, R.: Assessing a partial preorder of alternatives using ordinal regression and additive utility functions: A new UTA method. In: 58th Meeting of the EURO Working Group on MCDA, Moscow (2003)

Greco, S., Matarazzo, B., Słowiński, R.: Decision rule approach. In: Figueira, J., Greco, S., Ehrgott, M. (eds.) Multiple Criteria Decision Analysis: State of the Art Surveys, pp. 507–562. Springer Science + Business Media Inc., New York (2005)

Greco, S., Mousseau, V., Słowiński, R.: Ordinal regression revisited: Multiple criteria ranking with a set of additive value functions. European Journal of Operational Research 191(2), 416–436 (2008)

Jacquet-Lagrèze, E., Siskos, Y.: Assessing a set of additive utility functions for multicriteria decision making: The UTA method. European Journal of Operational Research 10(2), 151–164 (1982)

Jacquet-Lagrèze, E., Meziani, R., Słowiński, R.: MOLP with an interactive assessment of a piecewise linear utility function. European Journal of Operational Research 31, 350–357 (1987)

Kiss, L., Martel, J., Nadeau, R.: ELECCALC - an interactive software for modelling the decision maker's preferences. Decision Support Systems 12(4-5), 757–777 (1994)

March, J.: Bounded rationality, ambiguity and the engineering of choice. Bell Journal of Economics 9, 587–608 (1978)

Marichal, J., Roubens, M.: Determination of weights of interacting criteria from a reference set. European Journal of Operational Research 124(3), 641–650 (2000)

Mousseau, V., Slowinski, R.: Inferring an ELECTRE TRI model from assignment examples. Journal of Global Optimization 12(2), 157–174 (1998)

Pekelman, D., Sen, S.: Mathematical programming models for the determination of attribute weights. Management Science 20(8), 1217–1229 (1974)

Roy, B., Bouyssou, D.: Aide Multicritère à la Décision: Méthodes et Cas. Economica, Paris (1993)

Saaty, T.: The Analytic Hierarchy Process. McGraw Hill, New York (1980)

Saaty, T.: The analytic hierarchy and analytic network processes for the measurement of intangible criteria and for decision-making. In: Figueira, J., Greco, S., Ehrgott, M. (eds.) Multiple Criteria Decision Analysis: The State of the Art Surveys, pp. 345–407. Springer Science+Business Media, Inc., New York (2005)

Siskos, Y., Grigoroudis, V., Matsatsinis, N.: UTA methods. In: Figueira, F., Greco, S., Ehrgott, M. (eds.) Multiple Criteria Decision Analysis: State of the Art Surveys, pp. 297–343. Springer Science + Business Media Inc., New York (2005)

Słowiński, R., Greco, S., Matarazzo, B.: Rough set based decision support. In: Burke, E., Kendall, G. (eds.) Introductory Tutorials on Optimization, Search and Decision Support Methodologies, pp. 475–527. Springer Science + Business Media Inc., New York (2005)

Srinivasan, V., Shocker, A.: Estimating the weights for multiple attributes in a composite criterion using pairwise judgments. Psychometrika 38(4), 473–493 (1973)

Dominance-Based Rough Set Approach to Interactive Multiobjective Optimization

Salvatore Greco[1], Benedetto Matarazzo[1], and Roman Słowiński[2,3]

[1] Faculty of Economics, University of Catania, Corso Italia, 55,
    95129 Catania, Italy, salgreco@unict.it, matarazz@unict.it
[2] Institute of Computing Science, Poznań University of Technology,
    60-965 Poznan, Poland, roman.slowinski@cs.put.poznan.pl
[3] Systems Research Institute, Polish Academy of Sciences,
    01-447 Warsaw, Poland

Abstract. In this chapter, we present a new method for interactive multiobjective optimization, which is based on application of a logical preference model built using the Dominance-based Rough Set Approach (DRSA). The method is composed of two main stages that alternate in an interactive procedure. In the first stage, a sample of solutions from the Pareto optimal set (or from its approximation) is generated. In the second stage, the Decision Maker (DM) indicates relatively good solutions in the generated sample. From this information, a preference model expressed in terms of "if ..., then ..." decision rules is induced using DRSA. These rules define some new constraints which can be added to original constraints of the problem, cutting-off non-interesting solutions from the currently considered Pareto optimal set. A new sample of solutions is generated in the next iteration from the reduced Pareto optimal set. The interaction continues until the DM finds a satisfactory solution in the generated sample. This procedure permits a progressive exploration of the Pareto optimal set in zones which are interesting from the point of view of DM's preferences. The "driving model" of this exploration is a set of user-friendly decision rules, such as "if the value of objective i_1 is not smaller than α_{i_1} and the value of objective i_2 is not smaller than α_{i_2}, then the solution is good". The sampling of the reduced Pareto optimal set becomes finer with the advancement of the procedure and, moreover, a return to previously abandoned zones is possible. Another feature of the method is the possibility of learning about relationships between values of objective functions in the currently considered zone of the Pareto optimal set. These relationships are expressed by DRSA association rules, such as "if objective j_1 is not greater than α_{j_1} and objective j_2 is not greater than α_{j_2}, then objective j_3 is not smaller than β_{j_3} and objective j_4 is not smaller than β_{j_4}".

Reviewed by: José Rui Figueira, Technical University of Lisbon, Portugal
Hisao Ishibuchi, Osaka Prefecture University, Japan
Kaisa Miettinen, University of Jyväskylä, Finland

5.1 Introduction

We propose a new method to interactive multiobjective optimization permitting to use the preference model expressed in terms of easily understandable "*if ..., then ...*" decision rules, induced from information about preferences of the Decision Maker (DM) expressed in terms of a simple indication of relatively good solutions in a given sample. The method we propose complements well any multiobjective optimization method (see Chapter 1), which finds the Pareto optimal set or its approximation, such as Evolutionary Multiobjective Optimization methods (see Chapter 3).

Interactive multiobjective optimization (for a systematic introduction see Chapter 2) consists of organizing the search for the most preferred solution by alternating stages of calculation and dialogue with the DM. In many interactive methods, the first stage of calculation provides a first sample of candidate solutions from the Pareto optimal set or from its approximation. This sample is presented to the DM. In the dialogue stage, the DM reacts to this proposal by supplying additional information revealing his/her preferences. This information is taken into account in the search for a new sample of candidate solutions in the next calculation stage, so as to provide solutions which better fit DM's preferences. The search stops when, among the candidate solutions, the DM finds one which yields a satisfactory compromise between objective function values, or when the DM comes to the conclusion that there is no such a solution in the current problem setting. The convergence of interactive procedures is of psychological rather than mathematical nature. Information supplied by the DM in the dialogue stage is a critical information about presented candidate solutions – as such, it is *preference information*, which is an indispensable component of each method supporting Multiple Criteria Decision Making (MCDM) (for an updated collection of state of the art surveys see (Figueira *et al.*, 2005b)). Let us remark that from a semantic point of view, criterion and objective function mean the same, thus we will use them alternatively. The concept of criterion is more handy in the context of evaluation of a finite set of solutions, and the concept of objective function better fits the context of optimization.

Preference information permits building a *preference model* of the DM. The preference model induces a preference structure in the set of candidate solutions (objects, alternatives, actions); a proper exploitation of this structure leads to a recommendation consistent with the preferences of the DM – the recommendation may concern one of the following three main problems of multiple criteria decision:

- *sorting* of candidate solutions into pre-defined and preference ordered decision classes (also called *ordinal classification*),
- *choice* of the most preferred solution(s),
- *ranking* of the solutions from the most to the least preferred.

The interactive method proposed hereafter combines two multiple criteria decision problem settings. In the dialogue stage, it requires the DM to sort a sample of solutions into two classes: "good" and "others". Finally, it gives a recommendation for the choice. A similar combination of sorting and choice has been used by Jaszkiewicz and Ferhat (1999). Remark that most interactive optimization methods require the DM to select one (feasible or infeasible) solution as a reference point (see Chapter 2). Moreover, there exist interactive multiobjective optimization methods requiring preference information in terms of ranking of a set of reference solutions (see, e.g., (Jacquet-Lagrèze *et al.*, 1987) and Chapter 4).

Experience indicates that decision support methods requiring from a DM a lot of cognitive effort in the dialogue stage fail to be accepted. Preference information may be either direct or indirect, depending on whether it specifies directly values of some parameters used in the preference model (e.g. trade-off weights, aspiration levels, discrimination thresholds, etc.), or some examples of holistic judgments, called *exemplary decisions*, from which compatible values of the preference model parameters are induced.

Direct preference information is used in the traditional paradigm, according to which the preference model is first constructed and then applied on the set of candidate solutions.

Indirect preference information is used in the *regression paradigm*, according to which the holistic preferences on a subset of candidate solutions are known first, and then a consistent preference model is inferred from this information to be applied on the whole set of candidate solutions.

Presently, MCDM methods based on indirect preference information and the regression paradigm are of increasing interest for they require relatively weaker cognitive effort from the DM. Preference information given in terms of exemplary decisions is very natural and, for this reason, reliable. Indeed, the regression paradigm emphasizes the discovery of intentions as an interpretation of actions rather than as a prior position, which was called by March the *posterior rationality* (March, 1978). It is also consistent with the inductive learning used in artificial intelligence approaches (Michalski *et al.*, 1998). Typical applications of this paradigm in MCDM are presented in (Greco *et al.*, 1999b, 2008; Jacquet-Lagrèze *et al.*, 1987; Jacquet-Lagrèze and Siskos, 1982; Mousseau and Słowiński, 1998).

The form of exemplary decisions which constitute preference information depends on the multiple criteria problem setting. In multiple criteria sorting, an exemplary decision is an assignment of a selected solution to one of decision classes, because sorting is based on absolute evaluation of solutions. In multiple criteria choice and ranking, however, an exemplary decision is a pairwise comparison of solutions, because choice and ranking is based on relative evaluations of solutions. While it is relatively easy to acquire a set of exemplary decisions, they are rarely logically consistent. By *inconsistent exemplary decisions* we mean decisions which do not respect the dominance principle (called also Pareto principle). In multiple criteria sorting, decision examples concerning solutions x and y are inconsistent if one of the following two situations occurs:

α) x and y have the same evaluations on all criteria (x and y are indiscernible), however, they have been assigned to different decision classes,

β) x has not worse evaluations on all criteria than y (x dominates y), however, x has been assigned to a worse decision class than y.

In multiple criteria choice and ranking, decision examples concerning pairs of solutions (x, y) and (w, z) are inconsistent if one of the following two situations occur:

α') differences of evaluations of x and y are the same as differences of evaluations of w and z on all criteria ((x, y) and (w, z) are indiscernible), however, x has been compared to y differently than w to z,

β') differences of evaluations of x and y are not smaller than differences of evaluations of w and z on all criteria ((x, y) dominates (w, z)), however, x has been compared to y as being less preferred than w is preferred to z, i.e. even if with respect to all considered criteria the strength of preference of x over y is not smaller than the strength of preference of w over z, the overall strength of preference of x over y is smaller than the overall strength of preference of w over z.

The dominance principle is the only objective principle that is widely agreed upon in the multiple criteria decision analysis. Inconsistency of exemplary decisions may come from many sources. Examples include:

- incomplete set of criteria,
- limited clear discrimination between criteria,
- unstable preferences of decision makers.

Inconsistencies cannot be considered as error or noise to be simply eliminated from the preference information or amalgamated with the consistent part of this information by some averaging operators. Indeed, they can convey important information that should be taken into account in the construction of the DM's preference model.

The *rough set* concept proposed by Pawlak (1982, 1991) is intended to deal with inconsistency in information and this is a major argument to support its application to multiple criteria decision analysis.

Since its conception, rough set theory has proved to be an excellent mathematical tool for the analysis of inconsistent information. The rough set philosophy is based on the assumption that with every object (e.g. a solution of multiobjective optimization problem) of a universe U there is associated a certain amount of information (data, knowledge). This information can be expressed by means of a number of attributes. The attribute values describe the objects. Objects which have the same description are said to be indiscernible (similar) with respect to the available information. The *indiscernibility relation* thus generated constitutes the mathematical basis of rough set theory. It induces a partition of the universe into blocks of indiscernible objects, called elementary sets or granules, which can be used to build concepts (e.g. classes of acceptable, doubtful or non-acceptable solutions).

Any subset X of the universe may be expressed in terms of these granules either precisely (as a union of granules) or approximately. In the latter case, the subset X may be characterized by two ordinary sets, called the *lower* and *upper approximations*. A rough set is defined by means of these two approximations. The lower approximation of X is composed of all the granules included in X (whose elements, therefore, certainly belong to X), while the upper approximation of X consists of all the granules which have a non-empty intersection with X (whose elements, therefore, may belong to X). The difference between the upper and lower approximation constitutes the *boundary region* of the rough set, whose elements cannot be characterized with certainty as belonging or not to X (by using the available information). The information about objects from the boundary region is, therefore, inconsistent. The cardinality of the boundary region states, moreover, the extent to which it is possible to express X in terms of certainty, on the basis of the available information. In fact, these objects have the same description, but are assigned to different classes, such as patients having the same symptoms (the same description), but different pathologies (different classes). For this reason, this cardinality may be used as a measure of inconsistency of the information about X.

Some important characteristics of the rough set approach make it a particularly interesting tool in a variety of problems and concrete applications. For example, it is possible to deal with both quantitative and qualitative attributes (e.g., in case of diagnostics, the blood pressure and the temperature are quantitative attributes, while the color of eyes or state of consciousness are qualitative attributes) and inconsistencies need not to be removed before the analysis. In result of the rough set approach, it is possible to acquire a posteriori information regarding the relevance of particular attributes and their subsets (Greco *et al.*, 1999b, 2001b). Moreover, given a partition of U into disjoint decision classes, the lower and upper approximations of this partition give a structure to the available information such that it is possible to induce from this information certain and possible *decision rules*, which are logical "*if..., then...*" statements.

Several attempts have been made to employ rough set theory for decision support (Pawlak and Słowiński, 1994; Słowiński, 1993). The classical rough set approach is not able, however, to deal with preference ordered value sets of attributes, as well as with the preference ordered decision classes. In decision analysis, an attribute with a preference ordered value set (scale) is called criterion, and the multiattribute classification problem corresponds to a multiple criteria sorting problem (also called ordinal classification).

At this point, let us return to the issue of inconsistent preference information provided by the DM in the process of solving a multiple criteria sorting problem. As can be seen from the above brief description of the classical rough set approach, it is able to deal with inconsistency of type α) and α'), however, it is not able to recognize inconsistency of type β) and β').

In the late 90's, adapting the classical rough set approach to analysis of preference ordered information became a particularly challenging problem within the field of multiple criteria decision analysis. Why might it be so important? The answer is that the result of the rough set analysis in form of certain decision rules, having a syntax "if for x some conditions hold, then x certainly belongs to class Cl", and being induced from the lower approximation, as well as possible decision rules, having a syntax "if for x some conditions hold, then x possibly belongs to class Cl", and being induced from the upper approximation, is very attractive for its ability of representing DM's preferences hidden in exemplary decisions. Such a preference model is very convenient for decision support, because it is intelligible and it speaks the same language as the DM.

An extension of the classical rough set approach, which enables the analysis of preference ordered information was proposed by Greco, Matarazzo and Słowiński (Greco et al., 1999b,a; Słowiński et al., 2002b). This extension, called the *Dominance-based Rough Set Approach* (DRSA), is mainly based on the replacement of the indiscernibility relation by a dominance relation in the rough approximation. This change permits to recognize inconsistency of both types, α) and α') on one hand, and β) and β') on the other hand, and to build lower and upper approximations of decision classes, corresponding to consistent only or all available information, respectively. An important consequence of this fact is the possibility of inferring from these approximations the preference model in terms of decision rules. Depending upon whether they are induced from lower approximations or from the upper approximations of decision classes, one gets certain or possible decision rules, which represent certain or possible knowledge about the DM's preferences. Such a preference model is more general than the classical functional models considered within multiattribute utility theory (see Keeney and Raiffa, 1976; Dyer, 2005) or the relational models considered, for example, in outranking methods (for a general review of outranking methods see (Roy and Bouyssou, 1993; Figueira et al., 2005a; Brans and Mareschal, 2005; Martel and Matarazzo, 2005), while for their comparison with the rough set approach see (Greco et al., 2002b, 2004a; Słowiński et al., 2002a).

DRSA has been applied to all three types of multiple criteria decision problems, to decision with multiple decision makers, to decision under uncertainty, and to hierarchical decision making. For comprehensive surveys of DRSA, see (Greco et al., 1999b, 2001b, 2004c, 2005a,b; Słowiński et al., 2005).

The aim of this chapter is to extend further the range of applications of DRSA to interactive multiobjective optimization. It is organized as follows. In the next Section, we recall the main concepts of DRSA, which are also illustrated with an example in Section 5.3. In Section 5.4, we introduce association rules expressing relationships between objective function values in the currently considered zone of the Pareto optimal set. In Section 5.5, we present the new interactive multiobjective optimization method based on DRSA. In Section 5.6, we illustrate this method with an example, and, in Section 5.7, we discuss its characteristic features. Final Section contains conclusions.

5.2 Dominance-Based Rough Set Approach (DRSA)

DRSA is a methodology of multiple criteria decision analysis aiming at obtaining a representation of the DM's preferences in terms of easily understandable "*if ..., then ...*" decision rules, on the basis of some exemplary decisions (past decisions or simulated decisions) given by the DM. In this Section, we present the DRSA to sorting problems, because in the dialogue stage of our interactive method this multiple criteria decision problem is considered. In this case, exemplary decisions are *sorting examples*, i.e. objects (solutions, alternatives, actions) described by a set of criteria and assigned to preference ordered classes. The criteria and the class assignment considered within DRSA correspond to the *condition attributes* and the *decision attribute*, respectively, in the classical Rough Set Approach (Pawlak, 1991). For example, in multiple criteria sorting of cars, an example of decision is an assignment of a particular car evaluated on such criteria as maximum speed, acceleration, price and fuel consumption to one of three classes of overall quality: "bad", "medium", "good".

Let us consider a set of criteria $F = \{f_1, \ldots, f_n\}$, the set of their indices $I = \{1, \ldots, n\}$, and a finite universe of objects (solutions, alternatives, actions) U, such that $f_i : U \to \Re$ for each $i = 1, \ldots, n$. To explain DRSA, it will be convenient to assume that $f_i(\cdot)$, $i = 1, \ldots, n$, are gain-type functions. Thus, without loss of generality, for all objects $x, y \in U$, $f_i(x) \geq f_i(y)$ means that "x is at least as good as y with respect to criterion i", which is denoted as $x \succeq_i y$. We suppose that $\succeq_i$ is a complete preorder, i.e. a strongly complete and transitive binary relation, defined on U on the basis of evaluations $f_i(\cdot)$. Note that in the context of multiobjective optimization, $f_i(\cdot)$ corresponds to objective functions. Furthermore, we assume that there is a decision attribute d which makes a partition of U into a finite number of decision classes called sorting, $\pmb{Cl} = \{Cl_1, \ldots, Cl_m\}$, such that each $x \in U$ belongs to one and only one class $Cl_t, t = 1, \ldots, m$. We suppose that the classes are preference ordered, i.e. for all $r, s = 1, \ldots, m$, such that $r > s$, the objects from Cl_r are preferred to the objects from Cl_s. More formally, if $\succeq$ is a *comprehensive weak preference relation* on U, i.e. if for all $x, y \in U$, $x \succeq y$ reads "x is at least as good as y", then we suppose

$$[x \in Cl_r, \ y \in Cl_s, \ r > s] \Rightarrow x \succ y,$$

where $x \succ y$ means $x \succeq y$ and *not* $y \succeq x$. The above assumptions are typical for consideration of a multicriteria sorting problem.

In DRSA, the explanation of the assignment of objects to preference ordered decision classes is made on the base of their evaluation with respect to a subset of criteria $P \subseteq I$. This explanation is called *approximation* of decision classes with respect to P. Indeed, in order to take into account the order of decision classes, in DRSA the classes are not considered one by one but, instead, unions of classes are approximated: *upward union* from class Cl_t to class Cl_m denoted by $Cl_t^{\geq}$, and *downward union* from class Cl_t to class Cl_1, denoted by $Cl_t^{\leq}$, i.e.:

$$Cl_t^{\geq} = \bigcup_{s \geq t} Cl_s, \quad Cl_t^{\leq} = \bigcup_{s \leq t} Cl_s, \quad t = 1, ..., m.$$

The statement $x \in Cl_t^{\geq}$ reads "x belongs to at least class Cl_t", while $x \in Cl_t^{\leq}$ reads "x belongs to at most class Cl_t". Let us remark that $Cl_1^{\geq} = Cl_m^{\leq} = U$, $Cl_m^{\geq} = Cl_m$ and $Cl_1^{\leq} = Cl_1$. Furthermore, for $t=2,...,m$, we have:

$$Cl_{t-1}^{\leq} = U - Cl_t^{\geq} \quad \text{and} \quad Cl_t^{\geq} = U - Cl_{t-1}^{\leq}.$$

In the above example concerning multiple criteria sorting of cars, the upward unions are: $Cl_{medium}^{\geq}$, that is the set of all the cars classified at least "medium" (i.e. the set of cars classified "medium" or "good"), and $Cl_{good}^{\geq}$, that is the set of all the cars classified at least "good" (i.e. the set of cars classified "good"), while the downward unions are: $Cl_{medium}^{\leq}$, that is the set of all the cars classified at most "medium" (i.e. the set of cars classified "medium" or "bad"), and $Cl_{bad}^{\leq}$, that is the set of all the cars classified at most "bad" (i.e. the set of cars classified "bad"). Notice that, formally, also $Cl_{bad}^{\geq}$ is an upward union as well as $Cl_{good}^{\leq}$ is a downward union, however, as "bad" and "good" are extreme classes, the two unions boil down to the whole universe U.

The key idea of the rough set approach is explanation (approximation) of knowledge generated by the decision attributes, by *granules of knowledge* generated by condition attributes.

In DRSA, where condition attributes are criteria and decision classes are preference ordered, the knowledge to be explained is the assignments of objects to *upward* and *downward unions of classes* and the granules of knowledge are sets of objects in *dominance cones* in the criteria values space.

We say that x *dominates* y with respect to $P \subseteq I$ (shortly, x *P-dominates* y), denoted by xD_Py, if for every criterion $i \in P$, $f_i(x) \geq f_i(y)$. The relation of P-dominance is reflexive and transitive, that is it is a partial preorder.

Given a set of criteria $P \subseteq I$ and $x \in U$, the granules of knowledge used for approximation in DRSA are:

- a set of objects dominating x, called *P-dominating set*,
  $D_P^+(x) = \{y \in U : yD_Px\}$,
- a set of objects dominated by x, called *P-dominated set*,
  $D_P^-(x) = \{y \in U : xD_Py\}$.

Let us observe that we can write the P-dominating sets and the P-dominated sets as follows. For $x \in U$ and $P \subseteq I$ we define the *P-positive dominance cone* of x, $P^{\geq}(x)$, and the *P-negative dominance cone* of x, $P^{\leq}(x)$, as follows

$$P^{\geq}(x) = \{\mathbf{z} = (z_1, ..., z_{|P|}) \in \Re^{|P|} : z_j \geq f_{i_j}(x), \text{for all } i_j \in P\},$$

$$P^{\leq}(x) = \{\mathbf{z} = (z_1, ..., z_{|P|}) \in \Re^{|P|} : z_j \leq f_{i_j}(x), \text{for all } i_j \in P\}.$$

Thus, the *P-dominating set* and the *P-dominated set* of x can be formulated as

$$D_P^+(x) = \{y \in U : (f_i(y), i \in P) \in P^{\geq}(x)\},$$

$$D_P^-(x) = \{y \in U : (f_i(y), i \in P) \in P^{\leq}(x)\}.$$

Let us remark that we can also write:

$$D_P^+(x) = \{y \in U : P^{\geq}(y) \subseteq P^{\geq}(x)\},$$

$$D_P^-(x) = \{y \in U : P^{\leq}(y) \subseteq P^{\leq}(x)\}.$$

For the sake of simplicity, with an abuse of definition, in the following we shall speak of dominance cones also when we refer to the dominating and dominated sets.

Let us recall that the dominance principle requires that an object x dominating object y with respect to considered criteria (i.e. x having evaluations at least as good as y on all considered criteria) should also dominate y on the decision (i.e. x should be assigned to at least as good decision class as y).

The *P-lower approximation* of $Cl_t^{\geq}$, denoted by $\underline{P}(Cl_t^{\geq})$, and the *P-upper approximation* of $Cl_t^{\geq}$, denoted by $\overline{P}(Cl_t^{\geq})$, are defined as follows ($t=1,...,m$):

$$\underline{P}(Cl_t^{\geq}) = \{x \in U : D_P^+(x) \subseteq Cl_t^{\geq}\},$$

$$\overline{P}(Cl_t^{\geq}) = \{x \in U : D_P^-(x) \cap Cl_t^{\geq} \neq \emptyset\}.$$

Analogously, one can define the *P-lower approximation* and the *P-upper approximation* of $Cl_t^{\leq}$ as follows ($t=1,...,m$):

$$\underline{P}(Cl_t^{\leq}) = \{x \in U : D_P^-(x) \subseteq Cl_t^{\leq}\},$$

$$\overline{P}(Cl_t^{\leq}) = \{x \in U : D_P^+(x) \cap Cl_t^{\leq} \neq \emptyset\}.$$

The *P*-lower and *P*-upper approximations so defined satisfy the following *inclusion properties*, for each $t \in \{1, \ldots, m\}$ and for all $P \subseteq I$:

$$\underline{P}(Cl_t^{\geq}) \subseteq Cl_t^{\geq} \subseteq \overline{P}(Cl_t^{\geq}), \quad \underline{P}(Cl_t^{\leq}) \subseteq Cl_t^{\leq} \subseteq \overline{P}(Cl_t^{\leq}).$$

The *P*-lower and *P*-upper approximations of $Cl_t^{\geq}$ and $Cl_t^{\leq}$ have an important *complementarity property*, according to which,

$$\underline{P}(Cl_t^{\geq}) = U - \overline{P}(Cl_{t-1}^{\leq}) \text{ and } \overline{P}(Cl_t^{\geq}) = U - \underline{P}(Cl_{t-1}^{\leq}), \quad t=2,...,m,$$

$$\underline{P}(Cl_t^{\leq}) = U - \overline{P}(Cl_{t+1}^{\geq}) \text{ and } \overline{P}(Cl_t^{\leq}) = U - \underline{P}(Cl_{t+1}^{\geq}), \quad t=1,...,m-1.$$

The *P-boundaries* of $Cl_t^{\geq}$ and $Cl_t^{\leq}$, denoted by $Bn_P(Cl_t^{\geq})$ and $Bn_P(Cl_t^{\leq})$, respectively, are defined as follows ($t=1,...,m$):

$$Bn_P(Cl_t^{\geq}) = \overline{P}(Cl_t^{\geq}) - \underline{P}(Cl_t^{\geq}), \quad Bn_P(Cl_t^{\leq}) = \overline{P}(Cl_t^{\leq}) - \underline{P}(Cl_t^{\leq}).$$

Due to the above complementarity property, $Bn_P(Cl_t^{\geq}) = Bn_P(Cl_{t-1}^{\leq})$, for $t = 2, \ldots, m$.

For every $P \subseteq I$, the *quality of approximation* of sorting Cl by a set of criteria P is defined as the ratio of the number of objects P-consistent with the dominance principle and the number of all the objects in U. Since the P-consistent objects are those which do not belong to any P-boundary $Bn_P(Cl_t^{\geq})$ or $Bn_P(Cl_t^{\leq})$, $t = 1, \ldots, m$, the quality of approximation of sorting Cl by a set of criteria P, can be written as

$$\gamma_P(Cl) = \frac{\left| U - \left(\bigcup_{t \in \{1,\ldots,m\}} Bn_P(Cl_t^{\leq}) \right) \cup \left(\bigcup_{t \in \{1,\ldots,m\}} Bn_P(Cl_t^{\geq}) \right) \right|}{|U|}$$

$$= \frac{\left| U - \left(\bigcup_{t \in \{1,\ldots,m\}} Bn_P(Cl_t^{\geq}) \right) \right|}{|U|} = \frac{\left| U - \left(\bigcup_{t \in \{1,\ldots,m\}} Bn_P(Cl_t^{\leq}) \right) \right|}{|U|}.$$

$\gamma_P(Cl)$ can be seen as a degree of consistency of the sorting examples, where P is the set of criteria and Cl is the considered sorting.

Each minimal (in the sense of inclusion) subset $P \subseteq I$ such that $\gamma_P(Cl) = \gamma_I(Cl)$ is called a *reduct* of sorting Cl, and is denoted by RED_{Cl}. Let us remark that for a given set of sorting examples one can have more than one reduct. The intersection of all reducts is called the *core*, and is denoted by $CORE_{Cl}$. Criteria in $CORE_{Cl}$ cannot be removed from consideration without deteriorating the quality of approximation of sorting Cl. This means that, in set I, there are three categories of criteria:

- *indispensable* criteria included in the core,
- *exchangeable* criteria included in some reducts, but not in the core,
- *redundant* criteria, neither indispensable nor exchangeable, and thus not included in any reduct.

The dominance-based rough approximations of upward and downward unions of decision classes can serve to induce a generalized description of sorting decisions in terms of "*if* ..., *then* ..." decision rules. For a given upward or downward union of classes, $Cl_t^{\geq}$ or $Cl_s^{\leq}$, the decision rules induced under a hypothesis that objects belonging to $\underline{P}(Cl_t^{\geq})$ or $\underline{P}(Cl_s^{\leq})$ are positive examples (that is objects that have to be matched by the induced decision rules), and all the others are negative (that is objects that have to be not matched by the induced decision rules), suggest a *certain* assignment to "class Cl_t or better", or to "class Cl_s or worse", respectively. On the other hand, the decision rules induced under a hypothesis that objects belonging to $\overline{P}(Cl_t^{\geq})$ or $\overline{P}(Cl_s^{\leq})$ are positive examples, and all the others are negative, suggest a *possible* assignment to "class Cl_t or better", or to "class Cl_s or worse", respectively. Finally, the decision rules induced under a hypothesis that objects belonging to the intersection $\overline{P}(Cl_s^{\leq}) \cap \overline{P}(Cl_t^{\geq})$ are positive examples, and all the others are negative, suggest an assignment to some classes between Cl_s and Cl_t ($s < t$). These rules are matching inconsistent objects $x \in U$, which cannot be assigned without doubts to classes Cl_r, $s \leq r \leq t$, with $s < t$, because $x \notin \underline{P}(Cl_r^{\geq})$ and $x \notin \underline{P}(Cl_r^{\leq})$ for all r such that $s \leq r \leq t$, with $s < t$.

Given the preference information in terms of sorting examples, it is meaningful to consider the following five types of decision rules:

1) *certain* $D_{\geq}$-*decision rules*, providing lower profiles (i.e. sets of minimal values for considered criteria) of objects belonging to $\underline{P}(Cl_t^{\geq})$, $P = \{i_1, \ldots, i_p\} \subseteq I$:
   if $f_{i_1}(x) \geq r_{i_1}$ *and* ... *and* $f_{i_p}(x) \geq r_{i_p}$, *then* $x \in Cl_t^{\geq}$,
   $t = 2, \ldots, m$, $r_{i_1}, \ldots, r_{i_p} \in \Re$;

2) *possible* $D_{\geq}$-*decision rules*, providing lower profiles of objects belonging to $\overline{P}(Cl_t^{\geq})$, $P = \{i_1, \ldots, i_p\} \subseteq I$:
   if $f_{i_1}(x) \geq r_{i_1}$ *and* ... *and* $f_{i_p}(x) \geq r_{i_p}$, *then* x *possibly belongs to* $Cl_t^{\geq}$,
   $t = 2, \ldots, m$, $r_{i_1}, \ldots, r_{i_p} \in \Re$;

3) *certain* $D_{\leq}$-*decision rules*, providing upper profiles (i.e. sets of maximal values for considered criteria) of objects belonging to $\underline{P}(Cl_t^{\leq})$, $P = \{i_1, \ldots, i_p\} \subseteq I$:
   if $f_{i_1}(x) \leq r_{i_1}$ *and* ... *and* $f_{i_p}(x) \leq r_{i_p}$, *then* $x \in Cl_t^{\leq}$,
   $t = 1, \ldots, m-1$, $r_{i_1}, \ldots, r_{i_p} \in \Re$;

4) *possible* $D_{\leq}$-*decision rules*, providing upper profiles of objects belonging to $\overline{P}(Cl_t^{\leq})$, $P = \{i_1, \ldots, i_p\} \subseteq I$:
   if $f_{i_1}(x) \leq r_{i_1}$ *and* ... *and* $f_{i_p}(x) \leq r_{i_p}$, *then* x *possibly belongs to* $Cl_t^{\leq}$,
   $t = 1, \ldots, m-1$, $r_{i_1}, \ldots, r_{i_p} \in \Re$;

5) *approximate* $D_{\geq\leq}$-*decision rules*, providing simultaneously lower and upper profiles of objects belonging to $Cl_s \cup Cl_{s+1} \cup \ldots \cup Cl_t$, without possibility of discerning to which class:
   if $f_{i_1}(x) \geq r_{i_1}$ *and* ... *and* $f_{i_k}(x) \geq r_{i_k}$ *and* $f_{i_{k+1}}(x) \leq r_{i_{k+1}}$ *and* ... *and* $f_{i_p}(x) \leq r_{i_p}$, *then* $x \in Cl_s \cup Cl_{s+1} \cup \ldots \cup Cl_t$,
   $\{i_1, \ldots, i_p\} \subseteq I$ $s, t \in \{1, \ldots, m\}$, $s < t$, $r_{i_1}, \ldots, r_{i_p} \in \Re$.

In the premise of a $D_{\geq\leq}$-decision rule, we can have "$f_i(x) \geq r_i$" and "$f_i(x) \leq r_i'$", where $r_i \leq r_i'$, for the same $i \in I$. Moreover, if $r_i = r_i'$, the two conditions boil down to "$f_i(x) = r_i$".

Let us remark that the values $r_{i_1}, \ldots, r_{i_p}$ in the premise of each decision rule are evaluation profiles with respect to $\{i_1, \ldots, i_p\} \subseteq I$ of some objects in the corresponding rough approximations. More precisely,

1) in case
   if $f_{i_1}(x) \geq r_{i_1}$ *and* ... *and* $f_{i_p}(x) \geq r_{i_p}$, *then* $x \in Cl_t^{\geq}$,

is a certain $D_\geq$-decision rule, there exists some $y \in \underline{P}(Cl_t^{\geq})$, $P = \{i_1, \dots, i_p\}$, such that
$f_{i_1}(y) = r_{i_1}$ and $\dots$ and $f_{i_p}(y) = r_{i_p}$;

2) in case

if $f_{i_1}(x) \geq r_{i_1}$ and $\dots$ and $f_{i_p}(x) \geq r_{i_p}$, then x possibly belongs to $Cl_t^{\geq}$, is a possible $D_\geq$-decision rules, there exists some $y \in \overline{P}(Cl_t^{\geq})$, $P = \{i_1, \dots, i_p\}$, such that
$f_{i_1}(y) = r_{i_1}$ and $\dots$ and $f_{i_p}(y) = r_{i_p}$;

3) in case

if $f_{i_1}(x) \leq r_{i_1}$ and $\dots$ and $f_{i_p}(x) \leq r_{i_p}$, then $x \in Cl_t^{\leq}$, is a certain $D_\leq$-decision rules, there exists some $y \in \underline{P}(Cl_t^{\leq})$, $P = \{i_1, \dots, i_p\}$, such that
$f_{i_1}(y) = r_{i_1}$ and $\dots$ and $f_{i_p}(y) = r_{i_p}$;

4) in case

if $f_{i_1}(x) \leq r_{i_1}$ and $\dots$ and $f_{i_p}(x) \leq r_{i_p}$, then x possibly belongs to $Cl_t^{\leq}$, is a possible $D_\leq$-decision rules, there exists some $y \in \overline{P}(Cl_t^{\leq})$, $P = \{i_1, \dots, i_p\}$, such that
$f_{i_1}(y) = r_{i_1}$ and $\dots$ and $f_{i_p}(y) = r_{i_p}$;

5) in case

if $f_{i_1}(x) \geq r_{i_1}$ and $\dots$ and $f_{i_k}(x) \geq r_{i_k}$ and $f_{i_{k+1}}(x) \leq r_{i_{k+1}}$ and $\dots$ and $f_{i_p}(x) \leq r_{i_p}$, then $x \in Cl_s \cup Cl_{s+1} \cup \dots \cup Cl_t$, is an approximate $D_{\geq\leq}$-decision rules, there exists some $y \in \overline{P}(Cl_t^{\geq})$, and some $z \in \overline{P}(Cl_s^{\leq})$, $\{i_1, \dots, i_k\} \cup \{i_{k+1}, \dots, i_p\} = P$, such that
$f_{i_1}(y) = r_{i_1}$ and $\dots$ and $f_{i_k}(y) = r_{i_k}$, and $f_{k+1}(z) = r_{k+1}$ and $\dots$ and $f_{i_p}(z) = r_{i_p}$.

Note that in the above rules, each condition profile defines a dominance cone in $|P|$-dimensional condition space $\Re^{|P|}$, where $P = \{i_1, \dots, i_p\}$ is the set of criteria considered in the rule, and each decision defines a dominance cone in one-dimensional decision space $\{1, \dots, m\}$. Both dominance cones are positive for $D_\geq$-rules, negative for $D_\leq$-rules and partially positive and partially negative for $D_{\geq\leq}$-rules.

Let also point out that dominance cones corresponding to condition profiles can originate from any point of $\Re^{|P|}$, $P = \{i_1, \dots, i_p\}$, without the risk of being too specific. Thus, contrary to traditional granular computing, the condition space $\Re^n$ (i.e. the set of all possible vectors of evaluations of objects with respect to considered criteria) does not need to be discretized. This implies that the rules obtained from DRSA are also meaningful for analysis of vectors coming from a continuous multiobjective optimization problem where the concept of dominance cone is particularly useful (see e.g. Chapter 1 where the related concept of ordering cones is discussed).

Since a decision rule is a kind of implication, by a *minimal* rule we mean an implication such that there is no other implication with the premise of at least the same weakness (in other words, a rule using a subset of conditions and/or weaker conditions) and the conclusion of at least the same strength (in other words, a $D_\geq$- or a $D_\leq$-decision rule assigning objects to the same union or sub-union of classes, or a $D_{\geq\leq}$-decision rule assigning objects to the same or smaller set of classes).

The rules of type 1) and 3) represent certain knowledge extracted from data (sorting examples), while the rules of type 2) and 4) represent possible knowledge; the rules of type 5) represent doubtful knowledge, because they are supported by inconsistent objects only.

Moreover, the rules of type 1) and 3) are *exact* or *deterministic* if they do not cover negative examples, and they are *probabilistic* otherwise. In the latter case, each rule is characterized by a *confidence ratio*, representing the probability that an object matching the premise of the rule also matches its conclusion.

Given a certain or possible $D_\geq$-decision rule $r \equiv$ "*if* $f_{i_1}(x) \geq r_{i_1}$ *and* ... *and* $f_{i_p}(x) \geq r_{i_p}$, *then* $x \in Cl_t^\geq$", an object $y \in U$ *supports* r if $f_{i_1}(y) \geq r_{i_1}$ and ... and $f_{i_p}(y) \geq r_{i_p}$ and $y \in Cl_t^\geq$. Moreover, object $y \in U$ supporting decision rule r is a *base* of r if $f_{i_1}(y) = r_{i_1}$ and ... and $f_{i_p}(y) = r_{i_p}$. Similar definitions hold for certain or possible $D_\leq$-decision rules and approximate $D_{\geq\leq}$-decision rules. A decision rule having at least one base is called *robust*. Identification of supporting objects and bases of robust rules is important for interpretation of the rules in multiple criteria decision analysis perspective. The ratio of the number of objects supporting the premise of a rule and the number of all considered objects is called *relative support* of a rule. The relative support and the confidence ratio are basic characteristics of a rule, however, some *Bayesian confirmation measures* reflect much better the attractiveness of a rule (Greco *et al.*, 2004b).

A set of decision rules is *complete* if it covers all considered objects (sorting examples) in such a way that consistent objects are re-assigned to their original classes, and inconsistent objects are assigned to clusters of classes referring to this inconsistency. We call each set of decision rules that is complete and non-redundant *minimal*, i.e. exclusion of any rule from this set makes it incomplete.

One of three induction strategies can be adopted to obtain a set of decision rules (Stefanowski, 1998):

- generation of a *minimal* representation, i.e. a minimal set of rules,
- generation of an *exhaustive* representation, i.e. all rules for a given data table,
- generation of a *characteristic* representation, i.e. a set of rules covering relatively many objects, however, not necessarily all objects, from U.

Procedures for induction of decision rules from dominance-based rough approximations have been proposed by Greco *et al.* (2001a).

In (Giove *et al.*, 2002), a new methodology for the induction of monotonic decision trees from dominance-based rough approximations of preference ordered decision classes has been proposed.

5.3 Example Illustrating DRSA

In this section we present a didactic example which illustrates the main concepts of DRSA. Let us consider the following multiple criteria sorting problem. Students of a college must obtain an overall evaluation on the basis of their achievements in Mathematics, Physics and Literature. The three subjects are clearly criteria (condition attributes) and the comprehensive evaluation is a decision attribute. For simplicity, the value sets of the criteria and of the decision attribute are the same, and they are composed of three values: bad, medium and good. The preference order of these values is obvious. Thus, there are three preference ordered decision classes, so the problem belongs to the category of multiple criteria sorting. In order to build a preference model of the jury, we will analyze a set of exemplary evaluations of students (sorting examples) provided by the jury. They are presented in Table 5.1.

Note that the dominance principle obviously applies to the sorting examples, since an improvement of a student's score on one of three criteria, with other scores unchanged, should not worsen the student's overall evaluation, but rather improve it.

Table 5.1. Exemplary evaluations of students (sorting examples)

Student	Mathematics	Physics	Literature	Overall Evaluation
S1	good	medium	bad	bad
S2	medium	medium	bad	medium
S3	medium	medium	medium	medium
S4	good	good	medium	good
S5	good	medium	good	good
S6	good	good	good	good
S7	bad	bad	bad	bad
S8	bad	bad	medium	bad

Observe that student S1 has not worse evaluations than student S2 on all the considered criteria, however, the overall evaluation of S1 is worse than the overall evaluation of S2. This contradicts the dominance principle, so the two sorting examples are inconsistent. Let us observe that if we reduced the set of considered criteria, i.e. the set of considered subjects, then some more inconsistencies could occur. For example, let us remove from Table 5.1 the evaluation on Literature. In this way we get Table 5.2, where S1 is inconsistent not only with $S2$, but also with $S3$ and $S5$. In fact, student $S1$ has not worse evaluations than students $S2$, $S3$ and $S5$ on all the considered criteria (Mathematics and Physics), however, the overall evaluation of $S1$ is worse than the overall evaluation of $S2$, $S3$ and $S5$.

Table 5.2. Exemplary evaluations of students excluding Literature

Student	Mathematics	Physics	Overall Evaluation
$S1$	good	medium	bad
$S2$	medium	medium	medium
$S3$	medium	medium	medium
$S4$	good	good	good
$S5$	good	medium	good
$S6$	good	good	good
$S7$	bad	bad	bad
$S8$	bad	bad	bad

Observe, moreover, that if we remove from Table 5.1 the evaluations on Mathematics, we obtain Table 5.3, where no new inconsistencies occur, comparing to Table 5.1.

Table 5.3. Exemplary evaluations of students excluding Mathematics

Student	Physics	Literature	Overall Evaluation
$S1$	medium	bad	bad
$S2$	medium	bad	medium
$S3$	medium	medium	medium
$S4$	good	medium	good
$S5$	medium	good	good
$S6$	good	good	good
$S7$	bad	bad	bad
$S8$	bad	medium	bad

Similarly, if we remove from Table 5.1 the evaluations on Physics, we obtain Table 5.4, where no new inconsistencies occur, comparing to Table 5.1.

The fact that no new inconsistency occurs when Mathematics or Physics is removed, means that the subsets of criteria {Physics, Literature} or {Mathematics, Literature} contain sufficient information to represent the overall evaluation of students with the same quality of approximation as using the complete set of three criteria. This is not the case, however, for the subset {Mathematics, Physics}. Observe, moreover, that subsets {Physics, Literature} and {Mathematics, Literature} are minimal, because no other criterion

Table 5.4. Exemplary evaluations of students excluding Physics

Student	Mathematics	Literature	Overall Evaluation
$S1$	good	bad	bad
$S2$	medium	bad	medium
$S3$	medium	medium	medium
$S4$	good	medium	good
$S5$	good	good	good
$S6$	good	good	good
$S7$	bad	bad	bad
$S8$	bad	medium	bad

can be removed without new inconsistencies occur. Thus, {Physics, Literature} and {Mathematics, Literature} are the reducts of the complete set of criteria {Mathematics, Physics, Literature}. Since Literature is the only criterion which cannot be removed from any reduct without introducing new inconsistencies, it constitutes the core, i.e. the set of indispensable criteria. The core is, of course, the intersection of all reducts, i.e. in our example:

{Literature} = {Physics, Literature} ∩ {Mathematics, Literature}.

In order to illustrate in a simple way the concept of rough approximation, let us confine our analysis to the reduct {Mathematics, Literature}. Let us consider student $S4$. His positive dominance cone $D^+_{\{Mathematics,Literature\}}(S4)$ is composed of all the students having evaluations not worse than him on Mathematics and Literature, i.e. of all the students dominating him with respect to Mathematics and Literature. Thus, we have

$$D^+_{\{Mathematics,Literature\}}(S4) = \{S4, S5, S6\}.$$

On the other hand, the negative dominance cone of student $S4$, $D^-_{\{Mathematics,Literature\}}(S4)$, is composed of all the students having evaluations not better than him on Mathematics and Literature, i.e. of all the students dominated by him with respect to Mathematics and Literature. Thus, we have

$$D^-_{\{Mathematics,Literature\}}(S4) = \{S1, S2, S3, S4, S7, S8\}.$$

Similar dominance cones can be obtained for all the students from Table 5.4. For example, for $S2$ we have

$$D^+_{\{Mathematics,Literature\}}(S2) = \{S1, S2, S3, S4, S5, S6\}.$$

and

$$D^-_{\{Mathematics,Literature\}}(S2) = \{S2, S7\}.$$

Using dominance cones, we can calculate the rough approximations. Let us consider, for example, the lower approximation of the set of students having a "good" overall evaluation $\underline{P}(Cl_{good}^{\geq})$, with $P=\{$Mathematics, Literature$\}$. We have, $\underline{P}(Cl_{good}^{\geq}) = \{S4, S5, S6\}$, because positive dominance cones of students $S4$, $S5$ and $S6$ are all included in the set of students with an overall evaluation "good". In other words, this means that there is no student dominating $S4$ or $S5$ or $S6$ while having an overall evaluation worse than "good". From the viewpoint of decision making, this means that, taking into account the available information about evaluation of students on Mathematics and Literature, the fact that student y dominates $S4$ or $S5$ or $S6$ is a *sufficient* condition to conclude that y is a "good" student.

As to the upper approximation of the set of students with a "good" overall evaluation, we have $\overline{P}(Cl_{good}^{\geq}) = \{S4, S5, S6\}$, because negative dominance cones of students $S4$, $S5$ and $S6$ have a nonempty intersection with the set of students having a "good" overall evaluation. In other words, this means that for each one of the students $S4$, $S5$ and $S6$, there is at least one student dominated by him with an overall evaluation "good". From the point of view of decision making, this means that, taking into account the available information about evaluation of students on Mathematics and Literature, the fact that student y dominates $S4$ or $S5$ or $S6$ is a *possible* condition to conclude that y is a "good" student.

Let us observe that for the set of criteria $P=\{$Mathematics, Literature$\}$, the lower and upper approximations of the set of "good" students are the same. This means that sorting examples concerning this class are all consistent. This is not the case, however, for the sorting examples concerning the union of decision classes "at least medium". For this upward union we have $\underline{P}(Cl_{medium}^{\geq}) = \{S3, S4, S5, S6\}$ and $\overline{P}(Cl_{medium}^{\geq}) = \{S1, S2, S3, S4, S5, S6\}$. The difference between $\overline{P}(Cl_{medium}^{\geq})$ and $\underline{P}(Cl_{medium}^{\geq})$, i.e. the boundary $Bn_P(Cl_{medium}^{\geq}) = \{S1, S2\}$, is composed of students with inconsistent overall evaluations, which has already been noticed above. From the viewpoint of decision making, this means that, taking into account the available information about evaluation of students on Mathematics and Literature, the fact that student y is dominated by $S1$ and dominates $S2$ is a condition to conclude that y can obtain an overall evaluation "at least medium" with some doubts.

Until now we have considered rough approximations of only upward unions of decision classes. It is interesting, however, to calculate also rough approximations of downward unions of decision classes. Let us consider first the lower approximation of the set of students having "at most medium" overall evaluation $\underline{P}(Cl_{medium}^{\leq})$. We have, $\underline{P}(Cl_{medium}^{\leq}) = \{S1, S2, S3, S7, S8\}$, because the negative dominance cones of students $S1, S2, S3, S7$, and $S8$ are all included in the set of students with overall evaluation "at most medium". In other words, this means that there is no student dominated by $S1$ or $S2$ or $S3$ or $S7$ or $S8$ while having an overall evaluation better than "medium". From the viewpoint of decision making, this means that, taking into account the available information about evaluation of students on Mathematics and Literature, the fact that student y is dominated by $S1$ or $S2$ or $S3$ or $S7$ or $S8$ is a *sufficient* condition to conclude that y is an "at most medium" student.

As to the upper approximation of the set of students with an "at most medium" overall evaluation, we have $\overline{P}(Cl^{\leq}_{medium}) = \{S1, S2, S3, S7, S8\}$, because the positive dominance cones of students $S1, S2, S3, S7$, and $S8$ have a nonempty intersection with the set of students having an "at most medium" overall evaluation. In other words, this means that for each one of the students $S1, S2, S3, S7$, and $S8$, there is at least one student dominating him with an overall evaluation "at most medium". From the viewpoint of decision making, this means that, taking into account the available information about evaluation of students on Mathematics and Literature, the fact that student y is dominated by $S1$ or $S2$ or $S3$ or $S7$ or $S8$ is a *possible* condition to conclude that y is an "at most medium" student.

Finally, for the set of students having a "bad" overall evaluation, we have $\underline{P}(Cl^{\leq}_{bad}) = \{S7, S8\}$ and $\overline{P}(Cl^{\leq}_{bad}) = \{S1, S2, S7, S8\}$. The difference between $\overline{P}(Cl^{\leq}_{bad})$ and $\underline{P}(Cl^{\leq}_{bad})$, i.e. the boundary $Bn_P(Cl^{\leq}_{bad}) = \{S1, S2\}$ is composed of students with inconsistent overall evaluations, which has already been noticed above. From the viewpoint of decision making, this means that, taking into account the available information about evaluation of students on Mathematics and Literature, the fact that student y is dominated by $S1$ and dominates $S2$ is a condition to conclude that y can obtain an overall evaluation "bad" with some doubts. Observe, moreover, that $Bn_P(Cl^{\geq}_{medium}) = Bn_P(Cl^{\leq}_{bad})$.

Given the above rough approximations with respect to the set of criteria $P=\{\text{Mathematics, Literature}\}$, one can induce a set of decision rules representing the preferences of the jury. The idea is that evaluation profiles of students belonging to the lower approximations can serve as a base for some certain rules, while evaluation profiles of students belonging to the boundaries can serve as a base for some approximate rules. The following decision rules have been induced (between parentheses there are id's of students supporting the corresponding rule; the student being a rule base is underlined):

rule 1) if the evaluation on Mathematics is (at least) good, and the evaluation on Literature is at least medium, then the overall evaluation is (at least) good, $\{\underline{S4}, S5, S6\}$,

rule 2) if the evaluation on Mathematics is at least medium, and the evaluation on Literature is at least medium, then the overall evaluation is at least medium, $\{\underline{S3}, S4, S5, S6\}$,

rule 3) if the evaluation on Mathematics is at least medium, and the evaluation on Literature is (at most) bad, then the overall evaluation is bad or medium, $\{S1, \underline{S2}\}$,

rule 4) if the evaluation on Mathematics is at least medium, then the overall evaluation is at least medium, $\{S2, S3, \underline{S4}, S5, S6\}$, rule 5) if the evaluation on Literature is (at most) bad, then the overall evaluation is at most medium, $\{\underline{S1}, \underline{S2}, \underline{S7}\}$,

rule 6) if the evaluation on Mathematics is (at most) bad, then the overall evaluation is (at most) bad, $\{\underline{S7}, \underline{S8}\}$.

Analogously in rule 3) we wrote "if ... the evaluation in Literature is (at most) bad" and not simply "if ... the evaluation in Literature is bad" (as it was possible, since in the considered example "bad" is the worst possible evaluation). The same remark holds for rule 5) and rule 6). Notice that rules 1)–2), 4)–7) are certain, while rule 3) is an approximate one. These rules represent knowledge discovered from the available information. In the current context, the knowledge is interpreted as a preference model of the jury. A characteristic feature of the syntax of decision rules representing preferences is the use of expressions "at least" or "at most" a value; in case of extreme values ("good" and "bad"), these expressions are put in parentheses because there is no value above "good" and below "bad".

Even if one can represent all the knowledge using only one reduct of the set of criteria (as we have done using $P=\{$Mathematics, Literature$\}$), when considering a larger set of criteria than a reduct, one can obtain a more synthetic representation of knowledge, i.e. the number of decision rules or the number of elementary conditions, or both of them, can get smaller. For example, considering the set of all three criteria, {Mathematics, Physics, Literature}, we can induce a set of decision rules composed of the above rules 1), 2), 3) and 6), plus the following :

rule 7) if the evaluation on Physics is at most medium, and the evaluation on Literature is at most medium, then the overall evaluation is at most medium, $\{S1, S2, \underline{S3}, S7, S8\}$.

Thus, the complete set of decision rules induced from Table 5.1 is composed of 5 instead of 6 rules.

Once accepted by the DM, these rules represent his/her preference model. Assuming that rules 1)–7) in our example represent the preference model of the jury, it can be used to evaluate new students. For example, student $S9$ who is "medium" in Mathematics and Physics and "good" in Literature, would be evaluated as "medium" because his profile matches the premise of rule 2), having as consequence an overall evaluation at least "medium". The overall evaluation of $S9$ cannot be "good", because his profile does not match any rule having as consequence an overall evaluation "good" (in the considered example, the only rule of this type is rule 1) whose premise is not matched by the profile of $S9$).

5.4 Multiple Criteria Decision Analysis Using Association Rules

In interactive multiobjective optimization, and, more generally, in multiple criteria decision analysis, the DM is interested in relationships between attainable values of criteria. For instance, in a car selection problem, one can observe that in the set of considered cars, if the maximum speed is at least 200 km/h and the time to reach 100 km/h is at most 7 seconds, then the price is not less than 40,000\$ and the fuel consumption is not less than 9 liters per 100 km. These relationships are association rules whose general syntax, in case of minimization of criteria f_i, $i \in I$, is:

"if $f_{i_1}(x) \leq r_{i_1}$ and ... and $f_{i_p}(x) \leq r_{i_p}$, then $f_{i_{p+1}}(x) \geq r_{i_{p+1}}$ and ... and $f_{i_q}(x) \geq r_{i_q}$", where $\{i_1, \ldots, i_q\} \subseteq I$, $r_{i_1}, \ldots, r_{i_q} \in \Re$.

If criterion f_i, $i \in I$, should be maximized, the corresponding condition in the association rule should be reversed, i.e. in the premise, the condition becomes $f_i(x) \geq r_i$, and in the conclusion it becomes $f_i(x) \leq r_i$.

Given an association rule $r \equiv$ "if $f_{i_1}(x) \leq r_{i_1}$ and ... and $f_{i_p}(x) \leq r_{i_p}$, then $f_{i_{p+1}}(x) \geq r_{i_{p+1}}$ and ... and $f_{i_q}(x) \geq r_{i_q}$", an object $y \in U$ *supports* r if $f_{i_1}(y) \leq r_{i_1}$ and ... and $f_{i_p}(y) \leq r_{i_p}$ and $f_{i_{p+1}}(y) \geq r_{i_{p+1}}$ and ... and $f_{i_q}(y) \geq r_{i_q}$. Moreover, object $y \in U$ supporting decision rule r is a *base* of r if $f_{i_1}(y) = r_{i_1}$ and ... and $f_{i_p}(y) = r_{i_p}$ and $f_{i_{p+1}}(y) = r_{i_{p+1}}$ and ... and $f_{i_q}(y) = r_{i_q}$. An association rule having at least one base is called *robust*.

We say that an association rule $r \equiv$ "if $f_{i_1}(x) \leq r_{i_1}$ and ... and $f_{i_p}(x) \leq r_{i_p}$, then $f_{i_{p+1}}(x) \geq r_{i_{p+1}}$ and ... and $f_{i_q}(x) \geq r_{i_q}$" holds in universe U if:

1) there is at least one $y \in U$ supporting r,

2) r is not contradicted in U, i.e. there is no $z \in U$ such that $f_{i_1}(z) \leq r_{i_1}$ and ... and $f_{i_p}(z) \leq r_{i_p}$, while *not* $f_{i_{p+1}}(z) \geq r_{i_{p+1}}$ or ... or $f_{i_q}(z) \geq r_{i_q}$.

Given two association rules

- $r_1 \equiv$ "if $f_{i_1}(x) \leq r_{i_1}^1$ and ... and $f_{i_p}(x) \leq r_{i_p}^1$, then $f_{i_{p+1}}(x) \geq r_{i_{p+1}}^1$ and ... and $f_{i_q}(x) \geq r_{i_q}^1$",

- $r_2 \equiv$ "if $f_{j_1}(x) \leq r_{j_1}^2$ and ... and $f_{j_s}(x) \leq r_{j_s}^2$, then $f_{j_{s+1}}(x) \geq r_{j_{s+1}}^2$ and ... and $f_{j_t}(x) \geq r_{j_t}^2$",

we say that rule r_1 is not weaker than rule r_2, denoted by $r_1 \trianglerighteq r_2$, if:

α) $\{i_1, \ldots, i_p\} \subseteq \{j_1, \ldots, j_s\}$,

β) $r_{i_1}^1 \geq r_{i_1}^2, \ldots, r_{i_p}^1 \geq r_{i_p}^2$,

γ) $\{i_{p+1}, \ldots, i_q\} \supseteq \{j_{s+1}, \ldots, j_t\}$,

δ) $r_{j_{s+1}}^1 \geq r_{j_{s+1}}^2, \ldots, r_{j_t}^1 \geq r_{j_t}^2$.

Conditions β) and δ) are formulated for criteria f_i to be minimized. If criterion f_i should be maximized, the corresponding inequalities should be reversed, i.e. $r_i^1 \leq r_i^2$ in condition β) as well as in condition δ). Notice that $\unrhd$ is a binary relation on the set of association rules, which is a partial preorder, i.e. it is reflexive (each rule is not weaker than itself) and transitive. The asymmetric part of the relation $\unrhd$ is denoted by $\rhd$, and $r_1 \rhd r_2$ reads "r_1 is stronger than r_2.

For example, consider the following association rules:

- $r_1 \equiv$ "if the maximum speed is at least 200 km/h and the time to reach 100 km/h is at most 7 seconds, then the price is not less than 40,000\$ and the fuel consumption is not less than 9 liters per 100 km",

- $r_2 \equiv$ "if the maximum speed is at least 200 km/h and the time to reach 100 km/h is at most 7 seconds and the horse power is at least 175 kW, then the price is not less than 40,000\$ and the fuel consumption is not less than 9 liters per 100 km",

- $r_3 \equiv$ "if the maximum speed is at least 220 km/h and the time to reach 100 km/h is at most 7 seconds, then the price is not less than 40,000\$ and the fuel consumption is not less than 9 liters per 100 km",

- $r_4 \equiv$ "if the maximum speed is at least 200 km/h and the time to reach 100 km/h is at most 7 seconds, then the price is not less than 40,000\$",

- $r_5 \equiv$ "if the maximum speed is at least 200 km/h and the time to reach 100 km/h is at most 7 seconds, then the price is not less than 35,000\$ and the fuel consumption is not less than 9 liters per 100 km",

- $r_6 \equiv$ "if the maximum speed is at least 220 km/h and the time to reach 100 km/h is at most 7 seconds and the horse power is at least 175 kW, then the price is not less than 35,000\$".

Let us observe that rule r_1 is stronger than each of the other five rules for the following reasons:

- $r_1 \rhd r_2$ for condition α) because, all things equal elsewhere, in the premise of r_2 there is an additional condition: "the horse power is at least 175 kW",

- $r_1 \rhd r_3$ for condition β) because, all things equal elsewhere, in the premise of r_3 there is a condition with a worse threshold value: "the maximum speed is at least 220 km/h" instead of "the maximum speed is at least 200 km/h",

- $r_1 \rhd r_4$ for condition γ) because, all thing equal elsewhere, in the conclusion of r_4 one condition is missing: "the fuel consumption is not less than 9 liters per 100 km",

- $r_1 \rhd r_5$ for condition δ) because, all thing equal elsewhere, in the conclusion of r_5 there is a condition with a worse threshold value: "the price is not less than 35,000 \$" instead of "the price is not less than 40,000\$",

- $r_1 \triangleright r_6$ for conditions α), β), γ) and δ) because all weak points for which rules r_2, r_3, r_4 and r_5 are weaker than rule r_1 are present in r_6.

An association rule r is *minimal* if there is no other rule stronger than r with respect to $\triangleright$. An algorithm for induction of association rules from preference ordered data has been presented in (Greco *et al.*, 2002a).

5.5 Interactive Multiobjective Optimization Using Dominance-Based Rough Set Approach (IMO-DRSA)

In this section, we present a new method for Interactive Multiobjective Optimization using Dominance-based Rough Set Approach (IMO-DRSA). The method is composed of the following steps.

Step 1. Generate a representative sample of solutions from the currently considered part of the Pareto optimal set.

Step 2. Present the sample to the DM, possibly together with association rules showing relationships between attainable values of objective functions in the Pareto optimal set.

Step 3. If the DM is satisfied with one solution from the sample, then this is the most preferred solution and the procedure stops. Otherwise continue.

Step 4. Ask the DM to indicate a subset of "good" solutions in the sample.

Step 5. Apply DRSA to the current sample of solutions sorted into "good" and "others", in order to induce a set of decision rules with the following syntax "if $f_{j_1}(\mathbf{x}) \le \alpha_{j_1}$ and ... and $f_{j_p}(\mathbf{x}) \le \alpha_{j_p}$, then solution $\mathbf{x}$ is good", $\{j_1, \ldots, j_p\} \subseteq \{1, \ldots, n\}$.

Step 6. Present the obtained set of rules to the DM.

Step 7. Ask the DM to select the decision rules most adequate to his/her preferences.

Step 8. Adjoin the constraints $f_{j_1}(\mathbf{x}) \le \alpha_{j_1}$, ... , $f_{j_p}(\mathbf{x}) \le \alpha_{j_p}$ coming from the rules selected in *Step* 7 to the set of constraints imposed on the Pareto optimal set, in order to focus on a part interesting from the point of view of DM's preferences.

Step 9. Go back to *Step* 1.

In a sequence of iterations the method is exploring the Pareto optimal set of a multiobjective optimization problem or an approximation of this set. In the calculation stage (*Step* 1), any multiobjective optimization method (see Chapter 1), which finds the Pareto optimal set or its approximation, such as Evolutionary Multiobjective Optimization methods (see Chapter 3), can be used. In the dialogue stage of the method (*Step* 2 to 7), the DM is asked to select a decision rule induced from his/her preference information, which is equivalent to fixing some upper bounds for the minimized objective functions f_j.

In *Step* 1, the representative sample of solutions from the currently considered part of the Pareto optimal set can be generated using one of existing procedures, such as (Steuer and Choo, 1983; Wierzbicki, 1980; Jaszkiewicz and Słowiński, 1999). It is recommended to use a fine grained sample of representative solutions to induce association rules; however, the sample of solutions presented to the DM in *Step* 2 should be much smaller (about a dozen) in order to avoid an excessive cognitive effort of the DM. Otherwise, the DM would risk to give non reliable information (for a discussion about cognitive aspects of interactive multiobjective optimization methods see Chapter 2 and Chapter 15).

The association rules presented in *Step* 2 help the DM in understanding what (s)he can expect from the optimization problem. More precisely, any association rule

"if $f_{i_1}(x) \le r_{i_1}$ and ... and $f_{i_p}(x) \le r_{i_p}$, then $f_{i_{p+1}}(x) \ge r_{i_{p+1}}$ and ... and $f_{i_q}(x) \ge r_{i_q}$", where $\{i_1, \dots, i_q\} \subseteq I$, $r_{i_1}, \dots, r_{i_q} \in \Re$

says to the DM that, if (s)he wants attain the values of objective functions $f_{i_1}(x) \le r_{i_1}$ and ... and $f_{i_p}(x) \le r_{i_p}$, then (s)he cannot reasonably expect to obtain values of objective functions $f_{i_{p+1}}(x) < r_{i_{p+1}}$ and ... and $f_{i_q}(x) < r_{i_q}$.

In *Step* 2, it could be useful to visualize the currently considered part of Pareto optimal set using some techniques discussed in Chapter 8 and Chapter 9.

With respect to the sorting of solutions into the two classes of "good" and "others", observe that "good" means in fact "relatively good", i.e. better than the rest in the current sample. In case, the DM would refuse to classify as "good" any solution, one can ask the DM to specify some minimal requirements of the type $f_{j_1}(\mathbf{x}) \le \alpha_{j_1}$ and ... and $f_{j_p}(\mathbf{x}) \le \alpha_{j_p}$ for "good" solutions. These minimal requirements give some constraints that can be used in *Step* 8, in the same way as the analogous constraints coming from selected decisions rules.

The rules considered in *Step* 5 have a syntax corresponding to minimization of objective functions. In case of maximization of an objective function f_j, the condition concerning this objective in the decision rule should have the form $f_j(\mathbf{x}) \ge \alpha_j$.

Remark, moreover, that the Pareto optimal set reduced in *Step* 8 by constraints $f_{j_1}(\mathbf{x}) \le \alpha_{j_1}, \dots, f_{j_p}(\mathbf{x}) \le \alpha_{j_p}$ is certainly not empty if these constraints are coming from one decision rule only. Let us remember that we consider robust rules (see Section 2) and, therefore, the threshold values $\alpha_{j_1}, \dots, \alpha_{j_p}$ are values of objective functions of some solutions from the Pareto optimal set. If $\{j_1, \dots, j_p\} = \{1, \dots, n\}$, i.e. $\{j_1, \dots, j_p\}$ is the set of all objective functions, then the new reduced part of the Pareto optimal set contains only one solution $\mathbf{x}$ such that $f_1(\mathbf{x}) = \alpha_1, \dots, f_n(\mathbf{x}) = \alpha_n$. If

$\{j_1, \ldots, j_p\} \subset \{1, \ldots, n\}$, i.e. $\{j_1, \ldots, j_p\}$ is a proper subset of the set of all objective functions, then the new reduced part of the Pareto optimal set contains solutions satisfying conditions $f_{j_1}(\mathbf{x}) \leq \alpha_{j_1}$ and ... and $f_{j_p}(\mathbf{x}) \leq \alpha_{j_p}$. Since the considered rules are robust, then there is at least one solution $\mathbf{x}$ satisfying these constraints. When the Pareto optimal set is reduced in *Step* 8 by constraints $f_{j_1}(\mathbf{x}) \leq \alpha_{j_1}, \ldots, f_{j_p}(\mathbf{x}) \leq \alpha_{j_p}$ coming from more than one rule, then it is possible that the resulting reduced part of the Pareto optimal set is empty. Thus, before passing to *Step* 9, it is necessary to verify if the reduced Pareto optimal set is not empty. If the reduced Pareto optimal set is empty, then the DM is required to revise his/her selection of rules. The DM can be supported in this task, by information about minimal sets of constraints $f_j(\mathbf{x}) \leq \alpha_j$ coming from the considered decision rules to be removed in order to get a non-empty part of the Pareto optimal set.

The constraints introduced in *Step* 8 are maintained in the following iterations of the procedure, however, they cannot be considered as irreversible. Indeed, the DM can always come back to the Pareto optimal set considered in one of previous iterations and continue from this point. This is in the spirit of a learning oriented conception of interactive multiobjective optimization, i.e. it agrees with the idea that the interactive procedure permits the DM to learn about his/her preferences and about the "shape" of the Pareto optimal set (see e.g. Chapter 2 and Chapter 15).

5.6 Illustrative Example

To illustrate the interactive multiobjective optimization procedure based on DRSA, we consider a product mix problem. There are three products: A, B, C which are produced in quantities denoted by x_A, x_B, and x_C, respectively. The unit prices of the three products are $p_A = 20$, $p_B = 30$, $p_C = 25$. The production process involves two machines. The production times of A, B, C on the first machine are equal to $t_{1A} = 5$, $t_{1B} = 8$, $t_{1C} = 10$, and on the second machine they are equal to $t_{2A} = 8$, $t_{2B} = 6$, $t_{2C} = 2$. Two raw materials are used in the production process. The first raw material has a unit cost of 6 and the quantity required for production of one unit of A, B and C is $r_{1A} = 1$, $r_{1B} = 2$ and $r_{1C} = 0.75$, respectively. The second raw material has a unit cost of 8 and the quantity required for production of one unit of A, B and C is $r_{2A} = 0.5$, $r_{2B} = 1$ and $r_{2C} = 0.5$, respectively. We know, moreover, that the market cannot absorb a production greater than 10, 20 and 10 units for A, B and C, respectively. To decide how much of A, B and C should be produced, the following objectives have to be taken into account:

- Profit (to be maximized),
- Time (total production time on two machines – to be minimized),
- Production of A (to be maximized),
- Production of B (to be maximized),

- Production of C (to be maximized),
- Sales (to be maximized).

The above product mix problem can be formulated as the following multiobjective optimization problem:

Maximize
$20x_A + 30x_B + 25x_C - (1x_A + 2x_B + 0.75x_C)6 - (0.5x_A + 1x_B + 0.5x_C)8$
[Profit],

Minimize $5x_A + 8x_B + 10x_C + 8x_A + 6x_B + 2x_C$ [Time],

Maximize x_A [Production of A],

Maximize x_B [Production of B],

Maximize x_C [Production of C],

Maximize $20x_A + 30x_B + 25x_C$ [Sales],

subject to:

$x_A \leq 10, \quad x_B \leq 20, \quad x_C \leq 10$ [Market absorption limits],

$x_A \geq 0, \quad x_B \geq 0, \quad x_C \geq 0$ [Non-negativity constraints].

A sample of representative Pareto optimal solutions has been calculated and proposed to the DM. Let us observe that the problem we are considering is a Multiple Objective Linear Programming (MOLP) and thus representative Pareto optimal solutions can be calculated using classical linear programming, looking for the solutions optimizing each one of the considered objectives or fixing all the considered objective functions but one at a satisfying value, and looking for the solution optimizing the remaining objective function. The set of representative Pareto optimal solutions is shown in Table 5.5. Moreover, a set of potentially interesting association rules have been induced from the sample and presented to the DM. These rules represent strongly supported relationships between attainable values of objective functions. The association rules are the following (between parentheses there are id's of solutions supporting the rule):

rule 1)   if Time ≤ 140, then Profit ≤ 180.38 and Sales ≤ 280.77
($s1, s2, s3, s4, s12, s13$),

rule 2)   if Time ≤ 150, then Profit ≤ 188.08 and Sales ≤ 296.15
($s1, s2, s3, s4, s5, s6, s7, s8, s9, s12, s13$),

rule 3)   if $x_B \geq 2$, then Profit ≤ 209.25 and $x_A \leq 6$ and $x_C \leq 7.83$
($s4, s5, s6, s9, s10, s11, s12, s13$),

rule 4)   if Time $\leq$ 150, then $x_B \leq 3$
($s1, s2, s3, s4, s5, s6, s7, s8, s9, s12, s13$),

rule 5)   if Profit $\geq$ 148.38 and Time $\leq$ 150, then $x_B \leq 2$
($s1, s2, s3, s5, s7, s8$),

rule 6)   if $x_A \geq 5$, then Time $\geq$ 150
($s5, s6, s8, s9, s10, s11$),

rule 7)   if Profit $\geq$ 127.38 and $x_A \geq 3$, then Time $\geq$ 130
($s4, s5, s6, s8, s9, s10, s11, s12$),

rule 8)   if Time $\leq$ 150 and $x_B \geq 2$, then Profit $\leq$ 148.38
($s4, s5, s6, s9, s12, s13$),

rule 9)   if $x_A \geq 3$ and $x_C \geq 4.08$, then Time $\geq$ 130
($s4, s5, s8, s10, , s11, s12$),

rule 10) if Sales $\geq$ 265.38 , then Time $\geq$ 130
($s2, s3, s4, s5, s6, s7, s8, s9, s10, s11$).

Then, the DM has been asked if (s)he was satisfied with one of the proposed Pareto optimal solutions. Since his/her answer was negative, (s)he was requested to indicate a subset of "good" solutions which are indicated in the "Evaluation" column of Table 5.5.

Table 5.5. A sample of Pareto optimal solutions proposed in the first iteration

Solution	Profit	Time	Prod. A	Prod. B	Prod. C	Sales	Evaluation
s1	165	120	0	0	10	250	*
s2	172.69	130	0.769	0	10	265.38	*
s3	180.38	140	1.538	0	10	280.77	good
s4	141.13	140	3	3	4.92	272.92	good
s5	148.38	150	5	2	4.75	278.75	good
s6	139.13	150	5	3	3.58	279.58	*
s7	188.08	150	2.308	0	10	296.15	*
s8	159	150	6	0	6	270	*
s9	140.5	150	6	2	3.67	271.67	good
s10	209.25	200	6	2	7.83	375.83	*
s11	189.38	200	5	5	5.42	385.42	*
s12	127.38	130	3	3	4.08	252.08	*
s13	113.63	120	3	3	3.25	231.25	*

Taking into account the sorting of Pareto optimal solutions into "good" and "others", made by the DM, twelve decision rules have been induced from the lower approximation of "good" solutions. The frequency of the presence of objectives in the premises of the rules gives a first idea of the importance of the considered objectives. These frequencies are the following:

- Profit: $\frac{4}{12}$,
- Time: $\frac{12}{12}$,
- Production of A: $\frac{7}{12}$,
- Production of B: $\frac{4}{12}$,
- Production of C: $\frac{5}{12}$,
- Sales: $\frac{5}{12}$.

The following potentially interesting decision rules were presented to the DM:

rule 1) if Profit ≥ 140.5 and Time ≤ 150 and $x_B \geq 2$,

> then product mix is good $(s4, s5, s9)$,

rule 2) if Time ≤ 140 and $x_A > 1.538$ and $x_C > 10$,

> then product mix is good $(s3)$,

rule 3) if Time ≤ 150 and $x_B \geq 2$ and $x_C \geq 4.75$,

> then product mix is good $(s4, s5)$,

rule 4) if Time ≤ 140 and Sales ≥ 272.9167,

> then product mix is good $(s3, s4)$,

rule 5) if Time ≤ 150 and $x_B \geq 2$ and $x_C \geq 3.67$ and Sales ≥ 271.67,

> then product mix is good $(s4, s5, s9)$.

Among these decision rules, the DM has selected rule 1) as the most adequate to his/her preferences. This rule permits to define the following constraints reducing the feasible region of the production mix problem:

- $20x_A + 30x_B + 25x_C - (x_A + 2x_B + 0.75x_C)6 - (0.5x_A + x_B + 0.5x_C)8 \geq 140.5$
  [Profit ≥ 140.5],
- $5x_A + 8x_B + 10x_C + 8x_A + 6x_B + 2x_C \leq 150$ [Time ≤ 150],
- $x_B \geq 2$ [Production of $B \geq 2$].

These constraints have been considered together with the original constraints for the production mix problem, and a new sample of representative Pareto optimal solutions shown in Table 5.6 have been calculated and presented to the DM, together with the following potentially interesting association rules:

rule 1') if Time ≤ 140, then Profit ≤ 174 and $x_C \leq 9.33$ and Sales ≤ 293.33
$(s5', s6', s7', s8', s9', s10', s11', s12')$,

rule 2') if $x_A \geq 2$, then $x_B \leq 3$ and Sales ≤ 300.83
$(s2', s3', s4', s6', s7', s9')$,

rule 3') if $x_A \geq 2$, then Profit ≤ 172 and $x_C \leq 8$
$(s2', s3', s4', s6', s7', s9')$,

rule 4') if Time $\leq$ 140, then $x_A \leq 2$ and $x_B \leq 3$
($s5', s6', s7', s8', s9', s10', s11', s12'$),

rule 5') if Profit $\geq$ 158.25, then $x_A \leq 2$
($s1', s3', s4', s5', s6', s8'$),

rule 6') if $x_A \geq 2$, then Time $\geq$ 130
($s2', s3', s4', s6', s7', s9'$),

rule 7') if $x_C \geq 7.17$, then $x_A \leq 2$ and $x_B \leq 2$
($s1', s3', s5', s6', s8', s10'$),

rule 8') if $x_C \geq 6$, then $x_A \leq 2$ and $x_B \leq 3$
($s1', s3', s4', s5', s6', s7', s8', s9', s10', s11', s12'$),

rule 9') if $x_C \geq 7$, then Time $\geq$ 125 and $x_B \leq 2$
($s1', s3', s5', s6', s8', s10', s11'$),

rule 10') if Sales $\geq$ 280, then Time $\geq$ 140 and $x_B \leq 3$
($s1', s2', s3', s4', s5', s7'$),

rule 11') if Sales $\geq$ 279.17, then Time $\geq$ 140
($s1', s2', s3', s4', s5', s6', s7'$),

rule 12') if Sales $\geq$ 272, then Time $\geq$ 130
($s1', s2', s3', s4', s5', s6', s7', s8'$).

The DM has been asked again if (s)he was satisfied with one of the proposed Pareto optimal solutions. Since his/her answer was negative, (s)he was requested again to indicate a subset of "good" solutions, which are indicated in the "Evaluation" column of Table 5.6.

Table 5.6. A sample of Pareto optimal solutions proposed in the second iteration

Solution	Profit	Time	Prod. A	Prod. B	Prod. C	Sales	Evaluation
s1'	186.53	150	0.154	2	10	313.08	*
s2'	154.88	150	3	3	5.75	293.75	*
s3'	172	150	2	2	8	300	good
s4'	162.75	150	2	3	6.83	300.83	good
s5'	174	140	0	2	9.33	293.33	*
s6'	158.25	140	2	2	7.17	279.17	good
s7'	149	140	2	3	6	280	*
s8'	160.25	130	0	2	8.5	272	good
s9'	144.5	130	2	2	6.33	258.33	*
s10'	153.38	125	0	2	8.08	262.08	*
s11'	145.5	125	1	2	7	255	good
s12'	141.56	125	1.5	2	6.46	251.46	good

Taking into account the sorting of Pareto optimal solutions into "good" and "others", made by the DM, eight decision rules have been induced from the lower approximation of "good" solutions. The frequencies of the presence of objectives in the premises of the rules are the following:

- Profit: $\frac{2}{8}$,

- Time: $\frac{1}{8}$,

- Production of A: $\frac{5}{8}$,

- Production of B: $\frac{3}{8}$,

- Production of C: $\frac{3}{8}$,

- Sales: $\frac{2}{8}$.

The following potentially interesting decision rules were presented to the DM:

rule 1) if Time $\leq$ 125 and $x_A \geq 1$, then product mix is good
($s11'$, $s12'$),

rule 2) if $x_A \geq 1$ and $x_C \geq 7$, then product mix is good
($s3'$, $s6'$, $s11'$),

rule 3) if $x_A \geq 1.5$ and $x_C \geq 6.46$, then product mix is good
($s3'$, $s4'$, $s6'$, $s12'$),

rule 4) if Profit $\geq$ 158.25 and $x_A \geq 2$, then product mix is good
($s3'$, $s4'$, $s6'$),

rule 5) if $x_A \geq 2$ and Sales $\geq$ 300, then product mix is good
($s3'$, $s4'$).

Among these decision rules, the DM has selected rule 4) as the most adequate to his/her preferences. This rule permits to define the following constraints reducing the Pareto optimal set of the production mix problem:

- $20x_A+30x_B+25x_C-(x_A+2x_B+0.75x_C)6-(0.5x_A+x_B+0.5x_C)8 \geq 158.25$
  [Profit $\geq$ 158.25],

- $x_A \geq 2$ [Production of $A \geq 2$].

Let us observe that the first constraint is just strengthening an analogous constraint introduced in the first iteration (Profit $\geq$ 140.5).

Considering the new set of constraints, a new sample of representative Pareto optimal solutions shown in Table 5.7 has been calculated and presented to the DM, together with the following potentially interesting association rules:

rule 1″)   if Time $\leq$ 145, then $x_A \leq 2$ and $x_B \leq 2.74$ and Sales $\leq$ 290.2
($s2''$, $s3''$, $s4''$),

rule 2″)   if $x_C \geq 6.92$, then $x_A \leq 3$ and $x_B \leq 2$ and Sales $\leq$ 292.92
($s3''$, $s4''$, $s5''$),

rule 3″)   if Time $\leq$ 145, then Profit $\leq$ 165.13 and $x_A \leq 2$ and $x_C \leq 7.58$
($s2″, s3″, s4″$),

rule 4″)   if $x_C \geq 6.72$, then $x_B \leq 2.74$
($s2″, s3″, s4″, s5″$),

rule 5″)   if Sales $\geq$ 289.58, then Profit $\leq$ 165.13 and Time $\geq$ 145 and
$x_C \leq 7.58$ ($s1″, s2″, s3″, s5″$).

The DM has been asked again if (s)he was satisfied with one of the presented
Pareto optimal solutions shown in Table 5.7 and this time (s)he declared that
solution $s3″$ is satisfactory for him/her. This ends the interactive procedure.

Table 5.7. A sample of Pareto optimal solutions proposed in the third iteration

Solution	Profit	Time	Prod. A	Prod. B	Prod. C	Sales	Evaluation
$s1″$	158.25	150	2	3.49	6.27	301.24	*
$s2″$	158.25	145	2	2.74	6.72	290.20	*
$s3″$	165.13	145	2	2	7.58	289.58	selected
$s4″$	158.25	140	2	2	7.17	279.17	*
$s5″$	164.13	150	3	2	6.92	292.93	*
$s6″$	158.25	145.3	3	2	6.56	284.02	*

5.7 Characteristics of the IMO-DRSA

The interactive procedure presented in Section 5.5 can be analyzed from the
point of view of input and output information. As to the input, the DM gives
preference information by answering easy questions related to sorting of some
representative solutions into two classes ("good" and "others"). Very often, in
multiple criteria decision analysis, in general, and in interactive multiobjective
optimization, in particular, the preference information has to be given in terms
of preference model parameters, such as importance weights, substitution rates
and various thresholds (see (Fishburn, 1967) for the Multiple Attribute Utility Theory and (Roy and Bouyssou, 1993; Figueira *et al.*, 2005a; Brans and
Mareschal, 2005; Martel and Matarazzo, 2005) for outranking methods; for
some well known interactive multiobjective optimization methods requiring
preference model parameters, see the Geoffrion-Dyer-Feinberg method (Geoffrion *et al.*, 1972), the method of Zionts and Wallenius (1976, 1983) and
the Interactive Surrogate Worth Tradeoff method (Chankong and Haimes,
1978, 1983) requiring information in terms of marginal rates of substitution,
the reference point method (Wierzbicki, 1980, 1982, 1986) requiring a reference point and weights to formulate an achievement scalarizing function, the

Light Beam Search method (Jaszkiewicz and Słowiński, 1999) requiring information in terms of weights and indifference, preference and veto thresholds, being typical parameters of ELECTRE methods). Eliciting such information requires a significant cognitive effort on the part of the DM. It is generally acknowledged that people often prefer to make exemplary decisions and cannot always explain them in terms of specific parameters. For this reason, the idea of inferring preference models from exemplary decisions provided by the DM is very attractive. The output result of the analysis is the model of preferences in terms of "*if...*, *then...*" decision rules which is used to reduce the Pareto optimal set iteratively, until the DM selects a satisfactory solution. The decision rule preference model is very convenient for decision support, because it gives argumentation for preferences in a logical form, which is intelligible for the DM, and identifies the Pareto optimal solutions supporting each particular decision rule. This is very useful for a critical revision of the original sorting of representative solutions into the two classes of "good" and "others". Indeed, decision rule preference model speaks the same language of the DM without any recourse to technical terms, like utility, tradeoffs, scalarizing functions and so on.

All this implies that IMO-DRSA has a transparent feedback organized in a learning oriented perspective, which permits to consider this procedure as a "glass box", contrary to the "black box" characteristic of many procedures giving final result without any clear explanation. Note that with the proposed procedure, the DM learns about the shape of the Pareto optimal set using the association rules. They represent relationships between attainable values of objective functions on the Pareto optimal set in logical and very natural statements. The information given by association rules is as intelligible as the decision rule preference model, since they speak the language of the DM and permit him/her to identify the Pareto optimal solutions supporting each particular association rule.

Thus, decision rules and association rules give an explanation and a justification of the final decision, that does not result from a mechanical application of a certain technical method, but rather from a mature conclusion of a decision process based on active intervention of the DM.

Observe, finally, that the decision rules representing preferences and the association rules describing the Pareto optimal set are based on ordinal properties of objective functions only. Differently from methods involving some scalarization (almost all existing interactive methods), in any step the proposed procedure does not aggregate the objectives into a single value, avoiding operations (such as averaging, weighted sum, different types of distance, achievement scalarization) which are always arbitrary to some extent. Remark that one could use a method based on a scalarization to generate the representative set of Pareto optimal solutions; nevertheless, the decision rule approach would continue to be based on ordinal properties of objective functions only, because the dialogue stage of the method operates on ordinal comparisons only. In the proposed method, the DM gets clear arguments for his/her decision in terms of "*if ..., then...*" decision rules and the verification if a proposed solution satisfies these decision rules is particularly easy. This is not the case of interactive multiobjective optimization methods based on scalarization. For example, in the methods using an achievement scalarization function, it is not evident what does it mean for a solution to be "close" to the reference point. How to justify the choice of the weights used in the achievement function? What is their interpretation? Observe, instead, that the method proposed in this chapter operates on data using ordinal comparisons which would not be affected by any increasing monotonic transformation of scales, and this ensures the meaningfulness of results from the point of view of measurement theory (see, e.g., Roberts, 1979).

With respect to computational aspects of the method, notice that the decision rules can be calculated efficiently in few seconds only using the algorithms presented in (Greco *et al.*, 2001a, 2002a). When the number of objective functions is not too large to be effectively controlled by the DM (let us say seven plus or minus two, as suggested by Miller (1956)), then the decision rules can be calculated in a fraction of one second. In any case, the computational effort grows exponentially with the number of objective functions, but not with respect to the number of considered Pareto optimal solutions, which can increase with no particularly negative consequence on calculation time.

5.8 Conclusions

We presented a new interactive multiobjective optimization method using a decision rule preference model to support interaction with the DM. It is using the Dominance-based Rough Set Approach, being a well grounded methodology of reasoning about preference ordered data. Due to the transparency and intelligibility of the transformation of the input information into the preference model, the proposed method can be qualified as a "glass box", contrary to the "black box", characteristic for many methods giving a final result without any clear argumentation. Moreover, the method is purely ordinal, because it does not make any invasive operations on data, such as averaging or calculation of a weighted sum, of a distance or of an achievement scalarization.

We believe that the good properties of the proposed method will encourage further developments and specializations to different decision problems involving multiple objectives, such as portfolio selection, inventory management, scheduling and so on.

Acknowledgements

The third author wishes to acknowledge financial support from the Polish Ministry of Science and Higher Education.

References

Brans, J.P., Mareschal, B.: PROMETHEE methods. In: Figueira, J., Greco, S., Ehrgott, M. (eds.) Multiple Criteria Decision Analysis: State of the Art Surveys, pp. 163–195. Springer, Berlin (2005)

Chankong, V., Haimes, Y.Y.: The interactive surrogate worth trade-off (iswt) method for multiobjective decision-making. In: Zionts, S. (ed.) Multiple Criteria Problem Solving, pp. 42–67. Springer, Berlin (1978)

Chankong, V., Haimes, Y.Y.: Multiobjective Decision Making Theory and Methodology. Elsiever Science Publishing Co., New York (1983)

Dyer, J.S.: MAUT-multiattribute utility theory. In: Figueira, J., Greco, S., Ehrgott, M. (eds.) Multiple Criteria Decision Analysis: State of the Art Surveys, pp. 265–295. Springer, Berlin (2005)

Figueira, J., Mousseau, V., Roy, B.: ELECTRE methods. In: Figueira, J., Greco, S., Ehrgott, M. (eds.) Multiple Criteria Decision Analysis: State of the Art Surveys, pp. 265–295. Springer, Berlin (2005a)

Figueira, J., Greco, S., Ehrgott, M. (eds.): Multiple Criteria Decision Analysis: State of the Art Surveys. Springer, Berlin (2005b)

Fishburn, P.C.: Methods of estimating additive utilities. Management Science 13(7), 435–453 (1967)

Geoffrion, A., Dyer, J., Feinberg, A.: An interactive approach for multi-criterion optimization, with an application to the operation of an academic department. Management Science 19(4), 357–368 (1972)

Giove, S., Greco, S., Matarazzo, B., Słowiński, R.: Variable consistency monotonic decision trees. In: Alpigini, J.J., Peters, J.F., Skowron, A., Zhong, N. (eds.) RSCTC 2002. LNCS (LNAI), vol. 2475, pp. 247–254. Springer, Heidelberg (2002)

Greco, S., Matarazzo, B., Słowiński, R.: Rough approximation of a preference relation by dominance relations. European J. Operational Research 117, 63–83 (1999a)

Greco, S., Matarazzo, B., Słowiński, R.: The use of rough sets and fuzzy sets in MCDM. In: Gal, T., Stewart, T., Hanne, T. (eds.) Advances in Multiple Criteria Decision Making, pp. 14.1–14.59, Kluwer, Boston (1999b)

Greco, S., Matarazzo, B., Słowiński, R., Stefanowski, J.: An algorithm for induction of decision rules consistent with the dominance principle. In: Ziarko, W., Yao, Y. (eds.) RSCTC 2000. LNCS (LNAI), vol. 2005, pp. 304–313. Springer, Heidelberg (2001a)

Greco, S., Matarazzo, B., Słowiński, R.: Rough sets theory for multicriteria decision analysis. European J. of Operational Research 129, 1–47 (2001b)

Greco, S., Słowiński, R., Stefanowski, J.: Mining association rules in preference-ordered data. In: Hacid, M.-S., Raś, Z.W., Zighed, A.D.A., Kodratoff, Y. (eds.) ISMIS 2002. LNCS (LNAI), vol. 2366, pp. 442–450. Springer, Heidelberg (2002a)

Greco, S., Matarazzo, B., Słowiński, R.: Preference representation by means of conjoint measurement & decision rule model. In: Bouyssou, D., et al. (eds.) Aiding Decisions with Multiple Criteria–Essays in Honor of Bernard Roy, pp. 263–313. Kluwer Academic Publishers, Dordrecht (2002b)

Greco, S., Matarazzo, B., Słowiński, R.: Axiomatic characterization of a general utility function and its particular cases in terms of conjoint measurement and rough-set decision rules. European J. of Operational Research 158, 271–292 (2004a)

Greco, S., Pawlak, Z., Słowiński, R.: Can Bayesian confirmation measures be useful for rough set decision rules? Engineering Applications of Artificial Intelligence 17, 345–361 (2004b)

Greco, S., Matarazzo, B., Słowiński, R.: Dominance-based rough set approach to knowledge discovery – (I) general perspective, (II) extensions and applications. In: Zhong, N., Liu, J. (eds.) Intelligent Technologies for Information Analysis, pp. 513–612. Springer, Berlin (2004c)

Greco, S., Matarazzo, B., Słowiński, R.: Decision rule approach. In: Figueira, J., Greco, S., Ehrgott, M. (eds.) Multiple Criteria Decision Analysis: State of the Art Surveys, pp. 507–563. Springer, Berlin (2005a)

Greco, S., Matarazzo, B., Słowiński, R.: Generalizing rough set theory through dominance-based rough set approach. In: Ślęzak, D., Yao, J., Peters, J.F., Ziarko, W., Hu, X. (eds.) RSFDGrC 2005. LNCS (LNAI), vol. 3642, pp. 1–11. Springer, Heidelberg (2005b)

Greco, S., Mousseau, V., Słowiński, R.: Ordinal regression revisted: Multiple criteria ranking with a set of additive value functions. European Journal of Operational Research 191, 415–435 (2008)

Jacquet-Lagrèze, E., Siskos, Y.: Assessing a set of additive utility functions for multicriteria decision making: the UTA method. European J. of Operational Research 10, 151–164 (1982)

Jacquet-Lagrèze, E., Meziani, R., Słowiński, R.: MOLP with an interactive assessment of a piecewise linear utility function. European J. of Operational Research 31, 350–357 (1987)

Jaszkiewicz, A., Ferhat, A.B.: Solving multiple criteria choice problems by interactive trichotomy segementation. European Journal of Operational Research 113, 271–280 (1999)

Jaszkiewicz, A., Słowiński, R.: The "Light Beam Search" approach - an overview of methodology and applications. European Journal of Operational Research 113, 300–314 (1999)

Keeney, R.L., Raiffa, H.: Decision with Multiple Objectives: Preference and Value Tradeoffs. John Wiley and Sons, New York (1976)

March, J.G.: Bounded rationality, ambiguity and the engineering of choice. Bell Journal of Economics 9, 587–608 (1978)

Martel, J.M., Matarazzo, B.: Other outranking approaches. In: Figueira, J., Greco, S., Ehrgott, M. (eds.) Multiple Criteria Decision Analysis: State of the Art Surveys, pp. 197–262. Springer, Berlin (2005)

Michalski, R.S., Bratko, I., Kubat, M. (eds.): Machine learning and datamining – Methods and applications. Wiley, New York (1998)

Miller, G.A.: The magical number seven, plus or minus two: some limits in our capacity for processing information. The Psychological Review 63, 81–97 (1956)

Mousseau, V., Słowiński, R.: Inferring an ELECTRE TRI model from assignment examples. Journal of Global Optimization 12, 157–174 (1998)

Pawlak, Z.: Rough sets. International Journal of Computer and Information Sciences 11, 341–356 (1982)

Pawlak, Z.: Rough Sets. Kluwer, Dordrecht (1991)

Pawlak, Z., Słowiński, R.: Rough set approach to multi-attribute decision analysis. European J. of Operational Research 72, 443–459 (1994)

Roberts, F.: Measurement Theory, with Applications to Decision Making, Utility and the Social Sciences. Addison-Wesley, Boston (1979)

Roy, B., Bouyssou, D.: Aide Multicritère à la Décision: Méthodes et Cas. Economica, Paris (1993)

Słowiński, R.: Rough set learning of preferential attitude in multi-criteria decision making. In: Komorowski, J., Raś, Z.W. (eds.) ISMIS 1993. LNCS, vol. 689, pp. 642–651. Springer, Heidelberg (1993)

Słowiński, R., Greco, S., Matarazzo, B.: Axiomatization of utility, outranking and decision-rule preference models for multiple-criteria classification problems under partial inconsistency with the dominance principle. Control and Cybernetics 31, 1005–1035 (2002a)

Słowiński, R., Greco, S., Matarazzo, B.: Rough set analysis of preference-ordered data. In: Alpigini, J.J., Peters, J.F., Skowron, A., Zhong, N. (eds.) RSCTC 2002. LNCS (LNAI), vol. 2475, pp. 44–59. Springer, Heidelberg (2002b)

Słowiński, R., Greco, S., Matarazzo, B.: Rough set based decision support. In: Burke, E.K., Kendall, G. (eds.) Search Methodologies: Introductory Tutorials in Optimization and Decision Support Techniques, pp. 475–527. Springer, New York (2005)

Stefanowski, J.: On rough set based approaches to induction of decision rules. In: Polkowski, L., Skowron, A. (eds.) Rough Sets in Data Mining and Knowledge Discovery, vol. 1, pp. 500–529. Physica, Heidelberg (1998)

Steuer, R.E., Choo, E.-U.: An interactive weighted tchebycheff procedure for multiple objective programming. Mathematical Programming 26, 326–344 (1983)

Wierzbicki, A.P.: The use of reference objectives in multiobjective optimization. In: Fandel, G., Gal, T. (eds.) Multiple Criteria Decision Making, Theory and Applications. LNEMS, vol. 177, pp. 468–486. Springer, Berlin (1980)

Wierzbicki, A.P.: A mathematical basis for satisficing decision making. Mathematical Modelling 3, 391–405 (1982)

Wierzbicki, A.P.: On the completeness and constructiveness of parametric characterizations to vector optimization problems. OR Spektrum 8, 73–87 (1986)

Zionts, S., Wallenius, J.: An interactive programming method for solving the multiple criteria problem. Management Science 22, 652–663 (1976)

Zionts, S., Wallenius, J.: An interactive multiple objective linear programming method for a class of underlying nonlinear utility functions. Management Science 29, 519–523 (1983)

6

Consideration of Partial User Preferences in Evolutionary Multiobjective Optimization

Jürgen Branke

Institute AIFB, University of Karlsruhe, 76128 Karlsruhe, Germany
branke@aifb.uni-karlsruhe.de

Abstract. Evolutionary multiobjective optimization usually attempts to find a good approximation to the complete Pareto optimal front. However, often the user has at least a vague idea about what kind of solutions might be preferred. If such information is available, it can be used to focus the search, yielding a more fine-grained approximation of the most relevant (from a user's perspective) areas of the Pareto optimal front and/or reducing computation time. This chapter surveys the literature on incorporating partial user preference information in evolutionary multiobjective optimization.

6.1 Introduction

Most research in evolutionary multiobjective optimization (EMO) attempts to approximate the complete Pareto optimal front by a set of well-distributed representatives of Pareto optimal solutions. The underlying reasoning is that in the absence of any preference information, all Pareto optimal solutions have to be considered equivalent.

On the other hand, in most practical applications, the decision maker (DM) is eventually interested in only a single solution. In order to come up with a single solution, at some point during the optimization process, the DM has to reveal his/her preferences to choose between mutually non-dominating solutions. Following a classification by Horn (1997) and Veldhuizen and Lamont (2000), the articulation of preferences may be done either before (a priori), during (progressive), or after (a posteriori) the optimization process, see also Figure 6.1.

A priori approaches aggregate different objectives into a single auxilliary objective in one way or another, which allows to use standard optimization techniques (including single-objective evolutionary algorithms) and usually

Reviewed by: Carlos Coello Coello, CINEVESTAV-IPN, Mexico
Salvatore Greco, University of Catania, Italy
Kalyanmoy Deb, Indian Institute of Technology Kanpur, India

J. Branke et al. (Eds.): Multiobjective Optimization, LNCS 5252, pp. 157–178, 2008.
© Springer-Verlag Berlin Heidelberg 2008

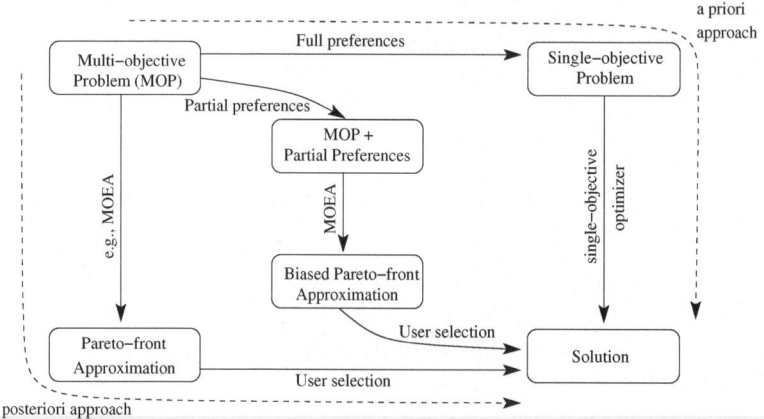

Fig. 6.1. Different ways to solve multiobjective problems.

results in a single solution. Many classical MCDM methodologies fall into this category. The most often used aggregation method is probably just a linear combination of the different objectives. Alternatives would be a lexicographic ordering of the objectives, or to use the distance from a specified target as objective. For an example of an approach based on fuzzy rules see Sait *et al.* (1999) or Sakawa and Yauchi (1999). As aggregation of objectives turn the multiobjective problem into a single objective problem, such evolutionary algorithms are actually out of scope of this chapter. A discussion of advantages and disadvantages of such aggregations can be found in Coello *et al.* (2002), Chapter 2.2. In any case, the aggregation of objectives into a single objective is usually not practical, because it basically requires to specify a ranking of alternatives before these alternatives are known. Classical MCDM techniques usually solve this predicament by repeatedly adjusting the auxilliary objective and re-solving the single objective problem until the DM is satisfied with the solution.

Most multiobjective evolutionary algorithms (MOEAs) can be classified as a posteriori. First, the EA generates a (potentially large) set of non-dominated solutions, then the DM can examine the possible trade-offs and choose according to his/her preferences. For an introduction to MOEAs, see Chapter 3. The most prominent MOEAs are the Non-Dominated Sorting Genetic Algorithm (NSGA-II, Deb *et al.*, 2002a) and the Strength-Pareto Evolutionary Algorithm (SPEA-II, Zitzler *et al.*, 2002).

Interactive approaches interleave the optimization with a progressive elicitation of user preferences. These approaches are discussed in detail in Chapter 7.

In the following, we consider an intermediate approach (middle path in Figure 6.1). Although we agree that it may be impractical for a DM to completely specify his or her preferences before any alternatives are known, we assume that the DM has at least a rough idea about what solutions might be preferred, and can specify partial preferences. The methods discussed here aim at integrating such imprecise knowledge into the EMO approach, biasing the search towards solutions that are considered as relevant by the DM. The goal is no longer to generate a good approximation to all Pareto optimal solutions, but a small set of solution that contains the DM's preferred solution with the highest probability. This may yield three important advantages:

1. **Focus:** Partial user preferences may be used to focus the search and generate a subset of all Pareto optimal alternatives that is particularly interesting to the DM. This avoids overwhelming the DM with a huge set of (mostly irrelevant) alternatives.

2. **Speed:** By focusing the search onto the relevant part of the search space, one may expect the optimization algorithm to find these solutions more quickly, not wasting computational effort to identify Pareto optimal but irrelevant solutions.

3. **Gradient:** MOEAs require some quality measure for solutions in order to identify the most promising search direction (gradient). The most important quality measure used in MOEA is Pareto dominance. However, with an increasing number of objectives, more and more solutions become incomparable, rendering Pareto dominance as fitness criterion less useful,resulting in a severe performance loss of MOEAs (e.g., Deb *et al.*, 2002b). Incorporating (partial) user preferences introduces additional preference relations, restoring the necessary fitness gradient information to some extend and ensuring MOEA's progress.

To reach these goals, the MOEA community can accomodate or be inspired by many of the classical MCDM methodologies covered in Chapters 1 and 2, as those generally integrate preference information into the optimization process. Thus, combining MOEAs, and their ability to generate multiple alternatives simultaneously in one run, and classical MCDM methodologies, and their ways to incorporate user preferences, holds great promise.

The literature contains quite a few techniques to incorporate full or partial preference information into MOEAs, and previous surveys on this topic include Coello (2000); Rachmawati and Srinivasan (2006), and Coello *et al.* (2002). In the following, we classify the different approaches based on the type of partial preference information they ask from the DM, namely objective scaling (Section 6.2), constraints (Section 6.3), a goal or reference point (Section 6.4), trade-off information (Section 6.5), or weighted performance measures (Section 6.6 on approaches based on marginal contribution). Some additional approaches are summarized in Section 6.7. The chapter concludes with a summary in Section 6.8.

6.2 Scaling

One of the often claimed advantages of MOEAs is that they do not require an a priori specification of user preferences because they generate a good approximation of the whole Pareto front, allowing the DM to pick his/her preferred solution afterwards. However, the whole Pareto optimal front may contain very many alternatives, in which case MOEAs can only hope to find a representative subset of all Pareto optimal solutions. Therefore, all basic EMO approaches attempt to generate a uniform distribution of representatives along the Pareto front. For this goal, they rely on distance information in the objective space, be it in the crowding distance of NSGA-II or in the clustering of SPEA-II[1]. Thus, what is considered uniform depends on the scaling of the objectives. This is illustrated in Figure 6.2. The left panel (a) shows an evenly distributed set of solutions along the Pareto front. Scaling the second objective by a factor of 100 (e.g., using centimeters instead of meters as unit), leads to a bias of the distribution and more solutions along the front parallel to the axis of the second objective (right panel). Note that depending on the shape of the front, this means that there is a bias towards objective 1 (as in the convex front in Figure 6.2), or objective 2 (if the front is concave). So, the user-defined scaling is actually a usually ignored form of user preference specification necessary also for MOEAs.

Many current implementations of MOEAs (e.g., NSGA-II and SPEA) scale objectives based on the solutions currently in the population (see, e.g., Deb (2001), S. 248). While this results in nice visualizations if the front is plotted with a 1:1 ratio, and relieves the DM from specifying a scaling, it assumes that ranges of values covered by the Pareto front in each objective are equally

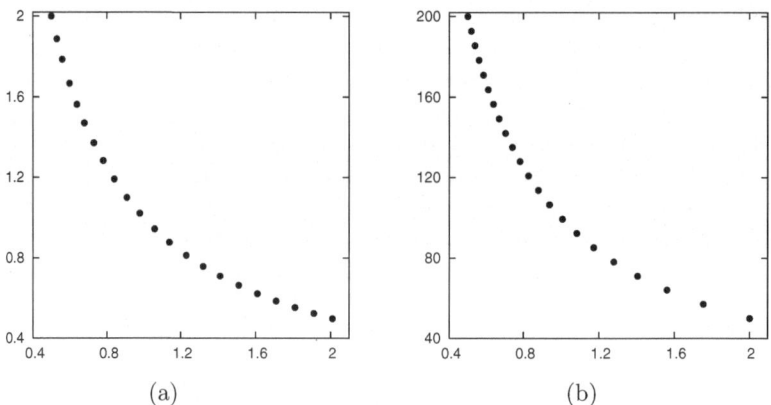

Fig. 6.2. Influence of scaling on the distribution of solutions along the Pareto front.

---

[1] The rest of this section assumes familiarity with the crowding distance concept. Readers unfamiliar with this concept are referred to, e.g., Deb (2001) or Chapter 3.

important. Whether this assumption is justified certainly depends strongly on the application and the DM's preferences.

In order to find a biased distribution anywhere on the Pareto optimal front, a previous study by Deb (2003) used a biased sharing[2] mechanism implemented on NSGA. In brief, the objectives are scaled according to preferences when calculating the distances. This allows to make distances in one objective appear larger than they are, with a corresponding change in the resulting distribution of individuals. Although this allows to focus on one objective or another, the approach does not allow to focus on a compromise region (for equal weighting of the objectives, the algorithm would produce no bias at all).

In Branke and Deb (2005), the biased sharing mechanism has been extended with a better control of the region of interest and a separate parameter controlling the strength of the bias. For a solution i on a particular front, the *biased crowding distance measure* D_i is re-defined as follows. Let η be a user-specified direction vector indicating the most probable, or central linearly weighted utility function, and let α be a parameter controlling the bias intensity. Then,

$$D_i = d_i \left(\frac{d_i'}{d_i}\right)^\alpha, \tag{6.1}$$

where d_i and d_i' are the original crowding distance and the crowding distance calculated based on the locations of the individuals projected onto the (hyper)plane with direction vector η. Figure 6.3 illustrates the concept.

As a result, for a solution in a region of the Pareto optimal front more or less parallel to the projected plane (such as solution 'a'), the original crowded distance d_a and projected crowding distance d_a' are more or less the same, thereby making the ratio d_a'/d_a close to one. On the other hand, for a solution in an area of the Pareto optimal front where the tangent has an orientation significantly different from the chosen plane (such as solution 'b'), the projected crowding distance d_b' is much smaller than the original crowding distance d_b. For such a solution, the biased crowding distance value D_i will be a small quantity, meaning that such a solution is assumed to be artificially crowded by neighboring solutions. A preference of solutions having a larger biased crowding distance D_i will then enable solutions closer to the tangent point to be found. The exponent α controls the extent of the bias, with larger α resulting in a stronger bias.

Note that biased crowding will focus on the area of the Pareto optimal front which is parallel to the iso-utility function defined by the provided direction vector η. For a *convex* Pareto optimal front, that is just the area around the optimal solution regarding a corresponding aggregate cost function. For a concave region, such an aggregate cost function would always prefer one of the edge points, while biased crowding may focus on the area in between.

---

[2] The sharing function in NSGA fulfills the same functionality as the crowding distance in NSGA-II, namely to ensure a diversity of solutions.

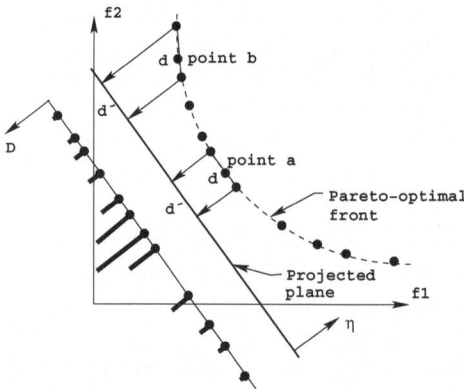

Fig. 6.3. The biased crowding approach is illustrated on a two-objective minimization problem (Branke and Deb, 2005).

Trautmann and Mehnen (2005) suggest an explicit incorporation of preferences into the scaling. They propose to map the objectives into the range $[0, 1]$ according to desirability functions. With one-sided sigmoid (monotone) desirability functions, the non-dominance relations are not changed. Therefore, the solutions found are always also non-dominated in the original objective space. What changes is the distribution along the front. Solutions that are in flat parts of the desirability function receive very similar desirability values and as MOEAs then attempt to spread solutions evenly in the desirability space, this will result in a more spread out distribution in the original objective space. However, in order to specify the desirability functions in a sensible manner, it is necessary to at least know the ranges of the Pareto front.

6.3 Constraints

Often, the DM can formulate preferences in the form of constraints, for example "Criterion 1 should be less than β". Handling constraints is a well-researched topic in evolutionary algorithms in general, and most of the techniques carry over to EMO in a straightforward manner. One of the simplest and most common techniques is probably to rank infeasible solutions according to their degree of infeasibility, and inferior to all feasible solutions (Deb, 2000; Jiménez and Verdegay, 1999). A detailed discussion of constraint handling techniques is out of the scope of this chapter. Instead, the interested reader is referred to Coello (2002) for a general survey on constraint handling techniques, and Deb (2001), Chapter 7, for a survey with focus on EMO techniques.

6.4 Providing a Reference Point

Perhaps the most important way to provide preference information is a reference point, a technique that has a long tradition in multicriteria decision making, see, e.g., Wierzbicki (1977, 1986) and also Chapter 2. A reference point consists of aspiration levels reflecting desirable values for the objective function, i.e., a target the user is hoping for. Such an information can then be used in different ways to focus the search. However, it should not lead to a dominated solution being preferred over the dominating solution.

The use of a reference point to guide the EMO algorithm has first been proposed in Fonseca and Fleming (1993). The basic idea there is to give a higher priority to objectives in which the goal is not fulfilled. Thus, when deciding whether a solution $\mathbf{x}$ is preferable to a solution $\mathbf{y}$ or not, first, only the objectives in which solution $\mathbf{x}$ does not satisfy the goal are considered, and $\mathbf{x}$ is preferred to $\mathbf{y}$ if it dominates $\mathbf{y}$ on these objectives. If $\mathbf{x}$ is equal to $\mathbf{y}$ in all these objectives, or if $\mathbf{x}$ satisfies the goal in all objectives, $\mathbf{x}$ is preferred over $\mathbf{y}$ either if $\mathbf{y}$ does not fulfill some of the objectives fulfilled by $\mathbf{x}$, or if $\mathbf{x}$ dominates $\mathbf{y}$ on the objectives fulfilled by $\mathbf{x}$. More formally, this can be stated as follows. Let $\mathbf{r}$ denote the reference point, and let there be m objectives without loss of generality sorted such that $\mathbf{x}$ fulfills objectives $k+1 \ldots m$ but not objectives $1 \ldots k$, i.e.

$$f_i(\mathbf{x}) > r_i \quad \forall i = 1 \ldots k \tag{6.2}$$

$$f_i(\mathbf{x}) \leq r_i \quad \forall i = k+1 \ldots m. \tag{6.3}$$

Then, $\mathbf{x}$ is preferred to $\mathbf{y}$ if and only if

$$\mathbf{x} \succ_{1 \ldots k} \mathbf{y} \vee$$
$$\mathbf{x} =_{1 \ldots k} \mathbf{y} \wedge [(\exists l \in [k+1 \ldots n] : f_l(\mathbf{y}) > r_k) \vee (\mathbf{x} \succ_{k+1 \ldots n} \mathbf{y})] \tag{6.4}$$

with $\mathbf{x} \succ_{i \ldots j} \mathbf{y}$ meaning that solution $\mathbf{x}$ dominates solution $\mathbf{y}$ on objectives i to j (i.e., for minimization problems as considered here, $f_k(\mathbf{x} \leq f_k(\mathbf{y} \forall k = i \ldots j$ with at least one strict inequality). A slightly extended version that allows the decision maker to additionally assign priorities to objectives has been published in Fonseca and Fleming (1998). This publication also contains the proof that the proposed preference relation is transitive. Figure 6.4 visualizes what part of the Pareto front remains preferred depending on whether the reference point is reachable (a) or not (b). If the goal has been set so ambitious that there is no solution which can reach the goal in even a single objective, the goal has no effect on search, and simply the whole Pareto front is returned.

In Deb (1999), a simpler variant has been proposed which simply ignores improvements over a goal value by replacing a solution's objective value $f_i(x)$ by $\max\{f_i(x), r_i\}$. If the goal vector $\mathbf{r}$ is outside the feasible range, the method is almost identical to the definition in Fonseca and Fleming (1993). However, if the goal can be reached, the approach from Deb (1999) will lose its selection

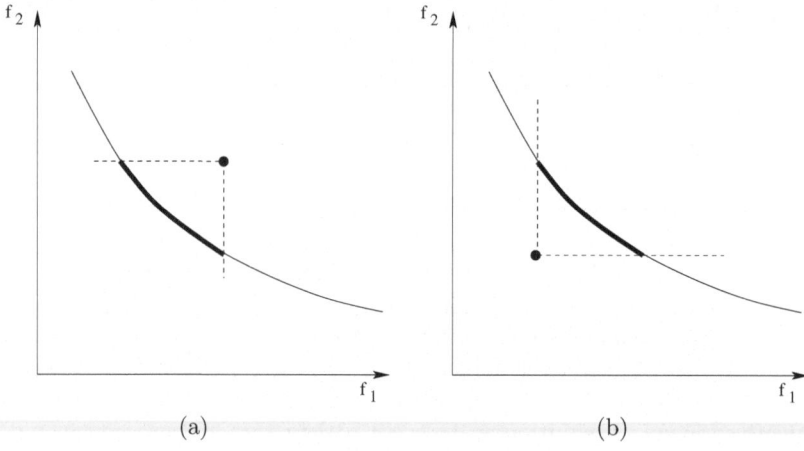

(a)   (b)

Fig. 6.4. Part of the Pareto optimal front that remains optimal with a given reference point **r** and the preference relation from Fonseca and Fleming (1993). The left panel (a) shows a reachable reference point, while the right panel (b) shows an unreachable one. Minimization of objectives is assumed.

pressure and basically stop search as soon as the reference point has been found, i.e., return a solution which is not Pareto optimal. On the other hand, the approach from Fonseca and Fleming (1993) keeps improving beyond the reference point. The goal-programming idea has been extended in Deb (2001) to allow for reference regions in addition to reference points.

Tan *et al.* (1999) proposed another ranking scheme which in a first stage prefers individuals fulfilling all criteria, and ranks those individuals according to standard non-dominance sorting. Among the remaining solutions, solution **x** dominates solution **y** if and only if **x** dominates **y** with respect to the objectives in which **x** does not fulfill the goal (as in Fonseca and Fleming (1993)), or if $|\mathbf{x} - \mathbf{r}| \succ |\mathbf{y} - \mathbf{r}|$. The latter corresponds to a "mirroring" of the objective vector along the axis of the fulfilled criteria. This may lead to some strange effects, such as non-transitivity of the preference relation (x is preferred to y, and y to z, but x and z are considered equal). Also, it seems odd to "penalize" solutions for largely exceeding a goal. What is more interesting in Tan *et al.* (1999) is the suggestion on how to account for multiple reference points, connected with AND and OR operations. The idea here is to rank the solutions independently with respect to all reference points. Then, rankings are combined as follows. If two reference points are connected by an AND operator, the rank of the solution is the maximum of the ranks according to the individual reference points. If the operator is an OR, the rank of the solution is the minimum of the ranks according to the individual reference points. This idea of combining the information of several reference points can naturally be combined with other preference relations using a reference point. The paper also presents a way to prioritize objectives by introducing additional goals. In effect, however, the priorization is equivalent to the one proposed in Fonseca and Fleming (1998).

In Deb and Sundar (2006); Deb *et al.* (2006), the crowding distance calculation in NSGA-II is replaced by the distance to the reference point, where solutions with a smaller distance are preferred. More specifically, solutions with the same non-dominated rank are sorted with respect to their distance to the reference point. Furthermore, to control the extent of obtained solutions, all solutions having a distance of ϵ or less between them are grouped. Only one randomly picked solution from each group is retained, while all other group members are assigned a large rank to discourage their use. As Fonseca and Fleming (1998) and Tan *et al.* (1999), this approach is able to improve beyond a reference point within the feasible region, because the non-dominated sorting keeps driving the population to the Pareto optimal front. Also, as Tan *et al.* (1999), it can handle multiple reference points simultaneously. With the parameter ϵ, it is possible to explicitly influence the diversity of solutions returned. Whether this extra parameter is an advantage or a burden may depend on the application.

Yet another dominance scheme was recently proposed in Molina *et al.* (2009), where solutions fulfilling all goals and solutions fulfilling none of the goals are preferred over solutions fulfilling only some of the goals. This, again, drives the search beyond the reference point if it is feasible, but it can obviously lead to situations where a solution which is dominated (fulfilling none of the goals) is actually preferred over the solution that dominates it (fulfilling some of the goals).

Thiele *et al.* (2007) integrate reference point information into the Indicator-Based Evolutionary Algorithm, see Section 6.6 for details.

The classical MCDM literature also includes some approaches where, in addition to a reference point, some further indicators are used to generate a set of alternative solutions. These include the reference direction method (Korhonen and Laakso, 1986) and light beam search (Jaszkiewicz and Slowinski, 1999). Recently, these methods have also been adopted into MOEAs.

In brief, the reference direction method allows the user to specify a starting point and a reference point, with the difference of the two defining the reference direction. Then, several points on this vector are used to define a set of achievement scalarizing functions, and each of these is used to search for a point on the Pareto optimal frontier. In Deb and Kumar (2007a), an MOEA is used to search for all these points simultaneously. For this purpose, the NSGA-II ranking mechanism has been modified to focus the search accordingly.

The light beam search also uses a reference direction, and additionally asks the user for some thresholds which are then used so find some possibly interesting neighboring solutions around the (according to the reference direction) most preferred solution. Deb and Kumar (2007b) use an MOEA to simultaneously search for a number of solutions in the neighborhood of the solution defined by the reference direction. This is achieved by first identifying the "most preferred" or "middle" solution using an achievement scalarizing function based on the reference point. Then, a modified crowding distance calculation is used to focus the search on those solutions which are not worse by more than the allowed threshold in all the objectives.

Summarizing, the first approach proposed in Fonseca and Fleming (1993) still seems to be a good way to include reference point information. While in most approaches the part of the Pareto optimal front considered as relevant depends on the reference point and the shape and location of the Pareto optimal front, in Deb and Sundar (2006) the desired spread of solutions in the vicinity of the Pareto optimal solution closest to the reference point is specified explicitly. The schemes proposed by Tan et al. (1999) and Deb and Sundar (2006) allow to consider several reference points simultaneously. The MOEAs based on the reference direction and light beam search (Deb and Kumar, 2007a,b) allow the user to specify additional information that influences the focus of the search and the set of solutions returned.

6.5 Limit Possible Trade-offs

If the user has no idea of what kind of solutions may be reachable, it may be easier to specify suitable trade-offs, i.e., how much gain in one objective is necessary to balance the loss in the other.

Greenwood et al. (1997) suggested a procedure which asks the user to rank a few alternatives, and from this derives constraints for linear weighting of the objectives consistent with the given ordering. Then, these are used to check whether there is a feasible linear weighting such that solution $\mathbf{x}$ is preferable to solution $\mathbf{y}$. More specifically, if the DM prefers a solution with objective values $\mathbf{f}(\mathbf{x})$ to a solution with objective values $\mathbf{f}(\mathbf{y})$, then, assuming linearly weighted additive utility functions and minimization of objectives, we know that

$$\sum_{k=1}^{n} w_k(f_k(\mathbf{x}) - f_k(\mathbf{y})) < 0 \tag{6.5}$$

$$\sum_{k=1}^{n} w_k = 1, \quad w_k \geq 0.$$

Let A denote the set of all pairs of solutions $(\mathbf{x}, \mathbf{y})$ ranked by the DM, and $\mathbf{x}$ preferred to $\mathbf{y}$. Then, to compare any two solutions $\mathbf{u}$ and $\mathbf{v}$, all linearly weighted additive utility functions are considered which are consistent with the ordering on the initially ranked solutions, i.e., consistent with Inequality 6.5 for all pairs of solutions $(\mathbf{x}, \mathbf{y}) \in A$. A preference of $\mathbf{u}$ over $\mathbf{v}$ is inferred if $\mathbf{u}$ is preferred to $\mathbf{v}$ for all such utility functions. A linear program (LP) is used to search for a utility function where $\mathbf{u}$ is not preferred to $\mathbf{v}$.

$$\min Z = \sum_{k=1}^{n} w_k (f_k(\mathbf{u}) - f_k(\mathbf{v})) \tag{6.6}$$

$$\sum_{k=1}^{n} w_k (f_k(\mathbf{x}) - f_k(\mathbf{y})) < 0 \qquad \forall (\mathbf{x}, \mathbf{y}) \in A \tag{6.7}$$

$$\sum_{k=1}^{n} w_k = 1, \quad w_k \geq 0.$$

If the LP returns a solution value $Z > 0$, we know there is no linear combination of objectives consistent with Inequality 6.7 such that $\mathbf{u}$ would be preferable, and we can conclude that $\mathbf{v}$ is preferred over $\mathbf{u}$. If the LP can find a linear combination with $Z < 0$, it only means that $\mathbf{v}$ is not preferred to $\mathbf{u}$. To test whether $\mathbf{u}$ is preferred to $\mathbf{v}$, one has to solve another LP and fail to find a linear combination of objectives such that $\mathbf{v}$ would be preferable. Overall, the method requires to solve 1 or 2 LPs for each pair of solutions in the population. Also, it needs special mechanisms to make sure that the allowed weight space does not become empty, i.e., that the user ranking is consistent with at least one possible linear weight assignment. The authors suggest to use a mechanism from White et al. (1984) which removes a minimal set of the DM's preference statements to make the weight space non-empty. Note that although linear combinations of objectives are assumed, it is possible to identify a concave part of the Pareto front, because the comparisons are only pair-wise. A more general framework for inferring preferences from examples (allowing for piecewise linear additive utility functions rather than linear additive utility functions) is discussed in Chapter 4.

In the guided MOEA proposed in Branke et al. (2001), the user is allowed to specify preferences in the form of maximally acceptable trade-offs like "one unit improvement in objective i is worth at most a_{ji} units in objective j". The basic idea is to modify the dominance criterion accordingly, so that it reflects the specified maximally acceptable trade-offs. A solution $\mathbf{x}$ is now preferred to a non-dominated solution $\mathbf{y}$ if the gain in the objective where $\mathbf{y}$ is better does not outweigh the loss in the other objective, see Figure 6.5 for an example. The region dominated by a solution is adjusted by changing the slope of the boundaries according to the specified maximal and minimal trade-offs. In this example, Solution A is now dominated by Solution B, because the loss in Objective 2 is too big to justify the improvement in Objective 1. On the other hand, Solutions D and C are still mutually non-dominated.

This idea can be implemented by a simple transformation of the objectives: It is sufficient to replace the original objectives with two auxiliary objectives Ω_1 and Ω_2 and use these together with the standard dominance principle, where

$$\Omega_1(x) = f_1(x) + \frac{1}{a_{21}} f_2(x)$$

$$\Omega_2(x) = \frac{1}{a_{12}} f_1(x) + f_2(x)$$

See Figures 6.7 and 6.6 for a visualization.

Because the transformation is so simple, the guided dominance scheme can be easily incorporated into standard MOEAs based on dominance, and it does not change the complexity nor the inner workings of the algorithm. However, an extension of this simple idea to more than two dimensions seems difficult.

Although developed independently and with a different motivation, the guided MOEA can lead to the same preference relation as the imprecise value function approach in Greenwood *et al.* (1997) discussed above. A maximally acceptable trade-off of the form "one unit improvement in objective i is worth at most a_{ji} units in objective j" could easily be transformed into the constraint

$$-w_i + a_{ji} \cdot w_j < 0 \qquad \text{or} \tag{6.8}$$

$$\frac{w_i}{w_j} > a_{ji} \tag{6.9}$$

The differences are in the way the maximally acceptable trade-offs are derived (specified directly by the DM in the guided MOEA, and inferred from a ranking of solutions in Greenwood *et al.* (1997)), and in the different implementation (a simple transformation of objectives in guided MOEA, and the solving of many LPs in the imprecise value function approach). While the guided MOEA is more elegant and computationally efficient for two objectives, the imprecise value function approach works independent of the number of objectives.

The idea proposed in Jin and Sendhoff (2002) is to aggregate the different objectives into one objective via weighted summation, but to vary the

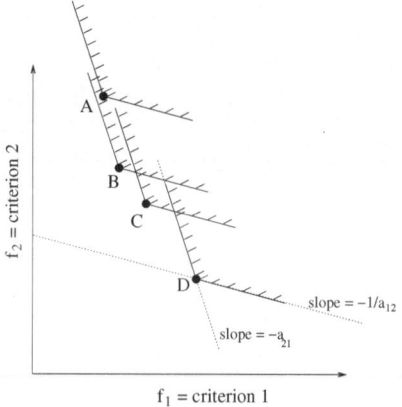

Fig. 6.5. Effect of the modified dominance scheme used by G-MOEA.

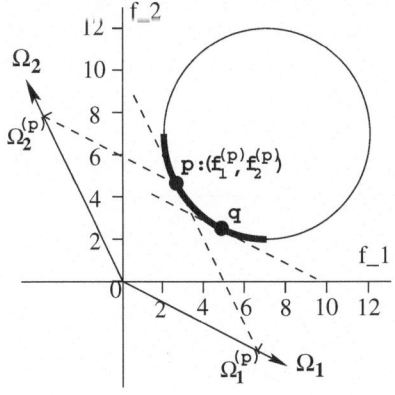
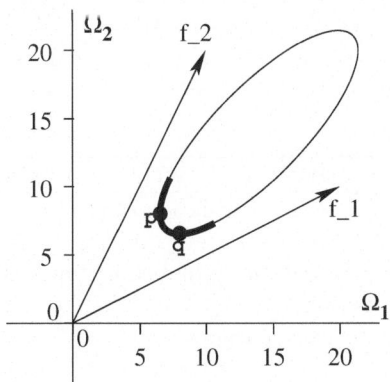

Fig. 6.6. When the guided dominance principle is used, non-dominated region of the Pareto optimal front is bounded by the two solutions p and q where the trade-off functions are tangent (Branke and Deb, 2005).

Fig. 6.7. The guided dominance principle is equivalent to the original dominance principle and appropriately transformed objective space (Branke and Deb, 2005).

weights gradually over time during the optimization. For two objectives, it is suggested to set $w_1(t) = |\sin(2\pi t/F)|$ and $w_2(t) = 1 - w_1(t)$, where t is the generation counter and F is a parameter to influence the oscillation period. The range of weights used in this process can be easily restricted to reflect the preferences of the DM by specifying a maximal and minimal weight $w_1^{\max}$ and $w_1^{\min}$, setting $w_1(t) = w_1^{\min} + (w_1^{\max} - w_1^{\min}) \cdot (\sin(2\pi t/F) + 1)/2$ and adjusting w_2 accordingly. The effect is a population moving along the Pareto front, covering the part of the front which is optimal with respect to the range of possible weight values. Because the population will not converge but keep oscillating along the front, it is necessary to collect all non-dominated solutions found in an external archive. Note also the slight difference in effect to restricting the maximal and minimal trade-off as do the other approaches in this section. While the other approaches enforce these trade-offs locally, on a one-to-one comparison, the dynamic weighting modifies the global fitness function. Therefore, the approach runs into problems if the Pareto front is concave, because a small weight change would require the population to make a big "jump".

6.6 Approaches Based on Marginal Contribution

Several authors have recently proposed to replace the crowding distance as used in NSGA-II by a solution's contribution to a given performance measure, i.e., the loss in performance if that particular solution would be absent from the population (Branke *et al.*, 2004; Emmerich *et al.*, 2005; Zitzler and Künzli, 2004). In the following, we call this a solution's marginal contribution. The algorithm then looks similar to Algorithm 2.

Algorithm 2 Marginal contribution MOEA

Initialize population of size μ
Determine Pareto-ranking
Compute marginal contributions
repeat
    Select parents
    Generate λ offspring by crossover and mutation and add them to the population
    Determine Pareto-ranking
    Compute marginal contributions
    while (population size $> \mu$) **do** {Environmental selection}
        From worst Pareto rank, remove individual with least marginal contribution
        Recompute marginal contributions
    end while
until termination condition

In Zitzler and Künzli (2004) and Emmerich *et al.* (2005), the performance measure used is the hypervolume. The hypervolume is the area (in 2D) or part of the objective space dominated by the solution set and bounded by a reference point **p**, see Chapter 14. Figure 6.8 gives an example for the hypervolume, and the parts used to rank the different solutions. The marginal contribution is then calculated only based on the individuals with the same Pareto rank. In the given example, Solution B has the largest marginal contribution. An obvious difficulty with hypervolume calculations is the determination of a proper reference point **p**, as this strongly influences the marginal contribution of the extreme solutions.

Zitzler *et al.* (2007) extend this idea by defining a weighting function over the objective space, and use the weighted hypervolume as indicator. This allows to incorporate preferences into the MOEA by giving preferred regions of the objective space a higher weight. In Zitzler *et al.* (2007), three different weighting schemes are proposed: a weight distribution which favors extremal solutions, a weight distribution which favors one objective over the other (but still keeping the best solution with respect to the less important objective), and a weight distribution based on a reference point, which generates a ridge-like function through the reference point (a, b) parallel to the diagonal. To calculate the weighted hypervolume marginal contributions, numerical integration is used.

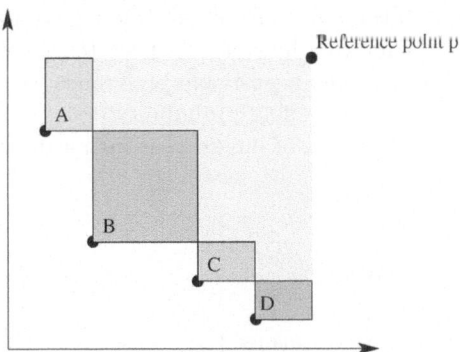

Fig. 6.8. Marginal contributions as calculated according to the hypervolume performance measure. The marginal contributions correspond to the respective shaded areas.

Another measure discussed in Zitzler and Künzli (2004) is the ϵ-Indicator. Basically, it measures the minimal distance by which an individual needs to be improved in each objective to become non-dominated (or can be worsened before it becomes dominated). Recently, Thiele *et al.* (2007) suggested to weight the ϵ-Indicator by an achievement scalarizing function based on a user specified reference point. The paper demonstrates that this allows to focus the search on the area around the specified reference point, and find interesting solutions faster.

Branke *et al.* (2004) proposed to use the "expected utility" as performance measure, i.e., a solution is evaluated by the expected loss in utility if this solution would be absent from the population. To calculate the expected utility, Branke *et al.* (2004) assumed that the DM has a linear utility function of the form $u(\mathbf{x}) = \lambda f_1(\mathbf{x}) + (1 - \lambda)f_2(\mathbf{x})$, and λ is unknown but follows a uniform distribution over $[0, 1]$. The expected marginal utility (emu) of a solution $\mathbf{x}$ is then the utility difference between the best and second best solution, integrated over all utility functions where solution $\mathbf{x}$ is best:

$$\text{emu}(\mathbf{x}) = \int_{\lambda=0}^{1} \max\{0, \min_{\mathbf{y}}\{u(\mathbf{y}) - u(\mathbf{x})\}\}d\lambda \qquad (6.10)$$

While the expected marginal utility can be calculated exactly in the case of two objectives, numerical integration is required for more objectives. The result of using this performance measure is a natural focus of the search on so-called "knees", i.e., convex regions with strong curvature. In these regions, an improvement in either objective requires a significant worsening of the other objective, and such solutions are often preferred by DMs (Das, 1999). An example of the resulting distribution of individuals along a Pareto front with a single knee is shown in Figure 6.9. Although this approach does not take into account individual user preferences explicitly, it favors the often preferred knee regions of the Pareto front. Additional explicit user preferences can be taken into account by allowing the user to specify the probability distribution for λ. For example, a probable preference for objective f_2 could be expressed by a linearly decreasing probability density of λ in the interval $[0..1]$, $p_\lambda(\alpha) = 2 - 2\alpha$. The effect of integrating such a preference information can be seen in Figure 6.10.

6.7 Other Approaches

The method by Cvetkovic and Parmee (2002) assigns each criterion a weight w_i, and additionally requires a minimum level for dominance τ, which corresponds to the concordance criterion of the ELECTRE method Figueira et al. (2005). Accordingly, the following weighted dominance criterion is used as dominance relation in the MOEA.

$$\mathbf{x} \succ_w \mathbf{y} \Leftrightarrow \sum_{i: f_i(\mathbf{x}) \leq f_i(\mathbf{y})} w_i \geq \tau.$$

To facilitate specification of the required weights, they suggest a method to turn fuzzy preferences into specific quantitative weights. However, since for every criterion the dominance scheme only considers whether one solution is better than another solution, and not by how much it is better, this approach allows only a very coarse guidance and is difficult to control. A somewhat similar dominance criterion has been proposed in Schmiedle et al. (2002). As

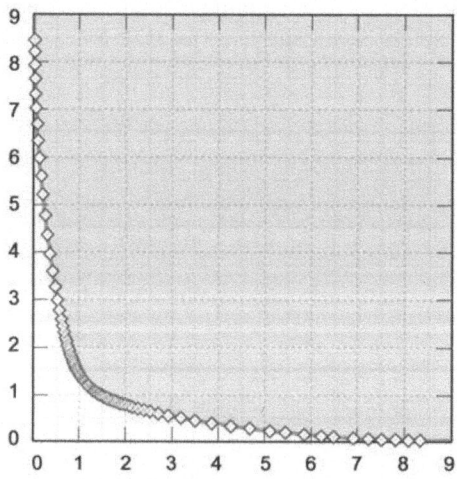

Fig. 6.9. Marginal contribution calculated according to expected utility result in a concentration of the individuals in knee areas.

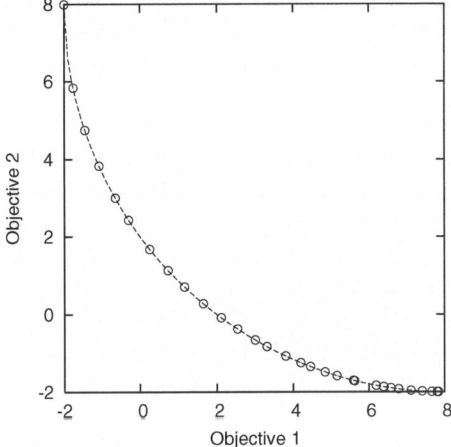

Fig. 6.10. Resulting distribution of individuals with the marginal expected utility approach and a linearly decreasing probability distribution for λ.

an additioanal feature, cycles in the preference relation graph are treated by considering all alternatives in a cycle as equivalent, and merging them into a single meta-node in the preference relation graph.

Hughes (2001) is concerned with MOEAs for noisy objective functions. The main idea to cope with the noise is to rank individuals by the sum of probabilities of being dominated by any other individual. To take preferences into account, the paper proposes a kind of weighting of the domination probabilities.

Some papers (Rekiek *et al.*, 2000; Coelho *et al.*, 2003; Parreiras and Vasconcelos, 2005) use preference flow according to Promethee II (Brans and Mareschal, 2005). Although this generates a preference order of the individuals, it does so depending on the different alternatives present in the population, not in absolute terms as, e.g., a weighted aggregation would do.

6.8 Discussions and Conclusions

If a single solution is to be selected in a multiobjective optimization problem, at some point during the process, the DM has to reveal his/her preferences. Specifying these preferences a priori, i.e., before alternatives are known, often means to ask too much of the DM. On the other hand, searching for all nondominated solutions as most MOEA do may result in a waste of optimization efforts to find solutions that are clearly unacceptable to the DM.

This chapter overviewed intermediate approaches, that ask for partial preference information from the DM a priori, and then focus the search to those regions of the Pareto optimal front that seem most interesting to the DM.

That way, it is possible to provide a larger number of relevant solutions. It seems intuitive that this should also allow to reduce the computation time, although this aspect has explicitly only been shown in Branke and Deb (2005) and Thiele *et al.* (2007).

Table 6.1 summarizes some aspects of some of the most prominent approaches. It lists the information required from the DM (Information), the part of the MOEA modified (Modification), and whether the result is a bounded region of the Pareto optimal front or a biased distribution (Influence). What method is most appropriate certainly depends on the application (e.g., whether the Pareto front is convex or concave, or whether the DM has a good conception of what is reachable) and on the kind of information the DM feels comfortable to provide. Many of the ideas can be combined, allowing the DM to provide preference information in different ways. For example, it would be straightforward to combine a reference point based approach which leads to sharp boundaries of the area in objective space considered as interesting with a marginal contribution approach which alters the distribution

Table 6.1. Comparison of some selected approaches to incorporate partial user preferences.

Name	Information	Modification	Influence
Constraints Coello (2002)	constraint	miscellaneous	region
Preference relation Fonseca and Fleming (1993)	reference point	dominance	region
Reference point based EMO, Deb *et al.* (2006)	reference point	crowding dist.	region
Light beam search based EMO, Deb and Kumar (2007b)	reference direction thresholds	crowding dist.	region
Imprecise value function Greenwood *et al.* (1997)	solution ranking	dominance	region
Guided MOEA Branke *et al.* (2001)	maximal/minimal trade-off	objectives	region
Weighted integration Zitzler *et al.* (2007)	weighting of objective space	crowding dist.	distribution
Marginal expected utility Branke *et al.* (2004)	trade-off probability distribution	crowding dist.	distribution
Biased crowding Branke and Deb (2005)	desired trade-off	crowding dist.	distribution

within this area. Furthermore, many of the ideas can be used in an interactive manner, which will be the focus of the following chapter (Chapter 7).

References

Branke, J., Deb, K.: Integrating user preference into evolutionary multi-objective optimization. In: Jin, Y. (ed.) Knowledge Incorporation in Evolutionary Computation, pp. 461–478. Springer, Heidelberg (2005)

Branke, J., Kaußler, T., Schmeck, H.: Guidance in evolutionary multi-objective optimization. Advances in Engineering Software 32, 499–507 (2001)

Branke, J., Deb, K., Dierolf, H., Osswald, M.: Finding knees in multi-objective optimization. In: Yao, X., et al. (eds.) PPSN 2004. LNCS, vol. 3242, pp. 722–731. Springer, Heidelberg (2004)

Brans, J. P., Mareschal, B.: PROMETHEE methods. In: Figueira, J., et al. (eds.) Multiple criteria decision analysis, pp. 163–196. Springer, Heidelberg (2005)

Coelho, R.F., Bersini, H., Bouillard, P.: Parametrical mechanical design with constraints and preferences: Application to a purge valve. Computer Methods in Applied Mechanics and Engineering 192, 4355–4378 (2003)

Coello Coello, C.A.: Handling preferences in evolutionary multiobjective optimization: A survey. In: Congress on Evolutionary Computation, vol. 1, pp. 30–37. IEEE Computer Society Press, Los Alamitos (2000)

Coello Coello, C.A.: Theoretical and numerical constraint-handling techniques used with evolutionary algorithms: A survey of the state of the art. Computer Methods in Applied Mechanics and Engineering 191(11-12), 1245–1287 (2002)

Coello Coello, C.A., van Veldhuizen, D.A., Lamont, G.B.: Evolutionary Algorithms for Solving Multi-Objective Problems. Kluwer Academic Publishers, Dordrecht (2002)

Cvetkovic, D., Parmee, I.C.: Preferences and their application in evolutionary multiobjective optimisation. IEEE Transactions on Evolutionary Computation 6(1), 42–57 (2002)

Das, I.: On characterizing the 'knee' of the pareto curve based on normal-boundary intersection. Structural Optimization 18(2/3), 107–115 (1999)

Deb, K.: Solving goal programming problems using multi-objective genetic algorithms. In: Proceedings of Congress on Evolutionary Computation, pp. 77–84 (1999)

Deb, K.: An efficient constraint handling method for genetic algorithms. Computer Methods in Applied Mechanics and Engineering 186(2-4), 311–338 (2000)

Deb, K.: Multi-Objective Optimization using Evolutionary Algorithms. Wiley, Chichester (2001)

Deb, K.: Multi-objective evolutionary algorithms: Introducing bias among Pareto-optimal solutions. In: Ghosh, A., Tsutsui, S. (eds.) Advances in Evolutionary Computing: Theory and Applications, pp. 263–292. Springer, Heidelberg (2003)

Deb, K., Kumar, A.: Interactive evolutionary multi-objective optimization and decision-making using reference direction method. In: Genetic and Evolutionary Computation Conference, pp. 781–788. ACM Press, New York (2007a)

Deb, K., Kumar, A.: Light beam search based multi-objective optimization using evolutionary algorithms. In: Congress on Evolutionary Computation, pp. 2125–2132. IEEE Computer Society Press, Los Alamitos (2007b)

Deb, K., Sundar, J.: Reference point based multi-objective optimization using evolutionary algorithms. In: Genetic and Evolutionary Computation Conference, pp. 635–642. ACM Press, New York (2006)

Deb, K., Agrawal, S., Pratap, A., Meyarivan, T.: A fast and elitist multi-objective genetic algorithm: NSGA-II. IEEE Transactions on Evolutionary Computation 6(2), 182–197 (2002a)

Deb, K., Thiele, L., Laumanns, M., Zitzler, E.: Scalable multi-objective optimization test problems. In: Congress on Evolutionary Computation (CEC), pp. 825–830 (2002b)

Deb, K., Sundar, J., Reddy, U.B., Chaudhuri, S.: Reference point based multi-objective optimization using evolutionary algorithms. International Journal of Computational Intelligence Research 2(3), 273–286 (2006)

Emmerich, M.T.M., Beume, N., Naujoks, B.: An EMO algorithm using the hypervolume measure as selection criterion. In: Coello Coello, C.A., Hernández Aguirre, A., Zitzler, E. (eds.) EMO 2005. LNCS, vol. 3410, pp. 62–76. Springer, Heidelberg (2005)

Figueira, J., Mousseau, V., Roy, B.: ELECTRE methods. In: Figueia, J., Greco, S., Ehrgott, M. (eds.) Multiple Criteria Decision Analysis: State of the Art Surveys, pp. 134–162. Springer, Heidelberg (2005)

Fonseca, C.M., Fleming, P.J.: Genetic algorithms for multiobjective optimization: Formulation, discussion, and generalization. In: International Conference on Genetic Algorithms, pp. 416–423 (1993)

Fonseca, C.M., Fleming, P.J.: Multiobjective optimization and multiple constraint handling with evolutionary algorithms - part I: A unified fomulation. IEEE Transactions on Systems, Man, and Cybernetics - Part A 28(1), 26–37 (1998)

Greenwood, G.W., Hu, X.S., D'Ambrosio, J.G.: Fitness functions for multiple objective optimization problems: combining preferences with Pareto rankings. In: Belew, R.K., Vose, M.D. (eds.) Foundations of Genetic Algorithms, pp. 437–455. Morgan Kaufmann, San Francisco (1997)

Horn, J.: Multicriterion Decision making. In: Bäck, T., Fogel, D., Michalewicz, Z. (eds.) Handbook of Evolutionary Computation, vol. 1, pp. F1.9:1–F1.9:15. Oxford University Press, Oxford (1997)

Hughes, E.J.: Constraint handling with uncertain and noisy multi-objective evolution. In: Congress on Evolutionary Computation, pp. 963–970. IEEE Computer Society Press, Los Alamitos (2001)

Jaszkiewicz, A., Slowinski, R.: The light beam search over a non-dominated surface of a multiple-objective programming problem. European Journal of Operational Research 113(2), 300–314 (1999)

Jiménez, F., Verdegay, J.L.: Evolutionary techniques for constrained optimization problems. In: Zimmermann, H.-J. (ed.) European Congress on Intelligent Techniques and Soft Computing, Verlag Mainz, Aachen (1999)

Jin, Y., Sendhoff, B.: Incorporation of fuzzy preferences into evolutionary multiobjective optimization. In: Asia-Pacific Conference on Simulated Evolution and Learning, Nanyang Technical University, Singapore, pp. 26–30 (2002)

Korhonen, P., Laakso, J.: A visual interactive method for solving the multiple criteria problem. European Journal of Operational Research 24, 277–287 (1986)

Molina, J., Santana, L.V., Hernandez-Diaz, A.G., Coello Coello, C.A., Caballero, R.: g-dominance: Reference point based dominance. European Journal of Operational Research (2009)

Parreiras, R.O., Vasconcelos, J.A.: Decision making in multiobjective optimization problems. In: Nedjah, N., de Macedo Mourelle, L. (eds.) Real-World Multi-Objective System Engineering, pp. 29–52. Nova Science Publishers, New York (2005)

Rachmawati, L., Srinivasan, D.: Preference incorporation in multi-objective evolutionary algorithms: A survey. In: Congress on Evolutionary Computation, pp. 3385–3391. IEEE Computer Society Press, Los Alamitos (2006)

Rekiek, B., Lit, P.D., Fabrice, P., L'Eglise, T., Emanuel, F., Delchambre, A.: Dealing with users's preferences in hybrid assembly lines design. In: Binder, Z., et al. (eds.) Management and Control of Production and Logistics Conference, pp. 989–994. Pergamon Press, Oxford (2000)

Sait, S.M., Youssef, H., Ali, H.: Fuzzy simulated evolution algorithm for multiobjective optimization of VLSI placement. In: Congress on Evolutionary Computation, pp. 91–97. IEEE Computer Society Press, Los Alamitos (1999)

Sakawa, M., Yauchi, K.: An interactive fuzzy satisficing method for multiobjective nonconvex programming problems through floating point genetic algorithms. European Journal of Operational Research 117, 113–124 (1999)

Schmiedle, F., Drechsler, N., Große, D., Drechsler, R.: Priorities in multi-objective optimization for genetic programming. In: Spector, L., et al. (eds.) Genetic and Evolutionary Computation Conference, pp. 129–136. Morgan Kaufmann, San Francisco (2002)

Tan, K.C., Lee, T.H., Khor, E.F.: Evolutionary algorithms with goal and priority information for multi-objective optimization. In: Congress on Evolutionary Computation, pp. 106–113. IEEE Computer Society Press, Los Alamitos (1999)

Thiele, L., Miettinen, K., Korhonen, P.J., Molina, J.: A preference-based interactive evolutionary algorithm for multiobjective optimization. Technical Report W-412, Helsinki School of Economics, Helsinki, Finland (2007)

Trautmann, H., Mehnen, J.: A method for including a-priori-preference in multicriteria optimization. Technical Report 49/2005, SFG 475, University of Dortmund, Germany (2005)

Van Veldhuizen, D., Lamont, G.B.: Multiobjective evolutionary algorithms: Analyzing the state-of-the-art. Evolutionary Computation Journal 8(2), 125–148 (2000)

White, C., Sage, A., Dozono, S.: A model of multiattribute decision-making and tradeoff weight determination under uncertainty. IEEE Transactions on Systems, Man, and Cybernetics 14, 223–229 (1984)

Wierzbicki, A.P.: Basic properties of scalarizing functions for multiobjective optimization. Optimization 8(1), 55–60 (1977)

Wierzbicki, A.P.: On the completeness and constructiveness of parametric characterizations to vector optimization problems. OR Spektrum 8(2), 73–87 (1986)

Zitzler, E., Künzli, S.: Indicator-based selection in multiobjective search. In: Yao, X., et al. (eds.) PPSN 2004. LNCS, vol. 3242, pp. 832–842. Springer, Heidelberg (2004)

Zitzler, E., Laumanns, M., Thiele, L.: SPEA2: Improving the Strength Pareto Evolutionary Algorithm for multiobjective optimization. In: Giannakoglou, K.C., et al. (eds.) Evolutionary Methods for Design, Optimisation and Control with Application to Industrial Problems (EUROGEN 2001), pp. 95–100. International Center for Numerical Methods in Engineering, CIMNE (2002)

Zitzler, E., Brockhoff, D., Thiele, L.: The hypervolume indicator revisited: On the design of pareto-compliant indicators via weighted integration. In: Obayashi, S., Deb, K., Poloni, C., Hiroyasu, T., Murata, T. (eds.) EMO 2007. LNCS, vol. 4403, pp. 862–876. Springer, Heidelberg (2007)

7

Interactive Multiobjective Evolutionary Algorithms

Andrzej Jaszkiewicz[1] and Jürgen Branke[2]

[1] Poznan University of Technology, Institute of Computing Science
jaszkiewicz@cs.put.poznan.pl
[2] Institute AIFB, University of Karlsruhe, 76128 Karlsruhe, Germany
branke@aifb.uni-karlsruhe.de

Abstract. This chapter describes various approaches to the use of evolutionary algorithms and other metaheuristics in interactive multiobjective optimization. We distinguish the traditional approach to interactive analysis with the use of single objective metaheuristics, the semi-a posteriori approach with interactive selection from a set of solutions generated by a multiobjective metaheuristic, and specialized interactive multiobjective metaheuristics in which the DM's preferences are interactively expressed during the run of the method. We analyze properties of each of the approaches and give examples from the literature.

7.1 Introduction

As already discussed in Chapters 1 and 6, in order to find the best compromise solution of a multiobjective optimization (MOO) problem, or a good approximation of it, MOO methods need to elicit some information about the DM's preferences. Thus, MOO methods may be classified with respect to the time of collecting the preference information as methods with either a priori, a posteriori, or progressive (interactive) articulation of preferences (Hwang et al., 1980; Słowiński, 1984). While the previous Chapter 6 discussed the use of (partial) a priori preference information in evolutionary MOO, here we will focus on interactive approaches. Note, however, that the two issues are closely related, as methods working with (partial) a priori information can be turned into interactive methods simply by allowing the DM to adjust preferences and re-start or continue the optimization interactively.

In general, interactive methods have the following main advantages:

- The preference information requested from the DM is usually much simpler than the preference information required by a priori methods.

Reviewed by: Francisco Ruiz, University of Málaga, Spain
Eckart Zitzler, ETH Zürich, Switzerland

J. Branke et al. (Eds.): Multiobjective Optimization, LNCS 5252, pp. 179–193, 2008.

- They have moderate computational requirements in comparison to a posteriori methods.
- As the DM controls the search process, he/she gets more involved in the process, learns about potential alternatives, and is more confident about the final choice.

Evolutionary algorithms and other metaheuristics may be used for interactive MOO in several ways. Traditional interactive methods usually assume that an underlying single objective, exact solver is available. This solver is used to solve a series of substitute single objective problems, whose solutions are guaranteed to be (weakly) Pareto optimal solutions. However, for many hard MOO problems, e.g., nonlinear or NP-hard combinatorial problems, no such efficient solvers are available. For such problems, one may use a single objective metaheuristic in place of an exact solver. This straightforward adaptation of the classical approach we will call the *traditional approach to interactive analysis with the use of single objective metaheuristics*.

In recent years, many multiobjective metaheuristics have been developed (Coello *et al.*, 2002; Deb, 2001). Most multiobjective metaheuristics aim at generating simultaneously, in a single run, a set of solutions being a good approximation of the whole Pareto optimal set. This set of solutions may then be presented to the DM to allow him/her choose the best compromise solution a posteriori. However, if the set of generated solutions and/or the number of objectives is large, the DM may be unable to perceive the whole set and to select the best solution without further support characteristic to interactive methods. Thus, the approach in which the DM interactively analyzes a large set of solutions generated by a multiobjective metaheuristic will be called the *semi-a posteriori approach*.

Furthermore, the DM may also interact with a metaheuristic during its run. Such interaction with evolutionary algorithms has been proposed for problems with (partially) subjective functions, and is known under the name of *interactive evolution* (Takagi, 2001). Finally, if at least the underlying objectives are known, we have *interactive multiobjective metaheuristics*, where the DM interacts with the method during its run, allowing it to focus on the desired region of the Pareto optimal set.

This chapter is organized as follows. In the next section, the traditional approach to interactive analysis with the use of single objective metaheuristics is analyzed. The semi-a posteriori approach is discussed in Section 7.3. Section 7.4 makes a short excursion to efficiency measures. Then, Section 7.5 contains the discussion of the interactive multiobjective metaheuristics. Future trends and research directions are discussed in the final section.

7.2 Traditional Approach to Interactive Analysis with the Use of Single Objective Metaheuristics

As was mentioned above, the traditional interactive methods rely on the use of an exact single objective solver (e.g., Miettinen, 1999, and also Chapter 2) In general, the single objective solver is used to solve a substitute single objective problem whose optimal solution is guaranteed to be (weakly) Pareto optimal. For example, in the simplest version of the reference point method (c.f., Chapter 1), in each iteration, the DM specifies a reference point in the objective space. The reference point defines an achievement scalarizing function that is optimized as substitute single objective problem with the use of an exact single objective solver. The single solution generated in this way is presented to the DM, an he/she decides whether this solution is satisfactory. If the DM is not satisfied he/she adjusts the reference point and the process is repeated. Typical examples of achievement scalarizing functions are the *weighted linear scalarizing function* defined as:

$$s_j(\mathbf{x}, \varLambda) = \sum_{j=1}^{J} \lambda_j f_j(\mathbf{x}),$$

and the *weighted Tchebycheff scalarizing function* defined as:

$$s_\infty(\mathbf{x}, \mathbf{z}^0, \varLambda) = \max_j \left\{ \lambda_j (f_j(\mathbf{x}) - \mathbf{z}_j^0) \right\}$$

where $\mathbf{z}^0$ is a *reference point*, and $\varLambda = [\lambda_1, \dots, \lambda_J]$ is a weight vector such that $\lambda_j \geq 0 \ \forall j$.

A well established theory is associated with the generation of Pareto-optima through solving substitute single objective problems. For example, it may be proved that each weighted linear scalarizing function has at least one global optimum (minimum) belonging to the set of Pareto-optimal solutions (Steuer, 1986). A Pareto-optimal solution that is a global optimum of a weighted linear scalarizing function is called *supported Pareto-optimal solution* (Ulungu and Teghem, 1994). Furthermore, each weighted Tchebycheff scalarizing function has at least one global optimum (minimum) belonging to the set of Pareto-optimal solutions. For each Pareto-optimal solution $\mathbf{x}$ there exists a weighted Tchebycheff scalarizing function s such that $\mathbf{x}$ is a global optimum (minimum) of s (Steuer, 1986, ch. 14.8). If, for a given problem, no efficient exact solver exists, a straightforward approach could be to use a single objective metaheuristic instead. One should be aware, however, that the mentioned above theoretical results are valid in the case of exact solvers only. If a metaheuristic solver is used several potential problems appear:

- Metaheuristics do not guarantee finding a global optimum. Thus, when applied to solve a substitute single objective problem, they may produce a dominated solution.

- Solutions generated in different iterations of the interactive method may dominate each other, which may be very misleading to the DM.
- No control mechanism allowing the DM to guide the search for the best compromise may be guaranteed. For example, the DM improving the value of one of the objectives in the reference point, expects that this objective will also improve in the newly generated Pareto optimal solution. Since metaheuristics do not give any guarantee on the quality of generated solutions, the change on this objective may even be opposite.
- Computational efficiency may be insufficient. In interactive methods, generation of new solutions needs to be done on-line. Thus, while a running time of several minutes may be fully acceptable for solving most single objective problems, it may be unacceptable when used within an interactive method.

Note that the potential non-optimality of metaheuristics may sometimes also be beneficial. In particular, a metaheuristic applied to the optimization of a linear scalarizing function may produce a non-supported Pareto-optimal solution as an approximate solution of the substitute problem. However, such results are due to randomness and can not be relied upon.

Several examples of this traditional approach may be found in the literature. Alves and Clímaco (2000) use either simulated annealing or tabu search in an interactive method for 0-1 programming. Kato and Sakawa (1998); Sakawa and Shibano (1998); Sakawa and Yauchi (1999) have proposed a series of interactive fuzzy satisficing methods using various kinds of genetic algorithms adapted to various kinds of problems. Miettinen and Mäkelä use two different versions of genetic algorithms with constraint handling (among other solvers) in NIMBUS (Miettinen, 1999; Miettinen and Mäkelä, 2006; Ojalehto et al., 2007). Gabrys et al. (2006) use a single objective EA within several classical interactive procedures, i.e., STEP, reference point approach, Tchebycheff method, and GDF.

7.3 Semi-a-Posteriori Approach – Interactive Selection from a Set of Solutions Generated by a Multiobjective Metaheuristic

Classical a posteriori approaches to MOO assume that the set of (approximately) Pareto optimal solutions is presented to the DM, who is then able to select the best compromise solution. In recent years, many multiobjective metaheuristics have been developed, see, e.g., Coello et al. (2002); Deb (2001) for overviews. The typical goal of these methods is to generate in a single run a set of solutions being a good approximation of the whole Pareto set. Because metaheuristics such as evolutionary algorithms or particle swarm optimization concurrently work with sets (populations) of solutions, they can simultaneously generate in one run a set of solutions which approximate the Pareto front. Thus, they are naturally suited as a posteriori MOO methods.

The set of potentially Pareto optimal solutions generated by a multiobjective metaheuristic may, however, be very large (Jaszkiewicz, 2002b), and even for relatively small problems may contain thousands of solutions. Obviously, most DMs will not be able to analyze such sets of solutions, in particular in the case of many objectives. Thus, they may require some further support which is typical for interactive methods.

In fact, numerous methods for interactive analysis of large finite sets of alternatives have been already proposed. This class of methods includes: Zionts method (Zionts, 1981), Korhonen, Wallenius and Zionts method (Korhonen et al., 1984), Köksalan, Karwan and Zionts method (Köksalan et al., 1988), Korhonen method (Korhonen, 1988), Malakooti method (Malakooti, 1989), Taner and Köksalan method (Taner and Köksalan, 1991), AIM (Lotfi et al., 1992), Light-Beam-Search-Discrete (Jaszkiewicz and Słowiński, 1997), Interactive Trichotomy Segmentation (Jaszkiewicz and Ferhat, 1999), and Interquad (Sun and Steuer, 1996). These methods are usually based on some traditional interactive methods (with on-line generation of solutions through optimization of the substitute problems). Thus, from the point of view of the DM the interaction may look almost the same, and he/she may be unaware whether the traditional or semi-a posteriori interaction is used. For example, using again the simplest version of the reference point method, in each iteration, the DM specifies a reference point in the objective space. The reference point defines an achievement scalarizing function. Then, the set of potentially Pareto optimal solutions is searched for the best solution with respect to this scalarizing function, which is presented to the DM. If the DM is not satisfied he/she adjusts the reference point and the process is repeated.

The advantages of this semi-a posteriori approach are:

- A large number of methods for both computational and interactive phases exist.
- All heavy calculations are done prior to interaction, thus interaction is very fast and saves the DM time.
- All solutions presented to the DM are mutually non-dominated. Note, however, that it does not necessarily mean that they are Pareto-optimal, since Pareto-optimality cannot be guaranteed by metaheuristics.
- Some control mechanisms allowing the DM to guide the search for the best compromise may be guaranteed.
- Additional various forms of statistical analysis of the pre-generated set of solutions (e.g. calculation of correlation coefficients between objectives) and graphical visualization may be used.

Generation of a high quality approximation of the whole Pareto set may, however, become computationally too demanding, especially in the case of realistic size combinatorial problems and a high number of objectives. Several studies (Jaszkiewicz, 2002b, 2003, 2004b; Purshouse, 2003) addressed the issue of computational efficiency of multiobjective metaheuristics compared to iteratively applied single objective metaheuristics to generate an approximation to the Pareto front. Below the idea of the *efficiency index* of a multiobjective metaheuristic with respect to a corresponding single objective metaheuristic introduced in Jaszkiewicz (2002b, 2003, 2004b) is presented.

Despite a large number of both multiobjective metaheuristics and interactive methods for analysis of large finite sets of alternatives, few examples of the semi-a posteriori approach may be found in the literature. Hapke *et al.* (1998) use an interactive Light Beam Search-discrete method over a set of solutions of project scheduling problems generated by Pareto simulated annealing. Jaszkiewicz and Ferhat (1999) use the interactive trichotomy segmentation method for analysis of a set of solutions of a personnel scheduling problem generated by the same metaheuristic. Tanaka *et al.* (1995) use an interactive procedure based on iterative evaluation of samples of points selected from a larger set generated by an MOEA. They use a radial basis function network to model the DM's utility function and suggest new solutions for evaluation.

7.4 Excursion: Efficiency Index

The general idea of the efficiency index presented here is to compare running times of single and multiobjective metaheuristics needed to generate solution sets of comparable quality. Many techniques have been proposed for evaluation of sets of solutions in the objective space, see Chapter14. Obviously, no single measure can cover all aspects of the quality. In this section, we will focus on the measures based on scalarizing functions, mainly because they can be directly applied to evaluate results of both single and multiobjective metaheuristics.

Assume that a single objective metaheuristic is applied to the optimization of an achievement scalarizing function. The quality of the approximate solution generated by the single objective metaheuristic is naturally evaluated with the achievement scalarizing function. Furthermore, one may run a single objective metaheuristic a number of times using a representative sample of achievement scalarizing functions defined, e.g., by a representative sample of weight vectors. Then, the average value of the scalarizing functions over the respective generated solutions measures the average quality of solutions generated.

Scalarizing functions are also often used for evaluation of the results of multiobjective metaheuristics (e.g. Czyzak and Jaszkiewicz, 1996; Viana and de Sousa, 2000). In Hansen and Jaszkiewicz (1998) and Jaszkiewicz (2002a) we have proposed a scalarizing functions-based quality measure fully compatible with the above outlined approach for evaluation of single objective metaheuristics. The quality measure evaluates sets of approximately Pareto optimal solutions with average value of scalarizing functions defined by a representative sample of weight vectors.

In order to compare the quality and the computational efficiency of a multiobjective metaheuristic and a single objective metaheuristic, a representative sample Λ_R of normalized weight vectors is used. Each weight vector $\lambda \in \Lambda_R$ defines a scalarizing function, e.g., a weighted Tchebycheff scalarizing function $s_\infty(\mathbf{z}, \mathbf{z}^0, \lambda)$. As the reference point $\mathbf{z}^0$ one may use an approximation of the ideal point. All scalarizing functions defined by vectors from set Λ_R constitute the set S_a of functions. $|\Lambda_R|$ is a sampling parameter, too low values of this parameter may increase the statistical error, but, in general, the results will not depend on it.

In order to evaluate the quality of solutions generated by a single objective metaheuristic the method solves a series of optimization problems for each achievement scalarizing function from set S_a. For each function $s_\infty(\mathbf{z}, \mathbf{z}^0, \lambda) \in S_a$, the single objective metaheuristic returns a solution corresponding to point $\mathbf{z}^\lambda$ in objective space. Thus, the average quality of solutions generated by the single objective metaheuristic is:

$$Q_s = \frac{\sum\limits_{\lambda \in \Lambda_R} s_\infty(\mathbf{z}^\lambda, \mathbf{z}^0, \lambda)}{|\Lambda_R|}$$

Note that due to the nature of metaheuristics, the best solution with respect to a particular achievement scalarizing function $s_\infty(\mathbf{z}, \mathbf{z}^0, \lambda') \in S_a$ may actually be found when optimizing another function $s_\infty(\mathbf{z}, \mathbf{z}^0, \lambda'')$ from set S_a. If desired, this can be taken into account by storing all solutions generated so far in all single objective runs, and using the respective best of the stored solutions instead of $\mathbf{z}^\lambda$ for each $\lambda \in \Lambda_R$.

In order to evaluate the quality of solutions generated by a multiobjective metaheuristic, a set $\mathbf{A}$ of solutions generated by the method in a single run is evaluated. For each function $s_\infty(\mathbf{z}, \mathbf{z}^0, \lambda) \in S_a$ the best point $\mathbf{z}^\lambda$ on this function is selected from set $\mathbf{A}$, i.e., $s_\infty(\mathbf{z}^\lambda, \mathbf{z}^0, \lambda) \leq s_\infty(\mathbf{z}, \mathbf{z}^0, \lambda) \forall \mathbf{z} \in \mathbf{A}$ Thus, the average quality of solutions generated by the multiobjective metaheuristic is:

$$Q_m = \frac{\sum\limits_{\lambda \in \Lambda_R} s_\infty(\mathbf{z}^\lambda, \mathbf{z}^0, \lambda)}{|\Lambda_R|}$$

The quality of solutions generated by both the single objective metaheuristic and the multiobjective metaheuristic may be assumed approximately the same if $Q_s = Q_m$.

The two quality measures Q_s and Q_m are used in order to compare computational efforts (running times) needed by a single objective metaheuristic and a multiobjective metaheuristic to generate solutions of the same quality. First, the single objective metaheuristic is run for each function $s_\infty(\mathbf{z}, \mathbf{z}^0, \lambda) \in S_a$. Let the average time of a single run of the single objective metaheuristic be denoted by T_s (T_s is rather independent of $|\Lambda_R|$ since it is an average time of a single objective run). Then, the multiobjective metaheuristic is run. During the run of the multiobjective metaheuristic the quality Q_m is observed using the same set S_a of functions. Q_m is applied to a set **A** of potentially Pareto optimal solutions generated up to a given time by the multiobjective meta-heuristic. This evaluation is repeated whenever set **A** changes. Note that the set S_a is used only to calculate Q_m, it has no influence on the work of the mul-tiobjective metaheuristic, which is guided by its own mechanisms (e.g., Pareto ranking). The multiobjective metaheuristic is stopped as soon as $Q_m \leq Q_s$ (or if some maximum allowed running time is reached). Let the running time of the multiobjective metaheuristic be denoted by T_m If condition $Q_m \leq Q_s$ was fulfilled one may calculate the efficiency index of the multiobjective meta-heuristic with respect to the single objective metaheuristic:

$$EI = \frac{T_m}{T_s}$$

In general, one would expect $EI > 1$, and the lower EI, the more efficient the multiple objective metaheuristic with respect to the single objective method.

The efficiency index may be used to decide what MOO method might be most efficient. Assume that a DM using an iterative single-objective approach has to do L iterations before choosing the best compromise solution[1]. If $EI < L$, then the off-line generation of approximately Pareto-optimal solutions with the use of a multiobjective metaheuristic would require less computational effort than on-line generation with the use of a single objective metaheuristic. In other words, EI compares efficiency of the traditional interactive approach and the semi-a posteriori approach.

The efficiency index has been used in several studies (Jaszkiewicz, 2002b, 2003, 2004b), similar techniques were also used by Purshouse (2003). Although the quantitative results depend on the test problem (e.g., TSP, knapsack, set covering) and method, the general observation is that the relative compu-tational efficiency of multiobjective metaheuristics reduces below reasonable levels with increasing number of objectives and instance size. For example, the computational experiment on multiobjective knapsack problems (Jaszkiewicz, 2004b) indicated that "the computational efficiency of the multiobjective metaheuristics cannot be justified in the case of 5-objective instances and can hardly be justified in the case of 4-objective instances". Thus, the semi-a posteriori would become very inefficient compared to an interactive approach in such cases.

7.5 Interactive Multiobjective Metaheuristics

Evolutionary algorithms and other metaheuristics are black box algorithms, i.e., it suffices to provide them with a quality measure for the solutions gen-erated. There are almost no restrictions regarding this quality measure; a

---

[1] Of course, in practice, this number is not known in advance and depends on the problem, the method and the DM, but often may be reasonably estimated on the basis of past experience

closed-loop description of the solution quality, or additional information such as gradients, is not necessary. This makes metaheuristics applicable to an enormous range of applications. In most applications, a solution's quality can be evaluated automatically. However, there are also applications where a solution's quality can not be computed automatically, but depends on user preferences. In this case, *interactive evolutionary computation* (Takagi, 2001) may be used. An interactive evolutionary algorithm relies on the user to evaluate and rank solutions during the run. These evaluations are then used to guide the further search towards the most promising regions of the solution space.

One popular example is the generation of a picture of a criminal suspect, with the only source of information being someone's memory (Caldwell and Johnston, 1991). The problem may be modeled as an optimization task where solutions are faces (built of some building blocks) and the objective function is a similarity to the suspect's face. Of course, the similarity may be evaluated only subjectively by the human witness. Another example may be optimization of aesthetics of a design (Kamalian *et al.*, 2004, 2006). An example in the realm of MCDM can be found in Hsu and Chen (1999).

Human fatigue is a crucial factor in such algorithms, as the number of solutions usually looked at by metaheuristics may become very large. Thus, various approaches based on approximate modeling (e.g., with a function learned from evaluation examples) of the DM's preferences have been proposed in this field (Takagi, 2001). The evolutionary algorithm tries to predict a DM's answers using this model, and asks the DM to evaluate only some of the new solutions.

There are apparent similarities of this field to interactive MOO. In both cases we are looking for solutions being the best from the point of view of subjective preferences. Thus, a very straightforward approach could be to apply an interactive evolutionary algorithm to a MOO problem asking the DM to evaluate presented solutions. However, in MOO, we assume to at least know the criteria that form the basis for the evaluation of a solution, and that these can be computed. Only how these objectives are combined to the overall utility of a solution is subjective. A direct application of interactive evolutionary algorithms to MOO would not take into account the fact that many solutions could be compared with the use of the dominance relations without consulting the DM. In other words, in an MOO problem, user evaluation is only necessary to compare mutually non-dominated solutions.

Interactive multiobjective metaheuristics are methods that are specifically adapted to interactive MOO, and use the dominance relation and the knowledge about the objectives to reduce the number of questions asked to the DM. Note that they may present to the DM intermediate solutions during the run, while in the case of the traditional or the semi-a priori approach only final solution(s) are presented to the DM.

As opposed to the traditional approach to interactive analysis, the interactive multiobjective metaheuristics do not make a complete run of a single objective method in each iteration. They rather modify the internal workings of a single or multiobjective metaheuristic, allowing interaction with the DM during the run.

Several methods belonging to this class may be found in the literature. Tanino et al. (1993) proposed probably the first method of this kind. The method is based on a relatively simple version of a Pareto-ranking based multiobjective evolutionary algorithm. The evaluation of solutions from the current population is based on both dominance relation and on preferences expressed iteratively by the DM. The DM has several options: He/she may directly point out satisfactory/unsatisfactory solutions, or specify aspiration/reservation levels that are used to identify satisfactory/unsatisfactory solutions.

Kamalian et al. (2004) suggest to use an a posteriori evolutionary MOO followed by an interactive evolutionary algorithm. First a Pareto ranking-based evolutionary algorithm is run to generate a rough approximation of the Pareto front. Then, the DM selects a sample of the most promising solutions that are subsequently used as starting population of a standard interactive evolutionary algorithm.

Kita et al. (1999) interleave generations of a Pareto ranking-based evolutionary algorithm with ranking of the solutions by a DM, while Kamalian et al. (2006) allow the user to modify the Pareto ranking computed automatically by changing the rank of some of the solutions.

Several authors allow the DM to set and adjust aspiration levels or reference points during the run, and thereby guide the MOEA towards the (from the DM's perspective) most promising solutions. For example, Fonseca and Fleming (1998) allow the user to specify aspiration levels in form of a reference point, and use this to modify the MOEA's ranking scheme in order to focus the search. This approach is discussed in more detail also in Chapter 6. The approach proposed by Geiger (2007) is based on Pareto Iterated Local Search. It first approximates the Pareto front by calculating some upper and lower bounds, to give the DM a rough idea of what can be expected. Based on this information, the DM may restrict the search to the most interesting parts of the objective space. Ulungu et al. (1998) proposed an interactive version of multiple objective simulated annealing. In addition to allowing to set aspiration levels, solutions may be explicitly removed from the archive, and weights may be specified to further focus the search. Thiele et al. (2007) also use DM's preferences interactively expressed in the form of reference points. They use an indicator-based evolutionary algorithm, and use the achievement scalarizing function to modify the indicator and force the algorithm to focus on the more interesting part of the Pareto front.

Deb and Chaudhuri (2007) proposed an interactive decision support system called I-MODE that implements an interactive procedure built over a number of existing EMO and classical decision making methods. The main idea of the interactive procedure is to allow the DM to interactively focus on interesting region(s) of the Pareto front. The DM has options to use several tools for generation of potentially Pareto optimal solutions concentrated in the desired regions. For example, he/she may use weighted sum approach, utility function based approach, Tchebycheff function approach or trade-off information. Note that the preference information may be used to define a number of interesting regions. For example, the DM may define a number of reference (aspiration) points defining different regions. The preference information is then used by an EMO to generate new solutions in (hopefully) interesting regions.

The interactive evolutionary algorithm proposed by Phelps and Köksalan (2003) allows the user to provide preference information about pairs of solutions during the run. Based on this information, the authors compute a "most compatible" weighted sum of objectives (i.e., a linear achievement scalarizing function) by means of linear programming, and use this as single substitute objective for some generations of the evolutionary algorithm. Note that the weight vector defines a single search direction and may change only when the user provides new comparisons of solutions. However, since only partial preference information is available, there is no guarantee that the weight vector obtained by solving the linear programming model defines the DM's utility function, even if the utility function has the form of a weighted sum. Thus, the use of a single weight vector may bias the algorithm towards some solutions not necessarily being the best for the DM. This bias may become even more significant when the DM's preferences cannot be modeled with a linear function.

Instead of using linear programming to derive a weighting of the objectives "most compatible" with the pairwise comparisons as in Phelps and Köksalan (2003), Barbosa and Barreto (2001) use two evolutionary algorithms, one to find the solutions, and one to determine the "most compatible" ranking. These EAs are run in turn: first, both populations (solutions and weights) are initialized, then the DM is asked to rank the solutions. After that, the population of weights is run for some generations to produce a weighting which is most compatible with the user ranking. Then, this weighting is used to evolve the solutions for some generations, and the process repeats. Todd and Sen (1999) also try to learn the user's utility function, but instead of only considering linear weightings of objectives, they use the preference information provided by the DM to train an artificial neural network, which is then used to evaluate solutions in the evolutionary algorithm.

The method of Jaszkiewicz (2007) is based on the Pareto memetic algorithm (PMA)(Jaszkiewicz, 2004a). The original PMA samples the set of scalarizing functions drawing a random weight vector for each single iteration and uses this during crossover and local search. In the proposed interactive version, preference information from pairwise comparisons of solutions is used to reduce the set of possible weight vectors. Note that different from the approaches above, it is not attempted to identify one most likely utility function, but simultaneously allows for a range of utility functions compatible with the preference information specified by the user.

7.6 Summary

In this chapter, we have described three principal ways for interactively using evolutionary algorithms or similar metaheuristics in MOO.

The traditional approach to interactive MOO with the use of single objective metaheuristics is a straightforward adaptation of the classical interactive methods. It suffers, however, from a number of weaknesses when metaheuristics are used in place of exact solvers, since many important theoretical properties are not valid in the case of heuristic solvers.

The semi-a posteriori approach allows combining multiobjective metaheuristics with methods for interactive analysis of large finite sets of alternatives. An important advantage of this approach is that various methods from both classes are available. The semi-a posteriori approach allows overcoming a number of weaknesses of the traditional approach. It may, however, become computationally inefficient for large problems with larger number of objectives.

Interactive multiobjective metaheuristics is a very promising class of methods specifically adapted to interactive solving of hard MOO problems. According to some studies (Phelps and Köksalan, 2003; Jaszkiewicz, 2007) they may be computationally efficient even for a large number of objectives and require relatively low effort from the DM. Such specifically designed methods may combine the main advantages of metaheuristics and interactive MOO avoiding weaknesses of the other approaches. Note that the way the DM interacts with such methods may be significantly different from the traditional approaches. For example, the DM may be asked to compare solutions being known to be located far from the Pareto optimal set.

Both interactive methods and the use of metaheuristics are among the most active research areas within MOO. Combination of these approaches may results in very effective methods for hard, multidimensional MOO problems. Despite of a number of proposals known from the literature, this field has not yet received appropriate attention from MOO researchers community.

References

Alves, M., Clímaco, J.: An interactive method for 0-1 multiobjective problems using simulated annealing and tabu search. Journal of Heuristics 6, 385–403 (2000)

Barbosa, H.J.C., Barreto, A.M.S.: An interactive genetic algorithm with co-evolution of weights for multiobjective problems. In: Spector, L., et al. (eds.) Genetic and Evolutionary Computation Conference, pp. 203–210. Morgan Kaufmann, San Francisco (2001)

Caldwell, C., Johnston, V.S.: Tracking a criminal suspect through "face-space" with a genetic algorithm. In: International Conference on Genetic Algorithms, pp. 416–421. Morgan Kaufmann, San Francisco (1991)

Coello Coello, C.A., Van Veldhuizen, D.A., Lamont, G.B.: Evolutionary Algorithms for Solving Multi-Objective Problems. Kluwer Academic Publishers, Dordrecht (2002)

Czyzak, P., Jaszkiewicz, A.: A multiobjective metaheuristic approach to the local-ization of a chain of petrol stations by the capital budgeting model. Control and Cybernetics 25(1), 177–187 (1996)

Deb, K.: Multi-Objective Optimization using Evolutionary Algorithms. Wiley, Chichester (2001)

Deb, K., Chaudhuri, S.: I-MODE: An interactive multi-objective optimization and decision-making using evolutionary methods. Technical Report KanGAL Report No. 2007003, Indian Institute of Technology Kanpur (2007)

Fonseca, C.M., Fleming, P.J.: Multiobjective optimization and multiple constraint handling with evolutionary algorithms - part I: A unified fomulation. IEEE Trans-actions on Systems, Man, and Cybernetics - Part A 28(1), 26–37 (1998)

Streichert, F., Tanaka-Yamawaki, M.: A new scheme for interactive multi-criteria decision making. In: Gabrys, B., Howlett, R.J., Jain, L.C. (eds.) KES 2006. LNCS (LNAI), vol. 4253, pp. 655–662. Springer, Heidelberg (2006)

Geiger, M.J.: The interactive Pareto iterated local search (iPILS) metaheuristic and its application to the biobjective portfolio optimization problem. In: IEEE Symposium on Computational Intelligence in Multicriteria Decision Making, pp. 193–199. IEEE Computer Society Press, Los Alamitos (2007)

Hansen, M.P., Jaszkiewicz, A.: Evaluating the quality of approximations to the nondominated set. Working paper, Institute of Mathematical Modelling Technical University of Denmark (1998)

Hapke, M., Jaszkiewicz, A., Słowiński, R.: Interactive analysis of multiple-criteria project scheduling problems. European Journal of Operational Research 107, 315–324 (1998)

Hsu, F.C., Chen, J.-S.: A study on multicriteria decision making model: Interactive genetic algorithms approach. In: Congress on Evolutionary Computation, vol. 3, pp. 634–639. IEEE Computer Society Press, Los Alamitos (1999)

Hwang, C.-L., Paidy, S.R., Yoon, K., Masud, A.S.M.: Mathematical programming with multiple objectives: A tutorial. Computers and Operations Research 7, 5–31 (1980)

Jaszkiewicz, A.: Genetic local search for multiple objective combinatorial optimiza-tion. European Journal of Operational Research 137(1), 50–71 (2002a)

Jaszkiewicz, A.: On the computational effectiveness of multiple objective metaheuris-tics. In: Trzaskalik, T., Michnik, J. (eds.) Multiple Objective and Goal Program-ming. Recent Developments, pp. 86–100. Physica-Verlag, Heidelberg (2002b)

Jaszkiewicz, A.: Do multiple-objective metaheuristics deliver on their promises? a computational experiment on the set-covering problem. IEEE Transactions on Evolutionary Computation 7(2), 133–143 (2003)

Jaszkiewicz, A.: A comparative study of multiple-objective metaheuristics on the bi-objective set covering problem and the pareto memetic algorithm. Annals of Operations Research 131(1-4), 135–158 (2004a)

Jaszkiewicz, A.: On the computational efficiency of multiobjective metaheuris-tics. the knapsack problem case study. European Journal of Operational Re-search 158(2), 418–433 (2004b)

Jaszkiewicz, A.: Interactive multiobjective optimization with the Pareto memetic algorithm. Foundations of Computing and Decision Sciences 32(1), 15–32 (2007)

Jaszkiewicz, A., Ferhat, A.B.: Solving multiple criteria choice problems by interac-tive trichotomy segmentation. European Journal of Operational Research 113(2), 271–280 (1999)

Jaszkiewicz, A., Słowiński, R.: The lbs-discrete interactive procedure for multiple-criteria analysis of decision problems. In: Climaco, J. (ed.) Multicriteria Analysis, International Conference on MCDM, Coimbra, Portugal, 1-6 August 1994, pp. 320–330. Springer, Heidelberg (1997)

Kamalian, R., Takagi, H., Agogino, A.M.: Optimized design of MEMS by evolutionary multi-objective optimization with interactive evolutionary computation. In: Deb, K., et al. (eds.) GECCO 2004. LNCS, vol. 3103, pp. 1030–1041. Springer, Heidelberg (2004)

Kamalian, R., Zhang, Y., Takagi, H., Agogino, A.M.: Evolutionary synthesis of micromachines using supervisory multiobjective interactive evolutionary computation. In: Yeung, D.S., Liu, Z.-Q., Wang, X.-Z., Yan, H. (eds.) ICMLC 2005. LNCS (LNAI), vol. 3930, pp. 428–437. Springer, Heidelberg (2006)

Kato, K., Sakawa, M.: An interactive fuzzy satisficing method for large scale multiobjective 0-1 programming problems with fuzzy numbers through genetic algorithms. European Journal of Operational Research 107(3), 590–598 (1998)

Kita, H., Shibuya, M., Kobayashi, S.: Integration of multi-objective and interactive genetic algorithms and its application to animation design. In: IEEE Systems, Man, and Cybernetics, pp. 646–651 (1999)

Köksalan, M., Karwan, M.H., Zionts, S.: An approach for solving discrete alternative multiple criteria problems involving ordinal criteria. Naval Research Logistics 35(6), 625–642 (1988)

Korhonen, P.: A visual reference direction approach to solving discrete multiple criteria problems. European Journal of Operational Research 34(2), 152–159 (1988)

Korhonen, P., Wallenius, J., Zionts, S.: Solving the discrete multiple criteria problem using convex cones. Management Science 30(11), 1336–1345 (1984)

Lotfi, V., Stewart, T.J., Zionts, S.: An aspiration-level interactive model for multiple criteria decision making. Computers and Operations Research 19, 677–681 (1992)

Malakooti, B.: Theories and an exact interactive paired-comparison approach for discrete multiple criteria problems. IEEE Transactions on Systems, Man, and Cybernetics 19(2), 365–378 (1989)

Miettinen, K.: Nonlinear Multiobjective Optimization. Kluwer Academic Publishers, Dordrecht (1999)

Miettinen, K., Mäkelä, M.M.: Synchronous approach in interactive multiobjective optimization. European Journal of Operational Research 170(3), 909–922 (2006)

Ojalehto, V., Miettinen, K., Mäkelä, M.M.: Interactive software for multiobjective optimization: IND-NIMBUS. WSEAS Transactions on Computers 6(1), 87–94 (2007)

Phelps, S., Köksalan, M.: An interactive evolutionary metaheuristic for multiobjective combinatorial optimization. Management Science 49(12), 1726–1738 (2003)

Purshouse, R.C.: On the evolutionary optimisation of many objectives. Ph.D. thesis, Department of Automatic Control and Systems Engineering, The University of Sheffield (2003)

Sakawa, M., Shibano, T.: An interactive fuzzy satisficing method for multiobjective 0-1 programming problems with fuzzy numbers through genetic algorithms with double strings. European Journal of Operational Research 107(3), 564–574 (1998)

Sakawa, M., Yauchi, K.: An interactive fuzzy satisficing method for multiobjective nonconvex programming problems through floating point genetic algorithms. European Journal of Operational Research 117(1), 113–124 (1999)

Słowiński, R.: Review of multiple objective programming methods (Part I in Polish). Przegląd Statystyczny 31, 47–63 (1984)

Steuer, R.E.: Multiple Criteria Optimization - Theory, Computation and Application. Wiley, Chichester (1986)

Sun, M., Steuer, R.E.: Interquad: An interactive quad tree based procedure for solving the discrete alternative multiple criteria problem. European Journal of Operational Research 89(3), 462–472 (1996)

Takagi, H.: Interactive evolutionary computation: fusion of the capabilities of ec optimization and human evaluation. Proceedings of the IEEE 89, 1275–1296 (2001)

Tanaka, M., Watanabe, H., Furukawa, Y., Tanino, T.: GA-based decision support system for multicriteria optimization. In: International Conference on Systems, Man and Cybernetics, vol. 2, pp. 1556–1561. IEEE Computer Society Press, Los Alamitos (1995)

Taner, O.V., Köksalan, M.: Experiments and an improved method for solving the discrete alternative multiple criteria problem. Journal of the Operational Research Society 42(5), 383–392 (1991)

Tanino, T., Tanaka, M., Hojo, C.: An interactive multicriteria decision making method by using a genetic algorithm. In: 2nd International Conference on Systems Science and Systens Engineering, pp. 381–386 (1993)

Thiele, L., Miettinen, K., Korhonen, P.J., Molina, J.: A preference-based interactive evolutionary algorithm for multiobjective optimization. Working papers w-412, Helsinki School of Economics, Helsinki (2007)

Todd, D.S., Sen, P.: Directed multiple objective search of design spaces using genetic algorithms and neural networks. In: Banzhaf, W., et al. (eds.) Genetic and Evolutionary Computation Conference, pp. 1738–1743. Morgan Kaufmann, San Francisco (1999)

Ulungu, B., Teghem, J., Ost, C.: Efficiency of interactive multi-objective simulated annealing through a case study. Journal of the Operational Research Society 49, 1044–1050 (1998)

Ulungu, E.L., Teghem, J.: Multi-objective combinatorial optimization problems: A survey. Journal of Multi-Criteria Decision Analysis 3, 83–101 (1994)

Viana, A., de Sousa, J.P.: Using metaheuristics in multiobjective resource constrained project scheduling. European Journal of Operational Research 120, 359–374 (2000)

Zionts, S.: Criteria method for choosing among discete alternatives. European Journal of Operational Research 7(1), 143–147 (1981)

8

Visualization in the Multiple Objective Decision-Making Framework

Pekka Korhonen and Jyrki Wallenius

Helsinki School of Economics, Department of Business Technology, P.O. Box 1210, FI-00101 Helsinki, Finland, pekka.korhonen@hse.fi, jyrki.wallenius@hse.fi

Abstract. In this paper we describe various visualization techniques which have been used or which might be useful in the multiple objective decision making framework. Several of the ideas originate from statistics, especially multivariate statistics. Some techniques are simply for illustrating snapshots of a single solution or a set of solutions. Others are used as an essential part of the human-computer interface.

8.1 Introduction

We describe various visualization techniques which have been proven useful or which we feel might prove useful in the multiple objective decision making framework. We focus on fundamental visualization techniques (see Chapter 9, for more specific techniques). Several of our ideas originate from statistics, especially multivariate statistics. Typically, in the multiple objectives framework, the decision maker (DM) is asked to evaluate a number of alternatives. Each alternative is characterized using an objective vector. From the perspective of visualization, the complexity of the decision problem depends on two dimensions: the number of objectives and the number of alternatives. A problem may be complex due to a large number of alternatives and a small number of objectives, or the other way round, although the nature of the complexity is different. Different visualization techniques are required for each case. The number of alternatives may also be uncountable, such as a subset of a feasible region in an objective space in multiobjective optimization.

In descriptive statistics, computer graphics is widely used to illustrate numerical information by producing standard visual representations (bar charts, line graphs, pie charts, etc.). More advanced visualization techniques, for example, Andrews (1972) curves and Chernoff (1973) faces have also been proposed. Especially Andrews curves and Chernoff faces were developed to illus-

Reviewed by: Julian Molina, University of Malaga, Spain
Mariana Vassileva, Bulgarian Academy of Sciences, Bulgaria
Kaisa Miettinen, University of Jyväskylä, Finland

J. Branke et al. (Eds.): Multiobjective Optimization, LNCS 5252, pp. 195–212, 2008.

trate multivariate data; a problem closely related to ours. These techniques have been developed for problems in which the main purpose is to obtain a holistic view of the data and/or to identify clusters, outliers, etc. In the multiple objective framework, an additional requirement is to provide the DM with information (value information) for articulating preferences.

In this chapter we review visualization techniques, which are useful for illustrating snapshots of a single solution or a set of solutions in discrete and continuous situations. The evaluation of alternatives is a key issue in multiple objective approaches. Although this book is devoted to continuous MCDM/EMO-problems, the set of alternatives which the DM is asked to evaluate is generally finite and its cardinality is small. An example of an exception is Lotov's Generalized Reachable Sets method (Lotov *et al.*, 2004). Therefore, graphical methods developed in statistics to illustrate discrete alternatives are essential, irrespective of whether the MCDM/EMO-problem is continuous or discrete. In addition, many continuous problems are approximated with discrete sets.

To facilitate DM's evaluations, graphical representation is an essential part of the human-computer interface. For more information, see Chapter 9. Interactive methods are described in Chapter 2. The organization of this chapter is as follows. In Section 8.2 we consider the use and misuse of standard statistical techniques for visual representation of numerical data. In Section 8.3 we describe visualization in the context of a multiple objective framework. Section 8.4 provides a discussion and conclusion.

8.2 Visual Representation of Numerical Data

Graphical techniques have been considered extremely useful by statisticians in analyzing data. However, they have not been utilized by the MCDM- or EMO-community to their full potential in, for example, interactive approaches. Statisticians have developed a number of graphical methods for data analysis. Standard graphical techniques, such as bar charts, value paths, line graphs, etc., have a common feature: they provide an alternative representation for numerical data and there is a one-to-one correspondence between a graphical representation and numerical data (see any basic textbook in statistics, for example, Levine *et al.* (2006).) The graphical representation can be transformed back into numerical form (with a certain accuracy) and conversely.

8.2.1 Standard Statistical Techniques

In this subsection we review some standard graphical techniques, such as bar charts, line graphs, and scatter plots, which are widely used to summarize information in statistical data sets. As an example, we use a small data set consisting of the unemployment rate (%) and the inflation rate (%) in nine countries.

The bar charts are a standard technique to summarize frequency data (see, e.g., Figure 8.10), and they are called histograms in this context. In Figure 8.1, we use a bar chart to represent the (values of) unemployment rate (%) and the inflation rate (%) for each of the nine countries. This is customary in a multiobjective framework, where instead of summary data the values of variables (objectives) of various alternatives are more interesting. Let us emphasize that in this subsection we follow the terminology of statistics and talk about variables, but in connection to multiobjective optimization variables correspond to objectives. In other words, we refer to the objective space and not to the variable space of multiobjective optimization.

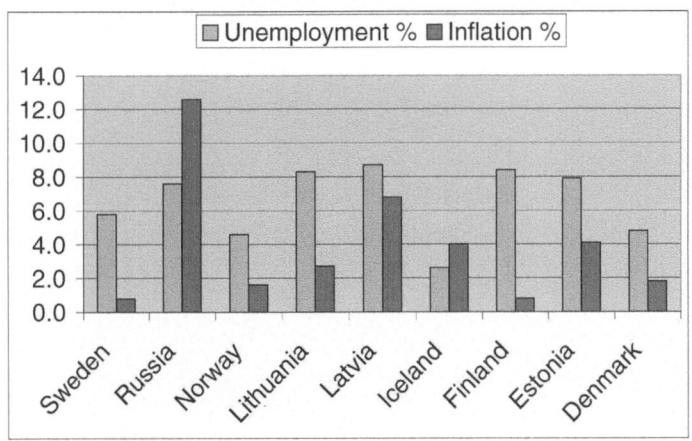

Fig. 8.1. Illustrating unemployment and inflation rates with a bar chart

Line graphs are another technique used like bar charts. (Line graphs (line charts) are called value paths in the multiple objective framework.) They are particularly appropriate, when the order of alternatives has a special meaning as in time series. In Figure 8.2, we ordered the alternatives in terms of decreasing inflation rates. This should make it easier to see how the unemployment rate and the inflation rate are related in different countries. The dependence between the unemployment rate and the inflation rate can alternatively be observed from the scatter diagram in Figure 8.3. Figure 8.3 is very useful in case where we are interested in recognizing the best (Pareto optimal) countries (Iceland, Norway, and Sweden).

There are also many other visualization techniques used in statistics, such as pie charts and boxplots, which may be useful in a specific multiple objective context. Pie charts are, for example, useful for visualizing probabilities and percentages. For further information, please consult any basic statistics textbook such as (Bowerman *et al.*, 2004).

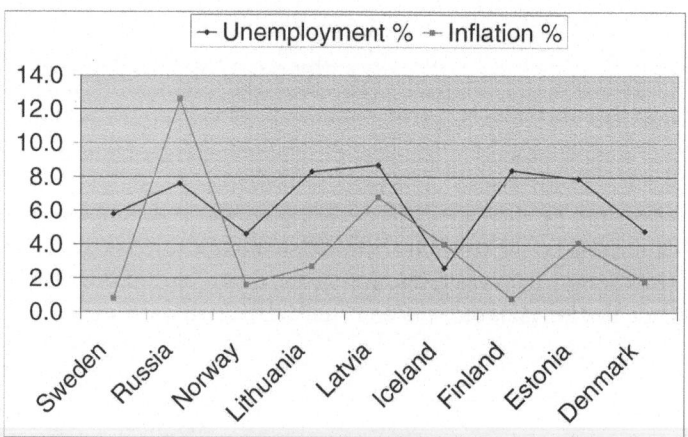

Fig. 8.2. Line graph

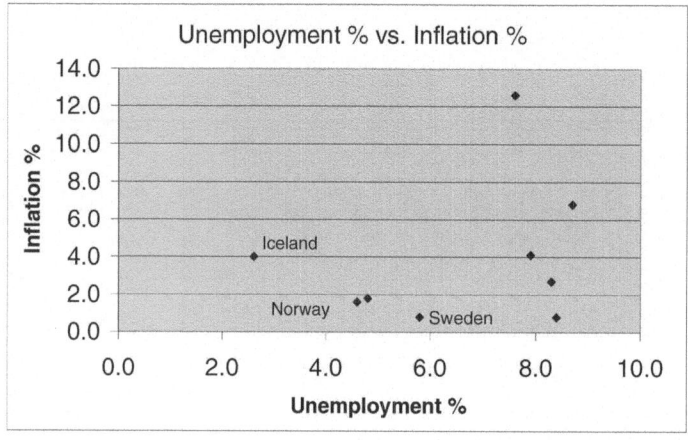

Fig. 8.3. Relationship between unemployment and inflation using a scatter diagram

8.2.2 Visualization of Multivariate Data

Visual representation is limited to two dimensions. Therefore, the main problem in visualizing multivariate data is to construct a two-dimensional representation, when the number of variables (i.e., objectives in the MCDM/EMO context) exceeds two. In statistics, two general principles have been applied to this problem:

1. reduce the dimensionality of a problem or
2. plot a multivariate observation as an object (an icon).

Principal component analysis and multidimensional scaling (MDS) (Mardia *et al.*, 1979) are two well-known techniques for obtaining a low-dimensional (specifically two-dimensional) representation of multivariate data, so that the data may be examined visually (see, for example, Everitt (1978)).

In principle, some standard graphical techniques, such as bar charts and line graphs can be used to illustrate more than two-variable data sets, but when the number of variables and/or alternatives increases the graphs quickly become unreadable. However, some standard techniques such as radar charts are more appropriate to illustrate multivariate data, provided the number of variables is reasonably small. As you can see from Figure 8.4, the radar chart is not very clear, even if we have only nine alternatives and two variables. However, the chart is readable if the number of variables is not large. A remedy to problems with a large number of variables is to represent them in different pictures.

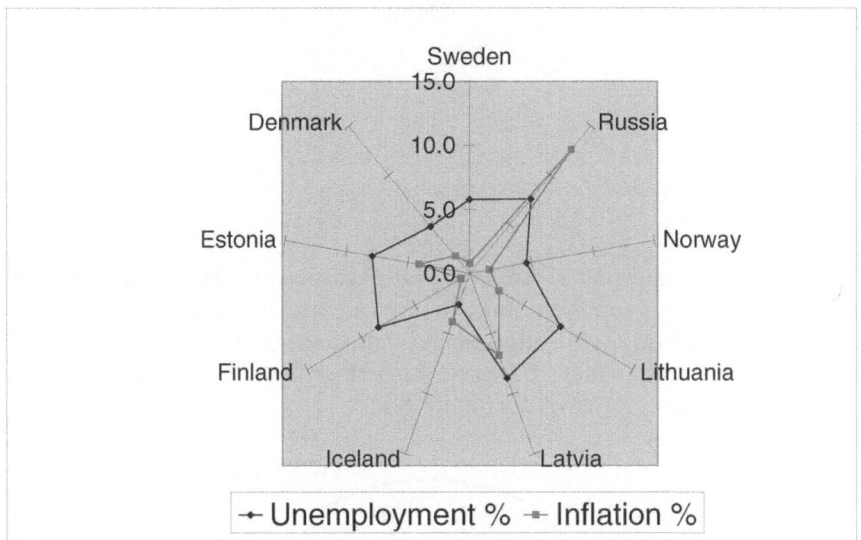

Fig. 8.4. Radar chart

In the early 1970's two promising techniques: Andrews (1972) curves and Chernoff (1973) faces were developed for visualizing multivariate data. Andrews plotted the following curve

$$g_i(t) = \frac{z_{i1}}{\sqrt{2}} + z_{i2} \sin t + z_{i3} \cos t + z_{i4} \sin 2t + \ldots$$

for each data point $z_i = (z_{i1}, z_{i2}, \ldots, z_{ip})$ over the interval $-\pi \leq t \leq \pi$, $i = 1, 2, \ldots, n$, where p refers to the number of variables (i.e., objectives in multiobjective optimization, denoted by k in this book). Thus, each observation is a harmonic curve in two dimensions. In this technique the number of variables is unlimited. The harmonic curves depend on the order in which the variables have been presented. Figure 8.5 reproduces the famous iris flower data (Fisher, 1936). The data set consists of three different species of iris flowers.

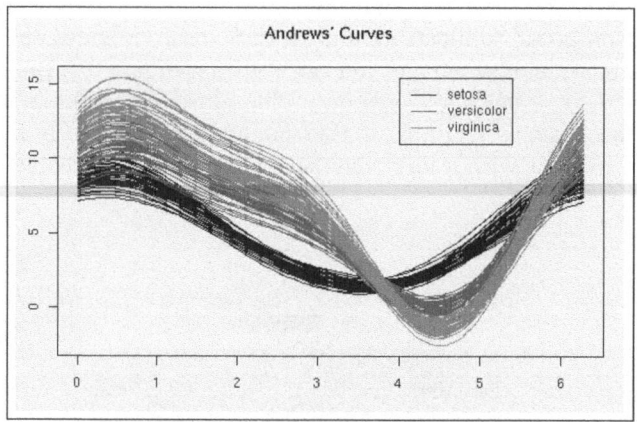

Fig. 8.5. Andrews curves (R Graph Gallery, 2007)

Chernoff used a human face to represent each observation graphically. The construction of Chernoff faces consists of geometrically well-defined elements, such as arcs of circles, arcs of ellipses, and straight lines. The values of variables are used as the parameters of these elements. Chernoff's original proposal consisted of 18 face parameters (Figure 8.6).

Fig. 8.6. Chernoff face

Andrews harmonic curves and Chernoff faces help us view similarities and dissimilarities between observations, identify clusters, outliers etc., but they are not very suitable for describing preference information. In Andrews curves, each curve stands for one observation, and the curves, which do not deviate much from each other, represent similar observations. However, Andrews curves are not good to illustrate the magnitude of variable values, and are thus not very practical to describing preference information. In a Chernoff face, it is easy to understand that a "smile" means something positive, but the length of the nose does not convey similar information. In addition, we have no knowledge about the joint effects of the face parameters. Big eyes and a long nose may make the face look silly in some user's mind, although big eyes are usually a positive feature.

In spite of the preceding disadvantages, the techniques provide us with new directions for developing visual techniques. Especially, there is a need for methods, which can also convey preference information.

8.2.3 A Careful Look at Using Graphical Illustrations

We should always present as truthful a representation of the data set as possible. However, it is possible to construct graphs that are misleading. We do not want that. One should always be aware of the ways statistical graphs and charts can be manipulated purposefully to distort the truth. In the multiple objective framework, where the DM is an essential part of the solution process, this may be even more important. For example, in interactive approaches, the DM is asked to react to graphical representations. A wrong illusion provided by a graph may lead to an undesirable final solution.

In the following, we present some common examples of (purposefully) misleading graphs. In Figure 8.7, we have described the development of traffic fatalities in Finland during selected years between 1980–2000 using histograms. The left-hand figure stands for the standard case, where the vertical axis starts from zero. In the right-hand figure, we present the same data, but start the vertical scale from 350 instead of 0. This makes the decrease in traffic fatalities appear more dramatic.

In Figure 8.8, we illustrate the per capita electricity consumption in Denmark and Finland. The consumption of electricity in Finland is about twice that of Denmark. It sounds appealing to illustrate the consumption by using an object somehow related to electricity such as light bulbs. Maintaining the shape of the light bulb in the left-hand figure makes the electricity consumption (height of the bulb) in Finland appear much larger than it actually is. In

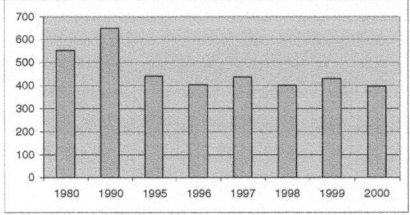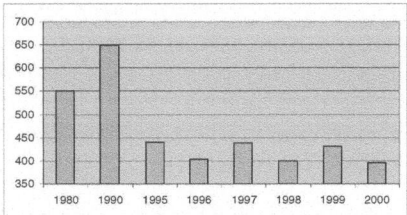

Fig. 8.7. Traffic fatalities in Finland: two representations

the right-hand figure, we have only stretched the height of Denmark's light bulb for Finland. This kind of illustration provides a correct impression.

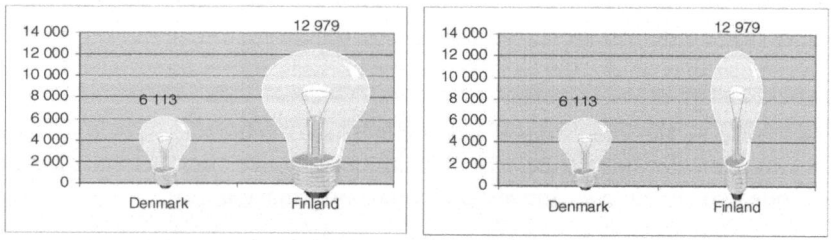

Fig. 8.8. Electricity consumption/capita in Denmark and Finland: two representations

In Figure 8.9, we illustrate the meaning of a stretched axis with a line graph. A neutral approach to describing, for example, the development of the unemployment rate by using a line graph is to start the percentage scale from zero and end it above the maximum, as is done in the left-hand figure. If we stretch the vertical axis and take the range roughly from the minimum to the maximum in the right-hand figure, it makes the downward trend look steeper, demonstrating "a dramatic improvement in the unemployment rate". Concerning the use of Figures 8.7–8.9 in an MCDM/EMO-context, it is not obvious which figure is better. For instance, sometimes it is useful to zoom in on some value range of objectives, especially, when we like to compare small differences in objective values. In other cases, a holistic figure is more desirable.

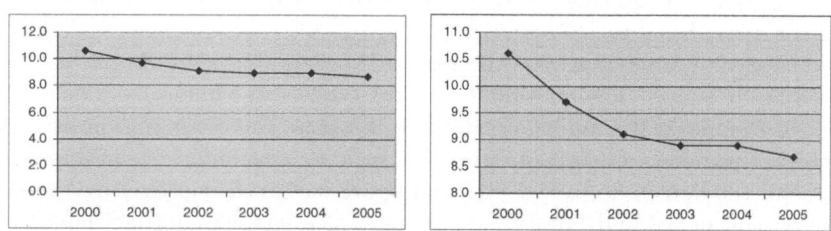

Fig. 8.9. Unemployment (%) in Finland during 2000-2005: two representations

Above, we have presented only some representative examples of misleading graphs. More information can be found in (Huff, 1954). In this classical book, the author shows how to take a graph and make it say anything you want. See also Wainer (1984).

8.3 Visualization in Multiple Objective Decision Making Approaches

Graphical techniques used with multivariate data have been of special interest for researchers working on MCDM problems, because of many similarities between these two problems. These techniques may also be used in MCDM/EMO problems to provide the DM with holistic information and to obtain a quick overall view of the relevant information, as well as detailed information for evaluation and comparison purposes.

Many authors have proposed the use of graphical techniques (bar charts, value paths, line graphs, etc.) to help evaluate alternatives (see, for example, Cohon (1978); Geoffrion et al. (1972); Grauer (1983); Grauer et al. (1984); Kok and Lootsma (1985); Korhonen and Laakso (1986); Korhonen and Wallenius (1988), Schilling et al. (1983); Silverman et al. (1985); Steuer (1986)). In Miettinen (1999), there is one chapter devoted to different visualization techniques. In most situations, the amount of information presented to the DM for evaluation may be considerable. A visual representation improves the readability of such information.

In statistics, the reduction of the dimensionality of a problem is a widely used technique to compress the information in multivariate data. If the number of objectives in the multiobjective context can be reduced to two (or three) without loosing essential information, then the data may be examined visually. Principal component analysis and multidimensional scaling (MDS) are also interesting from the point of view of MCDM. However, to our knowledge the principal component analysis has not been used for graphical purposes – with an exception of the paper by Mareschal and Brans (1988), in which they showed how to describe objectives and alternatives in the same picture. Korhonen et al. (1980) used MDS to reduce a four-objective problem into two dimensions and then described their search procedure in terms of a planar graph.

Graphical techniques have also been implemented as part of several computer systems developed for solving MCDM problems. Well known systems DIDASS (Dynamic Interactive Decision Analysis and Support Systems), Expert Choice, PREFCALC, NIMBUS, and Reachable Goals method are good examples. DIDASS has been developed by the System and Decision Sciences (SDS) research group at IIASA (Grauer et al., 1984). Expert Choice has been developed to implement the AHP (the Analytic Hierarchy Process) (Saaty, 1980). PREFCALC has been proposed by Jacquet-Lagreze and Siskos (1982) for assessing a set of additive utility functions. Miettinen and Mäkelä (1995, 2000, 2006) developed the WWW-NIMBUS system, the first web-based MCDM decision support system, for nonlinear multiobjective optimization (http://nimbus.it.jyu.fi/). Lotov and his colleagues proposed the Reachable Goals method, in which the authors slice and visualize the Pareto optimal set (Lotov et al., 2004). In addition, we would like to mention our systems VIG (Korhonen, 1987) and VIMDA (Korhonen, 1988; Korhonen and Karaivanova,

1999). For linear multiobjective optimization problems, VIG implements a free search in the Pareto optimal set by using a dynamic graphical interface called Pareto Race (Korhonen and Wallenius, 1988). VIMDA is a discrete version of the original Korhonen and Laakso (1986) method. The current version is able to deal with many millions of nondominated or Pareto optimal alternatives. In Figure 8.13 we display how a reference direction is projected into a set of randomly generated 500,000 alternatives.

8.3.1 Snapshots of a Single Solution

We can use graphical techniques to illustrate a single solution (objective vector, alternative). Classical techniques, such as bar charts, are mostly suitable for this purpose. The objectives are described on the x-axis and their values on the y-axis. Figure 8.10 illustrates unemployment rates in different population groups (total, male, female, youth, and long term) in Finland. This corresponds to a situation of one alternative and five objectives. When the number of objectives increases, the bar chart representation looses its ability to convey holistic information.

Chernoff faces (Figure 8.6) is one of the techniques, which makes it possible to provide information on an alternative with one icon. In the MCDM/EMO context, one has multiple alternatives to choose from. Sometimes, the choice is between two (or a few) alternatives (in what follows, we refer to such cases as discrete sets), sometimes the most preferred solution has to be chosen from an actually infinite number of alternatives (in what follows, we refer to such cases as infinite sets). In the following subsection we consider these situations.

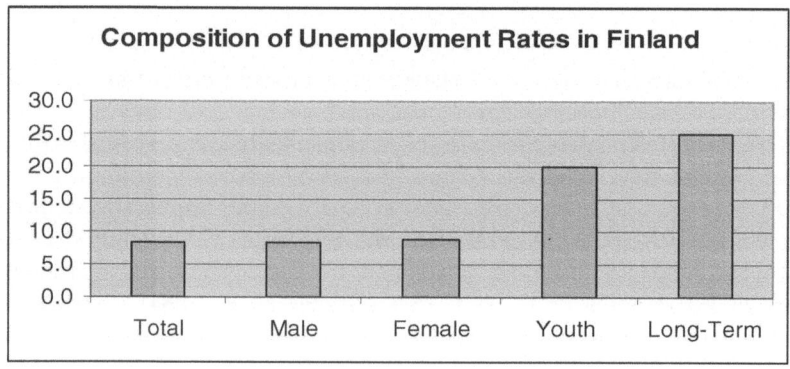

Fig. 8.10. Illustrating unemployment rates in different groups in Finland

8.3.2 Illustrating a Set of Solutions/Alternatives

We first consider finite sets and then extend the discussion to infinite sets (the continuous case).

Finite Sets

Bar charts, again, can be used to compare alternatives described with multiple objectives, when their number is small. In Figure 8.11 we compare the unemployment rates in five different population groups in Finland, Norway and Denmark. As we can see, the bar chart representation suits to objective-wise comparisons well. In this bar chart the country information is provided objective-wise. In fact one can think of Figure 8.11 as consisting of three separate bar charts presented in one graph. However, if one wants to compare across alternatives and choose the best one from the three countries, it is difficult to conclude whether the unemployment situation in Norway is better than in Denmark. (Finland is clearly the worst!)

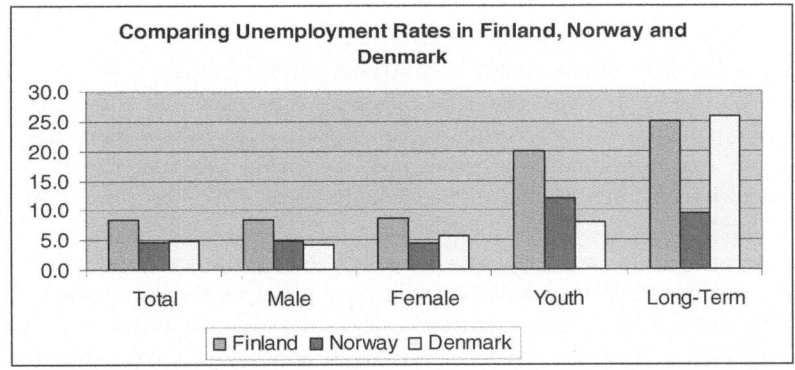

Fig. 8.11. Comparing unemployment rates in Finland, Norway, and Denmark

When the number of objectives is small, and we wish to compare alternatives, the bar chart representation is appropriate, but it would make sense to present the objectives pertaining to each alternative (country) together, as is done in Figure 8.1. A widely used alternative in MCDM is to use line graphs. Each line commonly stands for one objective. See Figure 8.2 for a typical two-objective example. The visual effect may be enhanced by ranking alternatives according to one objective, as is done in Figure 8.2.

A line graph may also be drawn in such a way that one line stands for one alternative, with the objectives on the x-axis. In the MCDM framework, this is often called a score profile (Belton and Stewart, 2001).

When standard graphical techniques (e.g., bar charts, value graphs, line graphs) are used to visualize alternative solutions, whether a solution is more or less preferred to another one in terms of some objective can be seen in the details. Short bars, lines with a negative slope, etc. stand for good values or improvements (in the minimization case). However, obtaining a holistic perception of these pieces of preference information, when there are a lot of details, is often impossible. On the other hand, advanced techniques, such as

Chernoff faces and Andrews harmonic curves, help the DM obtain a holistic perception of the alternatives, but do not provide a good basis for the purpose of evaluation. Therefore, Korhonen (1991) developed an approach, which transforms a vector into a picture in the spirit of Chernoff faces and Andrews curves, but which also enables a DM to see information that influences her/his preference on the basis of a visual representation. The underlying ideas are based on the use of two concepts: harmony and symmetry. For the icon, Korhonen chose a simple "harmonious house". The ideal (standard/normal) house is described in a harmonious (and symmetric) form. Deviations from this ideal are perceived as abnormal. The "degree" to which a house resembles the ideal serves as the basis for evaluation.

The first requirement for the icon is that it can be parametrized in such a way that by improving the value of an objective the icon becomes "better" or more "positive" in some sense. To convey this positive-negative information we can apply the concepts of harmony and symmetry.

The structure of the house is controlled by varying the positions of the corner points. Each corner point is in its default position and allowable moves of corner points are shown as squares. An objective can now be associated with the x- or y-coordinate of any corner point in such a way that the ideal value of the objective corresponds to the default value of the coordinate and the deviation from the ideal value is shown as a move in an x- or y-direction. The x- and y-coordinates of corner points are called house parameters. Two objectives can be associated with each corner point, one affecting the horizontal position and the other the vertical position of the corner point. When the value of the objective deviates much from the ideal, it has a dramatic effect on the position of a corner point.

A preliminary version of an interactive micro-computer decision support system known as VICO (A VIsual multiple criteria COmparison) has been developed to implement the above idea. A preliminary version of VICO has been used for experimental purposes with student subjects. The results were encouraging. Figure 8.12 refers to a test situation, where the subjects compared 20 companies using 11 objectives, consisting of detailed financial information. The purpose of the test was to identify the three companies, which had filed for bankruptcy. One of the companies (Yritys 19) in Figure 8.12 was one of the bankrupt companies. Note how disharmonious it is compared to a well-performing company (Yritys 1). In the system, the houses are compared in a pairwise fashion.

The method is dependent on how the house parameters are specified. It is important to associate key objectives with the house parameters that determine the structure of the house – thus influencing the degree of harmony. Of course, the directions of the changes also play an essential role. The method is subjective, and therefore it is important that a DM takes full advantage of this subjectivity and uses his/her knowledge of the problem and its relationships in the best possible way. VICO is quite suitable for MCDM problems, where the number of objectives is large and one uses pairwise comparisons to collect preference information.

An even more popular approach to visualizing a set of solutions is to use line graphs. This approach has been used, for example, in VIMDA (Korhonen, 1988). VIMDA is a "free search" type of approach that makes no assumptions, except monotonicity, about the properties of the DM's value function. It is a visual, interactive procedure for solving discrete multiple criteria decision problems. It is very suitable to continuous problems as well, when the number of objective vectors to be simultaneously evaluated by the DM is large. The search is controlled by varying the aspiration levels. The information is used to generate a set of discrete alternatives, which are provided for evaluation in a graphical form described in Figure 8.13.

The objective values in Figure 8.13 are shown on the ordinate. The current alternative is shown in the left-hand margin. The objective values of consecutive alternatives have been connected with lines using different colors and patterns. The cursor characterizes the alternative whose objective values are printed numerically at the top of the screen. The cursor moves to the right and to the left, and each time the objective values are updated. The DM is asked to choose his/her most preferred alternative from the screen by pointing the cursor.

Using this procedure, the DM is free to examine any Pareto optimal solution. Furthermore, this freedom is not restricted by previous choices. The currently implemented version of VIMDA does not include a stopping criterion based on a mathematical optimality test. The process is terminated when the DM is satisfied with the currently best solution.

Infinite Sets

The Geoffrion *et al.* (1972) interactive procedure was the first to present the idea of a (one dimensional) search in a projected direction in the context of continuous multiobjective optimization. They implemented the classic Frank-Wolfe (single objective) nonlinear programming algorithm for solving multiobjective problems. Korhonen and Laakso (1986), in their reference direction approach for solving continuous multiobjective optimization problems, adopted the idea of a visual line search from Geoffrion *et al.* (1972). In their approach, the DM provides his/her aspiration levels for the objectives, thereby defining a reference direction. This reference direction is projected onto the Pareto optimal set. Solutions along the projection are presented to the DM, who is assumed to choose his/her most preferred solution along the projection. The algorithm continues by updating the aspiration levels, forming a new reference direction, etc.

Lotov's "slices" represent an interesting visualization technique for continuous multiobjective problems (see, for example, Lotov *et al.* (1997, 2004)). The approach is based on a visualization of the feasible set in the objective space. For details, see Chapter 9.

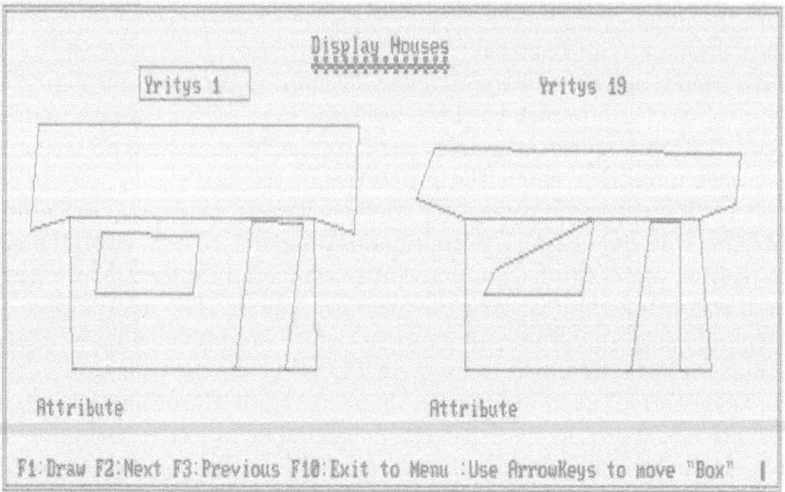

Fig. 8.12. Two harmonious houses

For other techniques providing a graphical illustration of Pareto optimal solutions, see, for example, Miettinen (2003).

8.3.3 Dynamic Representation of a Set of Solutions

Pareto Race (Korhonen and Wallenius, 1988) is a dynamic version of the Korhonen and Laakso (1986) reference direction approach. It enables a DM to move freely in the Pareto optimal set and, thus to work with the computer to find the most preferred values for the objectives (output variables). Figure 8.14 shows an example of the Pareto Race screen. In Pareto Race the DM sees the objective values on a display in numeric form and as bar graphs, as (s)he travels along the Pareto optimal set. The keyboard controls include an accelerator, gears and brakes. The search in the Pareto optimal set is analogous to driving an automobile. The DM can, for example, increase/decrease speed and brake at any time. It is also possible to change direction.

8.4 Discussion and Conclusion

In this chapter we have considered the use of graphics in the multiobjective decision making framework. The main issue is to enable the DM to view objective vectors (multidimensional alternatives) and to facilitate preference comparisons. The alternatives may be presented one at a time, two at a time, or many at a time for the DM's consideration. Based on the available information, the MCDM procedures ask the DM to choose the best (or the worst)

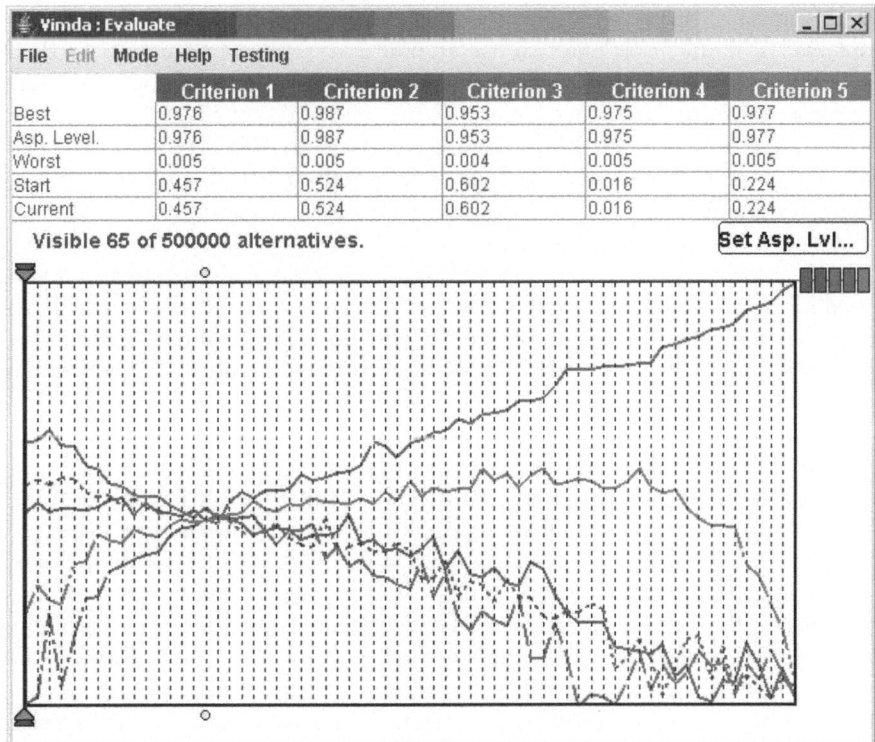

Fig. 8.13. VIMDA's visual interface

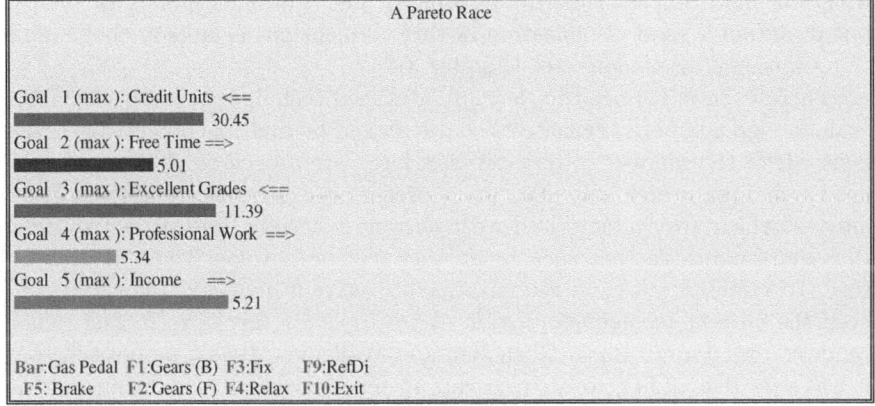

Fig. 8.14. Pareto Race interface

from the set of displayed alternatives, as well as to rank, or cluster the alternatives. In many cases one can use standard statistical techniques, or their variations, developed for visualizing numerical data. In statistics the main purpose is to provide a tool for classifying, clustering or identifying outliers – in short to obtain a holistic view of the data. In MCDM/EMO, the DM needs support in expressing preference information, based on displayed numerical data. Hence, in some cases more advanced techniques are needed for visualizing alternatives.

We have explored to what extent standard graphical techniques can be used in the MCDM/EMO framework, and reviewed some more advanced techniques specifically developed for the MCDM problem. Examples include the use of dynamic bar charts, such as the Pareto Race interface (Korhonen and Wallenius, 1988), and Korhonen's harmonious houses (Korhonen, 1991). Typically, the graphical representation is an essential part of the user interface in interactive MCDM procedures.

There exist several other ways to utilize graphics in MCDM procedures. For example, Salo and Hämäläinen (1992) developed a method allowing the DM to express approximate preference statements as interval judgments, which indicate a range for the relative importance of the objectives. The ranges are given as bar charts in an interactive manner. Moreover, Hämäläinen and his colleagues have implemented a system by name HIPRE 3+ (http://www.hipre.hut.fi/), which includes the above idea and other graphical interfaces. A typical example is a graphical assessment of a value function (Hämäläinen, 2004).

An important problem in MCDM/EMO is to provide a tool, with which one could obtain a holistic view of the Pareto optimal set. This is difficult in more than two or three dimensions in the objective space. Interestingly, in EMO the nondominated set evolves from one generation to the next; hence it would be important to visualize this process. The two-dimensional case is easy, but to obtain a good visualization in three dimensions is already challenging (for additional discussions, see Chapter 3).

There is clearly a need to develop advanced techniques, which help DMs evaluate and compare alternatives characterized by multiple objectives. Problems where the number of objectives is large are especially challenging. We need techniques which help DMs make preference comparisons between alternatives, which are characterized with tens of objectives (criteria, attributes). It is quite plausible that such techniques will be developed in the near future. To visualize search in the multiple objective framework is not easy, but even the current techniques provide useful tools for this case. In fact, when we developed Pareto Race (Korhonen and Wallenius, 1988), we noticed that it was very demanding to "drive a car" in ten dimensions, but using moving bar charts we were able to implement the idea. The visualization of the whole Pareto optimal set in more than three dimensions is a problem which may be too complicated to solve. However, we may surely invent partial solutions!

Acknowledgements

The authors wish to thank Kaisa Miettinen, Julian Molina, Alexander Lotov, and Mariana Vasileva for useful comments. Moreover, we would like to acknowledge the financial support of the Academy of Finland (grant 121980).

References

Andrews, D.: Plots of high dimensional data. Biometrics 28, 125–136 (1972)

Belton, V., Stewart, T.J.: Multiple Criteria Decision Analysis: An Integrated Approach. Kluwer Academic Publishers, Dordrecht (2001)

Bowerman, B.L., O'Connell, R.T., Orris, J.B.: Essentials in Business Statistics. McGraw-Hill, New York (2004)

Chernoff, H.: Using faces to represent points in k-dimensional space graphically. Journal of American Statistical Association 68, 361–368 (1973)

Cohon, J.L.: Multiobjective Programming and Planning. Academic Press, New York (1978)

Everitt, B.: Graphical Techniques for Multivariate Data. Heinemann Educational Books, London (1978)

Fisher, R.A.: The use of multiple measurements in taxonomic problems. Annual Eugenics 7, 179–188 (1936)

Geoffrion, A.M., Dyer, J.S., Feinberg, A.: An interactive approach for multi-criterion optimization, with an application to the operation of an academic department. Management Science 19, 357–368 (1972)

Grauer, M.: Reference point optimization – the nonlinear case. In: Hansen, P. (ed.) Essays and Surveys on Multiple Criteria Decision Making, vol. 209, pp. 126–135. Springer, Berlin (1983)

Grauer, M., Lewandowski, A., Wierzbicki, A.: DIDASS – theory, implementation and experiences. In: Grauer, M., Wierzbicki, A. (eds.) Interactive Decision Analysis, vol. 229, pp. 22–30. Springer, Berlin (1984)

Hämäläinen, R.P.: Reversing the perspective on the applications of decision analysis. Decision Analysis 1, 26–31 (2004)

Huff, D.: How to Lie with Statistics. Norton, New York (1954)

Jacquet-Lagreze, E., Siskos, J.: Assessing a set of additive utility functions for multicriteria decision making, the uta method. European Journal of Operational Research 10, 151–164 (1982)

Kok, M., Lootsma, F.: Pairwise-comparison methods in multiple objective programming, with applications in a long-term energy-planning model. European Journal of Operational Research 22, 44–55 (1985)

Korhonen, P.: VIG: A Visual Interactive Support System for Multiple Criteria Decision Making. Belgian Journal of Operations Research, Statistics and Computer Science 27, 3–15 (1987)

Korhonen, P.: A visual reference direction approach to solving discrete multiple criteria problems. European Journal of Operational Research 34, 152–159 (1988)

Korhonen, P.: Using harmonious houses for visual pairwise comparison of multiple criteria alternatives. Decision Support Systems 7, 47–54 (1991)

Korhonen, P., Karaivanova, J.: An algorithm for projecting a reference direction onto the nondominated set of given points. IEEE Transactions on Systems, Man, and Cybernetics 29, 429–435 (1999)

Korhonen, P., Laakso, J.: A visual interactive method for solving the multiple criteria problem. European Journal of Operational Research 24, 277–287 (1986)

Korhonen, P., Wallenius, J.: A Pareto race. Naval Research Logistics 35, 615–623 (1988)

Korhonen, P., Wallenius, J., Zionts, S.: A bargaining model for solving the multiple criteria problem. In: Fandel, G., Gal, T. (eds.) Multiple Criteria Decision Making Theory and Application, pp. 178–188. Springer, Berlin (1980)

Levine, D.M., Krehbiel, T.C., Berenson, M.L.: Business Statistics: A First Course. Prentice-Hall, Englewood Cliffs (2006)

Lotov, A.V., Bushenkov, V.A., Chernov, A.V., Gusev, D.V., Kamenev, G.K.: Internet, GIS and interactive decision maps. Journal of Geographic Information and Decision Analysis 1, 118–143 (1997)

Lotov, A.V., Bushenkov, V.A., Kamenev, G.K.: Interactive Decision Maps. Approximation and Visualization of Pareto Frontier. Kluwer Academic Publishers, Boston (2004)

Mardia, K.V., Kent, J.T., Bibby, J.M.: Multivariate Analysis. Academic Press, San Diego (1979)

Mareschal, B., Brans, J.: Geometrical representations for mcda. European Journal of Operational Research 34, 69–77 (1988)

Miettinen, K.: Nonlinear Multiobjective Optimization. Kluwer Academic Publishers, Boston (1999)

Miettinen, K.: Graphical illustration of Pareto optimal solutions. In: Tanino, T., Tanaka, T., Inuiguchi, M. (eds.) Multi-Objective Programming and Goal Programming: Theory and Applications, pp. 197–202. Springer, Berlin (2003)

Miettinen, K., Mäkelä, M.M.: Interactive bundle-based method for nondifferentiable multiobjective optimization: NIMBUS. Optimization 34, 231–246 (1995)

Miettinen, K., Mäkelä, M.M.: Interactive multiobjective optimization system WWW-NIMBUS on the Internet. Computers & Operations Research 27, 709–723 (2000)

Miettinen, K., Mäkelä, M.M.: Synchronous approach in interactive multiobjective optimization. European Journal of Operational Research 170, 909–922 (2006)

R Graph Gallery (2007),
http://addictedtor.free.fr/graphiques/RGraphGallery.php?graph=47

Saaty, T.: The Analytic Hierarchy Process. McGraw-Hill, New York (1980)

Salo, A., Hämäläinen, R.P.: Preference assessment by imprecise ratio statements. Operations Research 40, 1053–1061 (1992)

Schilling, D., Revelle, C., Cohon, J.: An approach to the display and analysis of multiobjective problems. Socio-Economic Planning Sciences 17, 57–63 (1983)

Silverman, J., Steuer, R.E., Whisman, A.: Computer graphics at the multicriterion computer/user interface. In: Haimes, Y., Chankong, V. (eds.) Decision Making with Multiple Objectives, vol. 242, pp. 201–213. Springer, Berlin (1985)

Steuer, R.E.: Multiple Criteria Optimization: Theory, Computation, and Application. John Wiley and Sons, New York (1986)

Wainer, H.: How to display data badly. The American Statistician 38, 137–147 (1984)

9

Visualizing the Pareto Frontier

Alexander V. Lotov[1] and Kaisa Miettinen[2]

[1] Dorodnicyn Computing Centre of Russian Academy of Sciences, Vavilova str., 40, Moscow 119333 Russia, lotov08@ccas.ru

[2] Department of Mathematical Information Technology, P.O. Box 35 (Agora), FI-40014 University of Jyväskylä, Finland, kaisa.miettinen@jyu.fi[*]

Abstract. We describe techniques for visualizing the Pareto optimal set that can be used if the multiobjective optimization problem considered has more than two objective functions. The techniques discussed can be applied in the framework of both MCDM and EMO approaches. First, lessons learned from methods developed for biobjective problems are considered. Then, visualization techniques for convex multiobjective optimization problems based on a polyhedral approximation of the Pareto optimal set are discussed. Finally, some visualization techniques are considered that use a pointwise approximation of the Pareto optimal set.

9.1 Introduction

Visualization of the Pareto optimal set in the objective space, to be called here *a Pareto frontier*, is an important tool for informing the decision maker (DM) about the feasible Pareto optimal solutions in biobjective optimization problems. This chapter is devoted to the visualization techniques that can be used in the case of more than two objective functions with both MCDM and EMO approaches for multiobjective optimization. In practice, we discuss methods for visualizing a large (or infinite) number of Pareto optimal solutions in the case of three and more objectives.

When discussing the visualization of the Pareto frontier, one has to note the difference between the decision space and the objective space (see definitions given in Preface). The techniques covered here are mainly aimed at the visualization of objective vectors (points) which form a part of the feasible objective region. Such an interest in visualizing the Pareto frontier is related mostly to the fact that in multiobjective optimization problems, the preferences of the DM are related to objective values, but not to decision variables

---

[*] In 2007 also Helsinki School of Economics, Helsinki, Finland
Reviewed by: Pekka Korhonen, Helsinki School of Economics, Finland
Sanaz Mostaghim, University of Karlsruhe, Germany
Roman Słowiński, Poznan University of Technology, Poland

J. Branke et al. (Eds.): Multiobjective Optimization, LNCS 5252, pp. 213–243, 2008.
© Springer-Verlag Berlin Heidelberg 2008

directly. Moreover, a possible relatively high dimension of the decision space, which can reach many thousands or more, prevents such a visualization. In contrast, the number of objectives in well-posed multiobjective optimization problems is usually not too high. Thus, one can hope that the visualization of the Pareto frontier would be possible.

Visualization techniques to be discussed in this chapter can be used in decision support in the framework of *a posteriori methods* (see, e.g., Chapter 1 and (Miettinen, 1999)). However, the techniques may also be used by analysts for evaluating Pareto frontiers or for comparing Pareto frontiers constructed using different methods. Usually, *a posteriori* methods include four steps: (1) approximation of the Pareto optimal set; (2) presentation of a representation of the whole Pareto optimal set; (3) specification of a preferred Pareto optimal objective vector by the DM; (4) selecting a (Pareto optimal) decision vector corresponding to the objective vector identified.

Approximation of the Pareto optimal set is computationally the most complicated step of *a posteriori* methods. Though modern computational methods often can approximate this set without human involvement, in complicated cases it must be performed by the analyst. After several decades of moderate development, *a posteriori* methods are now actively developed. This can be attributed to at least two reasons: better understanding of complications that arise in the process of constructing the value (utility) function and using of multiobjective optimization concepts by analysts or researchers deeply involved in real-life applications. These people are usually more interested in practical advantages of the methods than in their theoretical properties.

The analyst supporting the application of *a posteriori* methods has to answer two related questions: what kind of a technique to use to approximate the Pareto frontier and how to inform the DM concerning the resulting approximation. After the first Pareto frontier approximation method was proposed for biobjective linear problems more than 50 years ago by Gass and Saaty (1955), many different approximation techniques have been developed. Sometimes, they are called Pareto frontier generation techniques (Cohon, 1978) or vector optimization techniques (Jahn, 2004). Methods have been developed for problems involving nonlinear models and more than two objectives.

In case of more than two objectives, approximation methods are typically based on either a) approximation by a large but finite number of objective vectors or b) approximation by a convex polyhedral set. The first form can be used in general nonlinear problems and the second only in convex problems. In what follows, we refer to the approximation by a large number of objective vectors as a *pointwise approximation* of the Pareto optimal set.

As far as informing the DM about the Pareto optimal solutions is concerned, there are two possibilities: providing a list of the solutions to her/him or visualizing the approximation. Visualization of the Pareto frontier was introduced by Gass and Saaty (1955). They noted that in the case of two objectives, the Pareto frontier can easily be depicted because the objective space is a plane. Due to this, information concerning feasible objective values and objective tradeoffs can be provided to the DM in a graphic form. However, this convenient and straightforward form of displaying objective vectors forming the Pareto frontier on a plane is restricted to biobjective problems, only.

In the case of more objectives, the standard approach of *a posteriori* methods has been based on approximating the Pareto frontier by a large number of objective vectors and informing the DM about the Pareto frontier by providing a list of such points to her/him. However, selecting from a long list of objective vectors has been recognized to be complicated for human beings (Larichev, 1992). Yet, this approach is still applied (see interesting discussion as well as references in (Benson and Sayin, 1997)). We can attribute this to the fact that convenient human/computer interfaces and computer graphics that are needed in the case of more than two objectives have been absent (at least till the middle of the 1980's). In this respect, it is important to mention that the need for visualizing the Pareto frontier (for more than two objectives) in the form that gives the DM explicit information of objective tradeoffs has already been stated by Meisel (1973). Nowadays, this problem has been solved to a large extent: modern computers provide an opportunity to visualize the Pareto frontier for linear and nonlinear decision problems for three to about eight objectives; and computer networks are able to bring, for example, Java applets that display graphs of Pareto frontiers for different decision problems.

In what follows, we use the notation introduced in Preface. Remember that $P(Z)$ stands for the set of Pareto optimal objective vectors (Pareto frontier), which is a part of the feasible objective region Z. Another set that will often be used in this chapter is defined for multiobjective minimization problems as

$$Z_p = \{\mathbf{z} \in \mathbf{R}^k : z_i \geq z_i', \text{ for all } i, \ \mathbf{z}' \in Z\}.$$

Though the set Z_p has been considered in various theoretical studies of multiobjective optimization for many years, it was named as the Edgeworth-Pareto hull (EPH) of the set Z, in 1995, at the 12th International Conference on Multiple Criteria Decision Making in Hagen, Germany. EPH includes, along with the points of the set Z, all objective vectors dominated by points of the set Z. Importantly, EPH is the maximal set for which it holds $P(Z_p) = P(Z)$.

In this chapter, we concentrate on visualizing the Pareto frontier or an approximation of it as a whole whereas Chapter 8 was devoted mainly to visualizing individual alternatives or a small number of them. Let us here also mention so-called box indices suggested by Miettinen *et al.* (2008) that support the DM in comparing different Pareto optimal solutions by representing them in a rough enough scale in order to let her/him easily recognize the main characteristics of the solutions at a glance.

The rest of this chapter is organized as follows. General problems of visualizing the Pareto frontier are considered in Section 9.2. We discuss the application of visualization for informing DMs, consider lessons learned from methods developed for biobjective problems and briefly touch the possible instability of the Pareto frontier. Section 9.3 is devoted to visualization techniques for convex multiobjective optimization problems. These techniques are based on the polyhedral approximation of the Pareto frontier. We augment recent surveys of approximation methods (Ruzika and Wiecek, 2005) by discussing polyhedral approximation and theory of approximation, and then turn to visualization of the Pareto frontier in the convex case. In Section 9.4, visualization techniques are described that are based on the pointwise approximation of the Pareto frontier (including techniques based on enveloping a large number of approximating points). Such visualization techniques can be applied in both MCDM and EMO based approaches. Finally, we conclude in Section 9.5.

9.2 General Problems of Visualizing the Pareto Frontier

In this section, we discuss general problems of visualizing the Pareto frontier. Subsection 9.2.1 is devoted to the advantages of visualization. Then in Subsection 9.2.2 we discuss the lessons that can be learned from experiences with biobjective problems. Finally, issues related to the instability of the Pareto frontier with respect to disturbances of the parameters of the problem (that are often met in real-life problems, but usually not considered in multiobjective optimization textbooks) are briefly outlined in Subsection 9.2.3.

9.2.1 Why Is Visualization of the Pareto Frontier as a Whole Useful?

Visualization, that is, transformation of symbolic data into geometric information, can support human beings in forming a mental picture of the symbolic data. About one half of neurons in human brain are associated with vision in some way, and this evidence provides a solid basis for successful application of visualization for transforming data into knowledge. As a proverb says: "A picture is worth a thousand words." Another estimate of the role of visualization is given by Wierzbicki and Nakamori (2005). To their opinion, "a picture is worth a ten thousands words." In any case, visualization is an extremely effective tool for providing information to human beings. Visualization on the basis of computer graphics has proved to be a convenient technique that can help people to assess information. The question that we consider is how it can be effectively used in the field of multiobjective optimization, namely, with *a posteriori* methods.

As discussed earlier, *a posteriori* methods are based on informing the DM about the Pareto optimal set without asking for his/her preferences. The DM does not need to express his/her preferences immediately since expressing preferences in the form of the single-shot specification of the preferred Pareto optimal objective vector may be separated in time from the approximation phase. Thus, once the Pareto frontier has been approximated, the visualization of the Pareto frontier can be repeated as many times as the DM wants to and can last as long as needed.

The absence of time pressure is an important advantage of *a posteriori* methods. To prove this claim, let us consider psychological aspects of thinking. Studies in the field of psychology have resulted in a fairly complicated picture of a human decision making process. In particular, the concept of a mental model of reality that provides the basis of decision making has been proposed and experimentally proven (Lomov, 1984). The mental models have at least three levels that describe the reality in different ways: as logical thinking, as interaction of images and as sub-conscious processes. Preferences are connected to all three levels. A conflict between the mental levels may be one of the reasons of the well-known non-transitive behavior of people (both in experiments studying their preferences as well as in real-life situations).

A large part of human mental activities is related to the coordination of the levels. To settle the conflict between the levels, time is required. Psychologists assure that sleeping is used by the brain to coordinate the mental levels. (Compare with the proverb: "The morning is wiser than the evening"). In his famous letter on making a tough decision, Benjamin Franklin advised to spend several days to make a choice. It is known that group decision and brainstorming sessions are more effective if they last at least two days. Detailed discussion of this topic is given in (Wierzbicki, 1997).

Thus, to settle the conflict between the levels of one's mental models in finding a balance between different objectives in a multiobjective optimization problem, the DM needs to keep the Pareto optimal solutions in his/her brains for a sufficiently long time. It is well known that a human being cannot simultaneously handle very many objects (see (Miller, 1956) for the magical number seven plus or minus two). This statement is true in the case of letters, words, sentences and even paragraphs. Thus, a human being cannot think about hundreds or thousands objective vectors of the Pareto frontier approximation simultaneously. Even having such a long list in front of his/her eyes, the DM may be unable to find the best one (Larichev, 1992). Instead, the DM often somehow selects a small number of objective vectors from the list and compares them (for details see (Simon, 1983)). Though after some time, the most preferred one of these solutions will be selected, such an approach results in missing most of the Pareto optimal solutions, and one of them may be better than the selected one. Visualization can help to avoid this problem.

However, to be effective, a visualization technique must satisfy some requirements formulated, for example, by McQuaid *et al.* (1999). These requirements include

1. simplicity, that is, visualization must be immediately understandable,
2. persistence, that is, the graphs must linger in the mind of the beholder, and
3. completeness, that is, all relevant information must be depicted by the graphs.

If a visualization technique satisfies these requirements, the DM can consider the Pareto frontier mentally as long as needed and select the most preferred objective vector from the whole set of Pareto optimal solutions. If some features of the graphs may happen to be forgotten, (s)he can look at the frontier again and again.

9.2.2 Lessons Learned from Biobjective Problems

According to (Roy, 1972), "In a general bi-criterion case, it has a sense to display all efficient decisions by computing and depicting the associated criterion points; then, DM can be invited to specify the best point at the compromise curve". Indeed, the Pareto frontier, which is often called as a compromise curve in the biobjective case, can be depicted easily after its approximation was constructed. For complications that may arise in the process of approximating biobjective Pareto frontiers, we refer to (Ruzika and Wiecek, 2005).

As has already been said, the first method for approximating the Pareto frontier by Gass and Saaty (1955) used visualization of the Pareto frontier as the final output of the process of approximation. Since then, various studies of biobjective optimization problems have used visualization of the Pareto frontier. It is interesting to note that visualization of the feasible objective region or the EPH is sometimes used, as well. In this case, the set of Pareto optimal objective vectors can be recognized easily. For minimization problems, it is the lower left frontier of the feasible objective region. It can be recognized even in nonconvex problems. For further details, see Chapter 1. For an example, see Figure 9.1, where Pareto optimal solutions are depicted by bold lines.

It is extremely important that the graphs used provide, along with Pareto optimal objective vectors, information about objective tradeoffs. There exist different formulations of the intuitively clear concept of the objective tradeoff as a value that provides a comparison of two objective vectors (see, e.g.,

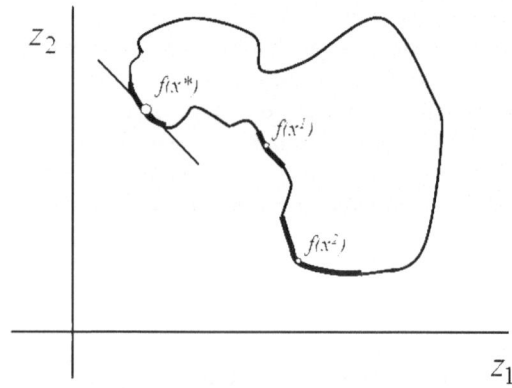

Fig. 9.1. Example of a feasible objective region and its Pareto frontier.

Chapter 2) or characterizes the properties of a movement along a curve in the objective space. Let us consider two of them for the biobjective case.

As an objective tradeoff (Chankong and Haimes, 1983; Miettinen, 1999) between any two objective functions f_1 and f_2 and decision vectors $\mathbf{x}^1$ and $\mathbf{x}^2$ (assuming $f_2(\mathbf{x}^2) - f_2(\mathbf{x}^1) \neq 0$) one can understand the value $T_{12}(\mathbf{x}^1, \mathbf{x}^2)$ defined in Chapter 2 as Definition 1. For any Pareto optimal decision vectors $\mathbf{x}^1$ and $\mathbf{x}^2$, the value of the objective tradeoff is negative because it describes the relation between the improvement of one objective and worsening of another. To estimate the objective tradeoff between vectors $\mathbf{x}^1$ and $\mathbf{x}^2$ visually, the DM has simply to compare the related objective values $\mathbf{z}^1 = \mathbf{f}(\mathbf{x}^1)$ and $\mathbf{z}^2 = \mathbf{f}(\mathbf{x}^2)$, respectively (see Figure 9.1). Such information is very important for the DM who can use it to decide, which of these two objective vectors is more preferable for her/him.

For a Pareto optimal objective vector $\mathbf{z}^*$, in which the Pareto frontier is smooth, the objective tradeoff can be expressed by the tradeoff rate $\frac{dz_2}{dz_1}(\mathbf{z}^*)$, where the derivative is taken along the curve which is the Pareto frontier in the biobjective case (see Figure 9.1). The value of the tradeoff rate informs the DM concerning the exchange between the objective values if one moves along the Pareto frontier. The tradeoff rate, which is given graphically by the tangent line to the Pareto frontier at $\mathbf{z}^*$, can easily be imagined at any point of the Pareto frontier. In case of a kink, tradeoff rates can be given by a cone of tangent lines (Henig and Buchanan, 1997; Miettinen and Mäkelä, 2002, 2003). Such cones can be estimated mentally by using the graph of the Pareto frontier, too. The importance of the role of the tradeoff rate has resulted in an alternative name for the Pareto optimal frontier: the tradeoff curve.

Tradeoff information is extremely important for the DM since it helps to identify the most preferred point along the tradeoff curve. This information is given by the graph of the Pareto frontier in a clear form, which can be accessed immediately. The graph of the Pareto frontier, if the frontier is not too complicated, can linger in the mind of the DM for a relatively long time. In any case, the DM can explore the curve as long as needed. Finally, the graph provides full information on the objective values and their mutual dependence along the tradeoff curve. Thus, such kind of visualization satisfies the requirements formulated above.

Visualization of the Pareto frontier helps transforming data on Pareto optimal objective values into knowledge of them. In other words, it helps in the formation of a mental picture of a multiobjective optimization problem (compare with learning processes discussed in Chapter 15). Since visualization can influence all levels of thinking, it can support the mental search for the most preferred Pareto optimal solution. Such a search may be logically imperfect, but acceptable for all levels of human mentality. Visualization of the Pareto frontier in biobjective problems helps to specify a preferable feasible objective vector as a feasible goal directly on the graph. This information is sufficient for finding the associated decision vector. In addition, the feasible goal (or its neighborhood) can be used as a starting point for various procedures, like identification of decision rules or in selecting a part of the Pareto frontier for a subsequent detailed study.

Thus, in biobjective cases, visualization of the Pareto frontier helps the DM

1. to estimate the objective tradeoff between any two feasible points and the tradeoff rate at any point of the Pareto frontier;
2. to specify the most preferred solution directly on the Pareto frontier.

The question can arise whether it is possible and profitable to visualize the Pareto frontier in the case of more than two objectives. To make visualization as effective as it is in the biobjective case, one needs to satisfy the general requirements of visualization. In addition, visualization must provide information on objective tradeoffs. Finally, the techniques must support the DM in identifying the most preferred solution.

Let us next consider the objective tradeoffs in the case of more than two objectives. The concept of an objective tradeoff introduced earlier can give rise to two different concepts in the multiobjective case: *partial objective tradeoff* and *total objective tradeoff*. Both these concepts are defined for objectives f_i and f_j and decision vectors $\mathbf{x}^1$ and $\mathbf{x}^2$, for which $f_j(\mathbf{x}^2) - f_j(\mathbf{x}^1) \neq 0$, by the same formula as earlier, that is, $T_{ij}(\mathbf{x}^1, \mathbf{x}^2)$ defined in Chapter 2 as Definition 1 (Miettinen, 1999). As discussed in Chapter 2, the value $T_{i,j}$ is said to be a *partial objective tradeoff* if other objective values are not taken into account. On the other hand, it is a *total objective tradeoff* if decision vectors $\mathbf{x}^1$ and $\mathbf{x}^2$ satisfy $f_l(\mathbf{x}^1) = f_l(\mathbf{x}^2)$ for all $l \neq i, j$. Thus, the total tradeoff can only be used for a small part of pairs of decisions. At first sight, a total tradeoff cannot play an important role. However, it is not so.

To give a geometric interpretation of the total tradeoff, it is convenient to consider *biobjective slices (cross-sections)* of the set Z (or the set Z_p). A biobjective slice (cross-section) of the set Z is defined as a set of such points in Z for which all objective values except two (i and j, in our case) are fixed. Then, the slice is a two-dimensional set containing only those objective vectors $\mathbf{z}^1 = f(\mathbf{x}^1)$ and $\mathbf{z}^2 = f(\mathbf{x}^2)$, for which it holds $z_l^1 = z_l^2$ for all $l \neq i, j$. Thus, since only the values of z_i and z_j change in the slice, the tradeoff can be estimated visually between any pair of points of the slice. In turn, it means that the total tradeoff can be estimated between objectives f_i and f_j for any pair of decision vectors $\mathbf{x}^1$ and $\mathbf{x}^2$ that may be unknown, but they certainly exist since they result in the objective vectors of the slice. Such a comparison is especially informative if both objective vectors belong to the Pareto frontier.

Application of biobjective slices is even more important while studying tradeoff rates between objective values. If the Pareto frontier is smooth in its point $\mathbf{z}^* = f(\mathbf{x}^*)$, a tradeoff rate becomes a *partial tradeoff rate* defined as $\frac{\partial z_i}{\partial z_j}(\mathbf{z}^*)$, where the partial derivative is taken along the Pareto frontier. Graphically, it is given by the tangent line to the frontier of the slice. The value of the partial tradeoff rate informs the DM about the tradeof rate between values of two objectives under study at the point $\mathbf{z}^*$, while other objectives are fixed at some values. Once again, the case of nonsmooth frontiers is discussed in (Henig and Buchanan, 1997; Miettinen and Mäkelä, 2002, 2003).

According to our knowledge, the idea of visualizing the biobjective slices of the Pareto frontier related to multiple objectives was introduced by Meisel (1973). He developed a method for visualizing biobjective *projections* of Pareto optimal solutions (compare the scatterplot techniques to be described in Section 9.4). At the same time, he argued that it would be much more useful to visualize the biobjective slices of the Pareto frontier instead of biobjective projections of particular objective vectors. The reason, according to Meisel, is related to the evidence that graphs of biobjective slices (in contrast to projections) can inform the DM on total objective tradeoffs and partial tradeoff rates. Moreover, they can support identification of the most preferred objective vector directly at the Pareto frontier.

The simplest approach to constructing and displaying the biobjective slices of the Pareto frontier can be based on a direct conversion of the multiobjective problem with $k > 2$ to a series of biobjective problems. Such a concept is close to the ε-constraint method (Chankong and Haimes, 1983) described in Chapter 1. The only difference is the following: One has to select any two objectives f_i and f_j (of k functions) to be minimized instead of only one (as in the ε-constraint method). Then, the following biobjective problem is considered

$$\begin{aligned} \text{minimize} \quad & \{f_i(\mathbf{x}), f_j(\mathbf{x})\} \\ \text{subject to} \quad & f_l(\mathbf{x}) \leq \varepsilon_l \text{ for all } l = 1, \ldots, k, \ l \neq i, j \\ & \mathbf{x} \in S. \end{aligned} \quad (9.1)$$

Here the values of ε_l, $l = 1, \ldots, k$, $l \neq i, j$ must be given in advance. As said, various methods have been proposed for constructing biobjective Pareto frontiers. These methods can be used for solving problem (9.1). As a result, one obtains the Pareto frontier for two selected objectives f_i and f_j for given values of ε_l of $k - 2$ other objectives. One has to note, however, that this is true only in the case if the plane does indeed cut the Pareto frontier.

To get a full picture of the Pareto frontier in this way, one needs to specify a grid in the space of $k - 2$ objectives and solve a biobjective problem for any point of the grid. For example, in case of five objective functions with 10 possible values of ε_l for any of $(k - 2 -)$ 3 objectives, we have to construct the Pareto frontier for 1000 biobjective problems, which naturally is a tremendous task. In addition, one has to somehow visualize these 1000 biobjective frontiers. For this reason, researchers usually apply this approach only in the case of $k = 3$ or, sometimes, $k = 4$ and restrict to a dozen biobjective problems. As to visualization, one usually can find a convenient way for displaying about a dozen tradeoff curves. Examples are given in (Mattson and Messac, 2005; Schuetze *et al.*, 2007).

One can prove that tradeoff curves obtained in this way do not intersect for $k = 3$ (though they can touch each other). Thus, in this case a system of Pareto frontiers displayed in the same graph is relatively simple (see, e.g.,

Figure 9.2 adopted from (Haimes *et al.*, 1990), p. 71, Fig. 4.2). Such graphs are known as *decision maps*.

If needed, a decision map can be considered as a collection of projections of biobjective slices on the objective plane of f_i and f_j, but this interpretation is not obligatory and does not bring additional information. Early examples of applying decision maps in multiobjective water management have been given by Louie *et al.* (1984) (p. 53, Fig. 7) and Jewel (1990) (p. 464, Fig. 13.10). In addition to tradeoff rates for any of several tradeoff curves, decision maps provide graphic information concerning objective tradeoffs between any pair of objective vectors that belong to different tradeoff curves. Thus, decision maps provide tradeoff information on the relations between Pareto optimal points for three objectives. In the next sections we describe methods for constructing decision maps that are more effective than using problem (9.1) with a large number of values for ε_3.

9.2.3 Comment on the Stability of the Pareto Frontier

Let us here pay attention to the issue of the stability of the Pareto frontier, that is, to the question whether the Pareto frontier depends continuously on the parameters of the problem. This issue is important for computer approximation and visualization. The problem consists of the evidence that a Pareto frontier is often instable to the disturbances of parameters of the model. To apply Pareto frontier approximation techniques correctly, one has to first prove that the Pareto frontier exists and is unique for some set of parameter values (which usually is true) and that the Pareto frontier is stable to their disturbances (which usually is unknown). A sufficient condition for the stability of the Pareto frontier follows from the theorems of Chapter 4 in (Sawaragi *et al.*, 1985): If some simple technical conditions are satisfied, the stability of the Pareto frontier is provided by the coincidence of an unbiased Pareto frontier and a weak Pareto frontier. If the class of disturbances is broad enough, this condition is also necessary.

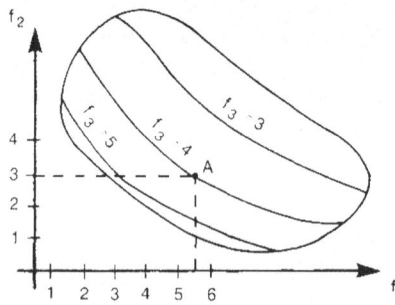

Fig. 9.2. A decision map.

Usually it is impossible to check whether Pareto and weak Pareto frontiers coincide before constructing the Pareto frontier. In the process of approximation, large parts of the Pareto frontier (and, hence, related decisions) can be lost. Or, vise versa, some parts of the frontier of the set Z, which do not belong to the Pareto frontier, can be considered to be Pareto optimal. The result may depend on the computer and even on its initial state. In this case, different users will get different approximations.

To solve this problem, robust algorithms for approximating the Pareto frontier have been proposed (Krasnoshchekov *et al.*, 1979). A more detailed list of references is given in Subsection 9.4.4. To avoid using these complicated algorithms, many researchers have tried to revise the problem of approximating the Pareto frontier to make it well posed. One of the approaches is based on squeezing the domination cone. As a result, a set is constructed which contains the Pareto frontier as its subset. However, such an approach may result in complications related to the instability of the broadened Pareto frontier. Another form of problem revision can be based on the approximation of the set Z or Z_p, thus avoiding the problem of the stability of the Pareto frontier. The idea of approximating sets Z or Z_p instead of the Pareto frontier itself has been known for years (Lotov, 1975, 1983; Yu, 1985; Kaliszewski, 1994; Benson, 1998). In the case of pointwise approximations, this idea has been used in (Evtushenko and Potapov, 1987; Reuter, 1990; Benson and Sayin, 1997). In Section 9.3 we show that visualizing the frontier of the EPH may help to solve the problem of an instable Pareto frontier.

9.3 Visualization of the Pareto Frontier Based on Approximation by Convex Polyhedra

An important class of nonlinear multiobjective optimization problems is the set of convex problems. Several definitions of convex multiobjective optimization problems can be given. We use the most general one: a multiobjective optimization problem is convex if the set Z_p is convex (Henig and Buchanan, 1994). Note that the feasible objective region may be nonconvex in this case. A sufficient condition for the convexity of an EPH is given, for example, as: the set Z_p is convex if the set S is convex and all objective functions are convex (Yu, 1974). One can see that this sufficient condition is not a necessary one because problems with nonconvex feasible regions do not satisfy the condition. At the same time, there exist important real-life problems which have nonconvex feasible regions, but a convex EPH. Note that linear multiobjective optimization (MOLP) problems are always convex.

In a convex case, in addition to the universal pointwise approximation of the Pareto frontier, its polyhedral approximation is possible. This can be given by hyperplanes or hyperfaces (i.e., faces of dimension $k-1$). A polyhedral approximation of the Pareto frontier provides a fast visualization of the Pareto frontier.

There exist two main polyhedral approximation approaches aimed at visualizing Pareto frontiers in convex problems. The first approach is based on the fact that the Pareto frontier is a part of the frontier of the convex feasible objective region (or its EPH). Thus, one can construct faces of the feasible objective region that approximate the Pareto frontier without approximating the feasible objective region by itself. In the second approach, one constructs a polyhedral set that approximates the feasible objective region or any convex set that has the same Pareto frontier as the feasible objective region (like EPH).

The first approach results in complications with more than two objectives. Note that the Pareto frontier of the convex feasible objective region is usually a nonconvex set by itself. Its approximation can include a large number of hyperfaces, which must be somehow connected to each other. Taking into account that, in addition, the Pareto frontier is often instable, it is extremely hard to develop a method and software that are able to approximate and visualize the Pareto frontier in this form if we have more than three objectives. Such complications are discussed in (Solanki et al., 1993; Das and Dennis, 1998).

The second approach is based on approximating the convex feasible objective region (or another set in the objective space that has the same Pareto frontier), for which a polyhedral approximation can be given by a system of a finite number of linear inequalities

$$Hz \leq h, \tag{9.2}$$

where H is a matrix and h is a vector, which are to be constructed by an approximation technique. A biobjective slice for the objective values z_i and z_j can be constructed very fast: it is sufficient to fix the values of other objectives in (9.2). Another advantage of the form (9.2) is the stability of compact polyhedral sets to errors in data and rounding errors (Lotov et al., 2004).

As far as the system of inequalities (9.2) is concerned, it is usually constructed when approximating any of the following convex sets:

1. set Z (i.e., the feasible objective region) (Lotov, 1975),
2. set Z_p (i.e., the EPH), which is the largest set that has the same Pareto frontier as Z (Lotov, 1983), or
3. a confined set that is the intersection of Z_p with such a box that the frontier of the intersection contains the Pareto frontier. Formally, this was proposed by Benson (1998) who called it an efficiency-equivalent polyhedron. However, such a set was used already in the 1980s in a software for approximating and visualizing Pareto frontiers, see, e.g., Figure 9.3.

From the point of view of visualization, approximating the EPH or an efficiency-equivalent polyhedron is preferable since its biobjective slices provide decision maps, in which the Pareto optimal frontier can be recognized by the DM immediately.

The structure of the rest of this section is as follows. Subsections 9.3.1 and 9.3.2 are devoted to visualization in the case of three and more than three objectives, respectively. In Subsection 9.3.3 we shortly outline the main methods for constructing a polyhedral approximation in the form (9.2) when $k > 2$. Various methods for approximating Pareto frontiers in convex problems with two objectives are described in (Ruzika and Wiecek, 2005).

9.3.1 Visualization in the Case of Three Objectives

If we have three objectives, two main approaches to visualizing the Pareto frontier can be applied: displaying the approximation of the Pareto frontier (or EPH) (i) as a three-dimensional graph; or (ii) using decision maps. Let us compare these two approaches with an example of a long-term dynamic economic model that takes environmental indicators into account (Lotov *et al.*, 2004). The objectives are a consumption indicator C^* (deviation from the balance growth – to be maximized), a pollution indicator Z^* (maximal pollution during a time period – to be minimized) and an unemployment indicator U^* (maximal unemployment during a time period – to be minimized). The three-dimensional graph of the EPH and the decision map for this problem are given in Figures 9.3 and 9.4, respectively. They were prepared manually to demonstrate differences between the two formats of visualization.

The Pareto frontier is hatched in a three-dimensional graph in Figure 9.3. This picture shows the general structure of the problem. Ten interesting objective vectors are specified in the graph. Some of them (vectors 1–6) belong to the plane $U^* = 0$. Due to this, it is possible to estimate the objective values in these vectors. At the same time, it is fairly complicated to check the objective values in the other objective vectors and to evaluate tradeoffs between them or tradeoff rates using such a graph (except the tradeoff curve for $U^* = 0$ that is, actually, a biobjective slice of the Pareto frontier).

Let us next compare the three-dimensional graph with the decision map in Figure 9.4. The decision map is of a style given in Figure 9.2. Tradeoff curves (biobjective slices of the EPH) were calculated for several values of

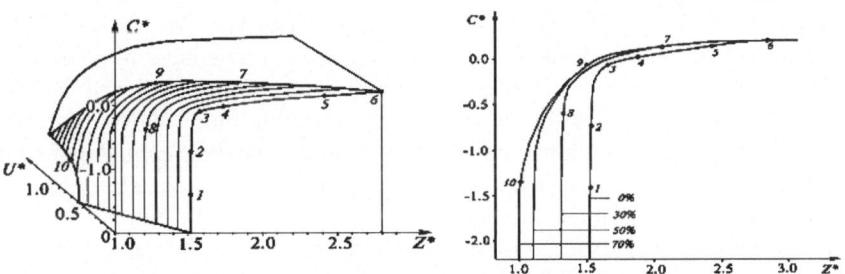

Fig. 9.3. A three-dimensional graph of the Pareto frontier.   **Fig. 9.4.** A decision map.

U^*, which are given in the graph near the associated tradeoff curves. The objective vectors 1–10 are also depicted in the figure.

One can see that the tradeoff curve corresponding to $U^* = 0$ has a kinked form: the tradeoff rate for Z^* and C^* changes substantially when $-0.4 < C^* < -0.2$. If $C^* < -0.40$, one can increase consumption without increasing pollution indicator Z^* (or with minimal increase). In contrast, while $C^* > 0$, a small additional growth of C^* results in a drastic growth of the pollution indicator Z^*. One can easily compare other points along this tradeoff curve, as well.

Four other tradeoff curves in Figure 9.4 have a similar form, but the kink is smoother than for $U^* = 0$. These tradeoff curves show in a clear way how it is possible to decrease pollution while maintaining the same value of C^* by switching from one tradeoff curve to another. In this way, a decision map provides more information on objective values and, especially, on tradeoffs, than a three-dimensional graph. One can see the total tradeoff between any two points belonging to the same tradeoff curve as well as the total tradeoff between any two points that have the same value of C^* or Z^*. The tradeoff rates are visible at any point of the tradeoff curves.

Let us next discuss how well decision maps meet the requirements on visualization techniques formulated in Subsection 9.2.1. First of all, let us note that tradeoff curves do not intersect in a decision map (though they may sometimes coincide). Due to this, they look like contour lines of topographic maps. Indeed, a value of a third objective (related to a particular tradeoff curve) plays the role of the height level related to a contour line of a topographic map. For example, a tradeoff curve describes such combinations of values of the first and the second objectives that are feasible for a given constraint imposed on the value of the third objective (like "places lower, than..." or "places higher, than..."). Moreover, one can easily estimate which values of the third objective are feasible for a given combination of the first and of the second objectives (like "height of this particular place is between..."). If the distance between tradeoff curves is small, this could mean that there is a steep ascent or descent in values, that is, a small move in the plane of two objectives is related to a substantial change in the value of the third objective.

Thus, decision maps are fairly similar to topographic maps. For this reason, one can use topographic maps for the evaluation of the effectiveness of the visualization given by decision maps. Topographic maps have been used for a long time and people usually understand information displayed without difficulties. Experience of application of topographic maps shows that they are *simple enough* to be immediately understood, *persistent enough* not to be forgotten by people after their exploration is over, and *complete enough* to provide information on the levels of particular points in the map.

The analogy between decision maps and topographic maps asserts that decision maps satisfy the requirements specified in Subsection 9.2.1. Note that the decision maps can be constructed in advance before they are studied, or on-line, after obtaining a request from a user. On-line calculation of decision maps provides additional options to change objectives located on axes, change the number of tradeoff curves on decision maps, zoom the picture and change graphic features of the display such as the color of the background, colors of the slices, etc. For example, it would be good to get an additional tradeoff curve for the values of U^* between zero and 30% for the example considered.

9.3.2 Visualization in the Case of More than Three Objectives

In the case of more than three objectives, the possibility of an interactive display of decision maps (if exists) is an important advantage. Visualization of the Pareto frontier with more than three objectives can be based on the exploration of a large number of decision maps, each of which describing objective tradeoffs between three objectives. Let as assume that an approximation of a convex feasible objective region or of another convex set in the objective space has already been constructed in the form (9.2). Then, a large number of decision maps can be constructed in advance and displayed on a computer screen, whenever asked for. However, the number of decision maps to be prepared increases drastically with the growth of the number of objectives. For example, in the case of four objectives, by using the modification of the ε-constraint method described in Subsection 9.2.2, one could prepare several dozens of decision maps for different values of the fourth objective. However, it is not clear in advance, whether these decision maps are informative enough for the DM. It may happen that different three-objective decision maps are more preferable for her/him (e.g., f_1 and f_4 on the axes, f_2 as "level of height" and f_3 to be used as a constraint). The situation is much worse in the case of five objectives. Thus, it seems to be reasonable to enable calculating and displaying the decision maps on-line.

To provide on-line displays of decision maps fast, one can approximate the Pareto frontier beforehand and then compute biobjective slices fast. In convex cases, one can use polyhedral approximations of the EPH for which biobjective slices can be computed and superimposed very fast. Decision maps corresponding to constraints imposed on the rest $k - 3$ objectives can be provided to the DM on request in different forms. Here we describe two of them: animation and matrices. These two forms compete in the case of four or five objectives. In the case of more than five objectives, these forms supplement each other.

Animation of decision maps requires generating and displaying a large number of biobjective slices fast (several hundred slices per second for four objectives). It is important that in the process of animation, the slices are constructed for the same two objectives whose values are located on the axes. Chernykh and Kamenev (1993) have developed a special fast algorithm for computing biobjective slices based on preprocessing the linear inequalities of the polyhedral approximation of the EPH. Due to this algorithm, thousands of decision maps can be computed and depicted in seconds. The control of animation is based on using such traditional computer visualization technique as scroll-bars. Figure 9.5 represents a gray scale copy of a color computer display for a real-life water quality problem (Lotov *et al.*, 2004) involving five objectives. The decision map consists of four superimposed biobjective differently colored slices. There is only one difference between decision maps given in Figures 9.5 and 9.4: the Pareto frontiers in Figure 9.5 are given as the frontiers of colored areas and not as curves. This form has proven to be more convenient for decision makers. A palette shows the relation between the values of the third objective and colors. Two scroll-bars are related to the values of the fourth and the fifth objectives.

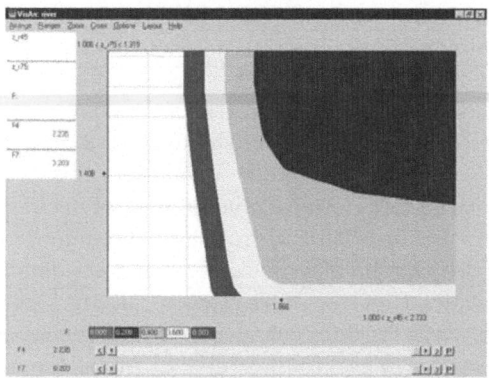

Fig. 9.5. Gray scale copy of a decision map and two scroll-bars.

A movement of a scroll-bar results in a change of the decision map. The DM can move the slider manually. However, the most effective form of displaying information to the DM is based on an automatic movement of the slider, that is, on a gradual increment (or decrement) in the constraint imposed on the value of an objective. A fast replacement of the decision maps offers the effect of animation. Because any reasonable number of scroll-bars can be located on the display, one can explore the influence of the fourth, the fifth (and maybe even the sixth and the seventh etc.) objectives on the decision map. Since the EPH has been approximated for the whole problem in advance, different forms of animation can be used, including a simultaneous movement of several sliders. However, such effects are not recommended since they are too complicated for DMs. Animation of only one slider at a time has turned out to be recommendable (by keeping the other sliders fixed).

Another form of displaying several decision maps is a matrix of decision maps, which can be interpreted as a collection of animation snap-shots. In Figure 9.6, one can see a matrix of decision maps for the same problem. Four values of an objective (associated with a scroll-bar in Figure 9.5) form four columns, and four values of another objective (associated with the other scroll-bar) form four rows. Note that such a matrix can be displayed without animation at all. The constraints on the objectives that define the decision maps used can be specified manually by the DM or automatically. The maximal number of rows and columns in such a matrix depends exclusively on the desires of the DM and on the quality of the computer display.

In the case of six or more objectives, animation of the entire matrix of decision maps is possible but may be cognitively very demanding. In this case, the values of the sixth, the seventh and the eights objectives can be related to scroll-bars, the sliders of which can be moved manually or automatically. It is important that the objectives in the matrix of decision map can be arranged in the order the DM wishes, that is, any objective can be associated with an axis, the color palette or a scroll-bar and the column or row of the matrix.

By shortening the ranges of the objectives the DM can zoom the decision maps. It is recommended to restrict the number of objective functions considered to five or six, because otherwise the amount of information becomes too ramified for a human being. However, in some real-life applications, environmental engineers have managed to apply matrices of decision maps even for nine objective functions.

The visualization technique for the Pareto frontier described in this subsection is called by the name interactive decision maps (IDM) technique. For further details, see (Lotov *et al.*, 2004). After exploring the Pareto frontier, the DM can specify a preferred combination of feasible objective values (feasible goal) directly on one of the decision maps. Since the goal identified is close to the precise Pareto frontier, the related decision can be found by solving an optimization problem using any distance function. For the same reason, the objective vector corresponding to the Pareto optimal decision vector found will be close to the goal identified and, thus, this approach can be called a feasible goals method (Lotov *et al.*, 2004).

9.3.3 Comment on Polyhedral Approximation

Note that the problem of the polyhedral approximation of the Pareto frontier is close to the classical problem of applied mathematics: polyhedral approximation of convex sets. Methods for polyhedral approximation of convex sets are based on iterative algorithms for constructing converging sequences of approximating polyhedra.

The main concepts of iterative methods for polyhedral approximation of convex sets were introduced already in the 1970s by McClure and Vitale (1975), who proposed the general idea of simultaneously constructing two polyhedral approximations, an internal one and an external one, where vertices of the internal approximation coincide with the points of contact of the external approximation. Note that by this an assessment of the quality of the approximation of the Pareto frontier is provided.

Then, both the approximations are used for iteratively decreasing the local discrepancy between them in a balanced adaptive way. It should result in a

230   A.V. Lotov and K. Miettinen

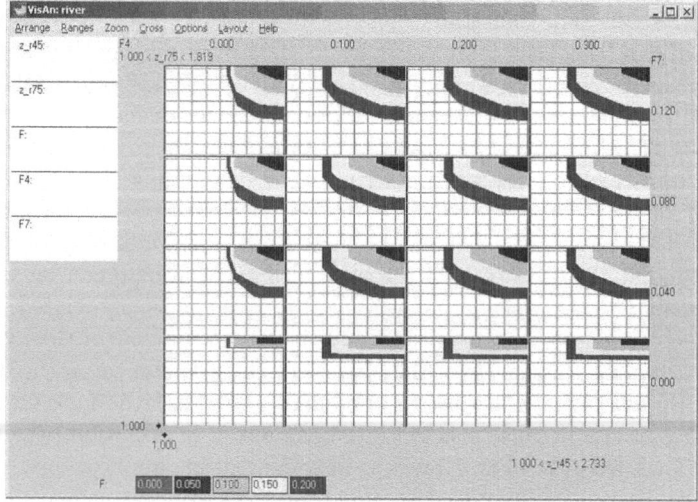

Fig. 9.6. Gray scale copy of a matrix of decision maps.

fast total convergence of the internal and external polyhedra providing by this convergence to the approximated convex set.

This idea was transformed by Cohon (1978); Cohon *et al.* (1979) into the first adaptive approximation method, the non-inferior set estimation (NISE) method. The NISE method was implemented as a software for approximating Pareto frontiers in linear biobjective optimization problems. Then, the estimation refinement (ER) method was proposed (Bushenkov and Lotov, 1982; Lotov, 1989) that integrated the concepts of the NISE method with methods of linear inequality theory. The ER method effectively constructs internal and external approximations of multi-dimensional ($3 < k < 8$) convex bodies. Its convergence can be proved to be asymptotically optimal. The ER method has been implemented as a software (downloadable from http://www.ccas.ru/mmes/mmeda/soft/) that has been applied for several real-life problems (Lotov *et al.*, 2004).

To illustrate approximation methods, we provide a simplified iteration of the ER method applied for approximating a convex and compact feasible objective region Z. Prior to the $(l+1)$-th iteration, a polyhedron P^l, for which vertices belong to the frontier of Z, has to be constructed in two forms: in the form of a list of its vertices as well as in the form of a system of linear inequalities

$$(\mathbf{u}^k, \mathbf{z}) \le u_0^k, \ k = 1, 2, ..., K(l) \tag{9.3}$$

where $K(l)$ is the number of inequalities in the system (9.3), $\mathbf{u}^k$ are vectors with unit norm and u_0^k are right sides of the inequalities.

- Step 1. Solve the optimization problems $(\mathbf{u}^k, \mathbf{z}) \to$ max over the set Z for all $k = 1, \ldots, K(l)$ except those solved at the previous iterations. Denote the maximal values by $g_Z(\mathbf{u}^k)$. Among $\mathbf{u}^k$ find a vector $\mathbf{u}^*$ that maximizes the increment, i.e., the value $g_Z(\mathbf{u}^k) - u_0^k$. If all increments are sufficiently small, then stop.
- Step 2. The new approximating polyhedron P^{l+1} is given by the convex hull of the polyhedron P^l and the point $\mathbf{z}^*$ of the boundary of Z that has $(\mathbf{u}^*, \mathbf{z}^*) = g_Z(\mathbf{u}^*)$. Construct a new linear inequality system of type (9.3) which describes P^{l+1} by using the stable algorithm of the beneath-beyond method described in (Lotov et al., 2004). Start the next iteration.

Note that the ER method provides assessment of the quality of the approximation of the Pareto frontier: the external polyhedron P_I^{l+1} that is given by the linear inequality system

$$(\mathbf{u}^k, \mathbf{z}) \leq g_Z(\mathbf{u}^k), k = 1, 2, ..., K(l+1)$$

contains Z. Due to this, at any iteration, we have internal and external estimates for Z. Therefore, it is possible to evaluate the accuracy visually or use the maximal value of $g_Z(\mathbf{u}^k) - u_0^k$ as the accuracy measure.

Later, various related methods have been developed, including methods which are dual to the ER method (Kamenev, 2002; Lotov et al., 2004). An interesting method has been proposed by Schandl et al. (2002a,b); Klamroth et al. (2002) for the case of more than two objectives. The method is aimed at constructing inner polyhedral approximations for convex, nonconvex, and discrete multiobjective optimization problems. In (Klamroth et al., 2002) the method for convex problems is described separately, and one can see that the algorithm for constructing the internal approximation is close to the ER method, while the algorithm for constructing an external approximation is close to the dual ER method. Other interesting ideas have been proposed, including (Voinalovich, 1984; Benson, 1998).

9.4 Visualization of Pointwise Approximations of the Pareto Frontier

In this section, we consider nonconvex multiobjective optimization problems. To be more specific, we discuss methods for visualizing approximations of the Pareto frontier given in the form of a list of objective vectors. We assume that this list has already been constructed. It is clear that a good approximation representing the Pareto optimal set should typically consist of a large number of solutions. This number may be hundreds, thousands or even much more. For this reason, methods developed for visualizing a small number of Pareto optimal points, as a rule, cannot be used for visualizing the Pareto frontier as a whole. Thus, we do not consider here such methods as bar and pie charts, value paths, star and web (radar) diagrams or harmonious houses, which are described in Chapter 8 and (Miettinen, 1999, 2003) and used, for example, in (Miettinen and Mäkelä, 2006). Let us point out that here we do not consider visualization techniques that are based on various transformations of the Pareto frontier, like GAIA, which is a part of the PROMETHEE method (Mareschal and Brans, 1988), BIPLOT (Lewandowski and Granat, 1991) or GRADS (Klimberg, 1992), or methods that utilize preference information (Vetschera, 1992).

There exists a very close, but different problem: visualization of objective vectors in finite selection problems involving multiple objectives (often named attributes in such problems) (Olson, 1996). If the number of alternatives is large, the problem is close to the problem of visualizing an approximation of the Pareto frontier.

Note that any point of a pointwise approximation of the Pareto optimal set usually corresponds to a feasible decision vector. Thus, along a pointwise approximation of the Pareto frontier, one usually obtains a pointwise approximation of the set of Pareto optimal decision vectors. However, we do not consider the set of Pareto optimal decision vectors since it is usually not visualized for the reasons which have been discussed in the introduction.

This section consists of four parts. Subsection 9.4.1 surveys visualization methods based on concepts of heatmap graphs and scatterplot matrices for the case of a large number of objective vectors. In Subsections 9.4.2 and 9.4.3, we describe visualization of pointwise Pareto frontier approximations by using biobjective slices of the EPH and application of enveloping, respectively. Finally, in Subsection 9.4.4, we discuss some topics related to pointwise approximations of Pareto frontiers (that were not touched in Chapter 1).

9.4.1 Heatmap Graphs and Scatterplots

A well-known technique for visualizing a small number of decision alternatives is the value path method proposed, for example, in (Geoffrion *et al.*, 1972). A value paths figure consists of k parallel lines related to attributes (objectives) with broken lines on them associated with the decision alternatives (Miettinen, 1999). See also Chapter 8. Such figures are not too helpful if we are interested in selecting the best alternative from a large number of them, but they can help to cluster such sets of alternatives (Cooke and van Noortwijk, 1999).

Interesting developments of value paths are heatmap graphs (Pryke *et al.*, 2007) that were adopted from visualization methods developed by specialists in data mining. In Figure 9.7, we provide a gray scale copy of a heatmap graph (which however, gives only a limited impression of the colorful graph). In the figure, there are 50 alternatives described by 11 decision variables (p1–p11) and two objective functions (o1–o2). An alternative is given by a straight line, while the deepness of gray provides information about the normalized values (from 0 to 1) of decision and objective values (black corresponds to zero, white corresponds to 1). Heatmap graphs are convenient for selecting clusters of alternatives, but can be used for selecting one alternative, as well.

Fig. 9.7. Gray scale copy of a heatmap graph

A scatterplot matrix is a technique for visualizing a finite number of objective vectors with biobjective projections. For clarity, we introduce it here for a small number of objective vectors even though it could be used for bigger sets of solutions. Hereby we augment the treatment given in Chapter 8 by showing that a visualization, like the scatterplot, can result in misinforming the DM and skipping the best alternatives even in the case of a small number of solutions.

A scatterplot matrix has been introduced in (Meisel, 1973), but it has become widely recognized since Cleveland (1985) discussed it. It has directly been inspired by a biobjective case, but is used in the case of three and more objectives. The technique displays objective vectors as their projections at all possible biobjective planes. These projections are displayed as a *scatterplot matrix*, which is a square matrix of panels each representing one pair of objective function. Thus, any panel shows partial objective tradeoffs for different pairs of objectives. The dimension of the matrix coincides with the number of objective functions. Thus, any pair of objective functions is displayed twice with the scales interchanged (with the diagonal panels being empty).

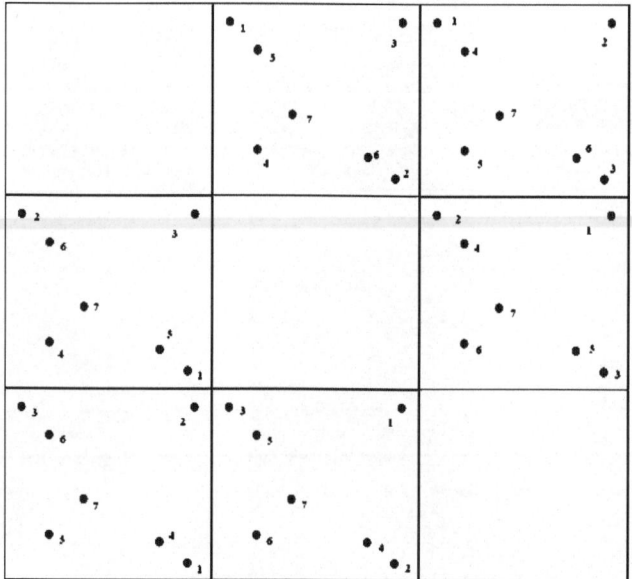

Fig. 9.8. Scatterplot matrix for three objectives.

A simple example of a scatterplot matrix is given in Figure 9.8. It is related to a simple decision problem involving seven Pareto optimal alternatives and three objectives to be minimized. Thus we have seven objective vectors

1. Alternative #1: (0, 1, 1)
2. Alternative #2: (1, 0, 1)
3. Alternative #3: (1, 1, 0)
4. Alternative #4: (0.2, 0.2, 0.8)
5. Alternative #5: (0.2, 0.8, 0.2)
6. Alternative #6: (0.8, 0.2, 0.2)
7. Alternative #7: (0.4, 0.4, 0.4)

The projections of the vectors are numbered according to their number in the list. One can see that in this example the vectors were specified in a way that the variety of their projections is the same for all biobjective planes (only the number of the original objective vector changes). It was done to illustrate the evidence that despite its simplicity, this technique can be misleading since only partial tradeoffs are displayed in the biobjective projections. In the list one can see that vector #7 is balanced (which means that it may be a preferred one). However, it is deep inside in all the projections in Figure 9.8. Thus, the user may not recognize its merits and,instead, select solutions for which projections look great, but the projected vectors have unacceptable values of some objectives. This example shows that simple projections on biobjective planes may result in misunderstanding the position of a solution in the Pareto optimal set. It is especially complicated in the case of a large number of alternatives.

For this reason, researchers try to improve the scatter plot technique by providing values of objectives, which are not given on the axes, in some form, like colors. However, such an approach may be effective only in the case of a small number of points and objectives. Furthermore, in some software implementations of the scatterplot technique, one can specify a region in one of the panels, and the software will color points belonging to that region in all other panels. Though such a tool may be convenient, balanced points may still be lost as in Figure 9.8.

Let us finally mention one more tool, so-called knowCube, by Trinkaus and Hanne (2005) that can be used when a large set of Pareto optimal solutions should be visualized. It is based on a radar chart (or spider-web chart) described in Chapter 8. However, in knowCube there can be so many solutions (in the beginning) that it is impossible to identify individual solutions. Then, the DM can iteratively study the solutions, for example, by fixing desirable values for some objective functions or filtering away undesirable solutions by specifying upper and/or lower bounds for objective function values.

9.4.2 Decision Maps in Visualizing Pointwise Approximations of the Pareto Frontier

Let us next consider visualization of an approximation of the Pareto frontier given by a finite number of objective vectors (from several hundreds to several thousands) using decision maps of biobjective slices. As described in Section 9.3, in convex cases decision maps visualize tradeoff curves and support a direct identification of the most preferred solution (goal) (for three to eight objectives). The same can be done if we provide a set of Pareto optimal points by visualizing biobjective slices of the EPH of these points. If we have a finite number of objective vectors, the description of the EPH is very simple: it is a union of domination cones $\mathbf{R}_k^+$ with vertices located in these points. Due to this simple explicit form, the biobjective slices of the EPH can be rapidly computed and displayed by computer graphics. Since the EPH of a finite set of points, in contrast to the set of points itself, is a bodily set, the frontier of its slice is given by a continuous line (frontier of the figure provided by superimposed cones). Collections of biobjective slices can be given in the form of decision maps, where the value of only one objective is changing from slice to slice. Visualization of finite sets of objective vectors by slices of their EPHs was introduced in (Bushenkov *et al.*, 1993) and the current state of the art is given in (Lotov *et al.*, 2004).

In Figure 9.9 (adopted from (Lotov *et al.*, 2005)), one can see a decision map involving 2879 objective vectors for a steel casting process described by a complicated nonlinear model with 325 decision variables. Note that in the case of a large number of Pareto optimal objective vectors, tradeoff curves are displayed.

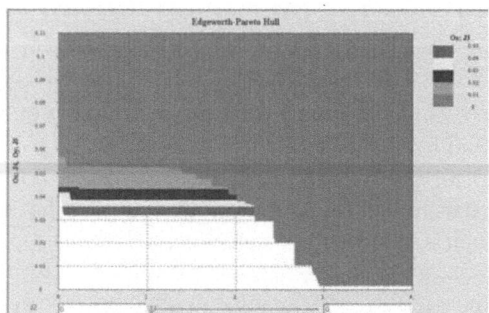

Fig. 9.9. Decision map and a scroll-bar for 2879 Pareto optimal objective vectors and four objectives.

Let us next pay some attention to the case of more than three objectives. As in the previous section, the ideas used in the IDM technique can be applied. Since biobjective slices of the EPH given by a system of cones can be constructed fairly fast, an interactive study of decision maps can be used. It can be implemented using the same tools as in the convex case (scroll-bars and matrices of the decision maps). For example, in Figure 9.9 one can see a scroll-bar that helps to animate decision maps. In this way, the animation helps to study a problem with four objectives.

It is important to stress that one can study the associated EPH in general and in details from various points of view by providing various decision maps on-line. One can see the values and tradeoffs for three objectives, while the influence of the fourth objective (or fifth, etc.) can be studied by moving sliders of scroll-bars. The DM can select different allocations of objectives among scrollbars, colors and axes. However, understanding interdependencies between many objectives may involve a lot of cognitive effort. As discussed, after studying the decision maps, the DM can identify the most preferred solution and obtain the closest Pareto optimal objective vector from the approximation as well as the associated decision. Examples of hybridizing this idea with the interactive NIMBUS method (Miettinen, 1999; Miettinen and Mäkelä, 2006) (for fine-tuning the solution) are given by Miettinen *et al.* (2003).

9.4.3 Application of Enveloping for Visualizing Alternatives

Finally, let us describe an alternative approach to visualizing objective vectors (where $k > 2$) based on approximating and visualizing the EPH of their envelope (convex hull). This has been proposed in (Bushenkov et al., 1993), and can be used in the case of a large number (from several hundreds to several millions) of objective vectors. Due to enveloping, the IDM technique can be applied for visualizing the Pareto frontier of a convex hull.

We assume that we have objective vectors ($k < 7$) either as a list of objective vectors or defined implicitly as objective vectors corresponding to integer-valued decision vectors satisfying some constraints. First, the EPH of the convex hull of these points must be approximated. One can use methods for the polyhedral approximation of the convex sets (see Section 9.3) or methods specially developed for constructing a convex hull of a system of multi-dimensional points (Barber et al., 1996). Then the IDM technique can be applied for visualization. An example of a decision map describing about 400 thousand objective vectors for an environmental problem is given in (Lotov et al., 2004). It is important that tradeoff rates can be seen more easily than in the case of visualizing the EPH for objective vectors without convex enveloping.

According to experiments, even non-professional users have been able to understand tradeoff rates and identify the most preferred solution. However, one must remember that it is the tradeoff rate of the envelope that is shown: the Pareto frontier of the convex hull is displayed, which includes additional infeasible points of the objective space that simplify the graph. Thus, the solution identified by the DM on the Pareto frontier of the envelope is only reasonable (the concept of a reasonable goal as the goal that is close to a feasible objective vector was introduced by Lotfi et al. (1992). Eventually, several alternatives can be selected which are close to the goal in some sense. For further details, see (Gusev and Lotov, 1994; Lotov et al., 2004).

9.4.4 Comment on Methods for Pointwise Approximations of the Pareto Frontier

Several basic methods have been proposed for generating a representation of the Pareto optimal set, that is, Pareto frontier approximation. Some scalarization based approaches were described in Chapter 1. That is why we do not consider them here. A more detailed overview of methods for pointwise approximation is provided in (Ruzika and Wiecek, 2005). Here we augment this overview.

There exist scalarization methods that take the possible instability of the Pareto frontier into account (Krasnoshchekov et al., 1979; Popov, 1982; Nefedov, 1984, 1986; Abramova, 1986; Smirnov, 1996) and, by solving a large number of scalarized parametric optimization problems they generate pointwise approximations. It is important that the methods are stable or insensitive to disturbances. If Lipschitz constants exist, a reasonable approximation accuracy can be achieved by solving a huge number of global optimization problems. Note that it is possible to avoid a huge number of solutions in the approximation by filtering (Reuter, 1990; Sayin, 2003; Steuer and Harris, 1980).

Methods based on covering the feasible region (in the decision space) by balls whose radius depends on the Lipschitz constants of objective functions (Evtushenko and Potapov, 1987) provide a theoretically justified approach to approximating the Pareto frontier. They use the EPH of a finite number of objective vectors as an external approximation of the precise EPH. The idea to use an approximation of the EPH for approximating the Pareto frontier in the nonconvex case has also been used in (Reuter, 1990; Kaliszewski, 1994; Benson and Sayin, 1997; Lotov et al., 2002).

As far as random search methods are concerned, they compute objective vectors at random decision vectors and choose the nondominated ones (Sobol and Statnikov, 1981). Though such methods are not theoretically justified (no estimate is given for the approximation quality in a general case), the convergence of the process is guaranteed when the number of random points tends to infinity. These methods are easy to implement (Statnikov and Matusov, 1995).

In simulated annealing (Chipperfield et al., 1999; Suppaptnarm et al., 2000; Kubotani and Yoshimura, 2003), physical processes are imitated. This approach is methodologically close to evolutionary (see Chapter 3) and other metaheuristic methods. Furthermore, various combinations of approximation methods considered above as well as EMO methods can be used in hybrid methods. One effective hybrid method, which combines random search, optimization and a genetic algorithm (Berezkin et al., 2006) was applied in a real-life problem with 325 decisison variables and four objective functions (see Figure 9.9). Some other examples of hybrid methods are discussed in Chapter 16.

Methods for the assessment of the quality of the approximation of the Pareto frontier for nonlinear models are intensively developed by specialists in evolutionary methods (see Chapter 14). In the framework of MCDM methods, in addition to the convex case (see the previous section), quality assessment is provided in methods described in (Evtushenko and Potapov, 1987; Schandl et al., 2002a,b; Klamroth et al., 2002). The quality assessment techniques play a major role in the hybrid methods by Lotov et al. (2002, 2004); Berezkin et al. (2006).

9.5 Conclusions

We have discussed visualization of the Pareto optimal set from different perspectives. Overall, the aim has been to help users of a posteriori methods to find the most preferred solutions for multiobjective optimization problems involving more than two objectives. Both cases of using polyhedral approximations of the Pareto optimal set as well as sets of Pareto optimal solutions as a starting point have been considered. It has been shown that visualization of the Pareto frontier can be carried out in nonlinear multiobjective optimization problems with up to four or five objectives, but this requires more cognitive effort if the number of objectives increases (e.g., till eight).

Acknowledgements

The work of A.V. Lotov was supported by the Russian Foundation for Basic Research (project # 07-01-00472). The work of K. Miettinen was partly supported by the Foundation of the Helsinki School of Economics.

References

Abramova, M.: Approximation of a Pareto set on the basis of inexact information. Moscow University Computational Mathematics and Cybernetics 2, 62–69 (1986)

Barber, C.B., Dobkin, D.P., Huhdanpaa, H.T.: The quickhull algorithm for convex hulls. ACM Transactions on Mathematical Software 2(4), 469–483 (1996)

Benson, H.P.: An outer–approximation algorithm for generating all efficient extreme points in the outcome set of a multiple-objective linear programming problem. Journal of Global Optimization 13, 1–24 (1998)

Benson, H.P., Sayin, S.: Toward finding global representations of the efficient set in multiple-objective mathematical programming. Naval Research Logistics 44, 47–67 (1997)

Berezkin, V.E., Kamenev, G.K., Lotov, A.V.: Hybrid adaptive methods for approximating a nonconvex multidimensional Pareto frontier. Computational Mathematics and Mathematical Physics 46(11), 1918–1931 (2006)

Bushenkov, V.A., Lotov, A.V.: Methods for the Constructing and Application of Generalized Reachable Sets (In Russian). Computing Centre of the USSR Academy of Sciences, Moscow (1982)

Bushenkov, V.A., Gusev, D.I., Kamenev, G.K., Lotov, A.V., Chernykh, O.L.: Visualization of the Pareto set in multi-attribute choice problem (In Russian). Doklady of Russian Academy of Sciences 335(5), 567–569 (1993)

Chankong, V., Haimes, Y.Y.: Multiobjective Decision Making Theory and Methodology. Elsevier, New York (1983)

Chernykh, O.L., Kamenev, G.K.: Linear algorithm for a series of parallel two-dimensional slices of multidimensional convex polytope. Pattern Recognition and Image Analysis 3(2), 77–83 (1993)

Chipperfield, A.J., Whideborn, J.F., Fleming, P.J.: Evolutionary algorithms and simulated annealing for MCDM. In: Gal, T., Stewart, T.J., Hanne, T. (eds.) Multicriteria Decision Making: Advances in MCDM Models, Algorithms, Theory and Applications, pp. 16-1–16-32, Kluwer Academic Publishers, Boston (1999)

Cleveland, W.S.: The Elements of Graphing Data. Wadsworth, Belmont (1985)

Cohon, J., Church, R., Sheer, D.: Generating multiobjective tradeoffs: An algorithm for bicriterion problems. Water Resources Research 15, 1001–1010 (1979)

Cohon, J.L.: Multiobjective Programming and Planning. Academic Press, New York (1978)

Cooke, R.M., van Noortwijk, J.M.: Generic graphics for uncertainty and sensitivity analysis. In: Schueller, G., Kafka, P. (eds.) Safety and Reliability, Proceedings of ESREL '99, pp. 1187–1192. Balkema, Rotterdam (1999)

Das, I., Dennis, J.: Normal boundary intersection: A new method for generating the Pareto surface in nonlinear multicriteria optimization problems. SIAM Journal on Optimization 8, 631–657 (1998)

Evtushenko, Y., Potapov, M.: Methods of Numerical Solutions of Multicriteria Problems. Soviet Mathematics Doklady 34, 420–423 (1987)

Gass, S., Saaty, T.: The Computational Algorithm for the Parametric Objective Function. Naval Research Logistics Quarterly 2, 39 (1955)

Geoffrion, A.M., Dyer, J.S., Feinberg, A.: An interactive approach for multi-criterion optimization, with an application to the operation of an academic department. Management Science 19(4), 357–368 (1972)

Gusev, D.V., Lotov, A.V.: Methods for decision support in finite choice problems (In Russian). In: Ivanilov, J. (ed.) Operations Research. Models, Systems, Decisions, pp. 15–43. Computing Center of Russian Academy of Sciences, Moscow (1994)

Haimes, Y.Y., Tarvainen, K., Shima, T., Thadathill, J.: Hierachical Multiobjective Analysis of Large-Scale Systems. Hemisphere Publishing Corporation, New York (1990)

Henig, M., Buchanan, J.T.: Generalized tradeoff directions in multiobjective optimization problems. In: Tzeng, G., Wang, H., Wen, U., Yu, P. (eds.) Multiple Criteria Decision Making – Proceedings of the Tenth International Conference, pp. 47–56. Springer, New York (1994)

Henig, M., Buchanan, J.T.: Tradeoff directions in multiobjective optimization problems. Mathematical Programming 78(3), 357–374 (1997)

Jahn, J.: Vector Optimization. Springer, Berlin (2004)

Jewel, T.K.: A Systems Approach to Civil Engineering, Planning, Design. Harper & Row, New York (1990)

Kaliszewski, I.: Quantitative Pareto Analysis by Cone Separation Technique. Kluwer, Dordrecht (1994)

Kamenev, G.K.: Dual adaptive algorithms for polyhedral approximation of convex bodies. Computational Mathematics and Mathematical Physics 42(8) (2002)

Klamroth, K., Tind, J., Wiecek, M.M.: Unbiased approximation in multicriteria optimization. Mathematical Methods of Operations Research 56, 413–437 (2002)

Klimberg, R.: GRADS: A new graphical display system for visualizing multiple criteria solutions. Computers & Operations Research 19(7), 707–711 (1992)

Krasnoshchekov, P.S., Morozov, V.V., Fedorov, V.V.: Decomposition in design problems (In Russian). Izvestiya Akademii Nauk SSSR, Series Technical Cybernetics (2), 7–17 (1979)

Kubotani, H., Yoshimura, K.: Performance evaluation of acceptance probability functions for multiobjective simulated annealing. Computers & Operations Research 30, 427–442 (2003)

Larichev, O.: Cognitive validity in design of decision-aiding techniques. Journal of Multi-Criteria Decision Analysis 1(3), 127–138 (1992)

Lewandowski, A., Granat, J.: Dynamic BIPLOT as the interaction interface for aspiration based decision support systems. In: Korhonen, P., Lewandowski, A., Wallenius, J. (eds.) Multiple Criteria Decision Support, pp. 229–241. Springer, Berlin (1991)

Lomov, B.F.: Philosophical and Theoretical Problems of Psychology. Science Publisher, Moscow (1984)

Lotfi, V., Stewart, T.J., Zionts, S.: An aspiration level interactive model for multiple criteria decision making. Computers & Operations Research 19(7), 671–681 (1992)

Lotov, A., Berezkin, V., Kamenev, G., Miettinen, K.: Optimal control of cooling process in continuous casting of steel using a visualization-based multi-criteria approach. Applied Mathematical Modelling 29(7), 653–672 (2005)

Lotov, A.V.: Exploration of economic systems with the help of reachable sets (In Russian). In: Proceedings of the International Conference on the Modeling of Economic Processes, Erevan, 1974, pp. 132–137. Computing Centre of USSR Academy of Sciences, Moscow (1975)

Lotov, A.V.: Coordination of Economic Models by the Attainable Sets Method (In Russian). In: Berlyand, E.L., Barabash, S.B. (eds.) Mathematical Methods for an Analysis of Interaction between Industrial and Regional Systems, pp. 36–44. Nauka, Novosibirsk (1983)

Lotov, A.V.: Generalized reachable sets method in multiple criteria problems. In: Lewandowski, A., Stanchev, I. (eds.) Methodology and Software for Interactive Decision Support, pp. 65–73. Springer, Berlin (1989)

Lotov, A.V., Kamenev, G.K., Berezkin, V.E.: Approximation and Visualization of Pareto-Efficient Frontier for Nonconvex Multiobjective Problems. Doklady Mathematics 66(2), 260–262 (2002)

Lotov, A.V., Bushenkov, V.A., Kamenev, G.K.: Interactive Decision Maps. Approximation and Visualization of Pareto Frontier. Kluwer Academic Publishers, Boston (2004)

Louie, P., Yeh, W., Hsu, N.: Multiobjective Water Resources Management Planning. Journal of Water Resources Planning and Management 110(1), 39–56 (1984)

Mareschal, B., Brans, J.-P.: Geometrical representation for MCDA. European Journal of Operational Research 34(1), 69–77 (1988)

Mattson, C.A., Messac, A.: Pareto frontier based concept selection under uncertainty, with visualization. Optimization and Engineering 6, 85–115 (2005)

McClure, D.E., Vitale, R.A.: Polygonal approximation of plane convex bodies. Journal of Mathematical Analysis and Applications 51(2), 326–358 (1975)

McQuaid, M.J., Ong, T.-H., Chen, H., Nunamaker, J.F.: Multidimensional scaling for group memory visualization. Decision Support Systems 27, 163–176 (1999)

Meisel, W.S.: Tradeoff decision in multiple criteria decision making. In: Cochrane, J., Zeleny, M. (eds.) Multiple Criteria Decision Making, pp. 461–476. University of South Carolina Press, Columbia (1973)

Miettinen, K.: Nonlinear Multiobjective Optimization. Kluwer Academic Publishers, Boston (1999)

Miettinen, K.: Graphical illustration of Pareto optimal solutions. In: Tanino, T., Tanaka, T., Inuiguchi, M. (eds.) Multi-Objective Programming and Goal Programming: Theory and Applications, pp. 197–202. Springer, Berlin (2003)

Miettinen, K., Mäkelä, M.M.: On generalized trade-off directions in nonconvex multiobjective optimization. Mathematical Programming 92, 141–151 (2002)

Miettinen, K., Mäkelä, M.M.: Characterizing general trade–off directions. Mathematical Methods of Operations Research 57, 89–100 (2003)

Miettinen, K., Mäkelä, M.M.: Synchronous approach in interactive multiobjective optimization. European Journal of Operational Research 170(3), 909–922 (2006)

Miettinen, K., Lotov, A.V., Kamenev, G.K., Berezkin, V.E.: Integration of two multiobjective optimization methods for nonlinear problems. Optimization Methods and Software 18, 63–80 (2003)

Miettinen, K., Molina, J., González, M., Hernández-Díaz, A., Caballero, R.: Using box indices in supporting comparison in multiobjective optimization. European Journal of Operational Research. To appear (2008), doi:10.1016/j.ejor.2008.05.103

Miller, G.A.: The magical number seven, plus or minus two: Some limits of our capacity for processing information. Psychological Review 63, 81–97 (1956)

Nefedov, V.N.: On the approximation of Pareto set. USSR Computational Mathematics and Mathematical Physics 24, 19–28 (1984)

Nefedov, V.N.: Approximation of a set of Pareto-optimal solutions. USSR Computational Mathematics and Mathematical Physics 26, 99–107 (1986)

Olson, D.: Decision Aids for Selection Problems. Springer, New York (1996)

Popov, N.: Approximation of a Pareto set by the convolution method. Moscow University Computational Mathematics and Cybernetics (2), 41–48 (1982)

Pryke, A., Mostaghim, S., Nazemi, A.: Heatmap Visualization of Population Based Multi Objective Algorithms. In: Obayashi, S., Deb, K., Poloni, C., Hiroyasu, T., Murata, T. (eds.) EMO 2007. LNCS, vol. 4403, pp. 361–375. Springer, Heidelberg (2007)

Reuter, H.: An approximation method for the efficiency set of multiobjective programming problems. Optimization 21, 905–911 (1990)

Roy, B.: Decisions avec criteres multiples. Metra International 11(1), 121–151 (1972)

Ruzika, S., Wiecek, M.M.: Survey paper: Approximation methods in multiobjective programming. Journal of Optimization Theory and Applications 126(3), 473–501 (2005)

Sawaragi, Y., Nakayama, H., Tanino, T.: Theory of Multiobjective Optimization. Academic Press, Orlando (1985)

Sayin, S.: A Procedure to Find Discrete Representations of the Efficient Set with Specified Coverage Errors. Operations Research 51, 427–436 (2003)

Schandl, B., Klamroth, K., Wiecek, M.M.: Introducing oblique norms into multiple criteria programming. Journal of Global Optimization 23, 81–97 (2002a)

Schandl, B., Klamroth, K., Wiecek, M.M.: Norm-based approximation in multicriteria programming. Computers and Mathematics with Applications 44, 925–942 (2002b)

Schütze, O., Jourdan, L., Legrand, T., Talbi, E.-G., Wojkiewicz, J.L.: A multiobjective approach to the design of conducting polymer composites for electromagnetic shielding. In: Obayashi, S., Deb, K., Poloni, C., Hiroyasu, T., Murata, T. (eds.) EMO 2007. LNCS, vol. 4403, pp. 590–603. Springer, Heidelberg (2007)

Simon, H.: Reason in Human Affairs. Stanford University Press, Stanford (1983)

Smirnov, M.: The logical convolution of the criterion vector in the problem of approximating Pareto set. Computational Mathematics and Mathematical Physics 36, 605–614 (1996)

Sobol, I.M., Statnikov, R.B.: Choice of Optimal Parameters in Multiobjective Problems (In Russian). Nauka Publishing House, Moscow (1981)

Solanki, R.S., Appino, P.A., Cohon, J.L.: Approximating the noninferior set in multiobjective linear programming problems. European Journal of Operational Research 68, 356–373 (1993)

Statnikov, R.B., Matusov, J.: Multicriteria Optimization and Engineering. Chapman and Hall, Boca Raton (1995)

Steuer, R.E., Harris, F.W.: Intra-set point generation and filtering in decision and criterion space. Computers & Operations Research 7, 41–58 (1980)

Suppaptnarm, A., Steffen, K.A., Parks, G.T., Clarkson, P.J.: Simulated annealing: An alternative approach to true multiobjective optimization. Engineering Optimization 33(1), 59–85 (2000)

Trinkaus, H.L., Hanne, T.: knowcube: a visual and interactive support for multicriteria decision making. Computers & Operations Research 32, 1289–1309 (2005)

Vetschera, R.: A preference preserving projection technique for MCDM. European Journal of Operational Research 61(1–2), 195–203 (1992)

Voinalovich, V.: External Approximation to the Pareto Set in Criterion Space for Multicriterion Linear Programming Tasks. Cybernetics and Computing Technology 1, 135–142 (1984)

Wierzbicki, A.P.: On the role of intuition in decision making and some ways of multicriteria aid of intuition. Journal of Multi-Criteria Decision Analysis 6(2), 65–76 (1997)

Wierzbicki, A.P., Nakamori, Y.: Creative Space. Springer, Berlin (2005)

Yu, P.L.: Cone convexity, cone extreme points and nondominated solutions in decision problems with multiple objectives. Journal of Optimization Theory and Applications 14(3), 319–377 (1974)

Yu, P.L.: Multiple-Criteria Decision Making Concepts, Techniques and Extensions. Plenum Press, New York (1985)

10

Meta-Modeling in Multiobjective Optimization

Joshua Knowles[1] and Hirotaka Nakayama[2]

[1] School of Computer Science, University of Manchester,
   Oxford Road, Manchester M13 9PL, UK
   j.knowles@manchester.ac.uk
[2] Konan University, Dept. of Information Science and Systems Engineering,
   8-9-1 Okamoto, Higashinada, Kobe 658-8501, Japan
   nakayama@konan-u.ac.jp

Abstract. In many practical engineering design and other scientific optimization problems, the objective function is not given in closed form in terms of the design variables. Given the value of the design variables, the value of the objective function is obtained by some numerical analysis, such as structural analysis, fluidmechanic analysis, thermodynamic analysis, and so on. It may even be obtained by conducting a real (physical) experiment and taking direct measurements. Usually, these evaluations are considerably more time-consuming than evaluations of closed-form functions. In order to make the number of evaluations as few as possible, we may combine iterative search with *meta-modeling*. The objective function is modeled during optimization by fitting a function through the evaluated points. This model is then used to help predict the value of future search points, so that high performance regions of design space can be identified more rapidly. In this chapter, a survey of meta-modeling approaches and their suitability to specific problem contexts is given. The aspects of dimensionality, noise, expensiveness of evaluations and others, are related to choice of methods. For the multiobjective version of the meta-modeling problem, further aspects must be considered, such as how to define improvement in a Pareto approximation set, and how to model each objective function. The possibility of interactive methods combining meta-modeling with decision-making is also covered. Two example applications are included. One is a multiobjective biochemistry problem, involving instrument optimization; the other relates to seismic design in the reinforcement of cable-stayed bridges.

10.1 An Introduction to Meta-modeling

In all areas of science and engineering, models of one type or another are used in order to help understand, simulate and predict. Today, numerical methods

Reviewed by: Jerzy Błaszczyński, Poznan University, Poland
Yaochu Jin, Honda Research Institute Europe, Germany
Koji Shimoyama, Tohoku University, Japan
Roman Słowiński, Poznan University of Technology, Poland

J. Branke et al. (Eds.): Multiobjective Optimization, LNCS 5252, pp. 245–284, 2008.

make it possible to obtain models or simulations of quite complex and large-scale systems, even when closed-form equations cannot be derived or solved. Thus, it is now a commonplace to model, usually on computer, everything from aeroplane wings to continental weather systems to the activity of novel drugs.

An expanding use of models is to optimize some aspect of the modeled system or process. This is done to find the best wing profile, the best method of reducing the effects of climate change, or the best drug intervention, for example. But there are difficulties with such a pursuit when the system is being modeled numerically. It is usually impossible to find an optimum of the system directly and, furthermore, iterative optimization by trial and error can be very expensive, in terms of computation time.

What is required, to reduce the burden on the computer, is a method of further *modeling the model*, that is, generating a simple model that captures only the relationships between the relevant input and output variables — not modeling any underlying process. *Meta-modeling*, as the name suggests, is such a technique: it is used to build rather simple and computationally inexpensive models, which hopefully replicate the relationships that are *observed* when samples of a more complicated, high-fidelity model or simulation are drawn.[1] Meta-modeling has a relatively long history in statistics, where it is called the response surface method, and is also related to the Design of Experiments (DoE) (Anderson and McLean, 1974; Myers and Montgomery, 1995).

Meta-modeling in Optimization

Iterative optimization procedures employing meta-models (also called surrogate models in this context) alternate between making evaluations on the given high-fidelity model, and on the meta-model. The full-cost evaluations are used to train the initial meta-model, and to update or re-train it, periodically. In this way, the number of full-cost evaluations can often be reduced substantially, whilst a high accuracy is still achieved. This is the main advantage of using a meta-model. A secondary advantage is that the trained meta-model may represent important information about the cost surface and the variables in a relatively simple, and easily interpretable manner. A schematic of the meta-modeling approach is shown in Figure 10.1.

The following pseudocode makes more explicit this process of optimization using models and meta-models:

1. Take an initial sample I of $(\boldsymbol{x}, y)$ pairs from the high-fidelity model.
2. From some or all the samples collected, build/update a model M of $p(y \in Y | \boldsymbol{x} \in X)$ or $f : X \to Y$.

---

[1] Meta-modeling need not refer to modeling of a computational *model*; real processes can also be meta-modeled.

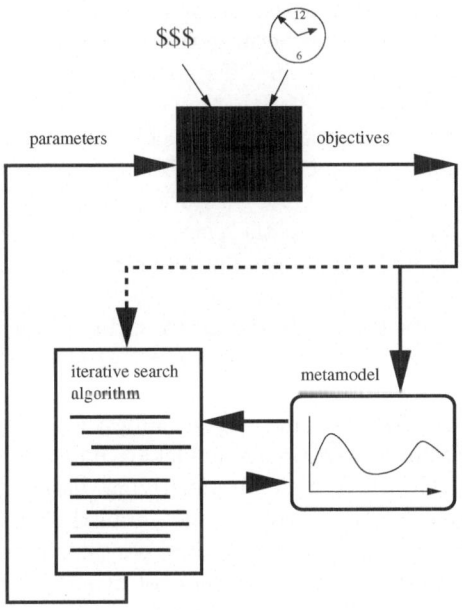

Fig. 10.1. A schematic diagram showing how meta-modeling is used for optimization. The high-fidelity model or function is represented as a black box, which is expensive to use. The iterative searcher makes use of evaluations on both the meta-model and the black box function.

3. Using M, choose a new sample P of points and evaluate them on the high-fidelity model.
4. Until stopping criteria satisfied, return to 2.

The pseudocode is intentionally very general, and covers many different specific strategies. For instance, in some methods the choice of new sample(s) P is made solely based on the current model M, e.g. by finding the optimum on the approximate model (cf. EGO, Jones *et al.* (1998) described in Section 3). Whereas, in other methods, P may be updated based on a memory of previously searched or considered points: e.g. an evolutionary algorithm (EA) using a meta-model may construct the new sample from its current population via the usual application of variation operators, but M is then used to screen out points that it predicts will not have a good evaluation (see (Jin, 2005) and Section 4).

The criterion for selecting the next point(s) to evaluate, or for screening out points, is not always based exclusively on their predicted value. Rather, the estimated *informativeness* of a point may also be accounted for. This can be estimated in different ways, depending on the form of the meta-model. There is a natural tension between choosing of samples because they are predicted to be high-performance points and because they would yield much information, and this tension can be resolved in different ways.

The process of actually constructing a meta-model from data is related to classical regression methods and also to machine learning. Where a model is built up from an initial sample of solutions *only*, the importance of placing those points in the design space in a theoretically well-founded manner is emphasized, a subject dealt with in the classical design of experiments (DoE) literature. When the model is updated using new samples, classical DoE principles do not usually apply, and one needs to look to machine learning theory to understand what strategies might lead to optimal performance. Here care must be taken, however. Although the *supervised learning* paradigm is usually taken as the default method used for training meta-models, it is worth noting that in supervised learning, it is usually a base assumption that the available data for training a regression model are independent and identically distributed samples drawn from some underlying distribution. But in meta-modeling, the samples are not drawn randomly in this way: they are *chosen*, and this means that training sets will often contain highly correlated data, which can affect the estimation of goodness of fit and/or generalization performance. Also, meta-modeling in optimization can be related to *active learning* (Cohn *et al.*, 1996), since the latter is concerned with the iterative choice of training samples; however, active learning is concerned only with maximising what is *learned* whereas meta-modeling in optimization is concerned mainly or partially with seeking optima, so neither supervised learning or active learning are identical with meta-modeling. Finally, in meta-modeling, samples are often added to a 'training set' incrementally (see, e.g. (Cauwenberghs and Poggio, 2001)), and the model is re-trained periodically; how this re-training is achieved also leads to a variety of methods.

Interactive and Evolutionary Meta-modeling

Meta-modeling brings together a number of different fields to tackle the problem of how to optimize expensive functions on a limited budget. Its basis in the DoE literature gives the subject a classical feel, but evolutionary algorithms employing meta-models have been emerging for some years now, too (for a comprehensive survey, see(Jin, 2005)).

In the case of multiobjective problems, it does not seem possible or desirable to make a clear distinction between interactive and evolutionary approaches to meta-modeling. Some methods of managing and using meta-models seek to add one new search point, derived from the model, at every iteration; others are closer to a standard evolutionary algorithm, with a population of solutions used to generate a set of new candidate points, which are then filtered using the model, so that only a few are really evaluated. We deal with both types of methods here (and those that lie in between), and also consider how interaction with a decision maker can be used.

Organization of the Chapter

We begin by considering the different settings in which (single-objective) meta-modeling may be used, and the consequences for algorithm design. In the succeeding section, we survey, in more detail, methods for constructing meta-models, i.e., different regression techniques and how models can be updated iteratively when new samples are collected. Section 10.4 explicitly considers how to handle meta-models in the case of multiobjective optimization. This section elaborates on the way in which a Pareto front approximation is gradually built up in different methods, and also considers how interaction with a DM may be incorporated. Practical advice for evaluating meta-models in multiobjective optimization is given in section 10.5, including ideas for performance measures as well as baseline methods to compare against. Two application sections follow, one from analytical chemistry and one from civil engineering.

10.2 Aspects of Managing Meta-models in Optimization

The combination of meta-modeling and optimization can be implemented in many different ways. In this section, we relate some of the ways of using meta-models to properties of the particular optimization scenario encountered: such properties as the cost of an evaluation, the number that can be performed concurrently, and the features of the cost landscape.[2]

10.2.1 Choice of Model Type

The whole concept of using a model of the cost landscape to improve the performance of an optimizer rests on the assumption that the model will aid in choosing worthy new points to sample, i.e., it will be predictive of the real evaluations of those points, to some degree. This assumption will hold only if the function being optimized is amenable to approximation by the selected *type* of meta-model. It is not realistic to imagine that functions of arbitrary form and complexity can be optimized more efficiently using meta-modeling. This important consideration is related to the 'no free lunch' theorems for search (Wolpert and Macready, 1997) and especially to the pithy observation

---

[2] The concept of a cost landscape relies on the fact that proximity of points in design space is defined, which in turn, assumes that a choice of problem representation has already been made. We shall not venture any further into the subject of choices of representation here.

of Thomas English that 'learning is hard and optimization is easy in the typical function' (English, 2000).

In theory, given a cost landscape, there should exist a type of model that approximates it fastest as data is collected and fitting progresses. However, we think it is fair to say that little is yet known about which types of model accord best with particular features of a landscape and, in any case, very little may be known to guide this choice. Nonetheless, some basic considerations of the problem do help guide in the choice of suitable models and learning algorithms. In particular, the dimension of the design space is important, as certain models cope better than others with higher dimensions. For example, naive Bayes' regression (Eyheramendy *et al.*, 2003) is used routinely in very high-dimensional feature spaces (e.g. in spam recognition where individual word frequencies form the input space). On the other hand, for low-dimensional spaces, where local correlations are important, naive Bayes' might be very poor, whereas a Gaussian process model (Schwaighofer and Tresp, 2003) might be expected to perform more effectively.

Much of the meta-modeling literature considers only problems over continuous design variables, as these are common in certain engineering domains. However, there is no reason why this restriction need prevail. Meta-modeling will surely become more commonly used for problems featuring discrete design spaces or mixed discrete and continuous variables. Machine learning methods such as classification and regression trees (C&RT) (Breiman, 1984), genetic programming (Langdon and Poli, 2001), and Bayes' regression may be more appropriate for modeling these high-dimensional landscapes than the splines and polynomials that are used commonly in continuous design spaces. Meta-modeling of cost functions in discrete and/or high-dimensional spaces is by no means the only approach to combining machine learning and optimization, however. An alternative is to model the distribution over the variables that leads to high-quality points in the objective space — an approach known as model-based search or estimation of distribution algorithms (Larranaga and Lozano, 2001; Laumanns and Ocenasek, 2002). The learnable evolution model is a related approach (Michalski, 2000; Jourdan *et al.*, 2005). These approaches, though interesting rivals to meta-models, do not predict costs or model the cost function, and are thus beyond the scope of this chapter.

When choosing the type of model to use, other factors that are less linked to properties of the cost landscape/design space should also be considered. Models differ in how they scale in terms of accuracy and speed of training as the number of training samples varies. Some models can be trained incrementally, using only the latest samples (some particular SVMs), whereas others (most multi-layer perceptrons) can suffer from 'catastrophic forgetting' if trained on new samples only, and need to use complete re-training over all samples when some new ones become available, or some other strategy of rehearsal(Robins, 1997). Some types of model need cross-validation to control overfitting, whereas others use regularization. Finally, some types of model, such as Gaussian random fields, model their own error, which can be a distinct advantage when deciding where to sample next (Jones *et al.*, 1998; Emmerich *et al.*, 2006). A more detailed survey of types of model, and methods for training them, is given in section 3.

10.2.2 The Cost of Evaluations

One of the most important aspects affecting meta-modeling for optimization is the actual cost of an evaluation. At one extreme, a cost function may only take the order of 1s to evaluate, but is still considered expensive in the context of evolutionary algorithm searches where tens of thousands of evaluations are typical. At the other extreme, when the design points are very expensive, such as vehicle crash tests (Hamza and Saitou, 2005), then each evaluation may be associated with financial costs and/or may take days to organize and carry out. In the former case, there is so little time between evaluations that model-fitting is best carried out only periodically (every few generations of the evolutionary algorithm) and new sample points are generated mainly as a result of the normal EA mechanisms, with the meta-model playing only a subsidiary role of filtering out estimated poor points. In the latter case, the overheads of fitting and cross-validating models is small compared with the time between evaluations, so a number of alternative models can be fitted and validated after every evaluation, and might be used very carefully to decide on the best succeeding design point, e.g. by searching over the whole design space using the meta-model(s) to evaluate points.

The cost of evaluations might affect the meta-modeling strategy in more complicated and interesting ways than the simple examples above, too. In some applications, the cost of evaluating a point is not uniform over the design space, so points may be chosen based partly on their expected cost to evaluate (though this has not been considered in the literature, to our knowledge). In other applications, the cost of determining whether a solution is feasible or not is expensive, while evaluating it on the objective function is cheap. This was the case in (Joslin *et al.*, 2006), and led to a particular strategy of only checking the constraints for solutions passing a threshold on the objective function.

In some applications, the cost of an evaluation can be high in time, yet many can be performed in parallel. This situation occurs, for example, in using optimization to design new drugs via combinatorial chemistry methods. Here, the time to prepare a drug sample and to test it can be of the order of 24 hours, but using high-throughput equipment, several hundred or thousands of different drug compounds can be made and tested in each batch (Corne *et al.*, 2002). Clearly, this places very particular constraints on the meta-modeling/optimization process: there is not always freedom to choose how frequently updates of the model are done, or how many new design points should be evaluated in each generation. Only future studies will show how to best deal with these scenarios.

10.2.3 Advanced Topics

Progress in meta-modeling seems to be heading in several exciting new directions, worth mentioning here.

Typically, when fitting a meta-model to the data, a single global model is learned or updated, using all available design points. However, some research is departing from this by using local models (Atkeson *et al.*, 1997), which are trained only on local subsets of the data (Emmerich *et al.*, 2006). Another departure from the single global model is the possibility of using an ensemble of meta-models (Hamza and Saitou, 2005). Ensemble learning has general advantages in supervised learning scenarios (Brown *et al.*, 2005) and may increase the accuracy of meta-models too. Moreover, Jin and Sendhoff (2004) showed that ensemble methods can be used to predict the quality of the estimation, which can be very useful in the meta-modeling approach.

Noise or stochasticity is an element in many systems that require optimization. Many current meta-modeling methods, especially those based on radial basis functions or Kriging (see next section) assume noiseless evaluation, so that the uncertainty of the meta-model at the evaluated points is assumed to be zero. However, with noisy functions, to obtain more accurate estimates of the expected quality of a design point, several evaluations may be needed. Huang *et al.* (2006) consider how to extend the well-known EGO algorithm (see next section) to account for the case of noise or stochasticity on the objective function. Other noisy optimization methods, such as those based on EAs (Fieldsend and Everson, 2005), could be combined with meta-models in future work.

A further exciting avenue of research is the use of *transductive learning*. It has been shown by Chapelle *et al.* (1999) that in supervised learning of a regression model, knowledge of the future test points (just in design space), at the time of training, can be used to improve the training and lead to better ultimate prediction performance on those points. This results has been imported into meta-modeling by Schwaighofer and Tresp (2003), which compares transductive Gaussian regression methods with standard, inductive ones, and finds them much more accurate.

10.3 Brief Survey of Methods for Meta-modeling

The Response Surface Method (RSM) is probably the most widely applied to meta-modeling (Myers and Montgomery, 1995). The role of RSM is to predict the response y for the vector of design variables $x \in R^n$ on the basis of the given sampled obsevation (x_i, y_i) $(i = 1, \ldots, p)$.

Usually, the Response Surface Method is a generic name, and it covers a wide range of methods. Above all, methods using experimental design are famous. However, many of them select sample points only on the basis of statisitical analysis of design variable space. They may provide a good approximation of black-box functions with a mild nonlineality. It is clear, however, that in cases in which the black-box function is highly nonlinear, we can obtain better performance by methods taking into account not only the statistical property of design variable space but also that of range space of the black-box function (in other words, the shape of function).

Moreover, machine learning techniques such as RBF (Radial Basis Function) networks and Support Vector Machines (SVM) have been recently applied for approximating the black-box function (Nakayama *et al.*, 2002, 2003).

10.3.1 Using Design of Experiments

Suppose, for example for simplicitly, that we consider a response function given by a quadratic polynomial:

$$y = \beta_0 + \sum_{i=1}^{n} \beta_i x_i + \sum_{i=1}^{n} \beta_{ii} x_i^2 + \sum_{i<j} \beta_{ij} x_i x_j \tag{10.1}$$

Since the above equation is linear with respect to β_i, we can rewrite the equation (10.1) into the following:

$$y = X\beta + \varepsilon, \tag{10.2}$$

where $E(\varepsilon) = 0$, $V(\varepsilon) = \sigma^2 I$.

The above (10.2) is well known as linear regression, and the solution β minimizing the squarred error is given by

$$\hat{\beta} = (X^T X)^{-1} X^T y \tag{10.3}$$

The variance covariance matrix $V(\hat{\beta}) = cov(\hat{\beta}_i, \hat{\beta}_j)$ of the least squarred error prediction $\hat{\beta}$ given by (10.3) becomes

$$V(\hat{\beta}) = cov(\hat{\beta}_i, \hat{\beta}_j) = E((\hat{\beta} - E(\hat{\beta}))(\hat{\beta} - E(\hat{\beta}))) \tag{10.4}$$
$$= (X^T X)^{-1} \sigma^2, \tag{10.5}$$

where σ^2 is the variance of error in the response y such that $E(\varepsilon \varepsilon^T) = \sigma^2 I$.

i) Orthogonal Design

Orthogonal design is usually applied for experimental design with linear polynomials. Selecting sample points in such a way that the set X is orthogonal, the matrix $X^T X$ becomes diagonal. It is well known that the orthogonal design with the first order model is effective for cases with only main effects or first order interaction effects. For the polynomial regressrion with higher order (≥ 2), orthogonal polynomials are usually used in order to make the design to be orthogonal (namely, $X^T X$ is diagonal). Then the coefficients of polynomials are easily evaluated by using orthogonal arrays.

Another kind of experimental design, e.g., CCD (Cetral Composite Design) is applied mostly for experiments with quadratic polynomials.

ii) D-optimality

Considering the equation (10.5), the matrix $(X^T X)^{-1}$ should be minimized so that the variance of the predicted β may decrease. Since each element of $(X^T X)^{-1}$ has $\det(X^T X)$ in the denomnator, we can expect to decrease not only variance but also covariance of β_i by maximizing $\det(X^T X)$. This is the idea of D-optimality in design of experiments. In fact, it is usual to use the moment matrix

$$M = \frac{X^T X}{p}, \tag{10.6}$$

where p is the number of sample points.

Other criteria are possible: to minimize the trace of $(X^T X)^{-1}$ (A-optimiality), to minimize the maximal value of the diagonal components of $(X^T X)^{-1}$ (minimax criterion), to maximize the minimal eigen value of $X^T X$ (E-optimality). In general, however, the D-optimality criterion is widely used for many practical problems.

10.3.2 Kriging Method

Consider the response $y(x)$ as a realization of a random function, $Y(x)$ such that

$$Y(x) = \mu(x) + Z(x). \tag{10.7}$$

Here, $\mu(x)$ is a global model and $Z(x)$ reflecting a deviation from the global model is a random function with zero mean and nonzero covariance given by

$$cov[Z(x), Z(x')] = \sigma^2 R(x, x') \tag{10.8}$$

where R is the correlation between $Z(x)$ and $Z(x')$. Usually, the stochastic process is supposed to be stationary, which implies that the correlation $R(x, x')$ depends only on $x - x'$, namely

$$R(x, x') = R(x - x'). \tag{10.9}$$

A commonly used example of such correlation functions is

$$R(x, x') = \exp[-\sum_{i=1}^{n} \theta_i |x_i - x'_i|^2], \tag{10.10}$$

where x_i and x'_i are i-th component of x and x', respectively.

Although a linear regressrion model $\sum_{j=1}^{k} \mu_j f_j(x)$ can be applied as a global model in (10.7) (*universal Kriging*), $\mu(x) = \mu$ in which μ is unknown but constant is commonly used in many cases (*ordinary Kriging*). In the ordinary Kriging, the best linear unbiased predictor of y at an untried x can be given by

$$\hat{y}(x) = \hat{\mu} + r^T(x)R^{-1}(y - 1\hat{\mu}), \qquad (10.11)$$

where $\hat{\mu} = (1^T R^{-1} 1)^{-1} 1^T R^{-1} y$ is the generalized least squares estimator of μ, $r(x)$ is the $n \times 1$ vector of correlations $R(x, x_i)$ between Z at x and sampled points x_i $(i = 1, \ldots, p)$, R is an $n \times n$ correlation matrix with (i, j)-element defined by $R(x_i, x_j)$ and 1 is a unity vector whose components are all 1.

Using Expected Improvement

Jones *et al.* (1998) suggested a method called EGO (Efficient Global Optimization) for black-box objective functions. They applied a stochastic process model (10.7) for predictor and the expected improvement as a figure of merit for additional sample points.

The estimated value of the mean of the stochastic process, $\hat{\mu}$, is given by

$$\hat{\mu} = \frac{1^T R^{-1} y}{1^T R^{-1} 1}. \qquad (10.12)$$

In this event, the variation σ^2 is estimated by

$$\hat{\sigma}^2 = \frac{(y - 1\hat{\mu})^T R^{-1}(y - 1\hat{\mu})}{n}. \qquad (10.13)$$

The mean squared error of the predictor is estimated by

$$s^2(x) = \sigma^2 [1 - r^T R^{-1} r + \frac{(1 - 1^T R^{-1} r)^2}{1^T R^{-1} 1}]. \qquad (10.14)$$

In the following $s = \sqrt{s^2(x)}$ is called a standard error.

Using the above predictor on the basis of stochastic process model, Jones *et al.* applied the expected improvemnet for adding a new sample point. Let $f^p_{min} = \min\{y_1, \ldots, y_p\}$ be the current best function value. They model the uncertainty at $y(x)$ by treating it as the realization of a normally distributed random variable Y with mean and standard deviation given by the above predictor and its standard error.

For minimization cases, the improvement at x is $I = [\max(f^p_{min} - Y, 0)$. Therefore, the expected improvement is given by

$$E[I(x)] = E[\max(f^p_{min} - Y, 0)].$$

It has been shown that the above formula can be expanded as follows:

$$E(I) = \begin{cases} (f^p_{min} - \hat{y})\Phi(\frac{f^p_{min} - \hat{y}}{s}) + s\phi(\frac{f^p_{min} - \hat{y}}{s}) & \text{if } s < 0 \\ 0 & \text{if } s = 0, \end{cases} \qquad (10.15)$$

where ϕ is the standard normal density and Φ is the distribution function.

We can add a new sample point which maximizes the expected improvement. Although Jones *et al.* proposed a method for maximizing the expected improvement by using the branch and bound method, it is possible to select the best one among several candidates which are generated randomly in the design variable space.

Furthermore, Schonlau (1997) extended the expected improvement as follows: Letting $I^g = \max((f_{\min}^p - Y)^g, \ 0)$, then

$$E(I^g) = s^g \sum_{i=0}^{g} (-1)^i \left(\frac{g!}{i!(g-i)!}\right)(f_{min}^{p'})^{g-i} T_i \qquad (10.16)$$

where

$$f_{min}^{p'} = \frac{f_{min}^p - \hat{y}}{s}$$

and

$$T_k = -\phi(f_{min}^{p'})(f_{min}^{p'})^{(k-1)} + (k-1)T_{k-2}.$$

Here

$$T_0 = \Phi(f_{min}^{p'})$$
$$T_1 = -\phi(f_{min}^{p'}).$$

It has been observed that larger value of g makes the global search, while smaller value of g the local search. Therefore, we can control the value of g depending upon the situation.

10.3.3 Computational Intelligence

Multi-layer Perceptron Neural Networks

The multi-layer perceptron (MLP) is used in several meta-modeling applications in the literature (Jin *et al.*, 2001; Gaspar-Cunha and Vieira, 2004). It is well-known that MLPs are universal approximators, which makes them attractive for modeling black box functions for which little information about their form is known. But, in practice, it can be difficult and time-consuming to train MLPs effectively as they still have biases and it is easy to get caught in local minima which give far from desirable performance. A large MLP with many weights has a large *capacity*, i.e. it can model complex functions, but it is also easy to over-fit it, so that generalization performance may be poor. The use of a regularization term to help control the complexity is necessary to ensure better generalization performance. Cross-validation can also be used during the training to mitigate overfitting.

Disadvantages of using MLPs may include the difficulty to train it quickly, especially if cross-validation with several folds is used (a problem in some applications). It is not easy to train incrementally (compare with RBFs). Moreover, an MLP does not estimate its own error (compare with Kriging), which means that it can be difficult to estimate the best points to sample next.

Radial Basis Function Networks

Since the number of sample points for predicting objective functions should be as few as possible, incremental learning techniques which predict black-box functions by adding learning samples step by step, are attractive. RBF Networks (RBFN) and Support Vector Machines (SVM) are effective to this end. For RBFN, the necessary information for incremental learning can be easily updated, while the information of support vector can be utilized in selecting additional samples as the sensitivity in SVM. The details of these approaches can be seen in (Nakayama et al., 2002) and (Nakayama et al., 2003). Here, we introduce the incremental learning by RBFN briefly in the following.

The output of an RBFN is given by

$$f(\boldsymbol{x}) = \sum_{j=1}^{m} w_j h_j(\boldsymbol{x}),$$

where h_j, $j = 1, \ldots, m$ are radial basis functions, e.g.,

$$h_j(\boldsymbol{x}) = e^{-\|\boldsymbol{x}-\boldsymbol{c}_j\|^2/r_j}.$$

Given the training data $(\boldsymbol{x}_i, \hat{y}_i)$, $i = 1, \cdots, p$, the learning of RBFN is usually made by solving

$$\min \quad E = \sum_{i=1}^{p} (\hat{y}_i - f(\boldsymbol{x}_i))^2 + \sum_{j=1}^{m} \lambda_j w_j^2$$

where the second term is introduced for the purpose of regularization.

In general cases with a large number of training data p, the number of basis functions m is set to be less than p in order to avoid overlearning. However, the number of training data is not so large in this paper, because it is desired to be as small as possible in applications under consideration. The value m is set, therefore, to be equal to p in later sections in this paper. Also, the center of radial basis function $\boldsymbol{c}_i$ is set to be $\boldsymbol{x}_i$. The values of λ_j and r_j are usually determined by cross-validation test. It is observed through our experience that in many problems we have a good performance with $\lambda_j = 0.01$ and a simple estimate for r_j given by

$$r = \frac{d_{max}}{\sqrt[n]{np}}, \tag{10.17}$$

where d_{max} is the maximal distance among the data; n is the dimension of data; p is the number of data.

Letting $A = (H_p^T H_p + \Lambda)$, we have

$$A\boldsymbol{w} = H_p^T \hat{\boldsymbol{y}},$$

as a necessary condition for the above minimization. Here

$$H_p^T = [\boldsymbol{h}_1 \; \cdots \; \boldsymbol{h}_p],$$

where $\boldsymbol{h}_j^T = [h_1(\boldsymbol{x}_j), \ldots, h_m(\boldsymbol{x}_j)]$, and Λ is a diagonal matrix whose diagonal components are $\lambda_1 \; \cdots \; \lambda_m$.

Therefore, the learning in RBFN is reduced to finding

$$A^{-1} = (H_p^T H_p + \Lambda)^{-1}.$$

The incremental learning in RBFN can be made by adding new samples and/or a basis function, if necesary. Since the learning in RBFN is equivalent to the matrix inversion A^{-1}, the additional learning here is reduced to the incremental calculation of the matrix inversion. The following algorithm can be seen in (Orr, 1996):

(i) Adding a New Training Sample

Adding a new sample $\boldsymbol{x}_{p+1}$, the incremental learning in RBFN can be made by the following simple update formula: Let

$$H_{p+1} = \begin{bmatrix} H_p \\ \boldsymbol{h}_{p+1}^T \end{bmatrix},$$

where $\boldsymbol{h}_{p+1}^T = [h_1(\boldsymbol{x}_{p+1}), \ldots, h_m(\boldsymbol{x}_{p+1})]$.
Then

$$A_{p+1}^{-1} = A_p^{-1} - \frac{A_p^{-1} \boldsymbol{h}_{p+1} \boldsymbol{h}_{p+1}^T A_p^{-1}}{1 + \boldsymbol{h}_{p+1}^T A_p^{-1} \boldsymbol{h}_{p+1}}.$$

(ii) Adding a New Basis Function

In those cases where a new basis function is needed to improve the learning for a new data, we have the following update formula for the matrix inversion: Let

$$H_{m+1} = \begin{bmatrix} H_m \; \boldsymbol{h}_{m+1} \end{bmatrix},$$

where $\boldsymbol{h}_{m+1}^T = [h_{m+1}(\boldsymbol{x}_1), \ldots, h_{m+1}(\boldsymbol{x}_p)]$.
Then

$$A_{m+1}^{-1} = \begin{bmatrix} A_m^{-1} & \boldsymbol{0} \\ \boldsymbol{0}^T & 0 \end{bmatrix}$$

$$+ \frac{1}{\lambda_{m+1} + \boldsymbol{h}_{m+1}^T (I_p - H_m A_m^{-1} H_m^T) \boldsymbol{h}_{m+1}} \times \begin{bmatrix} A_m^{-1} H_m^T \boldsymbol{h}_{m+1} \\ -1 \end{bmatrix} \begin{bmatrix} A_m^{-1} H_m^T \boldsymbol{h}_{m+1} \\ -1 \end{bmatrix}^T.$$

10.3.4 Support Vector Machines

Support vector machines (SVMs) were originally developed for pattern classification and later extended to regression (Cortes and Vapnik, 1995; Vapnik, 1998; Cristianini and Shawe-Tylor, 2000; B.Schölkopf and A.J.Smola, 2002). Regression using SVMs, called often support vector regression, plays an important role in meta-modeling. However, the essential idea of support vector regression lies in SVMs for classification. Therefore, we start with a brief review of SVM for classification problems.

Let X be a space of conditional attributes. For binary classification problems, the value of $+1$ or -1 is assigned to each pattern $\boldsymbol{x}_i \in X$ according to its class $\mathcal{A}$ or $\mathcal{B}$. The aim of machine learning is to predict which class newly observed patterns belong to on the basis of the given training data set $(\boldsymbol{x}_i, y_i)$ $(i = 1, \ldots, p)$, where $y_i = +1$ or -1. This is performed by finding a discriminant function $f(\boldsymbol{x})$ such that $f(\boldsymbol{x}) \geq 0$ for $\boldsymbol{x} \in \mathcal{A}$ and $f(\boldsymbol{x}) < 0$ for $\boldsymbol{x} \in \mathcal{B}$. Linear discriminant functions, in particular, can be expressed by the following linear form

$$f(\boldsymbol{x}) = \boldsymbol{w}^T \boldsymbol{x} + b$$

with the property

$$\boldsymbol{w}^T \boldsymbol{x} + b \geq 0 \quad \text{for} \quad \boldsymbol{x} \in \mathcal{A}$$
$$\boldsymbol{w}^T \boldsymbol{x} + b < 0 \quad \text{for} \quad \boldsymbol{x} \in \mathcal{B}.$$

In cases where training data set X is not linearly separable, we map the original data set X to a feature space Z by some nonlinear map ϕ. Increasing the dimension of the feature space, it is expected that the mapped data set becomes linearly separable. We try to find linear classifiers with maximal margin in the feature space. Letting $\boldsymbol{z}_i = \phi(\boldsymbol{x}_i)$, the separating hyperplane with maximal margin can be given by solving the following problem with the normalization $\boldsymbol{w}^T \boldsymbol{z} + b = \pm 1$ at points with the minimum interior deviation:

$$\min_{\boldsymbol{w}, b} \quad \|\boldsymbol{w}\| \qquad \qquad (\text{SVM}_{hard})_P$$

$$\text{s.t.} \quad y_i \left(\boldsymbol{w}^T \boldsymbol{z}_i + b \right) \geq 1, \ i = 1, \ldots, p.$$

Dual problem of $(\text{SVM}_{hard})_P$ with $\frac{1}{2}\|\boldsymbol{w}\|_2^2$ is

$$\max_{\alpha_i} \quad \sum_{i=1}^{p} \alpha_i - \frac{1}{2} \sum_{i,j=1}^{p} \alpha_i \alpha_j y_i y_j \phi(\boldsymbol{x}_i)^T \phi(\boldsymbol{x}_j) \qquad (\text{SVM}_{hard})_D$$

$$\text{s.t.} \quad \sum_{i=1}^{p} \alpha_i y_i = 0,$$

$$\alpha_i \geq 0, \ i = 1, \ldots, p.$$

Using the kernel function $K(\boldsymbol{x}, \boldsymbol{x}') = \phi(\boldsymbol{x})^T \phi(\boldsymbol{x}')$, the problem $(\text{SVM}_{hard})_D$ can be reformulated as follows:

$$\max_{\alpha_i} \quad \sum_{i=1}^{p} \alpha_i - \frac{1}{2} \sum_{i,j=1}^{p} \alpha_i \alpha_j y_i y_j K(\boldsymbol{x}_i, \boldsymbol{x}_j) \qquad (\text{SVM}_{hard})$$

$$\text{s.t.} \quad \sum_{i=1}^{p} \alpha_i y_i = 0,$$

$$\alpha_i \geq 0, \ i = 1, \ldots, p.$$

Although several kinds of kernel functions have been suggested, the Gaussian kernel

$$K(\boldsymbol{x}, \boldsymbol{x}') = \exp\left(-\frac{||\boldsymbol{x} - \boldsymbol{x}'||^2}{2r^2}\right)$$

is popularly used in many cases.

MOP/GP Approaches to Support Vector Classification

In 1981, Freed and Glover suggested to get just a hyperplane separating two classes with as few misclassified data as possible by using goal programming (Freed and Glover, 1981) (see also (Erenguc and Koehler, 1990)). Let ξ_i denote the exterior deviation which is a deviation from the hyperplane of a point $\boldsymbol{x}_i$ improperly classified. Similarly, let η_i denote the interior deviation which is a deviation from the hyperplane of a point $\boldsymbol{x}_i$ properly classified. Some of main objectives in this approach are as follows:

i) Minimize the maximum exterior deviation (decrease errors as much as possible)

ii) Maximize the minimum interior deviation (i.e., maximize the margin)

iii) Maximize the weighted sum of interior deviation

iv) Minimize the weighted sum of exterior deviation.

Introducing the idea iv) above, the well known soft margin SVM with slack variables (or, exterior deviations) ξ_i $(i = 1, \ldots, p)$ which allow classification errors to some extent can be formulated as follows:

$$\min_{\boldsymbol{w}, b, \xi_i} \quad \frac{1}{2}||\boldsymbol{w}||_2^2 + C \sum_{i=1}^{p} \xi_i \qquad (\text{SVM}_{soft})_P$$

$$\text{s.t.} \quad y_i\left(\boldsymbol{w}^T \boldsymbol{z}_i + b\right) \geqq 1 - \xi_i,$$

$$\xi_i \geqq 0, \quad i = 1, \ldots, p,$$

where C is a trade-off parameter between minimizing $||\boldsymbol{w}||_2^2$ and minimizing $\sum_{i=1}^{p} \xi_i$.

Using a kernel function in the dual problem yields

$$\max_{\alpha_i} \quad \sum_{i=1}^{p} \alpha_i - \frac{1}{2} \sum_{i,j=1}^{p} \alpha_i \alpha_j y_i y_j K(\boldsymbol{x}_i, \boldsymbol{x}_j) \qquad (\text{SVM}_{soft})$$

$$\text{s.t.} \quad \sum_{i=1}^{p} \alpha_i y_i = 0,$$

$$0 \leqq \alpha_i \leqq C, \quad i = 1, \ldots, p.$$

Lately, taking into account the objectives (ii) and (iv) of goal programming, we have the same formulation of ν-support vector algorithm developed by Schölkopf and Smola (1998):

$$\min_{\boldsymbol{w},b,\xi_i,\rho} \quad \frac{1}{2}\|\boldsymbol{w}\|_2^2 - \nu\rho + \frac{1}{p}\sum_{i=1}^{p}\xi_i \qquad (\nu-\text{SVM})_P$$

$$\text{s.t.} \quad y_i\left(\boldsymbol{w}^T\boldsymbol{z}_i + b\right) \geqq \rho - \xi_i,$$

$$\rho \geqq 0, \;\; \xi_i \geqq 0, \;\; i = 1,\ldots,p,$$

where $0 \leqq \nu \leqq 1$ is a parameter.

Compared with the existing soft margin algorithm, one of the differences is that the parameter C for slack variables does not appear, and another difference is that the new variable ρ appears in the above formulation. The problem $(\nu-\text{SVM})_P$ maximizes the variable ρ which corresponds to the minimum interior deviation (i.e., the minimum distance between the separating hyperplane and correctly classified points).

The Lagrangian dual problem to the problem $(\nu-\text{SVM})_P$ is as follows:

$$\max_{\alpha_i} \quad -\frac{1}{2}\sum_{i,j=1}^{p} y_i y_j \alpha_i \alpha_j K\left(\boldsymbol{x}_i, \boldsymbol{x}_j\right) \qquad (\nu-\text{SVM})$$

$$\text{s.t.} \quad \sum_{i=1}^{p} y_i \alpha_i = 0,$$

$$\sum_{i=1}^{\ell} \alpha_i \geqq \nu,$$

$$0 \leqq \alpha_i \leqq \frac{1}{p}, \;\; i = 1,\ldots,p.$$

Other variants of SVM considering both slack variables for misclassified data points (i.e., exterior deviations) and surplus variables for correctly classified data points (i.e., interior deviations) are possible (Nakayama and Yun, 2006a): Considering iii) and iv) above, we have the fomula of total margin SVM, while $\nu-$SVM can be derived from i) and iii).

Finally, $\mu-\nu-$SVM is derived by considering the objectives i) and ii) in MOP/GP:

$$\min_{\boldsymbol{w},b,\rho,\sigma} \quad \frac{1}{2}\|\boldsymbol{w}\|_2^2 - \nu\rho + \mu\sigma \qquad (\mu-\nu-\text{SVM})_P$$

$$\text{s.t.} \quad y_i\left(\boldsymbol{w}^T\boldsymbol{z}_i + b\right) \geqq \rho - \sigma, \;\; i = 1,\ldots,p,$$

$$\rho \geqq 0, \;\; \sigma \geqq 0,$$

where ν and μ are parameters.

The dual formulation is given by

$$\max_{\alpha_i} \quad -\frac{1}{2} \sum_{i,j=1}^{p} \alpha_i \alpha_j y_i y_j K\left(\boldsymbol{x}_i, \boldsymbol{x}_j\right) \qquad (\mu - \nu - \text{SVM})$$

$$\text{s.t.} \quad \sum_{i=1}^{p} \alpha_i y_i = 0,$$

$$\nu \leqq \sum_{i=1}^{p} \alpha_i \leqq \mu,$$

$$\alpha_i \geqq 0, \ \ i = 1, \ldots, p.$$

Letting $\boldsymbol{\alpha}^*$ be the optimal solution to the problem ($\mu-\nu-$SVM), the offset b^* can be chosen easily for any i satisfying $\alpha_i^* > 0$. Otherwise, b^* can be obtained by the similar way with the decision of the b^* in the other algorithms.

Support Vector Regression

Support Vector Machines were extended to regression by introducing the ε insensitive loss function by Vapnik (1998). Denote the given sample data by $(\boldsymbol{x}_i, y_i)$ for $i = 1, ..., p$. Suppose that the regression function on the Z space is expressed by $f(\boldsymbol{z}) = \sum_{i=1}^{p} w_i z_i + b$. The linear ε insensitive loss function is defined by

$$L^{\varepsilon}(\boldsymbol{z}, y, f) = |y - f(\boldsymbol{z})|_{\varepsilon} = \max(0, |y - f(\boldsymbol{z})| - \varepsilon).$$

For a given insensitivity parameter ε,

$$\min_{\boldsymbol{w}, b, \varepsilon, \xi_i, \acute{\xi}_i} \quad \frac{1}{2}\|\boldsymbol{w}\|_2^2 + C\left(\frac{1}{p}\sum_{i=1}^{p}(\xi_i + \acute{\xi}_i)\right) \qquad (soft-\text{SVR})_P$$

$$\text{s.t.} \quad \left(\boldsymbol{w}^T \boldsymbol{z}_i + b\right) - y_i \leqq \varepsilon + \xi_i, \ \ i = 1, \ldots, p,$$

$$y_i - \left(\boldsymbol{w}^T \boldsymbol{z}_i + b\right) \leqq \varepsilon + \acute{\xi}_i, \ \ i = 1, \ldots, p,$$

$$\varepsilon, \ \xi_i, \ \acute{\xi}_i \geqq 0$$

where C is a trade-off parameter between the norm of $\boldsymbol{w}$ and ξ ($\acute{\xi}$). The dual formulation to $(soft-\text{SVR})_P$ is given by

$$\max_{\alpha_i,\acute{\alpha_i}} \quad -\frac{1}{2}\sum_{i,j=1}^{p}(\acute{\alpha_i}-\alpha_i)(\acute{\alpha_j}-\alpha_j)K(\boldsymbol{x}_i,\boldsymbol{x}_j) \qquad (soft\text{--}SVR)$$

$$+\sum_{i=1}^{p}(\acute{\alpha_i}-\alpha_i)y_i-\varepsilon\sum_{i,j=1}^{p}(\acute{\alpha_i}+\alpha_i)$$

$$\text{s.t.} \quad \sum_{i=1}^{p}(\acute{\alpha_i}-\alpha_i)=0,$$

$$0\leqq\acute{\alpha_i}\leqq\frac{C}{p},\ \ 0\leqq\alpha_i\leqq\frac{C}{p},\ \ i=1,\ldots,p.$$

In order to decide ε automatically, Schölkopf and Smola proposed ν-SVR as follows (Schölkopf and Smola, 1998):

$$\min_{\boldsymbol{w},b,\varepsilon,\xi_i,\acute{\xi_i}} \quad \frac{1}{2}\|\boldsymbol{w}\|_2^2+C\Big(\nu\varepsilon+\frac{1}{p}\sum_{i=1}^{p}(\xi_i+\acute{\xi_i})\Big) \qquad (\nu\text{--}SVR)_P$$

$$\text{s.t.} \quad (\boldsymbol{w}^T\boldsymbol{z}_i+b)-y_i\leqq\varepsilon+\xi_i,\ \ i=1,\ldots,p,$$

$$y_i-(\boldsymbol{w}^T\boldsymbol{z}_i+b)\leqq\varepsilon+\acute{\xi_i},\ \ i=1,\ldots,p,$$

$$\varepsilon,\ \xi_i,\ \acute{\xi_i}\geqq 0,$$

where C and ν are trade-off parameters between the norm of $\boldsymbol{w}$ and ε and ξ_i ($\acute{\xi_i}$).

The dual formulation to $(\nu\text{--}SVR)_P$ is given by

$$\max_{\alpha_i,\acute{\alpha_i}} \quad -\frac{1}{2}\sum_{i,j=1}^{p}(\acute{\alpha_i}-\alpha_i)(\acute{\alpha_j}-\alpha_j)K(\boldsymbol{x}_i,\boldsymbol{x}_j) \qquad (\nu\text{--}SVR)$$

$$+\sum_{i=1}^{p}(\acute{\alpha_i}-\alpha_i)y_i$$

$$\text{s.t.} \quad \sum_{i=1}^{p}(\acute{\alpha_i}-\alpha_i)=0,$$

$$\sum_{i=1}^{p}(\acute{\alpha_i}+\alpha_i)\leqq C\cdot\nu,$$

$$0\leqq\acute{\alpha_i}\leqq\frac{C}{p},\ \ 0\leqq\alpha_i\leqq\frac{C}{p},\ \ i=1,\ldots,p.$$

In a similar fashion to classification, we can obtain $(\mu-\nu\text{--}SVR)$ as follows:

$$\min_{\boldsymbol{w},b,\varepsilon,\xi,\acute{\xi}} \quad \frac{1}{2}\|\boldsymbol{w}\|_2^2 + \nu\varepsilon + \mu(\xi + \acute{\xi}) \qquad (\mu - \nu\text{--SVR})_P$$

$$\text{s.t.} \quad \left(\boldsymbol{w}^T \boldsymbol{z}_i + b\right) - y_i \leqq \varepsilon + \xi, \quad i = 1, \ldots, p,$$

$$y_i - \left(\boldsymbol{w}^T \boldsymbol{z}_i + b\right) \leqq \varepsilon + \acute{\xi}, \quad i = 1, \ldots, p,$$

$$\varepsilon, \ \xi, \ \acute{\xi} \geqq 0,$$

where ν and μ are trade-off parameters between the norm of $\boldsymbol{w}$ and ε and $\acute{\xi}$. The dual formulation of $\mu - \nu\text{--SVR}$ is as follows:

$$\max_{\alpha_i,\acute{\alpha_i}} \quad -\frac{1}{2}\sum_{i,j=1}^{p} \left(\acute{\alpha_i} - \alpha_i\right)\left(\acute{\alpha_j} - \alpha_j\right) K\left(\boldsymbol{x}_i, \boldsymbol{x}_j\right) \qquad (\mu - \nu\text{--SVR})$$

$$+ \sum_{i=1}^{p} \left(\acute{\alpha_i} - \alpha_i\right) y_i$$

$$\text{s.t.} \quad \sum_{i=1}^{p} \left(\acute{\alpha_i} - \alpha_i\right) = 0,$$

$$\sum_{i=1}^{p} \acute{\alpha_i} \leqq \mu, \ \sum_{i=1}^{p} \alpha_i \leqq \mu,$$

$$\sum_{i=1}^{p} \left(\acute{\alpha_i} + \alpha_i\right) \leqq \nu,$$

$$\acute{\alpha_i} \geqq 0, \ \alpha_i \geqq 0, \ i = 1, \ldots, p.$$

10.4 Managing Meta-models of Multiple Objectives

Meta-modeling in the context of multiobjective optimization has been considered in several works in recent years (Chafekar *et al.*, 2005; Emmerich *et al.*, 2006; Gaspar-Cunha and Vieira, 2004; Keane, 2006; Knowles, 2006; Nain and Deb, 2002; Ray and Smith, 2006; Voutchkov and Keane, 2006). The generalization to multiple objective functions has led to a variety of approaches, with differences in what is modeled, and also how models are updated. These differences follow partly from the different possible methods that there are of building up a Pareto front approximation (see Figure 10.2).

In a modern multiobjective evolutionary algorithm approach like NSGA-II, selection favours solutions of low dominance rank and uncrowded solutions, which helps build up a diverse and converged Pareto set approximation. A straightforward way to obtain a meta-modeling-based multiobjective optimization algorithm is thus to take NSGA-II and simply plug in meta-models of each independent objective function. This can be achieved by running NSGA-II for several generations on the meta-model (initially constructed from a DoE sample), and then cycling through phases of selection, evaluation on the

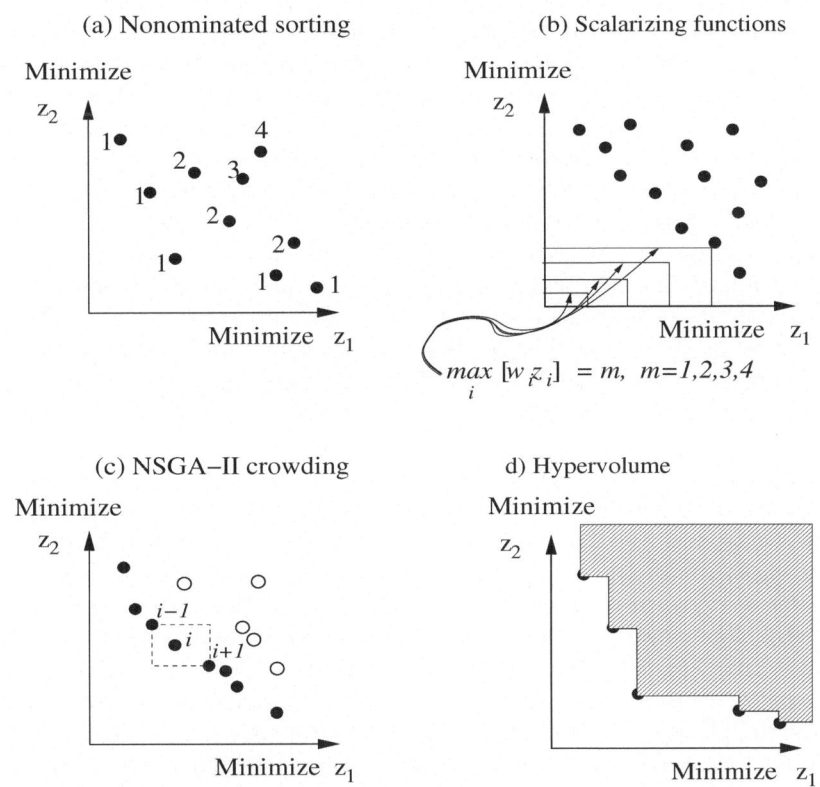

Fig. 10.2. Building up a Pareto front approximation can be achieved by different routes. Four are shown here: nondominated sorting, scalarizing, crowding, and maximising hypervolume.

real model, and update of the meta-model. This approach is the one taken by Voutchkov and Keane (2006) wherein a variety of response surface methods are compared, including splines, radial basis functions and polynomial regression. As the authors explain, an advantage of this approach is that each objective could, in theory, be modeled by a different type of response surface method (appropriate to it), or some objectives may be cheap to compute and may not need modeling at all.

An alternative to simply plugging meta-models into the existing selection step of an MOEA, is to use the meta-models, instead, to *pre-screen* points. This approach, much like EGO, may base the screening on both the predicted value of points and the confidence in these predictions. In a method still based on NSGA-II, Emmerich *et al.* (2006) proposed a number of different pre-screening criteria for multiobjective optimization, including the expected improvement and the probability of improvement. Note that in the case of multiobjective optimization, improvement is relative to the whole Pareto set approximation achieved so far, not a single value. Thus, to measure improvement, the estimated increase in hypervolume (Zitzler *et al.*, 2003) of the current approximation set (were a candidate point added to it) is used, based on a meta-model for each objective function. In experiments, Emmerich et al compared four different screening criteria on two and three-objective problems, and found improvement over the standard NSGA-II in all cases.

The approach of Emmerich et al is a sophisticated method of generalizing the use of meta-models to multiobjective optimization, via MOEAs, though it is as yet open whether this sophistication leads to better performance than the simpler method of Voutchkov and Keane (2006). Moreover, it does seem slightly unnatural to marry NSGA-II, which uses dominance rank and crowdedness to select its 'parents', with a meta-modeling approach that uses the hypervolume to estimate probable improvement. It would seem more logical to use evolutionary algorithms that themselves maximize hypervolume as the fitness assignment method, such as (Emmerich *et al.*, 2005). It remains to be seen whether such approaches would perform even better.

One worry with the methods described so far is that fitness assignments based on dominance rank (like NSGA-II) can perform poorly when the number of objectives is greater than three or four (Hughes, 2005). Hypervolume may be a better measure but it is very expensive to compute for large dimension, as the complexity of known methods for computing it is polynomial in the set size but exponential in d. Thus, scaling up objective dimension in methods based on either of the approaches described above might prove difficult.

A method that does not use either hypervolume or dominance rank is the ParEGO approach proposed by Knowles (2006). This method is a generalization to multiobjective optimization of the well-founded EGO algorithm (Jones *et al.*, 1998). To build up a Pareto front, ParEGO uses a series of weighting vectors to scalarize the objective functions. At each iteration of the algorithm, a new candidate point is determined by (i) computing the expected improvement (Jones *et al.*, 1998) in the 'direction' specified by the weighting vector drawn for that iteration, and (ii) searching for a point that maximizes this expected improvement (a single-objective evolutionary algorithm is used for this search). The use of such scalarizing weight vectors has been shown to scale well to many objectives, compared with Pareto ranking (Hughes, 2005). The ParEGO method has the additional advantage that it would be relatively straightforward to make it interactive, allowing the user to narrow down the set of scalarizing weight vectors to allow focus on a particular region of the Pareto front. This can further reduce the number of function evaluations it is necessary to perform.

Yet a further way of building up a Pareto front is exemplified in the final method we review here. (Chafekar *et al.*, 2005) proposes a genetic algorithm with meta-models OEGADO, based closely on their own method for single objective optimization. To make it work for the multiobjective case, a distinct genetic algorithm is run for each objective, with information exchange occurring between the algorithms at intervals, which helps the GAs to find the compromise solutions. The fact that each objective is optimized by its own genetic algorithm means that objective functions with different computational overhead can be appropriately handled — slow objectives do not slow down the evaluation of faster ones. The code may also be trivially implemented on parallel architectures.

10.4.1 Combining Interactive Methods and EMO for Generating a Pareto Frontier

Aspiration Level Methods for Interactive Multiobjective Programming

Since there may be many Pareto solutions in practice, the final decision should be made among them taking the total balance over all criteria into account. This is a problem of value judgment of DM. The totally balancing over criteria is usually called *trade-off*. Interactive multiobjective programming searches a solution in an interactive way with DM while making trade-off analysis on the basis of DM's value judgment. Among them, the aspiration level approach is now recognized to be effective in practice, because

(i) it does not require any consistency of DM's judgment,
(ii) aspiration levels reflect the wish of DM very well,
(iii) aspiration levels play the role of probe better than the weight for objective functions.

As one of aspiration level approaches, one of authors proposed the satisficing trade-off method (Nakayama and Sawaragi, 1984). Suppose that we have objective functions $f(x) := (f_1(x), \ldots, f_r(x))$ to be minimized over $x \in X \subset R^n$. In the satisficing trade-off method, the aspiration level at the k-th iteration $\overline{f}^k$ is modified as follows:

$$\overline{f}^{k+1} = T \circ P(\overline{f}^k).$$

Here, the operator P selects the Pareto solution nearest in some sense to the given aspiration level $\overline{f}^k$. The operator T is the trade-off operator which changes the k-th aspiration level $\overline{f}^k$ if DM does not compromise with the shown solution $P(\overline{f}^k)$. Of course, since $P(\overline{f}^k)$ is a Pareto solution, there exists no feasible solution which makes all criteria better than $P(\overline{f}^k)$, and thus DM has to trade-off among criteria if he wants to improve some of criteria. Based on this trade-off, a new aspiration level is decided as $T \circ P(\overline{f}^k)$. Similar process is continued until DM obtains an agreeable solution.

On the Operation P

The operation which gives a Pareto solution $P(\overline{\boldsymbol{f}}^k)$ nearest to $\overline{\boldsymbol{f}}^k$ is performed by some auxiliary scalar optimization. It has been shown in Sawaragi-Nakayama-Tanino (1985) that the only one scalarization technique, which provides any Pareto solution regardless of the structure of problem, is of the Tchebyshev norm type. However, the scalarization function of Tchebyshev norm type yields not only a Pareto solution but also a weak Pareto solution. Since weak Pareto solutions have a possibility that there may be another solution which improves a criteria while others being fixed, they are not necessarily "*efficient*" as a solution in decision making. In order to exclude weak Pareto solutions, the following scalarization function of the augmented Tchebyshev type can be used:

$$\max_{1 \leq i \leq r} \omega_i \left(f_i(\boldsymbol{x}) - \overline{f}_i \right) + \alpha \sum_{i=1}^{r} \omega_i f_i(\boldsymbol{x}), \tag{10.18}$$

where α is usually set a sufficiently small positive number, say 10^{-6}.

The weight ω_i is usually given as follows: Let f_i^* be an ideal value which is usually given in such a way that $f_i^* < \min \{ f_i(\boldsymbol{x}) \mid \boldsymbol{x} \in X \}$. For this circumstance, we set

$$\omega_i^k = \frac{1}{\overline{f}_i^k - f_i^*}. \tag{10.19}$$

The minimization of (10.18) with (10.19) is usually performed by solving the following equivalent optimization problem, because the original one is not smooth:

$$\begin{aligned}
\text{(AP)} \qquad & \underset{z, \, \boldsymbol{x}}{\text{minimize}} & & z + \alpha \sum_{i=1}^{r} \omega_i f_i(\boldsymbol{x}) \\
& \text{subject to} & & \omega_i^k \left(f_i(\boldsymbol{x}) - \overline{f}_i^k \right) \leq z \qquad (10.20) \\
& & & \boldsymbol{x} \in X.
\end{aligned}$$

On the Operation T

In cases that DM is not satisfied with the solution for $P(\overline{\boldsymbol{f}}^k)$, he/she is requested to answer his/her new aspiration level $\overline{\boldsymbol{f}}^{k+1}$. Let $\boldsymbol{x}^k$ denote the Pareto solution obtained by projection $P(\overline{\boldsymbol{f}}^k)$, and classify the objective functions into the following three groups:

(i) the class of criteria which are to be improved more,
(ii) the class of criteria which may be relaxed,
(iii) the class of criteria which are acceptable as they are.

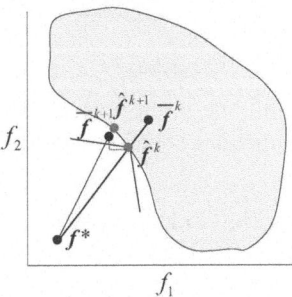

Fig. 10.3. Satisficing Trade-off Method

Let the index set of each class be denoted by I_I^k, I_R^k, I_A^k, respectively. Clearly, $\overline{f}_i^{k+1} < f_i(x^k)$ for all $i \in I_I^k$. Usually, for $i \in I_A^k$, we set $\overline{f}_i^{k+1} = f_i(x^{l_0})$. For $i \in I_R^k$, DM has to agree to increase the value of $\overline{f}_i^{k+1}$. It should be noted that an appropriate sacrifice of f_j for $j \in I_R^k$ is needed for attaining the improvement of f_i for $i \in I_I^k$.

Combining Satisficing Trade-off Method and Sequential Approximate Optimization

Nakayama and Yun proposed a method combining the satisficing trade-off method for interactive multiobjective programming and the sequential approximate optimization using $\mu - \nu$–SVR (Nakayama and Yun, 2006b). The procedure is summarized as follows:

Step 1. (Real Evaluation)
Evaluate actually the values of objective functions $f(x_1), f(x_2), \ldots, f(x_\ell)$ for sampled data $x_1, \ldots, x_\ell$ through computational simulation analysis or experiments.

Step 2. (Approximation)
Approximate each objective function $\hat{f}_1(x), \ldots, \hat{f}_m(x)$ by the learning of $\mu - \nu$–SVR on the basis of real sample data set.

Step 3. (Find a Pareto Solution Nearest to the Aspiration Level and Generate Pareto Frontier)
Find a Pareto optimal solution nearest to the given aspiration level for the approximated objective functions $\hat{f}(x) := (\hat{f}_1(x), \ldots, \hat{f}_m(x))$. This is performed by using GA for minimizing the augmented Tchebyshev scalarization function (10.18). In addition, generate Pareto frontier by MOGA for accumulated individuals during the procedure for optimizing the augmented Tchebyshev scalarization function.

Step 4. (Choice of Additional Learning Data)
Choose the additional ℓ_0-data from the set of obtained Pareto optimal solutions. Go to Step 1. (Set $\ell \leftarrow \ell + \ell_0$.)

how to choose the additional data

Stage 0.   First, add the point with highest achievement degree among Pareto optimal solutions obtained in Step 3. ($\leftarrow$ local information)

Stage 1.   Evaluate the ranks for the real sampled data of Step 1 by the ranking method (Fonseca and Fleming, 1993).

Stage 2.   Approximate the rank function associated with the ranks calculated in the Stage 1 by $\mu - \nu$–SVR.

Stage 3.   Calculate the expected fitness for Pareto optimal solutions obtained in Step 3.

Stage 4.   Among them, add the point with highest rank. ($\leftarrow$global information)

Next, we consider the following problem (Ex-1):

$$\text{minimize}\quad f_1 := x_1 + x_2$$
$$f_2 := 20\cos(15x_1) + (x_1 - 4)^4 + 100\sin(x_1 x_2)$$
$$\text{subject to}\ \ 0 \leqq x_1,\ x_2 \leqq 3.$$

The true function of each objective function f_1 and f_2 in the problem (Ex-1) are shown in Fig. 10.4.

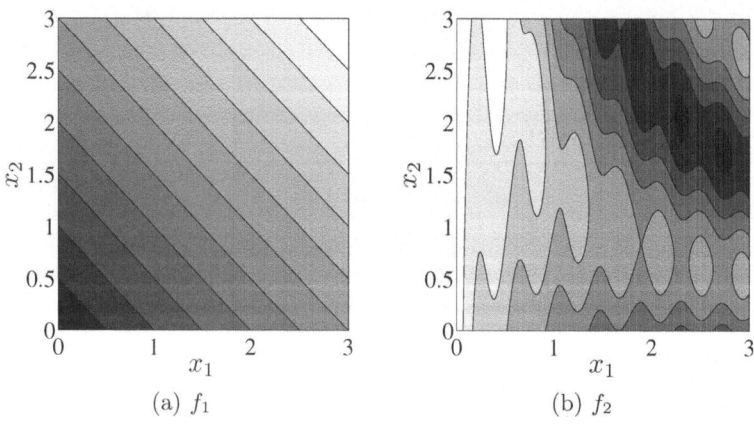

(a) f_1 (b) f_2

Fig. 10.4. The true contours to the problem

In our simulation, the ideal point and the aspiration level is respectively given by

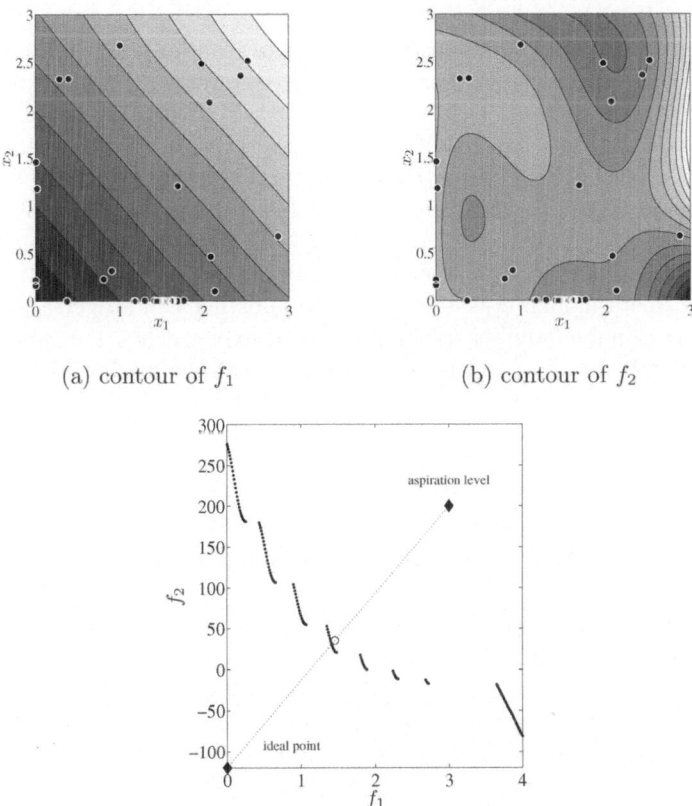

(a) contour of f_1 (b) contour of f_2

(c) population at the final generation

Fig. 10.5. # sample data : 50 points (Ex-1)

$$\left(f_1^*,\ f_2^*\right)\ =\ (0,\ -120),$$
$$\left(\overline{f}_1,\ \overline{f}_2\right)\ =\ (3,\ 200),$$

and the closest Pareto solution to the above aspiration level is as follows:

exact optimal solution $\left(\hat{x}_1,\ \hat{x}_2\right) = (1.41321, 0)$

exact optimal value $\left(\hat{f}_1,\ \hat{f}_2\right) = (1.41321, 30.74221)$

Starting with initial data 10 points randomly, we obtained the following approximate solution by proposed method after 50 real evaluations:

approximate solution $\left(x_1,\ x_2\right) = (1.45748, 0)$

approximate value $\left(f_1,\ f_2\right) = (1.45748, 35.34059)$

The final result is shown in Fig. 10.5

Additionally, a rough configuration of Pareto frontier is also obtained in Fig. 10.6 through the step 3. Although this figure shows a rough approximation of the whole Pareto frontier, it may be expected to provide a reasonable approximation to it in a neighborhood of the Pareto optimal point nearest to the aspiration level. This information of the approximation of Pareto frontier in a neighborhood of the obtained Pareto solution helps the decision maker to make trade-off analysis.

It has been observed that a method combining the satisficing trade-off method and meta-modeling is effective for supporting DM to get a final solution in a reasonable number of computational experiments. It is promising in practical problems since it has been observed that the method reduces the number of function evaluations up to less than $1/100$ to $1/10$ of usual methods such as MOGAs and usual aspiration level methods through several numerical experiments.

10.5 Evaluation of Meta-modeling Methods

Improvement in the design of heuristic search algorithms may sometimes be based on good theoretical ideas, but it is generally driven by the empirical testing and comparison of methods. Meta-model-based multiobjective optimization algorithms are relatively new, so, as yet, there has been little serious direct comparison of methods. In the coming years, this will become more of a focus.

10.5.1 What to Evaluate

In meta-modeling scenarios, full-cost evaluations are expensive and should be reduced. Therefore, it follows that whatever assessments of Pareto set approximation are used, these should generally be plotted against the number of full-cost function evaluations, so that it is possible to see how performance evolves over 'time'. In the case of multiobjective optimization, this is often overlooked, because of the desire to show Pareto front approximations. For two-objective problems, the use of snapshots of Pareto fronts can be informative, but for general dimension, plotting the value of an indicator, such as the hypervolume, against full-cost evaluation number (averaged over a number of runs) is a more advisable approach.

An alternative statistical method that has been used little to date, is the attainment function method of comparing two optimizers (da Fonseca *et al.*, 2001). With this, the number of full cost evaluations can be considered an additional objective. For the first-order attainment function, the method returns an overall significance result indicating whether the attainment function (which describes the probability of attaining a certain point by a certain time, for all points and times) differs significantly between the two algorithms.

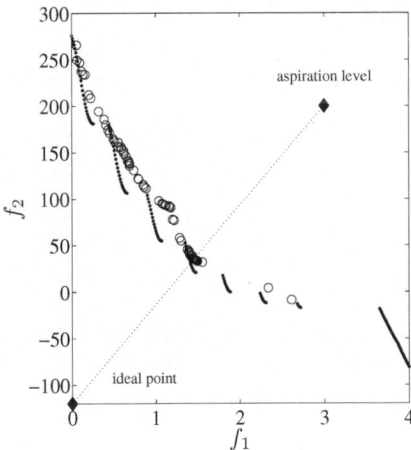

Fig. 10.6. The whole Pareto frontier with 50 sample points

In addition to the performance at optimizing a function, meta-modeling also involves its approximation. Therefore, it is sometimes desirable to show the time evolution of the accuracy of the model over all or some region of the design space.

In (Emmerich *et al.*, 2006), the accuracy of the model, as seen by the evolutionary algorithm optimizer was measured. This was done by computing the precision and recall of the pre-screening methods used in terms of being able to correctly rank solutions (rather than get their absolute evaluation correct). In a similar ilk, a number of measures of model quality that are based on evaluating the model's utility at making the 'correct' selection of individuals within an EA, were proposed in (Hüsken *et al.*, 2005).

Evaluation of interactive optimization and decision-making methods is even more of a challenge, and is dealt with in Chapter 7 of this book.

10.5.2 Adversaries

Assessment of progress in meta-modeling would be facilitated by using common, simple adversary algorithms or methods to compare against. Perhaps the simplest adversary is the random search. It is very interesting to compare with random search (as done in (Knowles, 2006) and see also (Hughes, 2006)) because it is not necessarily trivial to outperform the algorithm when the number of evaluations is small, and depending on the function. Moreover, when approximating a higher dimensional Pareto front, random search may be better than some multiobjective EAs, such as NSGA-II. The obvious additional adversary is the algorithm being proposed, with the meta-model removed (if this is possible).

10.6 Real Applications

10.6.1 Closed-Loop Mass-Spectrometer Optimization

Mass spectrometers are analytical instruments for determining the chemical compounds present in a sample. Typically, they are used for testing a hypothesis as to whether a particular compound is present or not. When used in this way, the instrument can be configured according to standard principles and settings provided by the instrument manufacturer. However, modern biological applications aim at using mass-spectrometry to mine data without a hypothesis, i.e. to measure/detect simultaneously the hundreds of compounds contained in complex biological samples. For such applications, the mass spectrometer will not perform well in a standard configuration, so it must be optimized.

(O'Hagan et al., 2006) describes experiments to optimize the configuration of a GCxGC-TOF mass spectrometer with the aim of improving its effectiveness at detecting all the metabolites (products of the metabolic system) in a human serum (blood) sample. To undertake this optimization, real experiments were conducted with the instrument in different configuration set-ups, as dictated by the optimization algorithm. Instrument configurations are specified by 15 continuous parameters. Typically, a single evaluation of an instrument configuration lasts of the order of one hour, although the throughput of the instrument (i.e. how long it takes to process the serum sample) was also one of the objectives and thus subject to some variation. The overall arrangement showing the optimization algorithm and how it is connected up to the instrument in a closed-loop is shown in Figure 10.7.

The experiments reported in (O'Hagan et al., 2006) used just 320 experiments (evaluations) to increase the number of metabolites observed (primary objective) substantially, as measured by the number of peaks in the mass-spectrometer's output. Simultaneously, the signal to noise ratio and the throughput of the instrument were optimized. A plot of the evolution of the three objectives is given in Figure 10.8.

The meta-modeling algorithm used for the optimization was ParEGO (Knowles, 2006), a multiobjective version of the efficient global optimization (EGO) algorithm (Jones et al., 1998). ParEGO uses a design and analysis of computer experiments (DACE) approach to model the objective function(s) (in this case, the instruments' behaviour under different operating configurations), based on an initial Latin hypercube sampling of the parameter space. Subsequently, the model is used to suggest the next experiment (set of instrumentation parameter values), such that the 'expected improvement' in the objective function(s) is maximized. The notion of expected improvement implicitly ensures that ParEGO balances exploration of new parameter combinations with exploitation and fine-tuning of parameter values that have led to good design points in previous experiments. The DACE model is updated after each fitness evaluation.

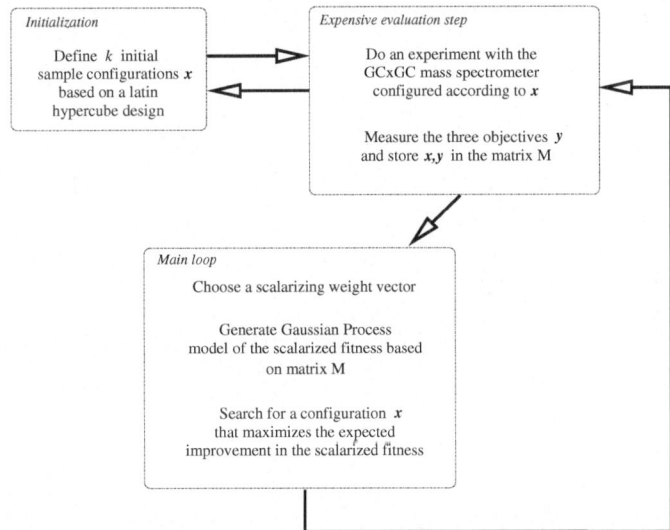

Fig. 10.7. Closed-loop optimization of mass spectrometer parameters using the ParEGO algorithm

10.6.2 Application to Reinforcement of Cable-Stayed Bridges

After the big earthquake in Kobe in 1995, many in-service structures were required to improve their anti-seismic property by law regulation in Japan. However, it is very difficult for large and/or complicated bridges, such as suspension bridges, cable-stayed bridges, arch bridges and so on, to be reinforced because of impractical executing methods and complicated dynamic responses. Recently, many kinds of anti-seismic device have been developed (Honda *et al.*, 2004). It is practical in the bridge to be installed a number of small devices taking into account of strength and/or space, and to obtain the most reasonable arrangement and capacity of the devices by using optimization technique. In this problem, the form of objective function is not given explicitly in terms of design variables, but the value of the function is obtained by seismic response analysis. Since this analysis needs much cost and long time, it is strongly desirable to make the number of analyses as few as possible. To this end, radial basis function networks (RBFN) are employed in predicting the form of objective function, and genetic algorithms (GA) in searching the optimal value of the predicted objective function (Nakayama *et al.*, 2006).

The proposed method was appplied to a problem of anti-seismic improvement of a cable-stayed bridge which typifies the difficulty of reinforcement of in-service structure. In this investigation, we determine an efficient arrangement and amount of additional mass for cables to reduce the seismic response of the tower of a cable-stayed bridge (Fig. 10.9).

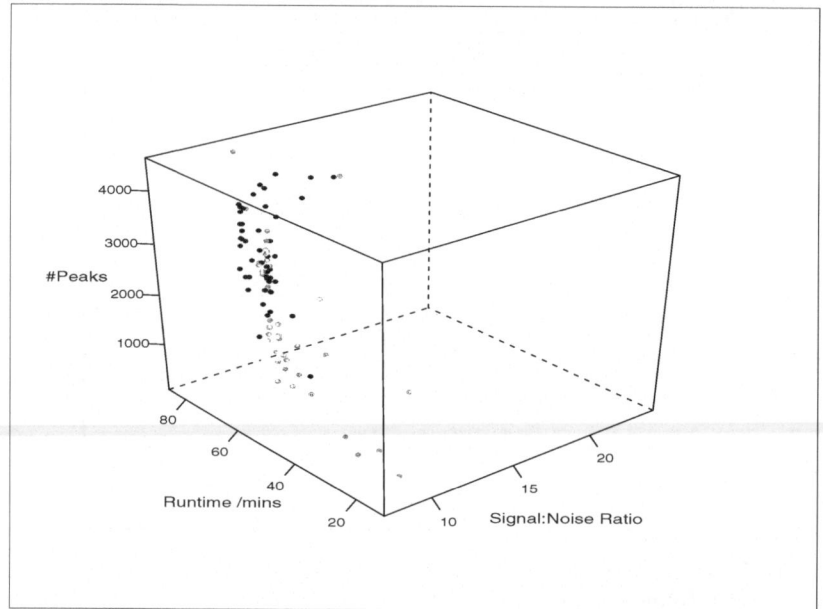

Fig. 10.8. Three-objective closed-loop optimization of GCxGC-TOF-mass spectrometry configurations for the analysis of human serum. The number of peaks and signal:noise ratio are maximized; the runtime of the mass spectrometer is minimized. The shading of the points represents the experiment number. Darker circles represent later experiments, and the six back dots represent replications of the same chosen final configuration. The number of peaks has risen to over 3000, whilst sufficient signal:noise has been maintained and runtime kept down to around 60 minutes.

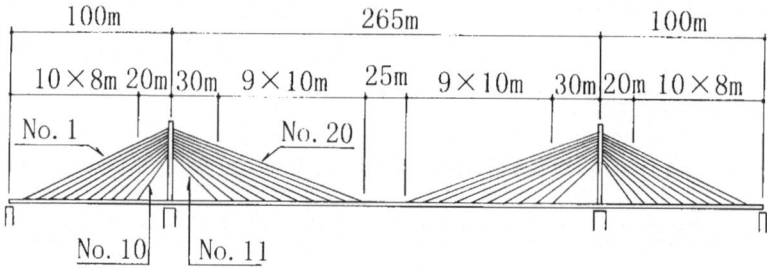

Fig. 10.9. Cable-stayed bridge

The influence of additional mass on cables was investigated by numerical sensitivity analysis. The analytical model shown in Fig. 10.9 is a 3-span continuous and symmetrical cable-stayed bridge whose 2 20 cables are in one plane and the towers stand freely in their transverse direction. The mass must be distributed over cables uniformly to prevent them from concentrating deformation.

The seismic response of interest was the stress at the fixed end of the tower when an earthquake occurs in the transverse direction. Seismic response analysis was carried out by a spectrum method. As there are a lot of modes whose natural frequencies were close to each other, the response was evaluated by the complete quadratic combination method. The input spectrum is given in the new Specifications for Highway Bridges in Japan.

The natural frequencies of modes accompanied with the bending of the tower (natural frequency of the tower alone is 1.4Hz) range from 0.79Hz to 2.31Hz, due to coupling with the cables.

As mentioned above, the seismic response of the tower can be controlled by additional mass to cables, but each cable influences other ones in a complex way. Thus, the most effective distribution of additional mass must be decided by optimization.

10.6.3 Case 1

The objective is to minimize the bending moment M at the base of tower. The variables are ratios of additional mass and mass of cables. The number of variables is 20. The lower bound and upper bound of each variable are 0.0, and 1.0, respectively. For comparison, we applied a quasi-Newton method based on approximated differentials as an existing method. We made five trials with different initial points in order to obtain a global optimum.

In applying our proposed method, we used BLX-α as a genetic algorithm which is observed to be effective for continuous variables. The population is 10, and the number of generation is 200. We set $\lambda = 0.01$, and decided the value of width r of Gaussian by the simple estimate given by (10.17).

We started the iteration with 60 sample points. The first 20 sample points are generated randomly with one of variables fixed at the upper bound 1 by turns; the next 20s are generated similarly with one of variables fixed at the lower bound 0 by turns; the last 20s similarly with one of variables fixed at the mid-value 0.5 by turns. The parameters for convergence are $C_x^0 = 20, C_f^0 = 20$ and $l_0 = 0.1$.

The result is shown in Table 10.1. It is seen that the proposed method can find out fairly good solutions within 1/10 or less times of analysis than the conventional optimization.

10.6.4 Case 2

Now, we take the number of cables to be added with masses, N, as another objective function in addition to the bending moment M. Namely, our objective function is

$$F = (M/M_0) + \alpha(N/N_0) \tag{10.21}$$

where α is a parameter for trade-off between the first term and the second one. M_0 and N_0 are used for normalization of the bending moment and the

Table 10.1. Result for Case 1

			exisiting method	RBF Network	
				best	average
		1	0.32	0.04	0.40
		2	1.00	0.69	0.84
		3	0.49	0.18	0.51
		4	0.62	0.82	0.80
		5	0.81	0.57	0.64
		6	0.52	0.43	0.56
		7	0.49	1.00	0.39
		8	0.52	0.44	0.66
		9	0.48	0.94	0.50
cable		10	0.48	0.50	0.56
No.		11	0.50	0.45	0.47
		12	0.55	1.00	0.74
		13	0.70	0.85	0.71
		14	0.61	0.50	0.30
		15	0.61	1.00	0.58
		16	0.46	0.24	0.37
		17	0.22	0.10	0.13
		18	1.00	0.95	0.91
		19	0.98	1.00	0.94
		20	1.00	1.00	0.91
bending moment (MN·m)			50.3	54.90	63.70
#analysis			1365	150.00	124.80

number of cables, respectively. In this experiment, we set $M_0 = 147.0$MN·m and $N_0 = 20$.

In this experiment, we used a simple GA, because some of variables are discrete. The parameters for calculation are the same as in Case 1. The result is given in Table 10.2. It should be noted that the number of analysis in our proposed method is reduced to about $1/20$ of the conventional method. Although the precision of solution in our method is behind the conventional method, it is sufficiently acceptable in practice.

Table 10.2. Result for Case 2

			existing method		RBF network	
			best	average	best	average
		1	0.00	0.00	0.00	0.83
		2	0.00	0.00	0.00	0.09
		3	0.00	0.00	0.00	0.00
		4	0.00	0.00	0.00	0.04
		5	0.00	0.00	0.00	0.00
		6	0.00	0.00	0.00	0.00
		7	0.00	0.00	0.00	0.00
		8	0.00	0.00	1.00	0.99
		9	0.00	0.00	0.00	0.10
cable		10	0.00	0.00	0.86	0.53
No.		11	0.00	0.00	0.00	0.00
		12	0.00	0.00	1.00	0.63
		13	0.00	0.00	0.00	0.13
		14	0.00	0.00	0.86	0.53
		15	0.00	0.00	0.00	0.00
		16	0.00	0.00	0.00	0.00
		17	0.00	0.00	0.00	0.00
		18	0.00	0.00	0.00	0.00
		19	0.71	0.74	1.00	1.00
		20	0.86	0.83	0.86	0.79
bending moment (MN·m)			62.6	62.8	67.1	69.7
#cable with additional mass			2	2	6	6.2
objective fn.			0.526	0.527	0.756	0.784
#analysis			4100	3780	199	193.3

The well known Response Surface Method (RSM, in short) is competitive with the proposed method. However, the proposed method has been observed to have advantage over RSM especially for highly nonlinear cases. On the other hand, EGO (Efficient Global Optimization) for black-box objective functions (Jones *et al.*, 1998; M.J. Sasena, 2000; Schonlau, 1997) takes time to calculate the expected improvement, while it is rather simple and easy to add two kinds of additional samples for global information and local information for approximation (Nakayama et al., 2002, 2003).

10.7 Concluding Remarks

The increasing desire to apply optimization methods in expensive domains is driving forward research in meta-modeling. Up to now, meta-modeling has been applied mainly in continuous, low dimensional design variable spaces, and methods from design of experiments and response surfaces have been used. High-dimensional discrete spaces may also arise in applications involving expensive evaluations and this will motivate research into meta-modeling of these domains too. Research in meta-modeling for multiobjective optimization is relatively young and there is still much to do. So far, there are few standards for comparisons of methods, and little is yet known about the relative performance of different approaches. The state of the art research surveyed in this chapter is beginning to grapple with the issues of incremental learning and the trade-off between exploitation and exploration within meta-modeling. In the future, scalability of methods in variable dimension and objective space dimension will become important, as will methods capable of dealing with noise or uncertainty. Interactive meta-modeling is also likely to be investigated more thoroughly, as the number of evaluations can be further reduced by these approaches.

References

Anderson, V.L., McLean, R.A.: Design of Experiments: A Realistic Approach. Marcel Dekker, New York (1974)

Atkeson, C.G., Moore, A.W., Schaal, S.: Locally weighted learning for control. Artificial Intelligence Review 11(1), 75–113 (1997)

Breiman, L.: Classification and Regression Trees. Chapman and Hall, Boca Raton (1984)

Brown, G., Wyatt, J.L., Tiňo, P.: Managing diversity in regression ensembles. The Journal of Machine Learning Research 6, 1621–1650 (2005)

Schölkopf, B., Smola, A.J.: Learning with Kernels: Support Vector Machines, Regularization, Optimization, and Beyond. MIT Press, Cambridge (2002)

Cauwenberghs, G., Poggio, T.: Incremental and decremental support vector machine learning. Advances in Neural Information Processing Systems 13, 409–415 (2001)

Chafekar, D., Shi, L., Rasheed, K., Xuan, J.: Multiobjective ga optimization using reduced models. IEEE Transactions on Systems, Man and Cybernetics, Part C: Applications and Reviews 35(2), 261–265 (2005)

Chapelle, O., Vapnik, V., Weston, J.: Transductive inference for estimating values of functions. Advances in Neural Information Processing Systems 12, 421–427 (1999)

Cohn, D.A., Ghahramani, Z., Jordan, M.I.: Active learning with statistical models. Journal of Artificial Intelligence Research 4, 129–145 (1996)

Corne, D.W., Oates, M.J., Kell, D.B.: On fitness distributions and expected fitness gain of mutation rates in parallel evolutionary algorithms. In: Guervós, J.J.M., Adamidis, P.A., Beyer, H.-G., Fernández-Villacañas, J.-L., Schwefel, H.-P. (eds.) PPSN 2002. LNCS, vol. 2439, pp. 132–141. Springer, Heidelberg (2002)

Cortes, C., Vapnik, V.: Support vector networks. Machine Learning 20, 273–297 (1995)

Cristianini, N., Shawe-Tylor, J.: An Introduction to Support Vector Machines and Other Kernel-based Learning Methods. Cambridge University Press, Cambridge (2000)

Grunert da Fonseca, V., Fonseca, C.M., Hall, A.O.: Inferential Performance Assessment of Stochastic Optimisers and the Attainment Function. In: Zitzler, E., Deb, K., Thiele, L., Coello Coello, C.A., Corne, D.W. (eds.) EMO 2001. LNCS, vol. 1993, pp. 213–225. Springer, Heidelberg (2001)

Emmerich, M.T.M., Beume, N., Naujoks, B.: An EMO algorithm using the hypervolume measure as selection criterion. In: Coello Coello, C.A., Hernández Aguirre, A., Zitzler, E. (eds.) EMO 2005. LNCS, vol. 3410, pp. 62–76. Springer, Heidelberg (2005)

Emmerich, M., Giannakoglou, K., Naujoks, B.: Single-and Multi-objective Evolutionary Optimization Assisted by Gaussian Random Field Metamodels. IEEE Transactions on Evolutionary Computation 10(4), 421–439 (2006)

English, T.M.: Optimization is easy and learning is hard in the typical function. In: Proceedings of the 2000 Congress on Evolutionary Computation (CEC00), pp. 924–931. IEEE Computer Society Press, Piscataway (2000)

Erenguc, S.S., Koehler, G.J.: Survey of mathematical programming models and experimental results for linear discriminant analysis. Managerial and Decision Economics 11, 215–225 (1990)

Eyheramendy, S., Lewis, D., Madigan, D.: On the naive Bayes model for text categorization. In: Proceedings Artificial Intelligence & Statistics 2003 (2003)

Fieldsend, J.E., Everson, R.M.: Multi-objective Optimisation in the Presence of Uncertainty. In: 2005 IEEE Congress on Evolutionary Computation (CEC'2005), Edinburgh, Scotland, September 2005, vol. 1, pp. 243–250. IEEE Computer Society Press, Los Alamitos (2005)

Fonseca, C.M., Fleming, P.J.: Genetic algorithms for multi-objective optimization: Formulation, discussion and generalization. In: Proceedings of the Fifth International Conference on Genetic Algorithms, pp. 416–426 (1993)

Freed, N., Glover, F.: Simple but powerful goal programming models for discriminant problems. European J. of Operational Research 7, 44–60 (1981)

Gaspar-Cunha, A., Vieira, A · A multi-objective evolutionary algorithm using neural networks to approximate fitness evaluations. International Journal of Computers, Systems, and Signals (2004)

Hamza, K., Saitou, K.: Vehicle crashworthiness design via a surrogate model ensemble and a co-evolutionary genetic algorithm. In: Proc. of IDETC/CIE 2005 ASME 2005 International Design Engineering Technical Conference, California, USA (2005)

Honda, M., Morishita, K., Inoue, K., Hirai, J.: Improvement of anti-seismic capacity with damper braces for bridges. In: Proceedings of the Seventh International Conference on Motion and Vibration Control (2004)

Huang, D., Allen, T.T., Notz, W.I., Zeng, N.: Global Optimization of Stochastic Black-Box Systems via Sequential Kriging Meta-Models. Journal of Global Optimization 34(3), 441–466 (2006)

Hughes, E.J.: Evolutionary Many-Objective Optimisation: Many Once or One Many? In: 2005 IEEE Congress on Evolutionary Computation (CEC'2005), vol. 1, pp. 222–227. IEEE Computer Society Press, Los Alamitos (2005)

Hughes, E.J.: Multi-Objective Equivalent Random Search. In: Runarsson, T.P., Beyer, H.-G., Burke, E.K., Merelo-Guervós, J.J., Whitley, L.D., Yao, X. (eds.) PPSN 2006. LNCS, vol. 4193, pp. 463–472. Springer, Heidelberg (2006)

Hüsken, M., Jin, Y., Sendhoff, B.: Structure optimization of neural networks for evolutionary design optimization. Soft Computing 9(1), 21–28 (2005)

Jin, Y.: A comprehensive survey of fitness approximation in evolutionary computation. Soft Computing - A Fusion of Foundations, Methodologies and Applications 9(1), 3–12 (2005)

Jin, Y., Sendhoff, B.: Reducing fitness evaluations using clustering techniques and neural network ensembles. In: Deb, K., et al. (eds.) GECCO 2004. LNCS, vol. 3102, pp. 688–699. Springer, Heidelberg (2004)

Jin, Y., Olhofer, M., Sendhoff, B.: Managing approximate models in evolutionary algorithms design optimization. In: Proceedings of the 2001 Congress on Evolutionary Computation, CEC2001, pp. 592–599 (2001)

Jones, D., Schonlau, M., Welch, W.: Efficient global optimization of expensive black-box functions. Journal of Global Optimization 13, 455–492 (1998)

Joslin, D., Dragovich, J., Vo, H., Terada, J.: Opportunistic fitness evaluation in a genetic algorithm for civil engineering design optimization. In: Proceedings of the Congress on Evolutionary Computation (CEC 2006), pp. 2904–2911. IEEE Computer Society Press, Los Alamitos (2006)

Jourdan, L., Corne, D.W., Savic, D.A., Walters, G.A.: Preliminary Investigation of the 'Learnable Evolution Model' for Faster/Better Multiobjective Water Systems Design. In: Coello Coello, C.A., Hernández Aguirre, A., Zitzler, E. (eds.) EMO 2005. LNCS, vol. 3410, pp. 841–855. Springer, Heidelberg (2005)

Keane, A.J.: Statistical improvement criteria for use in multiobjective design optimization. AIAA Journal 44(4), 879–891 (2006)

Knowles, J.: ParEGO: A hybrid algorithm with on-line landscape approximation for expensive multiobjective optimization problems. IEEE Transactions on Evolutionary Computation 10(1), 50–66 (2006)

Langdon, W.B., Poli, R.: Foundations of Genetic Programming. Springer, Heidelberg (2001)

Larranaga, P., Lozano, J.A.: Estimation of Distribution Algorithms: A New Tool for Evolutionary Computation. Kluwer Academic Publishers, Dordrecht (2001)

Laumanns, M., Očenášek, J.: Bayesian optimization algorithms for multi-objective optimization. In: Guervós, J.J.M., Adamidis, P.A., Beyer, H.-G., Fernández-Villacañas, J.-L., Schwefel, H.-P. (eds.) PPSN 2002. LNCS, vol. 2439, pp. 298–307. Springer, Heidelberg (2002)

Michalski, R.: Learnable Evolution Model: Evolutionary Processes Guided by Machine Learning. Machine Learning 38(1), 9–40 (2000)

Sasena, M.J., Papalambros, P.Y., Goovaerts, P.: Metamodeling sample criteria in a global optimization framework. In: 8th AIAA/NASA/USAF/ISSMO Symposium on Multidisciplinary Analysis and Optimization, Long Beach, AIAA-2000-4921 (2000)

Myers, R.H., Montgomery, D.C.: Response Surface Methodology: Process and Product Optimization using Designed Experiments. Wiley, Chichester (1995)

Nain, P., Deb, K.: A computationally effective multi-objective search and optimization technique using coarse-to-fine grain modeling. Kangal Report 2002005 (2002)

Nakayama, H., Sawaragi, Y.: Satisficing trade-off method for multi-objective programming. In: Grauer, M., Wierzbicki, A. (eds.) Interactive Decision Analysis, pp. 113–122. Springer, Heidelberg (1984)

Nakayama, H., Yun, Y.: Generating support vector machines using multiobjective optimization and goal programming. In: Jin, Y. (ed.) Multi-objective Machine Learning, pp. 173–198. Springer, Heidelberg (2006a)

Nakayama, H., Yun, Y.: Support vector regression based on goal programming and multi-objective programming. IEEE World Congress on Computational Intelligence, CD-ROM, Paper ID: 1536 (2006b)

Nakayama, H., Arakawa, M., Sasaki, R.: Simulation based optimization for unknown objective functions. Optimization and Engineering 3, 201–214 (2002)

Nakayama, H., Arakawa, M., Washino, K.: Optimization for black-box objective functions. In: Tseveendorj, I., Pardalos, P.M., Enkhbat, R. (eds.) Optimization and Optimal Control, pp. 185–210. World Scientific, Singapore (2003)

Nakayama, H., Inoue, K., Yoshimori, Y.: Approximate optimization using computational intelligence and its application to reinforcement of cable-stayed bridges. In: Zha, X.F., Howlett, R.J. (eds.) Integrated Intelligent Systems for Engineering Design, pp. 289–304. IOS Press, Amsterdam (2006)

O'Hagan, S., Dunn, W.B., Knowles, J.D., Broadhurst, D., Williams, R., Ashworth, J.J., Cameron, M., Kell, D.B.: Closed-Loop, Multiobjective Optimization of Two-Dimensional Gas Chromatography/Mass Spectrometry for Serum Metabolomics. Analytical Chemistry 79(2), 464–476 (2006)

Orr, M.: Introduction to radial basis function networks. Centre for Cognitive Science, University of Edinburgh (1996), http://www.cns.ed.ac.uk/people/mark.html

Ray, T., Smith, W.: Surrogate assisted evolutionary algorithm for multiobjective optimization. In: 47th AIAA/ASME/ASCE/AHS/ASC Structures, Structural Dynamics, and Materials Conference, pp. 1–8 (2006)

Robins, A.: Maintaining stability during new learning in neural networks. In: IEEE International Conference on Systems, Man, and Cybernetics, 1997, 'Computational Cybernetics and Simulation', vol. 4, pp. 3013–3018 (1997)

Schölkopf, B., Smola, A.: New support vector algorithms. Technical Report NC2-TR-1998-031, NeuroCOLT2 Technical report Series (1998)

Schonlau, M.: Computer Experiments and Global Optimization. Ph.D. thesis, Univ.of Waterloo, Ontario, Canada (1997)

Schwaighofer, A., Tresp, V.: Transductive and Inductive Methods for Approximate Gaussian Process Regression. Advances in Neural Information Processing Systems 15, 953–960 (2003)

Vapnik, V.N.: Statistical Learning Theory. John Wiley & Sons, Chichester (1998)

Voutchkov, I., Keane, A.J.: Multiobjective optimization using surrogates. Presented at Adaptive Computing in Design and Manufacture (ACDM 06), Bristol, UK (2006)

Wolpert, D.H., Macready, W.G.: No free lunch theorems for optimization. IEEE Transactions on Evolutionary Computation 1, 67–82 (1997)

Zitzler, E., Thiele, L., Laumanns, M., Fonseca, C.M., Grunert da Fonseca, V.: Performance Assessment of Multiobjective Optimizers: An Analysis and Review. IEEE Transactions on Evolutionary Computation 7(2), 117–132 (2003)

11

Real-World Applications of Multiobjective Optimization

Theodor Stewart[1], Oliver Bandte[2], Heinrich Braun[3], Nirupam Chakraborti[4], Matthias Ehrgott[5], Mathias Göbelt[3], Yaochu Jin[6], Hirotaka Nakayama[7], Silvia Poles[8], and Danilo Di Stefano[8]

[1] University of Cape Town, Rondebosch 7701, South Africa
  theodor.stewart@uct.ac.za
[2] Icosystem Corporation, Cambridge, MA 02138 oliver@icosystem.com
[3] SAP AG, Walldorf, Germany heinrich.braun@sap.com
[4] Indian Institute of Technology, Kharagpur 721 302, India
  nchakrab@iitkgp.ac.in
[5] The University of Auckland, Auckland 1142, New Zealand
  m.ehrgott@auckland.ac.nz
[6] Honda Research Institute Europe, 63073 Offenbach, Germany
  Yaochu.Jin@honda-ri.de
[7] Konan University, Higashinada, Kobe 658-8501, Japan nakayama@konan-u.ac.jp
[8] Esteco Research Labs, 35129 Padova, Italy silvia.poles@esteco.com

Abstract. This chapter presents a number of illustrative case studies of a wide range of applications of multiobjective optimization methods, in areas ranging from engineering design to medical treatments. The methods used include both conventional mathematical programming and evolutionary optimization, and in one case an integration of the two approaches. Although not a comprehensive review, the case studies provide evidence of the extent of the potential for using classical and modern multiobjective optimization in practice, and opens many opportunities for further research.

11.1 Introduction

The intention with this chapter is to provide illustrations of real applications of multiobjective optimization, covering both conventional mathematical programming approaches and evolutionary multiobjective optimization. These illustrations do cover a broad range of application, but do not attempt to provide a comprehensive review of applications.

Reviewed by: Alexander Lotov, Russian Academy of Sciences, Russia
Tatsuya Okabe, Honda Research and Development Inc., Japan
Kalyanmoy Deb, Indian Institute of Technology Kanpur, India

J. Branke et al. (Eds.): Multiobjective Optimization, LNCS 5252, pp. 285–327, 2008.
© Springer-Verlag Berlin Heidelberg 2008

In examining the case studies presented here, it may be seen that the applications may be distinguished along two primary dimensions, namely:

The number of objectives which may be:
- Few, i.e. 2 or 3 (impacts of which can be visualized graphically);
- Moderate, perhaps ranging from 4 to around 20;
- Large, up to hundreds of objectives

The level of interaction with decision makers, i.e. the involvement of policy makers, stakeholders or advisers outside of the technical team. The level of such interaction may be:
- Low, such as in many engineering design problems (but for an exception, see Section 11.7) where the analyst is part of the engineering team concerned with identifying a few potentially good designs;
- Moderate, such as in operational management or interactive design problems where solutions may need to be modified in the light of professional or technical experience from other areas of expertise;
- Intensive, such as in public sector planning or strategic management problems, where acceptable alternatives are constructed by interaction between decision makers and other stakeholders, facilitated by the analyst.

Not all combinations of number of objectives and level of interaction may necessarily occur. For example, the public sector planning or strategic management problems which require intensive interactions, also tend often to be associated with larger numbers of objectives. In the case studies reported in this chapter, we have attempted to provide summaries of a number of real case studies in which the authors have been involved, and which do between them illustrate all three levels for each dimension identified above. Table 11.1 summarizes the placement of each of the cases along the above two dimensions.

These studies exemplify the wide range of problems to which multiobjective optimization methods can and have been applied. Half of the case studies deal with engineering design problems, which is clearly an important area of application, but even within this category there is a wide diversity. For

Table 11.1. Classification of case studies

Number of Objectives	Level of Interaction	Case study
Few	Low	Aerodynamic design (Section 11.2)
		Industrial neural network design (Section 11.3)
		Molecular structures for drugs (Section 11.4)
Few	Moderate	Medical decision making (Section 11.5)
		Supply chain management (Section 11.6)
Moderate	Moderate	Interactive aircraft design (Section 11.7)
Moderate	Intensive	Land use planning (Section 11.8)
Large	Low	Lens and bridge designs (Section 11.9)

example, we have two examples from aircraft design, but one (Section 11.2) focuses on the trade-off between robustness and cost in aircraft design, while the other (Section 11.7) deals with the need to provide a broad holistic interactive decision support to aircraft designers. Although both applications relate to aircraft design, the issues raised are substantially different so that different sections in this chapter are devoted to each of them.

Other applications range over operational management of supply chains, effective treatment of cancers and conflicts between environmental, social and economic factors in regional planning.

A perhaps less usual application is that described in Section 11.4. Here the multiobjective optimization methods are applied not directly to design, operational or strategic decisions, but to the development of understanding of molecular processes in synthesizing drugs.

11.2 Aerodynamic Design Optimization

11.2.1 Problem Description

Although the number of objectives are typically low (two or three) if the geometrical constraints are not counted, aerodynamic design optimization is a challenging engineering task for a number of reasons. Firstly, aerodynamic optimization often needs to deal with a large number of design parameters. Secondly, no analytical function is available for evaluating the performance of a given design, and as a result many gradient-based optimization techniques are inapplicable. Thirdly, to evaluate the quality of designs, either computationally expensive computational fluid dynamics (CFD) simulations have to be performed or costly experiments have to be conducted. Finally, aerodynamic optimization involves multiple disciplines and more than one objective must be considered.

In recent years, evolutionary algorithms have successfully been applied to single and multiobjective aerodynamic optimization (Obayashi *et al.*, 2000; Olhofer *et al.*, 2000; Hasenjäger *et al.*, 2005). Despite the success that has been achieved in evolutionary aerodynamic optimization, several issues must be carefully addressed.

11.2.2 Methodology

Geometric Representation

Finding a proper representation scheme is the first and most important step toward successful optimization of aerodynamic structures. A few general criteria can be mentioned for choosing an appropriate geometric representation. Firstly, the representation should be sufficiently flexible to describe highly complex structures. An overly constrained representation will produce only suboptimal results. Secondly, the representation should be efficient, which means that the flexibility of representation can be achieved with a minimum number of free parameters. Inefficient representations may result in an unnecessarily large search space, which reduces the search efficiency of evolutionary algorithms. Finally, the representation should makes it possible to perform a local search. This requirement is important for refining the performance and for reducing search space.

Several methods are available for the geometric representation, such as B-Splines, Bezier curves, and T-Splines. Furthermore, constrained deformation instead of global deformation techniques can be used. An example of constrained deformation is free form deformation or simplified constrained deformation.

Nevertheless, it can happen that no single representation is able to satisfy all the above-mentioned properties. To solve this problem, adaptive representation techniques can be used. The basic idea of adaptive representation is to start the optimization with a relatively compact representation, so that the global search can be conducted first. After that, the number of search parameters can be increased, e.g., by inserting new control points in a B-Spline based representation. Encouraging results have been reported where an adaptive representation for evolutionary optimization of turbine blades has been adopted (Olhofer et al., 2001).

However, adaptive representation is not as straightforward as it appears. On the one hand, it is not trivial to establish when a new point should be inserted, or removed. On the other hand, the insertion or removal of a search point should be neutral to the fitness value. Moreover, an adaptation in representation may degrade the search performance of evolutionary algorithms, for example, for evolutionary strategies with a small population size (Jin et al., 2005).

In multiobjective optimization, even more complex situations can occur. For example, it has been found that in micro heat-exchanger optimization more than one representation is needed to achieve the whole Pareto front (Okabe et al., 2003).

Reduction of Computational Cost

Evolutionary algorithms (EAs) acquire strong search power at the cost of search efficiency. In contrast to gradient-based search methods, EAs often need a large number of quality evaluations to achieve acceptable solutions. This poses serious problems in aerodynamic optimization where each quality evaluation is costly. For example, a full three-dimensional CFD simulation may takes several hours on a high-end computer. To reduce the computational time for evolutionary optimization aerodynamic structures, the following approaches are adopted. Firstly, efficient and scalable evolutionary algorithms need to be developed. An efficient and scalable EA should be able to converge

to an acceptable non-dominated front within a small number of fitness evaluations, and should be insensitive to the increase in search dimension. Secondly,
both the CFD simulations and the fitness evaluations should be parallelized.
A further step is to take advantage the grid computing techniques that enable
us to use available computational resources as efficiently as possible. In addition, computational efficiency can further be improved if EAs are adapted to
the parallel or grid based computing hardware architecture. Finally, computationally expensive full simulations can be replaced partly by computationally
more efficient reduced simulations, or surrogates. In recent years, metamodeling techniques have been extensively investigated to reduce the computational
cost in evolutionary optimization of expensive problems, both for single objective (Jin *et al.*, 2002) and multiobjective optimization (Emmerich *et al.*,
2006). Refer to Chapter 10 on metamodeling for a more detailed discussion
of their use in multiobjective optimization.

Metamodels can introduce errors in quality evaluation, which may lead
to the convergence to a false minimum (Jin *et al.*, 2000). Therefore, a major
concern in using surrogates in evolutionary optimization is to reduce computational cost as much as possible without misleading the evolutionary algorithm. To this end, the meta-model should be properly interleaved with the
original fitness function, which is known as evolution control or model management (Jin *et al.*, 2002). A meta-model can be employed in different operations
of EAs, such as population initialization, crossover or mutation, evolutionary
fitness evaluation, or local search combined with evolutionary search. Different
model management techniques have been suggested. In the individual-based
techniques, all individuals in the current generation are evaluated with the
metamodel. Then, the most promising solutions according to the metamodel
are re-evaluated using the original fitness function. In a generation-based evolution control framework, the meta-model is used for fitness evaluation in
some of the generations, and the frequency at which the meta-model is used
can be adjusted according to the fidelity of the model (Jin *et al.*, 2002). If a
metamodel is employed in local search, the trust-region framework (Alexandrov *et al.*, 1998) can be adopted. A comprehensive survey of techniques for
using meta-models (surrogates) in evolutionary optimization can be found
in Jin (2005).

Robustness Considerations

In aerodynamic optimization, uncertainties in the environment must be taken
into account. For example, the Mach number may deviate from the normal
condition during the flight. In this case, a robust optimal solution is very
much desired. By robustness, it is meant in general that the performance of
an optimal solution should be insensitive to small perturbations of the design
variables or environmental parameters. In multiobjective optimization, the
robustness of a solution can be an important factor for a user in choosing the
final solution. Robust solutions can be achieved in evolutionary optimization

by a number of means. One simple approach is to add perturbations to the design variables or environmental parameters before the fitness is evaluated, which is known as implicit averaging, see e.g., Tsutsui and Ghosh (1997). An alternative to implicit averaging is explicit averaging, which means that the fitness value of a given design is averaged over a number of designs generated by adding random perturbations to the original design. One drawback of the explicit averaging method is the number of additional quality evaluations needed, which may make the approach impractical. Partly to solve this problem, metamodeling techniques have been considered (Ong *et al.*, 2006; Paenke *et al.*, 2006). A slightly different approach is to find the solution with the maximal allowed deviation given the allowed performance deterioration (Lim *et al.*, 2007). One potential advantage of this methods is that no assumptions need to be made concerning the noise distribution (as needed in the averaging based approaches).

Search for robust solutions can also be treated as a multiobjective task, i.e., to maximize the performance and the robustness simultaneously. These two tasks are very likely conflicting, and therefore, Pareto-based multiobjective methods can be employed to find a number of trade-off solutions. Refer to Jin and Branke (2005) for a more detailed discussion on evolutionary search for robust solutions.

11.2.3 An Example

We present here an example of evolutionary multiobjective optimization of a three-dimensional (3D) turbine stator blade used in gas turbines. Two objectives are taken into account in the optimization. The first objective is the average pressure loss, which indicates the energy efficiency of the blade. The second objective, as suggested in Hasenjäger *et al.* (2005), is the variation of the pitch-wise static outlet pressure.

The 3D geometry of the blade is represented by two sections of closed cubic B-splines, namely, a hub section and a tip section, each consisting of 25 control points, as illustrated in Fig. 11.1. In the representation, the first three and the last three control points of the closed B-splines are overlapping, resulting in 22 control points. In addition, since the hub and tip sections are supposed to lie on a cylindrical surface, the z-coordinate of the control points is fixed and not optimized during the evolution. As a result, 88 design parameters in total (x and y coordinates of 22 control points) are to be optimized by the evolutionary algorithm.

To evaluate the performance of a given design, 3D CFD simulations have to be performed. In our work, a 3D Navier-Stokes flow solver, HSTAR3D (Arima *et al.*, 1999) is employed, which usually takes from two to four hours on an AMD Opteron 2 GHz double processor depending on the convergence speed of the fluid dynamics. To reduce computation time, a two-level parallel computing architecture has been adopted. At the first level, fitness evaluations for each individual in the population are parallelized using the master-slave

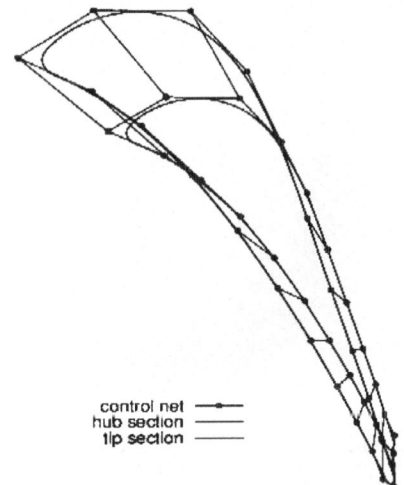

Fig. 11.1. Three-dimensional representation of a blade using B-splines.

model. At the second level, each CFD simulation is again parallelized on four computers using the node-only model. Consequently, if the population size is P, the needed number of computing nodes will be $4P+1$.

An efficient model-based evolutionary multiobjective optimization algorithm, the regularity modeling multiobjective estimation of distribution algorithm (RM-MEDA) (Zhang *et al.*, 2008), has been employed for the optimization of the 3D turbine blade. RM-MEDA is in principle a variant of estimation of distribution algorithms (EDA) (Larranaga and Lozano, 2001). Instead of using Gaussian models, a first-order principle curve has been used to model the regularly distributed Pareto-optimal solutions complemented by a Gaussian model. As demonstrated in Jin *et al.* (2008), by modeling the regularity in the distribution of Pareto-optimal solutions, the scalability of the EDA can be greatly improved. Furthermore, unlike most EDAs, which require a large population size, RM-MEDA performs well even with a small population size. In this example, a population size of 20 has been used.

The optimization results from two independent runs are plotted in Fig. 11.2, in each of which the population has been evolved for 100 generations. Note, however, that the population was initialized not randomly, but with solutions from previous optimization runs using weighted aggregation approaches (Hasenjäger *et al.*, 2005). Compared to the results reported in Hasenjäger *et al.* (2005), we see that the non-dominated solutions obtained by the RM-MEDA are better in terms of both coverage and accuracy.

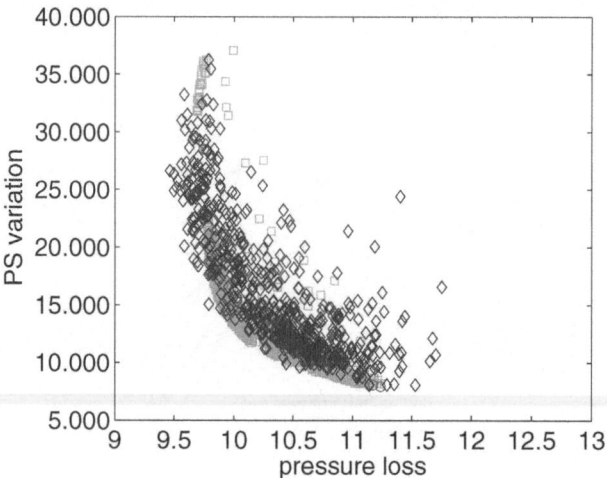

Fig. 11.2. Solutions obtained from two independent evolutionary runs. The diamonds denote the solutions from the first run, and the squares from the second run.

11.3 Design of a Neural Network for an Industrial Blast Furnace

11.3.1 Problem Description

Intuitively one can conceive of a neural network that is simultaneously attempting to satisfy two different requirements: it should be able to reproduce the data in an accurate manner and simultaneously, it should engage a small number of neural connections, primarily to avoid over-training of the data set. These two objectives in many cases could be conflicting with each other. As expected, as the number of nodes becomes smaller, the training error tends to shoot up, and the converse usually remains true as well. In term of these two objectives, one can thus think of working out a Pareto tradeoff, where each solution in the Pareto frontier denotes a neural network of a unique architecture with a unique set of weights. A procedure for evolving such frontiers through a Predator-Prey type multi-objective Genetic Algorithm (Li, 2003) has been demonstrated in a recent article (Pettersson *et al.*, 2007a) and was elaborately tested on the highly nonlinear data from an industrial iron making blast furnace shown schematically in Figure 11.3.

The aim of the study reported in Pettersson *et al.* (2007a) was to evolve a neural network that would optimally predict the carbon, sulfur and silicon contents of the hot metal produced in the blast furnace as a function of a number of process parameters. What was attempted there was simultaneously to minimize (i) the training error of the network (E) and (ii) the required number of active connections in the lower part of it (N). The idea was to tinker with the architecture of the lower part of the network, and to treat their corresponding weights as variables influencing the objective functions, as further elaborated in Figure 11.4. The trade off situation between E and N is expected to be represented as a Pareto-frontier.

11.3.2 Methodology

The evolutionary process: Here the crossing over was done between two entities both of which are essentially self-sustaining neural networks. This process is further elaborated in Figure 11.5. A self-adaptive real-coded mutation was performed on the weights, which draws its inspirations from Differential Evolution (Price *et al.*, 2005).

The multi-objective algorithm used in this study utilized a Moore neighbourhood inhabited by two distinct species: the predators and the preys. The preys are a family of sparse neural networks, initiated randomly as a population, and they evolved in the usual genetic way. The members of the prey population differed from each other both by the topology of the lower part connections and the corresponding weight values. The predators in this algorithm are a family of externally induced entities, which do not evolve, and the major purpose of their presence is to prune the prey populations based upon the fitness values. A two dimensional lattice was constructed as a computational space and both the predators and the prey were randomly introduced there, where each of them would have its own neighbourhood. The basic idea propagated in this algorithm inherits some of the concepts of cellular automata in Moore's neighbourhood. However, unlike cellular automata, here the lattice here does not denote the discretized physical space; it is just a mathematical construction that facilitates a smooth implementation of this algorithm. Further details are available in the original work (Pettersson *et al.*, 2007a).

The method seems to have worked better when the initial population is deliberately generated in the vicinity of the estimated nadir region. The progress of the rank-one members is captured in Figure 11.6 and a computed Pareto frontier is shown in Figure 11.7. Each discrete point in the frontier denotes a neural net with a different ability of prediction than the others. Some typical examples are shown in Figure 11.8. As the ultimate choice between them remains the task of the decision maker, the conservative middle ground 'B' shown in Figure 11.7 should be adequate for most applications.

This novel method of multi-objective analysis is not just to benefit the steel industries: basically it is robust enough to handle noisy data irrespective of their sources. Very recently this methodology has been augmented further through the use of Kalman filters (Saxén *et al.*, 2007), and it has also been effectively utilized for identifying the most important in-signal in a very large network (Pettersson *et al.*, 2007b), rendering it of further interest to the soft computing researchers at large.

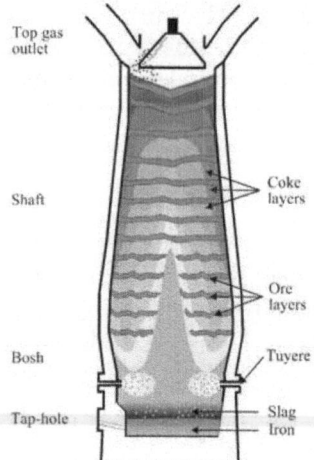

Fig. 11.3. Schematic diagram of an iron making blast furnace.

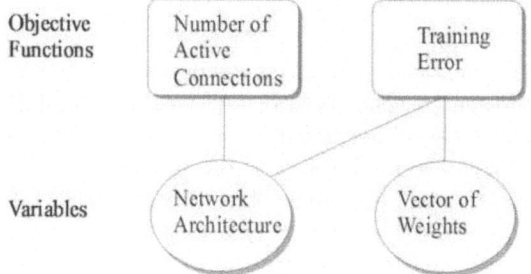

Fig. 11.4. Multi-objective formulation of the neural network.

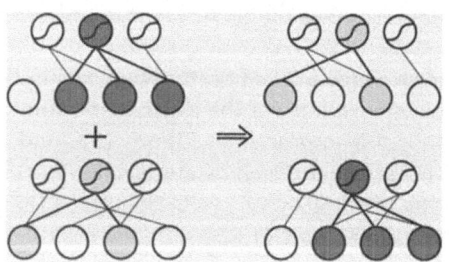

Fig. 11.5. The crossover scheme. The shaded regions are participating in the crossover process.

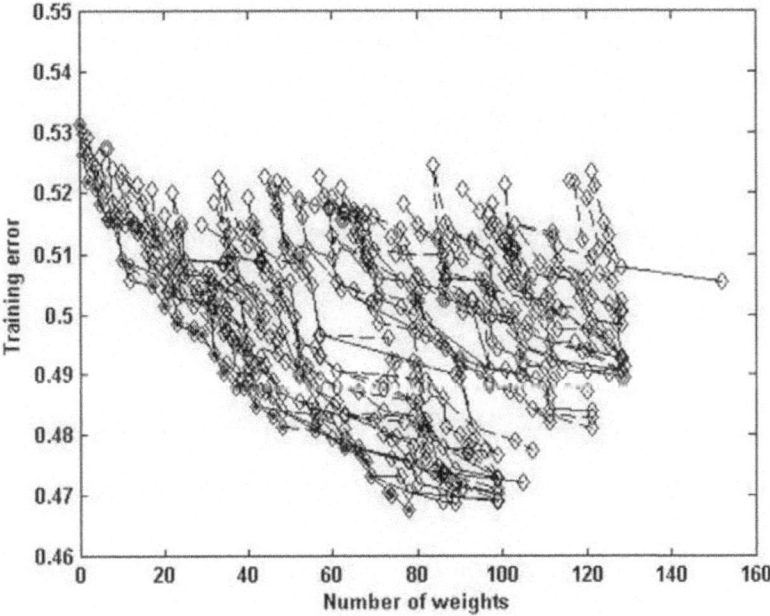

Fig. 11.6. Movement of rank 1 population in different generations.

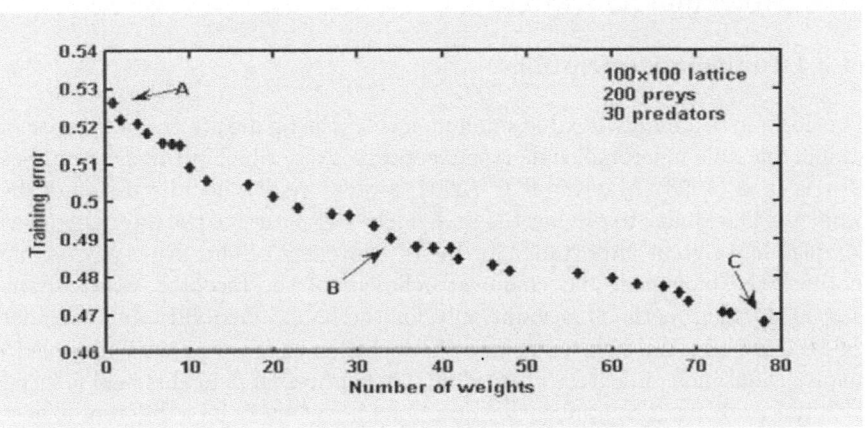

Fig. 11.7. A typical Pareto frontier presented in (Pettersson *et al.*, 2007a).

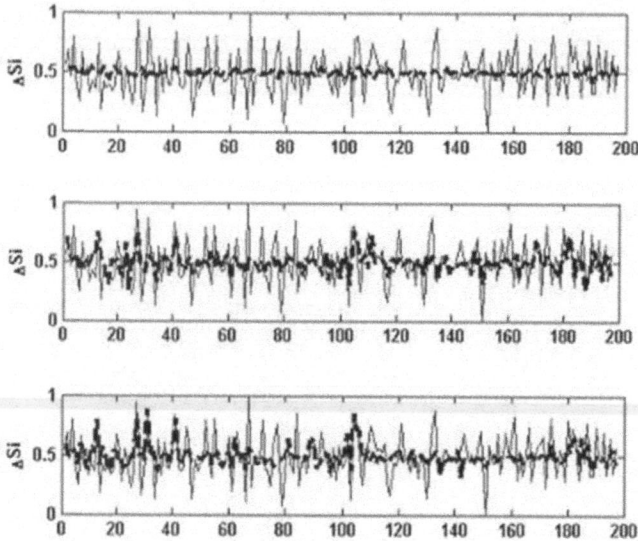

Fig. 11.8. Data prediction through three networks A (top), B (middle) and C (bottom). The lighter lines denote actual observations for a period of 200 days and the darker lines are the predicted values provided in (Pettersson *et al.*, 2007a).

11.4 Molecular Docking

11.4.1 Problem Description

The docking of a highly flexible small molecule (the ligand) to the active site of a highly flexible macromolecule (the receptor) is described in this Section. See Morris *et al.* (1998); MacKerell Jr. (2004) for a more detailed discussion of the problem. The ability to predict the final docked structure of the intermolecular complex has a great importance for the development of new drugs as docking modifies the biological and chemical behavior of the receptor. Most of the current docking methods account only for the ligand flexibility and consider the receptor as a rigid body because the inclusion of receptor flexibility could involve thousands of degrees of freedom. Current research in this field is faced with this problem. The application described here focuses on a different aspect of the docking procedure: the optimization methodology applied to find the best docked structure. The application of a multi-objective approach to the docking problem based on the Pareto optimization of different features of a docked structure is proposed. It is shown that this approach allows for the identification of the dominating interactions that drives the global process.

A drug performs its activity by binding itself to the receptor molecule, usually a protein. In their bounded structure, the molecules gain complementary chemical and geometrical properties that are essential for the therapeutic function of the drug. The computational process that searches for a ligand that best fits the binding site of the receptor from a geometrical and chemical point of view is known as molecular docking.

A molecule is represented by its atoms and the bonds connecting them. Atoms are described mainly by their Van der Waals radius that roughly defines their volume; bonds are described by their lengths (the distance between atoms), by the angle between two consecutive bonds and by their conformational state (the dihedral angle between three consecutive bonds). Molecules are not static systems. At room temperature they perform a variety of motions each one having a characteristic time scale. Since the time scales of stretching (changes in bond lengths) and bending (changes in bond angles) have greater time scales than conformational motions (changes in dihedral angles), bond lengths and bond angles can be considered fixed. Thus, from the docking point of view, only conformational degrees of freedom are important.

Typically, ligands have from 3 to 15 conformational degrees of freedom; their values define the conformational state of the ligand. Receptors have typically from 1000 to 3000 conformational degrees of freedom, so the dimension of the complete search space for best docked conformation becomes computationally unaffordable even for routine cases. The most widely used simplification is to consider only the ligand flexibility, so reducing the complexity of the search space.

The different possible ligand conformations are ranked according to their fitness with the receptor. What this fitness stands for is one of the key aspects of molecular docking and differentiates various docking methodologies. Most of the docking fitness functions are based on the calculation of the total energy of the docked structure. Energy based fitness functions are built starting from force fields which represent a functional form of the potential energy of a molecule. They are composed of a combination of different terms that can be classified in bonded terms (regarding bond energies, bond angles, bond conformations) and non-bonded terms (Van der Waals and electrostatic). This energy can be calculated in various ways, ranging from quantum mechanics to empirical methods. Obviously, a more "exact" fitness function as derived from quantum mechanical simulations strongly impacts on the computational complexity and is applicable only for small systems on massive parallel computers; the opportunity to use "rough" empirical models creates the possibility of treating more realistic cases.

In summary, a docking procedure is composed by two main elements: a fitness function to score different conformations for the molecular complex and a search procedure to explore the space of possible conformations. In current docking approaches, the bonded and non-bonded terms both contribute to the fitness function and the optimization has a single objective equal to their weighted sum. The weights are determined by statistical analysis of experimental data. The proposed multi-objective optimization approach incorporates two conflicting objectives, i.e. the concurrent minimization of internal and intermolecular energy terms, derived from a suitable scoring function, each one corresponding to an objective for the optimization algorithm.

11.4.2 Methodology

MOGA-II

MOGA-II is an improved version of the MOGA (Multi-Objective Genetic Algorithm) of Poloni (Poloni and Pediroda, 1997). It uses smart multi-search elitism for robustness and directional crossover for fast convergence. The efficiency of MOGA-II is controlled by its operators (classical crossover, directional crossover, mutation and selection) and by the use of elitism. The internal encoding of MOGA-II is implemented as in classical genetic algorithms. Elitism plays a crucial role in multi-objective optimization because it helps preserving the individuals that are closest to the Pareto front and the ones that have the best dispersion. MOGA-II uses four different operators for reproduction: one point crossover, directional crossover, mutation and selection. At each step of the reproduction process, one of the four operators is chosen with regard to the predefined operator probabilities.

A strong characteristic of this algorithm is the directional crossover that is slightly different from other crossover operators and assumes that a direction of improvement can be detected comparing the fitness of individuals. A novel operator called evolutionary direction crossover is introduced and it is shown that even in the case of a complex multi-modal function this operator outperforms classical crossover. The direction of improvement is evaluated by comparing the fitness of the individual Ind_i from generation t with the fitness of its parents belonging to generation $t - 1$. The new individual is then created by moving in a randomly weighted direction that lies within the ones individuated by the given individual and his parents.

Multi-objective Ligand-Receptor Docking

In this example, the MOGA-II implementation in modeFRONTIER® is used to optimize the docking towards each of the different contributions of the docking program Autodock v. 3.05 (http://autodock.scripps.edu) scoring function:

$$\Delta G = C_{\mathrm{CVDW}} \Delta G_{\mathrm{GVDW}} + C_{\mathrm{hbond}} \Delta G_{\mathrm{hbond}} + C_{\mathrm{elec}} \Delta G_{\mathrm{elec}}$$
$$+ C_{\mathrm{tor}} N_{\mathrm{tor}} + C_{\mathrm{sol}} \Delta G_{\mathrm{sol}} \quad (11.1)$$

that tries to estimate the change in Gibbs free energy G involved in passing from the system (receptor + ligand) to the docked system (receptor-ligand). The coefficients C are parametrized from experimental data and set to proper values. *VDW* stands for Van der Waals contribution, *hbond* for hydrogen bonds contribution, *elec* for electrostatic contributions, *tor* for the entropy change if N_{tor} rotatable bonds are connected with heavy atoms, and *sol* for the change in solvation energy.

11.4.3 Results

A "bound docking" experiment was performed: on the basis of the x-ray structure of the complex, the receptor coordinates were separated from those of the ligand, and then an attempt made to reconstruct the original x-ray structure by docking the ligand to the receptor. Starting from the scoring function of equation (11.1), Autodock gives the values for the internal energy of the ligand and for the intermolecular ligand-receptor interaction energy. These two outputs were assigned as the objective of the optimization.

The tests were conducted on PDB code 1KV3 chain A co-crystallized with GDP (http://www.rcsb.org/pdb). The resulting Pareto front is reported in Figure 11.9 (in which the units for the axes are Kcal/mol).

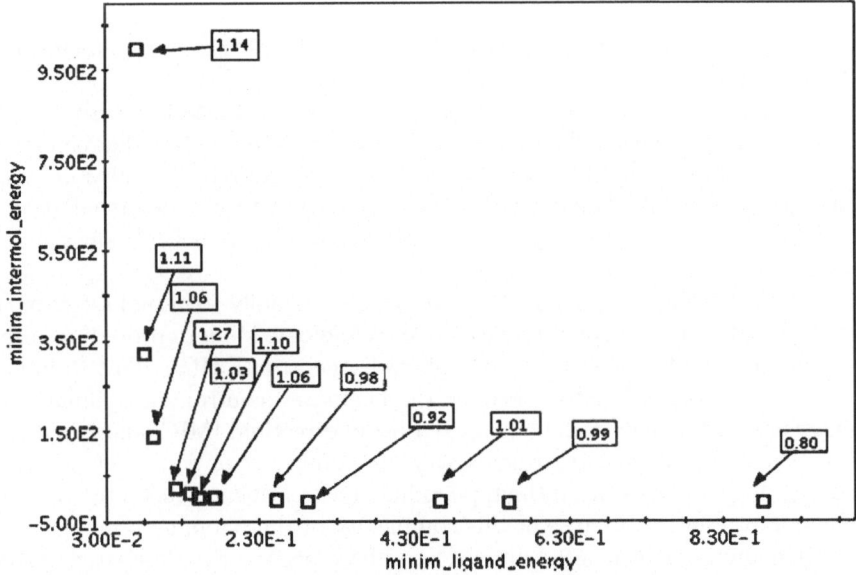

Fig. 11.9. Pareto frontier for the molecular docking problem. Energies are in Kcal/mol. Boxed values represent RMSD in Angstrom between the candidate solution and the original x-ray structure.

The squared values represent the root mean squared deviation (RMSD) in $\mathring{A}$ (angstrom) between the candidate solution and the original x-ray structure. Typically, RMSD values less than 1.5 $\mathring{A}$ are considered as good solutions. It is possible to note that in this case the docking process is mainly driven by the intermolecular energy. This information could be useful for a deeper understanding of the effective relative influence of the contributions of the scoring function to this particular docking process. From a practical point of view, it could also be useful for the design of a tailored scoring function for the docking of similar drug candidates. Also note the presence of a "knee" point (RMSD=1.27 Å). This is a particularly interesting solution of the docking problem in which a small improvement in the minimization of the ligand energy leads to a large deterioration of the intermolecular energy.

11.5 Radiotherapy Treatment Planning

11.5.1 Problem Description

Cancer is one of the most significant health problems worldwide. In industrialised countries it is the second most common cause of death and more than 5 in every 1,000 people are diagnosed with some type of cancer every year. The main treatment form besides surgery and chemotherapy is radiation therapy. It is estimated that 50% of all patients diagnosed with cancer would currently benefit from radiotherapy.

Ionising radiation is used to damage the DNA and interfere with division and growth of cancer cells. Radiation therapy exploits the fact that cancerous cells are more susceptible to radiation than healthy cells. The goal of radiotherapy treatment planning is therefore to ensure that enough radiation is delivered to the targeted region to kill the cancerous cells while surrounding anatomical structures are spared.

In the past it was possible for a physician manually to design a treatment that took full advantage of the available technology. Modern procedures use a technique called Intensity Modulated Radiotherapy (IMRT). This technique uses a multileaf collimator to shape the beam and control, or modulate, the intensity that is delivered along a fixed beam direction. IMRT allows patients to receive complicated treatments and the number of options that are available in IMRT places the optimal planning of a treatment outside the realm of human awareness. Because of this complexity of the planning process, treatment planning is segmented into a three-phase process that first selects beam directions then decides an intensity map (exposure times, fluence) for the directions selected in phase one, and finally chooses a delivery sequence that efficiently administers the treatment. Computer assisted optimisation methods are needed in each phase and since the end of the 20th century these problems have attracted the interest of the Operations Research community. Surveys on optimisation methods for the three phases can be found in Ehrgott et al. (2008a); Shao (2005); Ehrgott et al. (2008b), respectively. In the following we will concentrate on the intensity optimisation problem and assume that beam directions are given.

In 2000, we started a collaboration with the Physics Section of the Oncology Department at Auckland City Hospital to work on treatment planning problems. Treatment planners spend between 30 minutes and several hours on one single case. This is because the available planning system (like almost all commercially available ones worldwide) requires a trial and error approach.

Apart from desired dose levels in the tumour and surrounding critical structures, so called "importance factors" for these entities need to be specified as input. The software then employs heuristics to find a good treatment plan, which is presented to the planner. If it is unsatisfactory the importance factors have to be changed and the process will be repeated. Treatment planners are aware of the inefficiency of this approach. So the goal was to investigate the possibility of a planning system that would calculate several plans right away and provides decision support for choosing an appropriate one.

11.5.2 Methodology

Mathematical models for the intensity optimisation problem are based on the discretisation of the body and the beams. The body is divided into volume elements (voxels) represented by dose points. Voxels are cubic and their edge length is defined by the slice thickness and resolution of the patient's CAT images and is in the range of a few mm. at most. Deposited dose is calculated for one dose point in every voxel and assumed to be the same throughout the voxel. A beam is discretised into beam elements (bixels). Their size is defined by the number of leafs of the collimator and the number of stops for each leaf. The number of voxels may be tens or hundreds of thousands and the number of bixels can be up to 1,000 per beam. The relationship between intensity and dose is linear, i.e., $d = Ax$ where x is a vector of bixel intensities. The entries a_{ij} of A represent the rate at which dose is deposited in voxel i by bixel j. Finally, d is a dose vector that represents the discretised dose distribution in the patient. The computation of the values a_{ij} is referred to as dose calculation.

While most optimisation models in the medical physics literature have a single objective, they do try to accommodate the conflicting goals of destroying tumour cells and sparing healthy tissue. Almost all can be interpreted as weighted sum scalarisations of multi-objective programming models, where the weights are the importance factors mentioned above. Almost all of these multi-objective models are convex problems, so that their efficient sets can be mapped to one another. We decided to use a multi-objective version of the model of Holder (2003), which has some nice mathematical properties. Here A is decomposed by rows into A_T, A_C, and A_N depending on whether a voxel belongs to the tumour, critical structures, or normal tissue. Accordingly, TUB and TLB are vectors of upper and lower bounds on the dose delivered to the tumour voxels; CUB is a vector of upper bounds for the critical structure voxels; and NUB a vector of upper bounds for the remaining normal tissue voxels. The objectives of the model are to minimise the violation of any of the lower and upper bounds and can be stated as shown in (11.2). αUB, βUB, and γUB are parameters to restrict the deviations to clinically relevant values.

$$\min\{(\alpha,\beta,\gamma) : TLB - \alpha e \leqq A_T x \leqq TUB, A_C x \leqq CUB + \beta e,$$
$$A_N x \leqq NUB + \gamma e, 0 \leqq \alpha \leqq \alpha UB,$$
$$\min_i CUB_i \leqq \beta \leqq \beta UB, 0 \leqq \gamma \leqq \gamma UB, 0 \leqq x\}, \tag{11.2}$$

where e denotes a vector of ones of appropriate dimension.

We will denote the feasible set of (11.2) by X and its image in the objective space by Y. In (11.2) we have a multi-objective linear programme with three objectives, a large number of variables (order of thousands), and a very large number of constraints (order of hundred thousands). Under these circumstances we never did try to solve the problem with simplex methods, as it is known that the number of efficient basic feasible solutions can be very large, even for moderately sized problems. Moreover, treatment planners will never use the intensity maps to decide on a treatment, but always look at the dose distribution. The obvious choice was to try and solve the problem in the three-dimensional objective space.

To that end Benson's outer approximation algorithm (Benson, 1998) was implemented. With this method 2D planning problems (i.e. on a single CAT slice) could be solved, but the experiments indicated that 3D problems would require prohibitive computation times. It was therefore necessary to consider approximate solution of the problem. Discussions with physicists on whether that would be acceptable from a radiotherapy point of view provided valuable insights. We discovered that dose calculation is imprecise because the mathematical models to compute the entries of A are inexact since they cannot exactly capture the specific tissue composition in individual patients. The medical physicists assured us that it is acceptable to work with precision of about 0.1 Gy (Gy, for Gray, is the physical unit for radiation dose).

This allowed us to consider ε-efficient solutions of (11.2). It was possible to adapt Benson's algorithm in such a way that it does guarantee the construction of an ε-nondominated set, the modified algorithm is described in Shao and Ehrgott (2008). Solving the problems approximately reduced the computation times dramatically. Figure 11.10 (a) and (b) show the ε-nondominated set of a 2D problem with 986 voxels and 1140 bixels. For $\varepsilon = 0.1$ the problem had 152 nondominated extreme points and was solved in 20 minutes. For $\varepsilon = 0.005$ it took 9 hours to compute 1,989 nondominated extreme points.

11.5.3 Interactive Scheme

From the planners' point of view the whole set of nondominated points is not very useful, since it is infinite. Also, for the same reason of imprecision in dose calculation mentioned above, planners would not distinguish between plans if they differ only by very small amounts. It is necessary to select a finite set of nondominated points (efficient solutions). The nondominated extreme points and associated basic solutions have only mathematical relevance, but no clini-

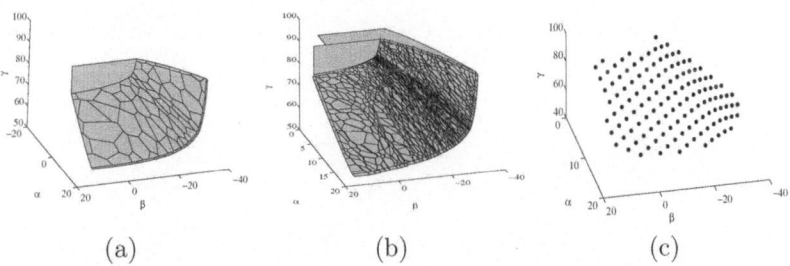

Fig. 11.10. ε-nondominated set with $\varepsilon = 0.1$ (a) and $\varepsilon = 0.005$ (b) and set of representative nondominated points (c).

cal meaning. The selection of plans should represent the whole nondominated set, but guarantee a certain minimal difference between the points. We developed a method to determine a representative subset of nondominated points in Shao and Ehrgott (2007). The method first constructs an equidistant lattice of points placed with distance d on a simplex S (the reference plane) that supports Y at the minimiser of $e^T y$ over Y and such that $Y \subset S + \mathbb{R}^3_{\geqq}$. For each lattice point q an LP

$$\min\{t : q + te \in Y, t \geqq 0\}$$

is solved. If the optimal value is $\hat{t}$, the point $q + \hat{t}e$ is tested for nondominance. It can be shown that the distance between remaining nondominated points is between d and $\sqrt{3}d$. Figure 11.10 (c) shows a representative set for the same example shown in Figure 11.10 (a) and (b).

Since the representative points are all nondominated the planners now have a choice between several plans. By the theory of linear programming, we know that they are all optimal solutions of some weighted sum problem using importance factors as used in current practice. Moreover, the whole range of such solutions is represented. To support planners in the choice of a plan, visual aids are necessary within a decision suppor system. Planners are used to judging the quality of a plan by looking at isodose curves and dose volume histograms (DVH). The former are colour-wash pictures showing curves of equal dose superimposed on CAT pictures. The latter are plots of the percentage of tumour and critical structures against dose levels, see Figure 11.11.

The representative set of solutions (treatment plans) is stored in a database and input to the software CARINA (Ehrgott and Winz, 2008) which first proposes a balanced solution of (11.2) (with as equal as possible values of α, β, γ) displaying the corresponding DVH and isodose plots as well as some information on available trade-offs. The planner can then specify changes (going to a neighbouring solution, searching for solutions with specific values, or for solu-

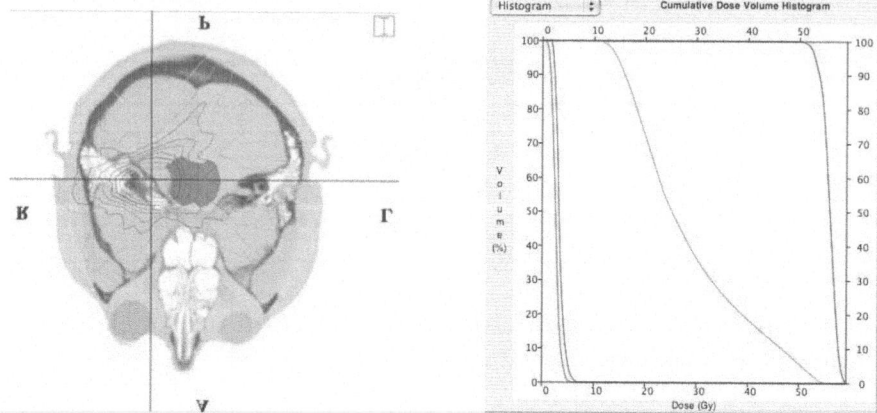

Fig. 11.11. Isodose curves and dose-volume histogram for a brain tumour treatment plan.

tions satisfying some thresholds). This process is continued until the planner accepts a treatment plan.

The interaction with the treatment planner is therefore ex-post, allowing the likely time consuming plan calculation to be decoupled from plan selection. As a consequnce, plan selection becomes faster as it is based on information retrieval from databse, a real-time operation. Moreover, the specification of dose levels is more natural than the "guessing" of importance factors.

11.5.4 Remarks

It is interesting to note that the optimisation model (11.2) tries to characterise dose distribution by a few numbers, whereas the quality is actually judged by the whole DVH. This, of course, is not part of the model. Attempts to specify some points on the DVH curves as constraints in optimisation models exist. But they lead to mixed integer programming models that at this time cannot be solved as multi-objective models.

Throughout the project radiotherapists have been involved in the project. This had several advantages. We obtained valuable information on radiotherapy practice and could ensure to develop usfeul tools that would be accepted by the actual users. Fears that we intend to replace people by software could easily be laid to rest once the radiotherapists understood that we never thought it is possible to replace their role in the treatment design, but that we could improve the planning process. Finally, we made sure to use the tools they are accustomed to work with every day.

The work on this project has thus far resulted in an academic software that allows solution of the multi-objective linear programme (11.2) for 2D and smaller sized 3D problems. The software has been developed in close consultation with treatment planners at Auckland City Hospital. Further work needs to address numerical issues arising in large 3D problems. Before actual use in clinical practice a lengthy and costly approval process needs to be completed for which the support of a medical software company will be required.

11.6 Supply Chain Management

11.6.1 Problem Description

In supply chains often thousands of individual decisions need to be made and coordinated. Due to the high degree of complexity successive planning approaches are therefore often chosen in practice.

Figure 11.12 shows typical planning tasks that arise in supply chains. These planning tasks are arranged in two dimensions. The first dimension is the "supply chain process". In this dimension, planning tasks are arranged focusing on the most important processes following the flow of goods in supply chains. These are procurement, production, distribution and sales. The second dimension is the "planning horizon". In this dimension the planning tasks are distinguished by their temporal impact on the supply chain. These may be strategic decisions with a long-term impact or operational decisions, which have only an immediate impact in the near future (short-term).

The strategic network planning module covers the long-term decisions across all supply chain processes. It supports the user to determine the structure of the supply network (plant location, distribution system) as well as the product program. Although its results are important for the long-term

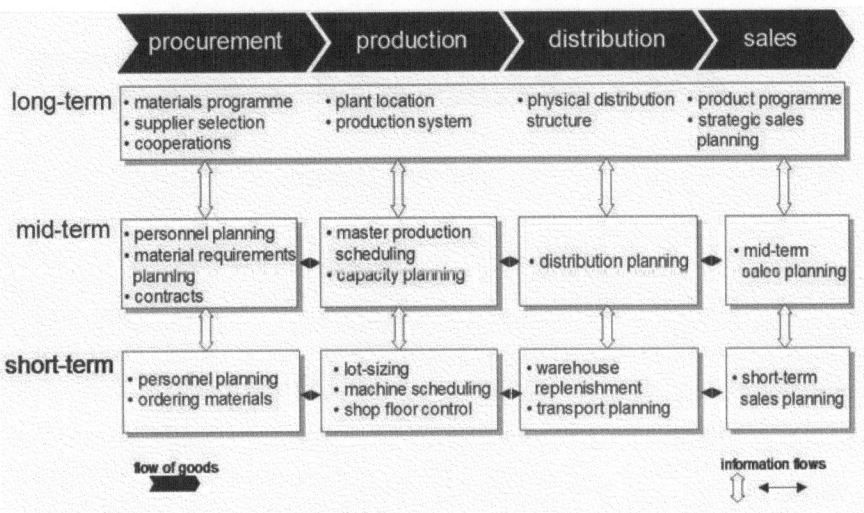

Fig. 11.12. Supply Chain Planning Matrix Fleischmann *et al.* (2005), p. 87.

profitability of supply chains, it is often not a core functionality of Advanced
Planning Systems (APS). This is because APS are primarily built to support
daily business, whereas strategic decisions are only reviewed periodically and
most often not within the regular organization, but rather on a project basis.

The master planning module coordinates procurement, production and
distribution on a mid-term level. Its major decision support is about sourcing:
Which product is produced at which location and when? Thus, in this module
the master production schedule is fixed. However, it is important to anticipate
the key characteristics of the lower (short-term) planning levels within this
module, because otherwise inconsistent plans (for procurement, production
and distribution) will result on the lower planning level.

In the area of distribution and transport planning, distribution related
planning tasks are addressed, the latter on a more detailed level (e.g., scheduling of transports, vehicle loading and routing). Production planning and
scheduling on the other hand are the two modules that support production related issues in the short-term planning horizon. Finally, purchasing and material requirements planning support the (short-term) procurement of materials
and components.

The capabilities offered within mySAP Supply Chain Management (SCM)
extend far beyond the scope of this article. The key functionalities we will
describe in the following are highlighted in Figure 11.13, which is based on
the generic supply chain planning matrix (Figure 11.12). They are part of the
SAP Advanced Planner and Optimizer (SAP APO), which is the advanced
planning component within the mySAP SCM solution. For more information
on SAP APO refer to Bartsch and Bickenbach (2001); Dickersbach (2005).

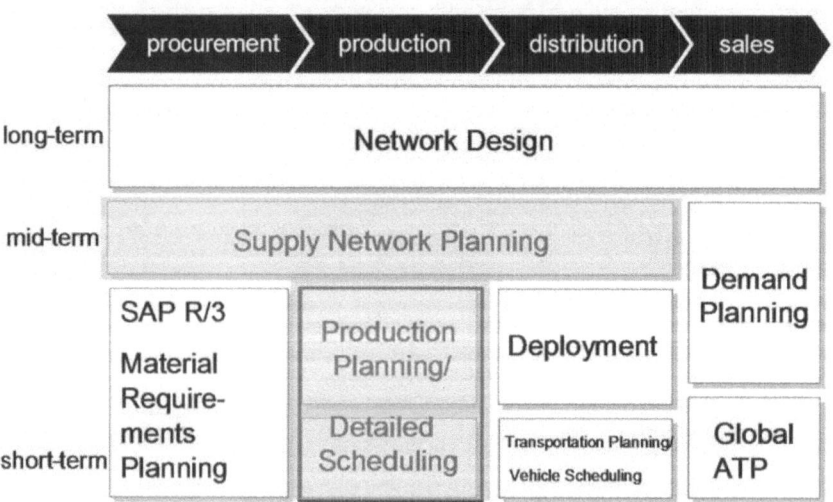

Fig. 11.13. Supply chain planning matrix using mySAP SCM terminology.

Solving a planning problem of this complexity in its entirety within one planning step is neither feasible from an algorithmic perspective nor sensible from a planning process point of view. A hierarchical decomposition of the complete planning problem into a master planning and a production planning and scheduling part addresses planning complexity as well as planning process design issues. In the following, we describe how the planning problem is partitioned into a master planning and a production planning part and which business requirements are addressed on which planning level.

11.6.2 Methodology

Multi Location Optimization in Supply Network Planning (SNP)

In SAP APO, the master planning process is implemented in the Supply Network Planning (SNP) module. SNP offers a multitude of functionalities, not all of which can be described in the limited scope of this article. More details on the SNP module can be found, among others, in Dickersbach (2005). The SNP model contains all relevant locations, i.e. production plants and distribution centres, in the supply network. SNP determines which of the plants produces which quantities of which products in which time periods. On a rough level, SNP also determines which production alternative is used at a specific plant, for instance with regard to ingredients and general process characteristics.

To reduce the complexity of the master planning model, not all products are considered in the SNP optimization run. The selection is made by flagging specific products as not relevant for SNP planning. SNP planning takes into account all products produced in a location, all products for which a stock transfer between locations is possible, externally procured active ingredients, goods for resale and selected forming auxiliaries. Not relevant to SNP are most raw materials, most forming auxiliaries as well as packaging materials. A similar logic is used for resources. Only bottleneck resources are selected for SNP planning.

The concentration on key products and bottleneck resources also results in a significant simplification of the recipes[1] used in SNP, which are derived from the more detailed recipes used in Production Planning and Detailed Scheduling (PP/DS) and the attached enterprise resource planning (ERP) system. Furthermore, compared to the recipes used on the PP/DS level, not all setup and cleaning operations are considered in SNP recipes. Small setup operations are normally neglected while key setup activities which are relevant for campaign planning on bottleneck resources due to their long duration or high costs are considered. To account for the resource capacity consumed by small setup and cleaning operations, a loss factor is applied to calculate the resource capacity for SNP planning.

---

[1] In APO, a recipe is commonly referred to as PPM (production process models) or PDS (production data structure).

One of the main aspects of the SNP planning process is the cost-based plan determination. The following cost types are used to build a cost model which represents the business scenario of value base planning:

- Penalties for not meeting customer demand / forecast,
- Penalties for late satisfaction of customer demand / forecast (location product specific)
- Penalties for not meeting safety stock / safety days' supply requirements (location product specific, linear or piecewise linear)
- Storage cost (location product specific)
- Penalty for exceeding maximum stock level / maximum coverage (location product specific, linear or piecewise linear)
- External procurement cost (linear or piecewise linear, location product specific)
- Handling in / out cost (location product specific)
- Transportation cost (transportation lane, product and means of transport specific, linear or piecewise linear)
- Variable production cost (production process specific, linear or piecewise linear)
- Fixed production cost / setup cost (production process specific)
- Resource utilization cost (resource specific)
- Costs for additional resource utilization (e.g. use of additional shifts, resource specific)
- Cost for falling below minimum resource utilization.

The definition of the cost model is of crucial importance for controlling the behaviour of the SNP optimizer. One of the central questions is whether to maximize service level – which usually means using high penalties for non and late delivery – or to maximize profits – which requires use of realistic sale prices. In the case study scenario, the non delivery cost levels reflect real sale prices sufficiently close to enable a profit maximization logic.

Another important feature of the case study scenario and the resulting cost model is inventory control. High seasonality effects and long campaign durations necessitate considerable build-up of stocks. To avoid an unbalanced build-up of stock, soft constraints for safety stock and maximum stock levels are used. To achieve an even better inventory levelling across products and locations, piecewise linear cost functions for falling below safety stock as well as for exceeding maximum stock levels are employed. In SNP optimization all revelevant constraints can be considered, including

- capacities for production, transportation, handling and storage resources,
- maximum location product specific storage quantities,
- minimum, maximum and fixed production lot sizes,
- minimum, maximum and fixed transportation lot sizes,
- minimum production campaign lot sizes.

An optimization model which considers all these constraints – especially those which can only be modelled using binary or general integer variables – can be highly complex.

Production Planning and Detailed Scheduling (PP/DS)

The short term planning process is dealt with in the Production Planning and Detailed Scheduling module within SAP APO.

PP/DS focuses on determining an optimal production sequence on key resources. In PP/DS, a more detailed modelling than on the SNP planning level is chosen. On the basis of the results determined in SNP optimization, a detailed schedule which considers additional resources and products is created. This schedule is fully executable and there is no need for manual planner intervention, even though manual re-planning and adjustments are fully supported within the PP/DS module. An executable plan can only be ensured by considering additional complex constraints in PP/DS optimization. These additional constraints include:

- Time-continuous planning
- Sequence-dependent setup and cleaning operations.

As the value based planning part is handled within SNP, the PP/DS optimizer uses a different objective function than the SNP optimizer. The following goals can be weighted in the objective function, which is subject to minimization:

- Sum of delays and maximum delay against given due dates
- Setup time and Setup cost
- Makespan (i.e. time interval between first and last activity for optimizing the compactness of the plan)
- Resource cost (i.e. costs associated with the selection of alternative resources)

The main objective of the PP/DS optimizer run in the scenario at hand is to minimize setup times and costs on resources without incurring too much delay against the order due dates. For some resource groups, resource costs are also used to ensure that priority is given to the 'best' (i.e. fastest, cheapest, etc.) resources.

11.6.3 Remarks

We have seen that both in Supply Network Planning and in Detailed Scheduling there are a huge number of objectives to be minimized. However these objectives can be mastered by forming a 4-level hierarchy.

On the root or top level, two dimensions of the second level can be differentiated: Service degree and real costs. The objective of real costs differentiates at the third level between for example:

- storage costs: the minimization of inventory by weighting the inventory of each storage location by an estimated cost factor
- safety costs: the minimization of the risk of getting our of stock by weighting the risk of each storage location by a cost factor
- setup costs: the minimization the overhead of change over for each resource by weighting each change over by a cost factor.

The objective of service degree differentiates at the third level between for example:

- delay penalties: the minimization of delay for each demand or customer order by weighting the priority of the customer
- non delivery penalties: the minimization of non delivery for each demand or customer order by weighting the priority of the customer.

Only for the top level are weighting factors not appropriate. The planner wants to see several alternative solutions of the Pareto front of these two objectives: By how much would costs increase if we wish to achieve a better service level? A high service level is clearly an important objective, but there is no direct cost measure for a delay. Summarizing in Advanced Planning for Supply Chain Management we can focus on an optimization problem with just two objectives maximizing service degree while minimizing the costs. In particular, two dimensional visualization of the Pareto front and representative solutions of this front are needed and sufficient.

11.7 Interactive Processes for Aircraft Design

11.7.1 Problem Description

Interactive Evolutionary Computation (IEC) has started to capture the fascination of researchers from fields as diverse as art, architecture, data mining, geophysics, medicine, psychology, robotics, and sociology. Takagi (2001) outlined many of these applications in his overview paper. However, to this day only very few researchers have applied IEC to the problem of engineering and design of complicated artifacts. While the main reason for the slow pace of adoption in engineering is mostly open for speculation, it is partially a result of the field's reluctance to accept new methods, like Genetic Algorithms, as well as the field's already heavy reliance on automated optimization processes that leave decidedly little room for subjectivity. While the reliance of engineers on analysis tools requires interactive evolutionary techniques to utilize them in the fitness generation, it is also true that many design decisions in practice are made through gut feel and intuition rather then analysis. Recognizing that fact, this example identifies an IEC approach to design that allows for automatic fitness calculation through analysis as well as selection and fitness assignment by the human designers directly.

There are few things humans build that are more complicated than aircraft. Not only are the reliability requirements enormous, given the fatal consequences of failure, but the system itself strides a multitude of areas in physics, such as aerodynamics, thermodynamics, mechanics, and materials. This convolution of disciplines has led historically to a very sequential design process, tackling the various disciplinary issues separately: aerodynamicists only tried to maximize the performance of the wing (or even just an airfoil), propulsion engineers tried to build the largest engines, structural engineers tried to build the sturdiest airframe, while material scientists attempted to only utilize the lightest and sturdiest materials. As a consequence, the design process itself was a highly inefficient iterative process of ever changing airplane configurations, only reconciled by rare, experienced individuals that were proficient in all (or at least many) disciplines. As these people retired, and significant computational power became available, a new design process emerged, attempting to satisfy the concerns of all disciplines concurrently: Multidisciplinary (Design) Optimization. MDO is inherently a multicriteria optimization problem, since each discipline contributes at least one objective function that potentially conflicts with the objective(s) of the other disciplines. The following example demonstrates the ability of one MCDM technique, Interactive Evolutionary Design, to address the difficult task of balancing the different disciplinary objectives when determining the preliminary design configuration of a Supersonic Business Jet.

Figure 11.14 outlines the interactive evolutionary design process employed for this application example (see Bandte and Malinchik, 2004, for background discussion). After the problem is set-up by defining design variables, objectives and constraints and sufficient feasibility has been established, a GA is being interrupted after several generations to display the current population via spider-graphs and Pareto Frontier displays for objective values as well as visualizations of the aircraft configurations. Based on this information the

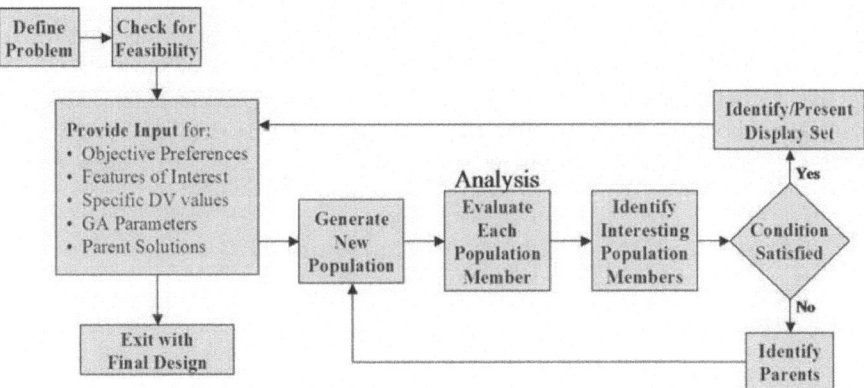

Fig. 11.14. Integrated interactive evolutionary design process.

designer can make some choices regarding objective preferences and features of interest, redirecting search and influencing selection respectively. To limit the scope of this example, only the redirection of the search through objective preferences is being implemented here. However, as exemplified later, designer selection of features of interest is an important part of interactive evolutionary design and should not be neglected in general. The following sections lay out in detail all tasks performed over several iterations for this example.

As in any design problem, the first step is to define the independent parameters, objectives and constraints, as well as evaluation functions that describe the objectives' dependencies on the independent variables. For this interactive evolutionary design environment, this step also identifies the genotype representation of a design alternative, the fitness evaluation function, influenced by the objectives, and how to handle design alternatives that violate constraints.

The supersonic business jet is described by five groups of design variables, displayed in a screen shot presented in Figure 11.15. The first group, general, consists of variables for the vehicle, some of which could be designated design requirements. The other four groups contain geometric parameters for the wing, fuselage, empennage, and engine. The engine group also entails propulsion performance parameters relevant to the design. All in all, the chromosome for this supersonic business jet contains 35 variables that can be varied to identify the best solution.

A mix of economic, size, and vehicle performance parameters were chosen as objectives in this example, with a special emphasis on noise generation, since it is anticipated to be a primary concern for a supersonic aircraft. Hence, for the initial loop the boom loudness and, as a counter weight, the acquisition cost are given slightly higher importance of 20%, while all other objectives are set at 10%. Furthermore, certain noise levels could be prohibitively large and prevent the design from getting regulatory approval. Hence, some of the noise objectives have to have constraint values imposed on them. In addition to these constraints, the design has to fulfill certain FAA requirements regarding take-off and landing distances as well as approach speed. Furthermore, the amount of available fuel has to be more than what is needed for the design mission. Finally, fitness is calculated via a weighted sum of normalized objective values, penalized by a 20% increase in value whenever at least one constraint is violated. Note that the "best solution" is identified as the one with the lowest fitness, i.e. objective function values. All constraints, objectives, normalization values and preferences are also displayed in Figure 11.15.

11.7.2 Methodology and Interactive Scheme

Run GA

Since the initial objective preferences were already specified at problem definition, the GA can be executed next without requiring further input from the designer. The GA chosen for this example is one of the most general found

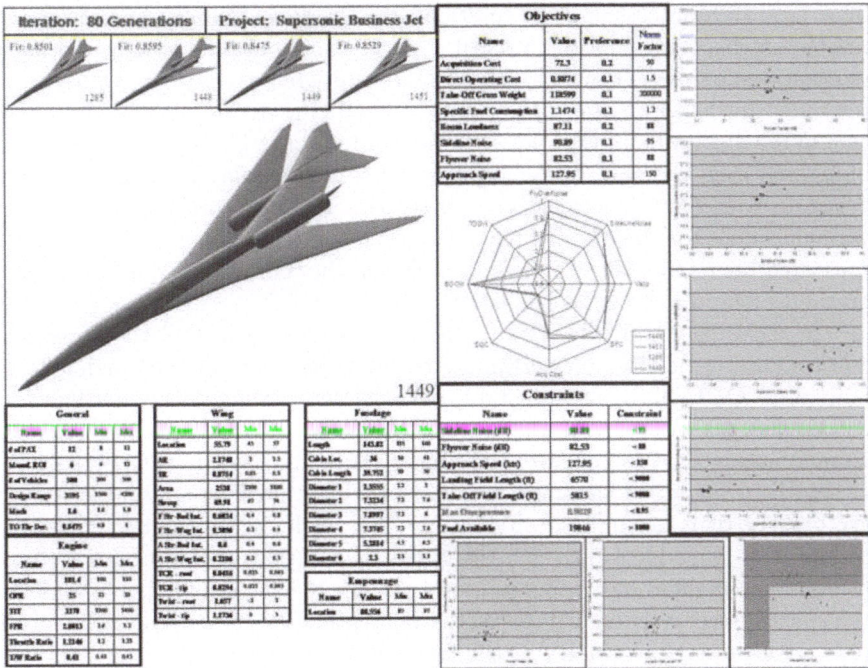

Fig. 11.15. Screen shot: Display of information after 80 generations.

in the literature Holland (1975); Mitchell (1996); Haupt and Haupt (1998). It has a population size of 20 and makes use of a real valued 35-gene representation, limited to the ranges selected at problem set-up. New generations are created from a population with a two individuals elite pool and proportionate probabilistic selection for crossover. The crossover algorithm utilizes a strategy with one splice point that is selected at random within the chromosome. Since the design variables are grouped in logical categories, this crossover algorithm enables, for example, a complete swap of the engine or fuselage-engine assembly between parents. Parent solutions are being replaced with offspring. Each member of the new population has a 15% probability for mutation at ten random genes, sampling a new value from a uniform distribution over the entire range of design variable values. The GA in this example is used for demonstration purposes only and therefore employs just a small population. A population size of 50 to 100 seems more appropriate for a more elaborate version of the presented interactive evolutionary design approach.

Display Information

Once the GA has executed 80 generations, it is interrupted for the first time to display the population of design alternatives found to this point. The designers are presented with information that is intended to provide maximum insight into the search process and the solutions it is yielding. In order to allow for a reasonable display size of the aircraft configuration, only the four best design alternatives, based on fitness and highest diversity in geometrical features, are presented in detail on the top of the left hand side of the display. A screen shot of the displayed information is presented in Figure 11.15, highlighting the individual with the best/lowest fitness, which is also enlarged to provide the designers with a more detailed view of the selected configuration. The design variable values for the highlighted alternative and their respective ranges are presented below this larger image, completing the "chromosome information" on the left hand side of the pane.

On the right hand side of the pane, the designers can find the objective and constraint information pertaining to the population and the highlighted individual. On the top, a simple table outlines the specific objective values for the highlighted alternative, as well as the objective preferences and normalization factors used to generate the fitness values for the current population. Below the table, a spider graph compares the four presented alternatives on the basis of their normalized objective values, while to the right four graphs display the objective values for the entire population, including its Pareto frontier (highlighted individual in black). Below the spider chart, a table lists the constraint parameter values for the highlighted alternative as well as the respective constraint values. A green font represents constraint parameter values near, orange font right around, and red font way beyond the constraint value. Finally at the bottom, three graphs display the population with respect to its member's constraint parameter values as well as the infeasible region, superimposed. These graphs in particular indicate the level of feasibility in the current population.

Provide Input

This step represents the central interaction point of the human with the IEC environment. Here they process the information displayed and communicate preferences for objectives, features of interest in particular designs, whether specific design variable values should be held constant in future iterations, what parameter setting the GA should run with in the next iteration (e.g. a condition that identifies the end of the GA iteration), or whether specific design alternatives should serve as parents for the next generation.

Analyzing the data provided, it is noticeable that all objectives except for the boom loudness are being satisfied well. Consequently, in an attempt to achieve satisfactory levels for the boom loudness in the next iteration, its preference is increased to 30%, reducing the acquisition cost's importance to 10%. This feedback is provided to the GA via a pop-up screen (not displayed here) that allows the designer to enter the new preference values for the next iteration. With this new preference information the GA is executed for another 80 generations.

Second Iteration, 160 Generations

After the GA has executed an additional 80 generations, it is apparent from the objective values that the last set of preferences did not emphasize the boom loudness and sideline noise enough, since boom loudness did not improve (from 87.11 to 87.12 dB) and the sideline noise got worse (from 90.89 to 93.65 dB). Consequently, for the next iteration the importance of both is increased to 35% while all other objectives are reduced to 5%.

Third Iteration, 240 Generations

Examining the population after another 80 generations yields that the last set of preferences still did not emphasize the boom loudness enough, since boom loudness improved only marginally (from 87.12 to 86.96 dB). On the other hand, sideline noise did improve significantly (from 93.65 to 90.94 dB), so that for the next iteration all emphasis can be given to boom loudness. To keep the score even for all other objectives, they are being kept at 5% with boom loudness at 65%.

Fourth Iteration, 400 Generations

For this iteration 160 more generations were executed to produce the population displayed in the screen shot presented in Figure 11.16. In part due to the longer GA run, a very good solution, #7172, is found after 400 generations with largely improved values for almost all the objectives. This result is somewhat surprising, considering most objectives had only a 5% level of preference and the one objective with 65%, boom loudness, improved only marginally.

This result can be attributed to an exemplified effect from summing correlated objective function values that are caused by similar design alternatives (with similar design variable settings) exhibiting similarly good (or bad) values for all objectives except boom loudness. The fact that boom loudness is a conflicting objective, specifically with sideline noise, can also be observed from the pronounced Pareto frontier in the second objective chart from the top.

However, the presented solution after 400 generations seems to satisfy the objectives better than the published solution in Buonanno *et al.* (2002), generated by MATLAB©'s Fmincon function (The MathWorks Inc. (2008)). So it could be concluded that none of these objective values are dramatically out of sync or range and the presented individual is the final solution.

11.8 Land Use Planning

11.8.1 Problem Description

The work described here was motivated by problems of land use allocation or re-allocation in the Netherlands. Land which is already intensely developed

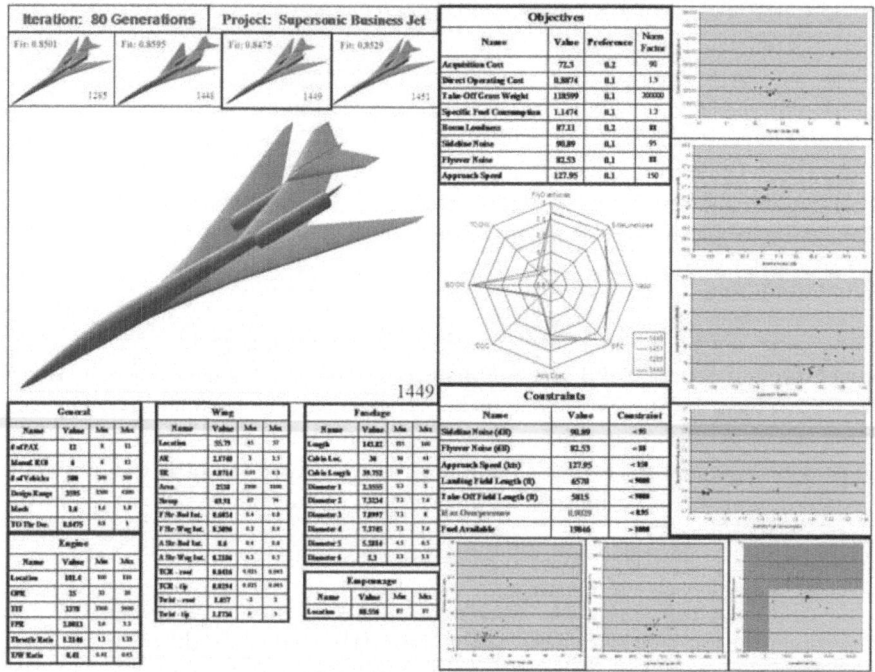

Fig. 11.16. Screen shot: Fourth iteration after 400 generations.

has often to be redeveloped to meet current needs for agriculture, residence, recreation and industry, while at the same time recognizing conservation goals (including possible restoration of some land to approximately pristine conditions). The initial model development was based on a specific region near Amsterdam (the Jisperveld), as briefly described in Section 5 of Stewart *et al.* (2004). However, the longer term intention is to build the model into a general land use planning decision support system (LUDSS). The function of such an LUDSS would be to generate a small number of plans which can be evaluated holistically by decision makers or planners. They would then indicate which features of the plan they like or dislike, which would lead to a readjustment of goals in the LUDSS and the generation of a new solution. This process may be repeated until planners are satisfied that no substantial further improvements are likely.

The model represents the region under consideration by a rectangular grid of (say) $R \times C$ equal-sized cells. It is then assumed that one and only one land use (from a set of Λ possible uses) is allocated to each grid. Formally, we define binary variables $x_{rc\ell}$, such that $x_{rc\ell} = 1$ if land use ℓ is allocated to cell (r, c) and $x_{rc\ell} = 0$ otherwise. For ease of notation we shall denote the three dimensional array of all $x_{rc\ell}$ values by $\mathbf{x}$. Typical constraints on $\mathbf{x}$ would include exclusions of certain land uses from certain cells (corresponding

$x_{rc\ell}$ set to zero) and upper and lower bounds on the total area allocated to a particular land use (i.e. on $\sum_{r=1}^{R} \sum_{c=1}^{C} x_{rc\ell}$).

Some objectives relate to directly quantifiable costs and benefits, and tend to be additive in nature. For example, if all such objectives (without loss in generality) are expressed as costs then:

$$f_i(\mathbf{x}) = \sum_{r=1}^{R} \sum_{c=1}^{C} \sum_{\ell=1}^{\Lambda} \beta_{rc\ell}^i x_{rc\ell}$$

where $\beta_{rc\ell}^i$ is the cost in terms of objective i associated with allocating land use ℓ to cell (r, c).

As initially described by Aerts *et al.* (2005), however, a critical management objective is to ensure that land uses are sufficiently compact to allow integrated planning and management. Aerts *et al.* (2005) introduce essentially one measure of compactness, related to the numbers of cells adjacent to cells of the same land use. This concept was extended in our work by means of a more detailed evaluation of the fundamental underlying management objectives. Defining a *cluster* of cells as a connected set of cells allocated to a single land use, three measures of performance for each land use type were identified as follows:

- *Numbers of clusters for each land use, C_ℓ*: These measure the degree of fragmentation of land uses, and minimization of the number of clusters would seek to ensure that areas of the same land use are connected as far as possible.
- *Relative magnitude of the largest cluster for each land use*: Maximization of the ratio $L_\ell = n_\ell^L / N_\ell$ is sought, where n_ℓ^L and N_ℓ are respectively the number of cells in the largest cluster and the total numbers of cells allocated for land use ℓ. If multiple clusters are formed, then it would often be better to have at least one large consolidated cluster, than for all clusters to be relatively small.
- *Compactness of land uses, denoted by R_ℓ*, defined by a weighted average across all clusters for land use ℓ of the ratio of the perimeter to the square root of the area of the cluster. This measure should be minimized as a compact area for one land use (e.g. a square or circular region) may be easier to manage than a long thread-like cluster.

The above measures define an additional 3Λ objectives, as the compactness goals need to be achieved for each land use individually. Furthermore, the calculation of C_ℓ, L_ℓ and R_ℓ require the execution of a clustering algorithm, so that these additional objectives are non-linear and computationally expensive. The total number of objectives is thus $k = k_0 + 3\Lambda$, where k_0 is the number of additive objectives.

11.8.2 Methodology

In view of the large number of objectives, it is not practical to seek to represent the efficient frontier in full. The approach adopted was thus based on sampling the efficient frontier by optimizing an aggregate measure of performance subject to the constraints on $\mathbf{x}$, for each of a number of different aggregations. The aggregation chosen was that of the *scalarizing function* introduced by Wierzbicki (1999) in the context of his *reference point* methodology, except that we chose a smoother function than that based on the maximum operator. Thus for any given *reference point* (which can be viewed as a set of goals or aspiration levels for each objective), say $g_1, g_2, \ldots, g_k$, the scalarizing which is to be minimized is defined by:

$$\sum_{i=1}^{k} \left[\frac{f_i(\mathbf{x}) - I_i}{g_i - I_i} \right]^4 \tag{11.3}$$

where I_i is the ideal (best achievable measure of performance) for objective i.

Constrained minimization of (11.3) with respect to $\mathbf{x}$ is a non-linear combinatorial optimization problem, with the added complexity that most of the functions cannot be evaluated explicitly in closed form (but are are derived as outputs from a clustering algorithm). Stewart *et al.* (2004) describe a special purpose genetic algorithm (GA), designed to exploit a number of special characteristics of the land use planning problem models, both in the generation of population elements and in the execution of cross-overs (see cited reference for details).

It is interesting to emphasize at this point that the chosen methodology includes elements from conventional multiobjective optimization and from evolutionary optimization, thus representing an integration of the two themes of the present volume.

11.8.3 Interactive Scheme and Results

In implementing the algorithm within an LUDSS, the process starts by selecting one or more tentative reference points, perhaps a central reference point (all goals positioned midway between ideals and worst performance levels) and a selection of reference points which favour each individual goal in turn. Each individual solution generated will be efficient (to the level of optimization accuracy achieved by the GA), and so represents a point on the efficient frontier. In response to assessments by the user as to the direction in which it is desired to move the solution, the reference points are adjusted and the optimizations repeated. In this way, the user is able incrementally to explore the efficient frontier until such time as a satisfactory solution, or short list of possible solutions, is found.

A detailed case study in the use of this system is given in Janssen *et al.* (2007). An illustration of the manner in which the interactions may progress is given in Figure 11.17, which presents three land use maps. The numbers in the maps indicate nine potential land use types, namely: 1. Intensive agriculture; 2. Extensive agriculture; 3. Residence; 4. Industry; 5. Day recreation; 6. Overnight recreation; 7. Wet natural area; 8. Water (recreational); and 9. Water (limited access).

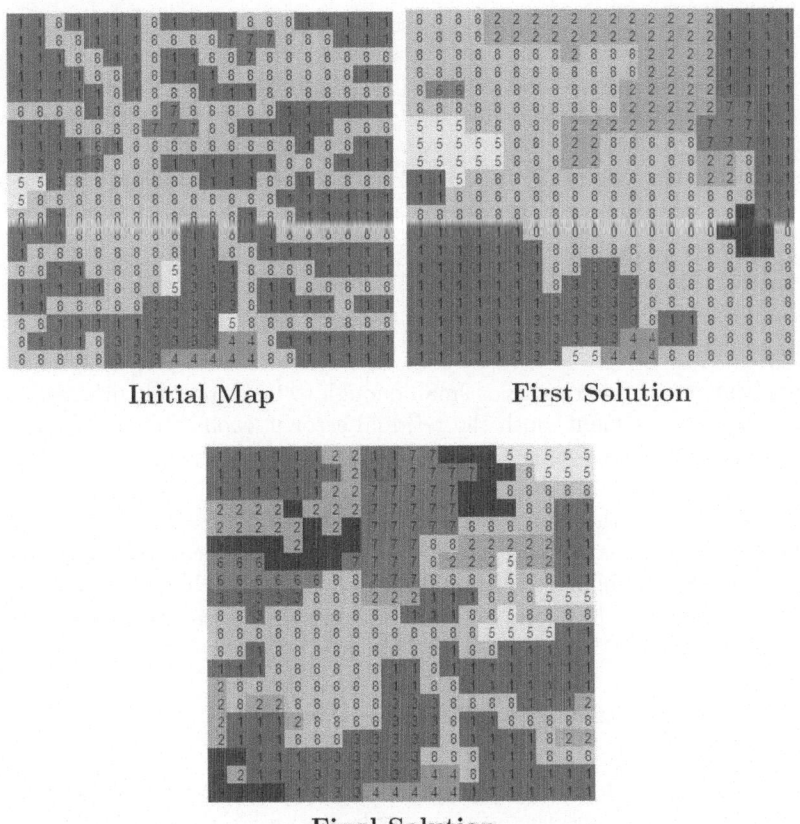

Initial Map   First Solution

Final Solution

Fig. 11.17. Three land use maps generated from the LUDSS.

The upper left hand map displays the original land use pattern. The upper right hand map was generated in the first optimization step, and provides a very compact allocation of land uses. However, the costs were deemed to be too high, largely because of the extent of agricultural land reclaimed from wet areas. For this reason, the priority on the cost attribute was increased. Some more fragmentation of the land uses was then re-introduced, much of which was acceptable except for that of the agricultural land. Also, the values associated with conservation goals were found to be unsatisfactory. Adjustment of these priorities led after the 8th iteration to the lower map in Figure 11.17, and this was found to represent a satisfactory compromise.

11.9 Engineering Design Problems with Large Numbers of Objectives

11.9.1 Problem Description: Cable-Stayed Bridges

Cable-stayed bridges are gaining much popularity due to their beautiful shape. During and after construction, this kind of bridge needs to have the cable length adjusted in order to attain errors of cable tension and camber (the configuration of the girders of the bridge) within some allowable range.

To this end, the following criteria are considered (Nakayama *et al.* (1995)):

- residual error in each cable tension,
- residual error in camber at each node,
- amount of shim adjustment for each cable,
- number of cables to be adjusted.

Since the change of cable rigidity is small enough to be neglected with respect to cable length adjustment, both the residual error in each cable tension and that in each camber are assumed to be linear functions of the amount of shim adjustment.

Let us define n as the number of cables in use, ΔT_i $(i = 1, \ldots, n)$ as the difference between the designed tension values and the measured ones, and x_{ik} as the tension change of i-th cable caused from the change of the k-th cable length by a unit. The residual error in cable tension caused by the shim adjustment is given by

$$p_i = |\Delta T_i - \sum_{k=1}^{n} x_{ik} \Delta l_k| \qquad (i = 1, \ldots, n)$$

Let m be the number of nodes, Δz_j $(j = 1, \ldots, m)$ the difference between the designed camber values and the measured ones, and y_{jk} the camber change at j-th node caused from the change of the k-th cable length by a unit. Then the residual error in the camber caused by the shim adjustments of $\Delta l_1, \ldots, \Delta l_n$ is given by

$$q_j = |\Delta Z_j - \sum_{k=1}^{n} y_{jk} \Delta l_k| \qquad (j = 1, \ldots, m)$$

In addition, the amount of shim adjustment can be treated as objective functions of

$$r_i = |\Delta l_i| \qquad (i = 1, \ldots, n)$$

The upper and lower bounds of shim adjustment inherent in the structure of the cable anchorage are as follows:

$$\Delta l_{Li} \leq \Delta l_i \leq \Delta l_{Ui} \qquad (i = 1, \ldots, n). \tag{11.4}$$

11.9.2 Methodology

Now we have a multi-objective optimization problem in which $p_1, \ldots, p_n$, $q_1, \ldots, q_m$ and $r_1, \ldots, r_n$ are to be minimized under the constraint (11.4). Some large scale bridges have around 100 cables at each side, so that the problem results easily in a very large number of objective functions. For this multi-objective optimization problem, engineers in bridge construction have tried to apply goal programming (Charnes and Cooper, 1961), in which they want to get a desirable solution by adjusting weights imposed on criteria. However, it has been pointed out in the literature (e.g. see Nakayama, 1995) that this task is very difficult even for simple problems. In addition, the shim adjustment must ususally be done during a relatively short period (say, 2:00 am to 8:00 am) with a stable temperature, because the cable length is greatly affected by change of temperature. Therefore, the decision of cable length adjustment must be made very quickly. Also, due to this reason, the traditional goal programming approach is not satisfactory for practical use in this problem.

On the other hand, an interactive multi-objective programming technique has been developed, called the satisficing trade-off method (Nakayama and Sawaragi, 1984). The method is one of aspiration level approaches to multi-objective optimization, which are observed to be effective in many practical problems because they are very simple and easy to implement and do not require any mathematical consistency of decision makers' judgment, and in addition take aspiration levels of decision makers as a probe rather than weights imposed on criteria.

Figure 11.18 shows the graphical user interface (GUI) for the erection management system of a cable stayed bridge using the satisficing trade-off method. The residual error of each criterion and the amount of shim adjustment are represented by bar graphs. The aspiration level is inputted by a mouse on the graph. After solving the auxiliary min-max problem, the Pareto solution according to the aspiration level is represented by another bar graph in a similar fashion. If the designer is not satisfied with the Pareto solution displayed, he/she can revise the aspiration level by means of mouse operations, and the process repeated.

This procedure is continued until the designer obtains a desirable shim-adjustment. The interactive operation using the GUI is very easy for the designer, and the visual information on trade-offs among criteria is user-friendly. The software has been used for real bridge construction, for example tje Tokiwa Swan Bridge (Ube City) and the Karasuo Harp Bridge (Kita-Kyusyu City) in 1992.

One of the important aspects of such a problem with a large number of objective functions is the graphical user interface. As can be easily seen in Fig.11.18, it is not too difficult for designers to make a trade-off analysis on the basis of the displayed visual information, even in cases with hundreds of objective functions. However, it might be difficult or even impossible to

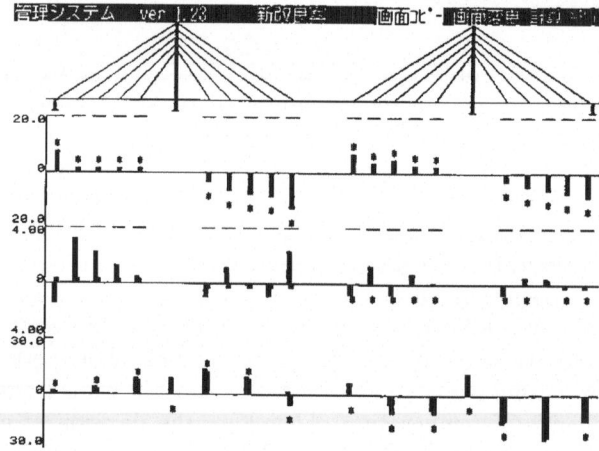

Fig. 11.18. Erection Management System of Cable-stayed Bridge.

grasp the total trade-off context on the basis of numerical information only, for problems with such a large number of objective functions.

11.9.3 A Further Application: Lens Design

Another good example with a large number of objective functions is lens design. There are many kinds of lenses such as copier, camera, medical instruments and so on. Above all, lenses in semiconductor chip production are very expensive (of the order of million dollars), and hence have to be designed very carefully.

In lens design, there are around 200 design variables such as

- kinds of glass,
- number of lenses,
- diameter,
- curvature,
- distance between lenses,

and around 400 criteria such as

- cost
- weight
- criteria for images:
  - aberration
  - chromatic
  - spherical
  - astigmatism
  - coma

- distortion
- curvature of field
- color balance
- resolution
- MTF
- CCI

In lens design, there is the further difficulty of nonlinear optimization in addition to the large number of objective functions: Scalarized optimization problems are usually highly nonlinear and highly multi-modal. Moreover, those functional forms are not given explicitly in terms of design variables. Those function values are evaluated on the basis of some kind of simulation (ray trace). Therefore, it is difficult to obtain a global minimum for the objective function.

So far, engineers use specific software in lens design. Their main attention has been directed to how they obtain a global optimum for the scalarized objective function, while the linearly weighted sum scalization function is applied. It will be surely a good subject to investigate how interactive multiobjective optimization techniques can work in lens design.

11.10 Concluding Comments

In what sense are the above applications different to other optimization studies? Clearly, the distinction lies in the multiplicity of objectives which are central to the applications discussed here. In common with more general approaches to multiple criteria decision making (MCDM), those applying the multiobjective methods start by careful problem structuring to identify the underlying objectives, and to represent these in meaningful manner, much as has been described in Chapter 3 of Belton and Stewart (2002).

Some of the approaches reported in the case studies do ultimately seek to identify an overall mathematical objective function as a surrogate measure of performance to be "maximized" or "minimized", but:

- This is done only after careful attention to tradeoffs between objectives and clear recognition that these tradeoffs may change as one explores the decision space;
- The methods are applied interactively, with systematic changes in formulation (revising goals or value tradeoffs) in the light of preference information (as described in other Chapters of this book concerning interactive methods), and thus providing a means of implicit exploration of the Pareto frontier.

Other reported approaches avoid the use of surrogate objective functions, by seeking to identify the Pareto frontier explicitly, leaving the user or ultimate client to explore the options visually before making the final selection. Unfortunately, it is difficult to provide an unambiguous visualization of the frontier for more more than two or three objectives, although the Chapter 9 in this book seeks to extend such opportunities.

The clear challenge to future research lies precisely in the interface between these implicit and explicit methods of searching the Pareto Frontier. An opportunity may lie in using the interactive methods using surrogate measures of performance for an initial exploration, but using the explicit search methods (linked to appropriate visualization) to refine the exploration across those objectives which are found most critical to the final decisions in the most promising regions of the decision space.

References

Aerts, J.C.J.H., van Herwijnen, M., Janssen, R., Stewart, T.J.: Evaluating spatial design techniques for solving land-use allocation problems. Journal of Environmental Planning and Management 48(1), 121–142 (2005)

Alexandrov, N.M., Dennis, J.E., Lewis, R.M., Torczon, V.: A trust region framework for managing use of approximation models in optimization. Journal on Structural Optimization 15(1), 16–23 (1998)

Arima, T., Sonoda, T., Shirotori, M., Tamura, A., Kikuchi, K.: A numerical investigation of transonic axial compressor rotor flow using a low-Reynolds-number $k - \epsilon$ turbulence model. ASME Journal of Turbomachinery 121(1), 44–58 (1999)

Bandte, O., Malinchik, S.: A broad and narrow approach to interactive evolutionary design – an aircraft design example. In: Deb, K., et al. (eds.) GECCO 2004. LNCS, vol. 3103, pp. 883–895. Springer, Heidelberg (2004)

Bartsch, H., Bickenbach, P.: Supply Chain Management mit SAP APO. Galileo Press, Bonn (2001)

Belton, V., Stewart, T.J.: Multiple Criteria Decision Analysis: An Integrated Approach. Kluwer Academic Publishers, Boston (2002)

Benson, H.: An outer approximation algorithm for generating all efficient extreme points in the outcome set of a multiple objective linear programming problem. Journal of Global Optimization 13(1), 1–24 (1998)

Buonanno, M., Lim, C., Mavris, D.N.: Impact of configuration and requirements on the sonic boom of a quiet supersonic jet. Presented at World Aviation Congress, Phoenix, AZ (2002)

Charnes, A., Cooper, W.: Management Models and Industrial Applications of Linear Programming, vol. 1. John Wiley, New York (1961)

Dickersbach, J.T.: Supply Chain Management with APO, 2nd edn. Springer, Berlin (2005)

Ehrgott, M., Winz, I.: Interactive decision support in radiation therapy treatment planning. OR Spectrum 30, 311–329 (2008)

Ehrgott, M., Holder, A., Reese, J.: Beam selection in radiotherapy design. In: Linear Algebra and Its Applications, vol. 428, pp. 1272–1312 (2008a)

Ehrgott, M., Hamacher, H.W., Nußbaum, M.: Decomposition of matrices and static multileaf collimators: A survey. In: Alves, C.J.S., Pardalos, P.M., Vicente, L.N. (eds.) Optimization in Medicine. Springer Series in Optimization and Its Applications, vol. 12, pp. 25–46. Springer Science & Business Media, New York (2008b)

Emmerich, M., Giannakoglou, K., Naujoks, B.: Single and multi-objective evolutionary optimization assisted by Gaussian random field meta-models. IEEE Transactions on Evolutionary Computation 10(4), 421–439 (2006)

Fleischmann, B., Meyr, H., Wagner, M.: Advanced planning. In: Stadtler, H., Kilger, C. (eds.) Supply Chain Management and Advanced Planning. Concepts, Models, Software and Case Studies, 3rd edn., pp. 81–106. Springer, Berlin (2005)

Hasenjäger, M., Sendhoff, B., Sonoda, T., Arima, T.: Three dimensional evolutionary aerodynamic design optimization using single and multi-objective approaches. In: Schilling, R., Haase, W., Periaux, J., Baier, H., Bugeda, G. (eds.) Evolutionary and Deterministic Methods for Design, Optimization and Control with Applications to Industrial and Societal Problems EUROGEN 2005, Munich, FLM (2005)

Haupt, R.L., Haupt, S.E.: Practical Genetic Algorithms. John Wiley & Sons, New York (1998)

Holder, A.: Designing radiotherapy plans with elastic constraints and interior point methods. Health Care Management Science 6, 5–16 (2003)

Holland, J.: Adaptation in Natural and Artificial Systems. The University of Michigan Press, Ann Arbor (1975)

Janssen, R., van Herwijnen, M., Stewart, T.J., Aerts, J.C.J.H.: Multiobjective decision support for land use planning. Environment and Planning B, Planning and Design. To appear (2007)

Jin, Y.: A comprehensive survey of fitness approximation in evolutionary computation. Soft Computing 9(1), 3–12 (2005)

Jin, Y., Branke, J.: Evolutionary optimization in uncertain environments – A survey. IEEE Transactions on Evolutionary Computation 9(3), 303–317 (2005)

Jin, Y., Olhofer, M., Sendhoff, B.: On evolutionary optimization with approximate fitness functions. In: Genetic and Evolutionary Computation Conference, pp. 786–792. Morgan Kaufmann, San Francisco (2000)

Jin, Y., Olhofer, M., Sendhoff, B.: A framework for evolutionary optimization with approximate fitness functions. IEEE Transactions on Evolutionary Computation 6(5), 481–494 (2002)

Jin, Y., Olhofer, M., Sendhoff, B.: On evolutionary optimization of large problems using small populations. In: Wang, L., Chen, K., Ong, Y.S. (eds.) ICNC 2005. LNCS, vol. 3611, pp. 1145–1154. Springer, Heidelberg (2005)

Jin, Y., Zhou, A., Zhang, Q., Tsang, E.: Modeling regularity to improve scalability of model-based multi-objective optimization algorithms. In: Multiobjective Problem Solving from Nature. Natural Computing Series, pp. 331–356. Springer, Heidelberg (2008)

Larranaga, P., Lozano, J.A. (eds.): Estimation of Distribution Algorithms: A New Tool for Evolutionary Computation. Kluwer Academic Publishers, Dordrecht (2001)

Li, X.-D.: A real-coded predator-prey genetic algorithm for multiobjective optimization. In: Fonseca, C.M., Fleming, P.J., Zitzler, E., Deb, K., Thiele, L. (eds.) EMO 2003. LNCS, vol. 2632, pp. 207–221. Springer, Heidelberg (2003)

Lim, D., Ong, Y.-S., Jin, Y., Sendhoff, B., Lee, B.S.: Inverse multi-objective robust evolutionary optimization. Genetic Programming and Evolvable Machines 7(4), 383–404 (2007)

MacKerell Jr., A.D.: Empirical force fields for biological macromolecules: Overview and issues. Journal of Computational Chemistry 25(13), 1584–1604 (2004)

Mitchell, M.: An Introduction to Genetic Algorithms. The MIT Press, Cambridge (1996)

Morris, G.M., Goodsell, D.S., Halliday, R.S., Huey, R., Hart, W.E., Belew, R.K., Olson, A.J.: Automated docking using a Lamarckian genetic algorithm and an empirical binding free energy function. Journal of Computational Chemistry 19(14), 1639–1662 (1998)

Nakayama, H.: Aspiration level approach to interactive multi-objective programming and its applications. In: Pardalos, P.M., Siskos, Y., Zopounidis, C. (eds.) Advances in Multicriteria Analysis, pp. 147–174. Kluwer Academic Publishers, Dordrecht (1995)

Nakayama, H., Sawaragi, Y.: Satisficing trade-off method for interactive multiobjective programming methods. In: Grauer, M., Wierzbicki, A.P. (eds.) Interactive Decision Analysis – Proceedings of an International Workshop on Interactive Decision Analysis and Interpretative Computer Intelligence, pp. 113–122. Springer, Heidelberg (1984)

Nakayama, H., Kaneshige, S., Takemoto, S., Watada, Y.: An application of a multiobjective programming technique to construction accuracy control of cable-stayed bridges. European Journal of Operational Research 87, 731–738 (1995)

Obayashi, S., Sasaki, D., Takaguchi, Y., Hirose, N.: Multi-objective evolutionary computation for supersonic wing-shape optimization. IEEE Transactions on Evolutionary Computation 4(2), 182–187 (2000)

Okabe, T., Foli, K., Olhofer, M., Jin, Y., Sendhoff, B.: Comparative Studies on Micro Heat Exchanger Optimisation. In: Proceedings of IEEE Congress on Evolutionary Computation (CEC-2003), pp. 647–654. IEEE Computer Society Press, Los Alamitos (2003)

Olhofer, M., Arima, T., Sonoda, T., Sendhoff, B.: Optimization of a stator blade used in a transonic compressor cascade with evolution strategies. In: Parmee, I. (ed.) Adaptive Computing in Design and Manufacture, pp. 45–54. Springer, Heidelberg (2000)

Olhofer, M., Jin, Y., Sendhoff, B.: Adaptive encoding for aerodynamic shape optimization using evolution strategies. In: Congress on Evolutionary Computation (CEC), Seoul, Korea, May 2001, vol. 2, pp. 576–583. IEEE Computer Society Press, Los Alamitos (2001)

Ong, Y.-S., Nair, P.B., Lim, K.Y.: Max-min surrogate-assisted evolutionary algorithms for robust design. IEEE Transactions on Evolutionary Computation 10(4), 392–404 (2006)

Paenke, I., Branke, J., Jin, Y.: Efficient search for robust solutions by means of evolutionary algorithms and fitness approximation. IEEE Transactions on Evolutionary Computation 10, 405–420 (2006)

Pettersson, F., Chakraborti, N., Saxén, H.: A genetic algorithms based multiobjective neural net applied to noisy blast furnace data. Applied Soft Computing 7, 387–397 (2007a)

Pettersson, F., Chakraborti, N., Singh, S.B.: Neural networks analysis of steel plate processing augmented by multi-objective genetic algorithms. Steel Research International 78, 890–898 (2007b)

Poloni, C., Pediroda, V.: GA coupled with computationally expensive simulations: tools to improve efficiency. In: Genetic Algorithms and Evolution Strategies in Engineering and Computer Science, pp. 267–288. John Wiley and Sons, Chichester (1997)

Price, K., Storn, R.N., Lampinen, J.A. (eds.): Differential Evolution: A Practical Approach to Global Optimizations. Springer, Berlin (2005)

Saxén, H., Pettersson, F., Gunturu, K.: Evolving nonlinear time-series models of the hot metal silicon content in the blast furnace. Materials and Manufacturing Processes 22, 577–584 (2007)

Shao, L.: A survey of beam intensity optimization in IMRT. In: Halliburton, T. (ed.) Proceedings of the 40th Annual Conference of the Operational Research Society of New Zealand, Wellington, 2-3 December 2005, pp. 255–264 (2005), Available online at http://secure.orsnz.org.nz/conf40/content/paper/Shao.pdf

Shao, L., Ehrgott, M.: Finding representative nondominated points in multiobjective linear programming. In: IEEE Symposium on Computational Intelligence in Multi-Criteria Decision Making, pp. 245–252. IEEE Computer Society Press, Los Alamitos (2007)

Shao, L., Ehrgott, M.: Approximately solving multiobjective linear programmes in objective space and an application in radiotherapy treatment planning. Mathematical Methods of Operations Research (2008)

Stewart, T.J., Janssen, R., van Herwijnen, M.: A genetic algorithm approach to multiobjective land use planning. Computers and Operations Research 32, 2293–2313 (2004)

Takagi, H.: Interactive evolutionary computation: Fusion of the capacities of EC optimization and human evaluation. Proceedings of the IEEE 89, 1275–1296 (2001)

The MathWorks Inc. (2008)

Tsutsui, S., Ghosh, A.: Genetic algorithms with a robust solution searching scheme. IEEE Transactions on Evolutionary Computation 1(3), 201–208 (1997)

Wierzbicki, A.P.: Reference point approaches. In: Gal, T., Stewart, T.J., Hanne, T. (eds.) Multicriteria Decision Making: Advances in MCDM Models, Algorithms, Theory, and Applications, Kluwer Academic Publishers, Boston (1999)

Zhang, Q., Zhou, A., Jin, Y.: RM-MEDA: A regularity model-based multi-objective estimation of distribution algorithm. IEEE Transactions on Evolutionary Computation 12(1), 41–63 (2008)

12

Multiobjective Optimization Software

Silvia Poles[1], Mariana Vassileva[2], and Daisuke Sasaki[3]

[1] ESTECO - Research Labs, Via Giambellino, 7 35129 Padova, ITALY
   silvia.poles@esteco.com
[2] Institute of Information Technologies, Bulgarian Academy of Sciences,
   BULGARIA mvassileva@iinf.bas.bg
[3] CFD Laboratory, Department of Engineering, University of Cambridge,
   Trumpington Street, Cambridge CB2 1PZ, UK ds432@eng.cam.ac.uk

Abstract. This chapter provides a description of multiobjective optimization software with a general overview of selected few available tools developed in the last decade. This chapter can be considered a revision of previous valid papers and chapters on nonlinear multiobjective optimization software such as the ones written by Weistroffer *et al.* (2005) and Miettinen (1999) that lists existing software packages up to the year 1999. More precisely, this chapter is focused on the tools and features that advisable multiobjective optimization software should contain.

12.1 Introduction

The main topic to be discussed in this chapter is available multiobjective optimization software. The main concern is devoted to software developed for nonlinear problems. Several questions may be raised when discussing multiobjective optimization software, but among the most recurring questions we may list the following:

- What do experts think about multiobjective optimization tools and what are the most important features good software should always possess?
- What is the current state-of-the-art of multiobjective optimization software?
- What are the advantages and gaps of all these optimization tools?

The description of an ideal software is very close to a complex integrated environment such as a "Process Integration and Design Optimization" (PIDO) or a "Problem Solving Environment" (PSE) (Gallopoulos *et al.*, 1991; Houstis *et al.*, 1997). PIDO and PSE are integrated computing environments which

Reviewed by: Oliver Bandte, Icosystem Cooperation, USA
Jyrki Wallenius, Helsinki School of Economics, Finland
Kaisa Miettinen, University of Jyväskylä, Finland

J. Branke et al. (Eds.): Multiobjective Optimization, LNCS 5252, pp. 329–348, 2008.

provide the users all the necessary tools for solving multiobjective optimization problems and for supporting decision making.

An ideal tool should have: an easy-to-use graphical user interface, a good set of optimization methods, a good tool for visualizing the results and choosing the final solutions. Moreover meta-modeling and validation of models are fundamental when dealing with time-consuming function evaluations. Last but not least, robustness and reliability of solutions are of primary importance for selecting the best design.

There are many attributes and characteristics that can be used to measure software quality as seen by end-users. Leaving out all the problems related to reliability, absence of bugs, extensibility and maintainability of each tool, we here refer to requirements that a decision maker may have for a multiobjective optimization software.

In the following sections a list of advisable program specifications is explained. Next, a list of software is described and their conformance to requirements and specifications is analyzed.

12.2 Software Features and Quality

12.2.1 Graphical User Interface

One of the most evident characteristics of a software is always a flexible, complete and easy-to-use graphical user interface (GUI). Even with multi-objective optimization tools, the GUI plays an important role. In this case, the GUI should give to the users of the software being it analysts or decision makers (e.g. engineers and managers) the ability to define and modify a problem, to define input, output, objectives and constraints. Moreover, the GUI should give to the decision makers the ability to choose optimization strategies, manage software and hardware resources, describe how the processes are synchronized and visualize and analyze results. Moreover, the GUI should be suitable for introducing decision maker's preferences in order to solve multiobjective decision making problems with an intelligent guidance.

For example, a multiobjective optimization problem can be described using graph-based formalisms as shown in Fig. 12.1. The figure describes a standard mechanical design problem, the design of a welded beam structure with the aim to minimize cost and displacement subject to constraints on shear.

12.2.2 Optimization Methods

Problems related to one or more than one conflicting objective functions, originate in several disciplines; their solution has been a challenge for a long time. Typically, using a single optimization technology is not sufficient to deal with real-life problems.

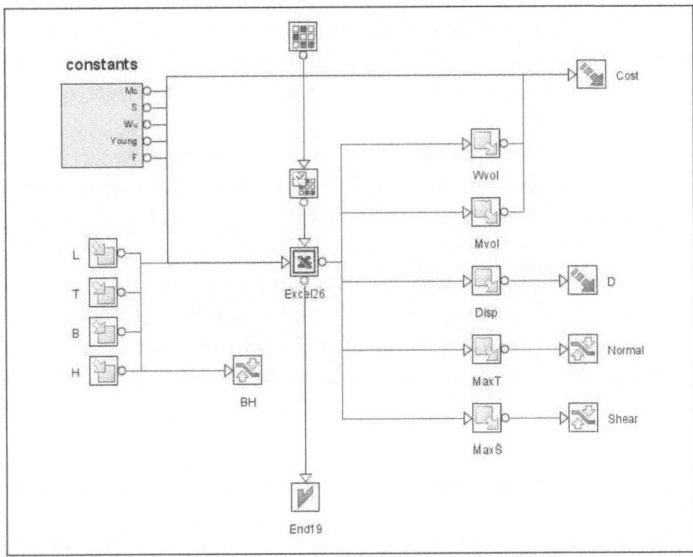

Fig. 12.1. An example of how a workflow can describe input, output, constraints and objectives of a multiobjective optimization problem.

In order to help engineers and decision makers, old and new multiobjective optimization techniques are studied in industries, project and portfolio management, military and governmental fields. The importance of managing more than one objective at once as opposed to just optimizing one outcome is well recognized, for example, in portfolio management. In fact, constructing a balanced bond portfolio must deal with uncertainty in the future price of bonds and several other aspects. Despite what is reported in (Kaliszewski, 2004), multiobjective optimization has recently started to gain attention within the engineering and scientific communities since many real world optimization problems in numerous disciplines and application areas, contain more than one outcome that should be optimized.

Each optimization technique is qualified by its search strategy that implies the robustness and/or the accuracy of the method. An indication of the robustness of an optimization method is the ability to reach the absolute extremes of the functions even when starting far away from the final solutions. On the contrary, the accuracy measures the capability of the optimization algorithm to get as close as possible to function extremes. There are hundreds or thousands of optimization methods in the literature: each numerical method can solve a specific or more generic problem. Different algorithms are intended to solve different types of multiobjective optimization problems such as linear, nonlinear, continuous, discrete, mixed, and so on. Different strategies can be selected for different problems. Unfortunately, real world applications often include one or more difficulties which make many of these methods inapplicable. Many engineering problems involve highly non-linear objective functions or even may not have an analytic expression in terms of the variables. A general overview of basic and recent approaches to multiobjective optimization has been given in Chapters 1–7.

Therefore, a multipurpose software that can be used in several fields and contests should include the most widely used and state-of-the-art methods using both *MCDM based* and *metaheuristics* approaches to multiobjective optimization. Obviously, some specific problems can be solved with software that contain only few mathematical programming based methods. Unfortunately, decision makers or analysts do not necessarily know the mathematical formulation of the problem at hand and the problem can change time after time. These are the main reasons why a really multipurpose software represents a viable solution.

12.2.3 Visualization, Post-processing and Statistical Charts

Visualization is the key in understanding the results coming out from large simulations in computational science and engineering. After a multiobjective optimization, we typically wish to visualize the entire set of results, rather than simply analyzing each single result. Understanding the results of a multiobjective process can be quite hard, particularly in higher dimensional spaces. Even though there are plenty of generic visualization tools (such as 2D and 3D scatter plots as explained in Chapter 8), an ad hoc visualization tool for Pareto optimal solutions is needed. Visualizing the objective space and the Pareto points is quite easy with 2 or 3 objectives. For a higher number of objectives, some more complex techniques should be implemented. For example, a common way of visualizing multivariate problems is using a parallel coordinates chart (Inselberg and Dimsdale, 1990). Some more complex techniques can be really useful with high dimensional spaces. Two important multi-dimensional visualizing tools are *self organizing maps* (SOMs) (Kohonen, 1982) that can really speed up the optimization phase as reported in (Obayashi and Sasaki, 2004) and *heatmaps* as described by Pryke *et al.* (2007).

These visualizing tools should be considered even as tools for *data management* and preliminary exploration. In multiobjective optimization, an initial explorative phase, called as a learning phase in Chapter 2, is important in order to determine the behavior and the main characteristics of the problem at hand. The principal aim of a preliminary exploration is to get the most relevant qualitative information from a problem making the smallest possible number of evaluations. This can be done by using a smart positioning of points in the space. This methodology provides a strong tool to design and analyze functions; it eliminates redundant observations and reduces the time and resources to make evaluations and experiments (Fig. 12.2).

Moreover, traditionally, visualization and statistical charts have been used as post-processing operations to visualize results. Anyhow, visualization can also be used to show the quality of the solutions. This kind of visualization can

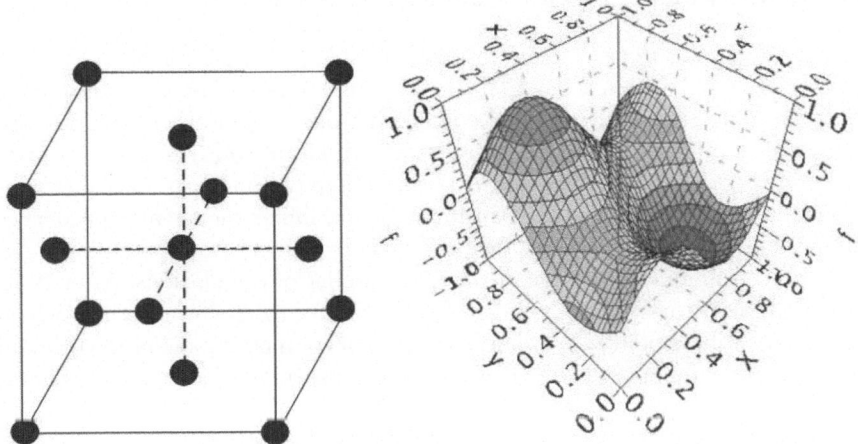

Fig. 12.2. Data management and preliminary exploration methods. A smart positioning of points in a 3-dimensional space (left) and a reliable meta-model (right)

give an important feedback during runtime and a good chart can support in deciding whether the optimization is going in the right direction or not. Based on visual feedback, the decision maker can stop and re-run the optimization using different parameters.

More detailed discussion of visualizing multiobjective optimization results is given in Chapters 8 and 9.

12.2.4 Decision Support Tool

In the absence of preference information, all Pareto optimal solutions can be regarded as equally desirable in the mathematical sense. Ranking a long list of Pareto optimal or nondominated alternatives is a difficult task especially when several solutions are available or when several conflicting goals are involved.

In several cases, more than one decision maker can be involved in selecting the best solution. In these cases, each person may even reflect different competencies and roles. Therefore, making coherent choices, with rational and transitive preferences, can be a very difficult task.

A decision support tool can assist the decision maker(s) in finding the best solution from among a set of reasonable alternatives. Moreover, a decision support tool can even allow the correct grouping of objectives into a single utility function by identifying possible relations between the objectives. A decision support tool can even guide the DM(s) in specifying preferences which leads to constructing a scalarized function that results to be coherent with the given preferences (see Chapters 1 and 2).

12.2.5 Meta-modeling and Validation of Models

In real life applications, it is not always possible to reduce the complexity of the problem and obtain a function that can be evaluated quickly. As reported in Chapter 11, in many practical engineering design and other scientific optimization problems, every single function evaluation can take hours or even days. In these cases, the time to run a single step of an algorithm makes running more than a few evaluations prohibitive and some other smart approaches are needed. In these situations, decision makers can turn to a preliminary exploration technique to perform a reduced number of calculations. After that, it is possible to use these well-distributed results to create a surface which interpolates these points. This surface represents a meta-model of the original problem and can be used to perform the optimization without costly computations. The use of mathematical and statistical tools to approximate, analyze and simulate complex real world systems is widely applied in many scientific domains. These kinds of interpolation and regression methodologies are now becoming common even for solving complex optimization problems where they are also known as *response surface methods* (RSMs). For example, RSMs are becoming very popular offering a surrogate model with a second generation of improvements in speed and accuracy in computer aided engineering. This approach allows direct optimization otherwise impossible.

Constructing a useful meta-model starting from a reduced number of real evaluations is not a trivial task. Mathematical and physical soundness, computational costs and prediction errors are not the only points to take into account when developing meta-models. Ergonomics of the software has to be considered in a wide sense. The users would like to grasp the general trends in the phenomena, especially when the behavior is nonlinear. Moreover, de-

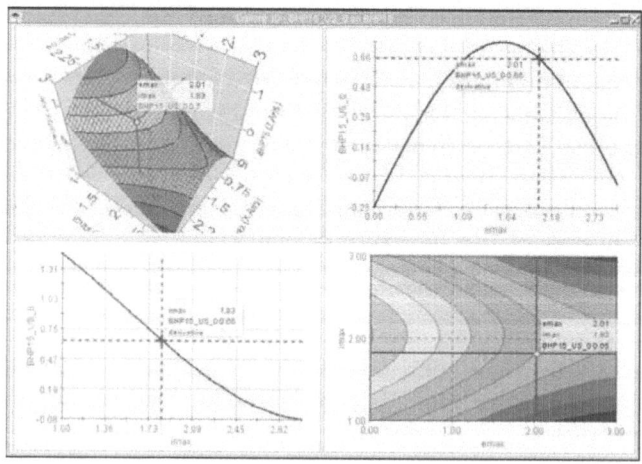

Fig. 12.3. A tool for meta-models: 3D-exploration

cision makers and engineers would like to re-use the experience accumulated, in order to spread the possible advantages to different projects. When using meta-models, the users should always keep in mind that this instrument allows a faster analysis than the complex models, but interpolation and extrapolation introduce a new element of error that must be managed carefully.

For these reasons, in the last years, different approximation strategies have been developed to provide inexpensive meta-models of the simulation models to substitute computationally expensive modules. As reported in Chapter 11, there is not a unique meta-model that is valid for any kind of situations. For this reason a good multiobjective optimization software should contain several different interpolation techniques such as, for example, neural networks, radial basis functions, kriging and Gaussian processes (see Chapter 11).

Once the meta-model has been constructed, it is really important to certify its fidelity. This is the reason why a tool for exploring (Fig. 12.3) and measuring the quality (Fig. 12.4) of meta-models in terms of statistical reliability would be appreciated together with all approximation strategies.

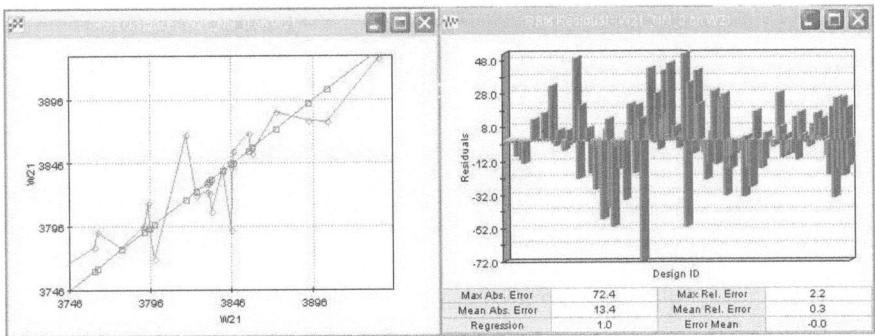

Fig. 12.4. Tools for measuring the quality of meta-models. Distance chart that points out the differences between real values and values calculated using the meta-model, (left) and residual chart (right). The residuals are the amounts which the meta-model has not been able to explain (approximation errors). These charts help to determine whether a meta-model is an acceptable representation of the original problem.

12.2.6 Robustness and Reliability

When dealing with uncertainty, conventional optimization techniques tend to "over-optimize", producing solutions that may perform well at the optimal point but have poor characteristics against the dispersion of design variables or environmental variables. As reported in Chapter 16, it is quite possible that the optimal solution will not be the most stable solution. For example, the function in Fig. 12.5 has a global optimum at point A, and a local optimum at point B. However, any small variations in the input parameters will cause the performance to drop off markedly around A. The performance of B may not be as good in absolute terms, but it is much more robust, since small changes in the input values do not cause drastic performance degradation.

For this reason, a tool that allows the user to perform a robust design analysis to check on the system's sensitivity to manufacturing tolerances or small changes in operating conditions can be really useful.

Fig. 12.5. Robustness of solutions

12.2.7 Parallelization

In those cases where each single function evaluation can be really time-consuming, *parallel computing* can be an important resource. In few words, *parallel computing* refers to evaluating simultaneously the same function on several processors in order to obtain results faster. The simple idea of parallelization is based on the fact that the optimization process usually can be divided into smaller steps. These smaller steps can be carried out simultaneously on multiple computers with some special coordination. The coordination can be done by a central manager that manages all the computers of the pool, collecting the requests and moving the computations accordingly to the current load of each computer. In this way, the whole optimization, or a part of it, can even be submitted to a queuing system and executed taking advantages of several different remote computers. This concept has been well described in Chapter 13.

This approach can really speed up the optimization because a parallel optimization algorithm can be much faster than a corresponding sequential algorithm. Parallel optimization methods can be developed by redesigning serial algorithms to make effective use of parallel hardware. Unfortunately, not all algorithms can be parallelized: for example, evolutionary algorithms can be parallelized more easily than many MCDM approaches.

12.2.8 Plug-in

A very good quality for a software is to be a completely open platform where anyone can contribute. The complexity of multiobjective optimization is becoming too big to design monolithic platforms. That is the reason why an open platform where scientists and software engineers can introduce their own methodologies and algorithms may represent a good solution.

Open platforms usually provide application programming interfaces (APIs) allowing third parties to create plug-ins that interact with the main application. In this kind of an open platform, the users can contribute with their own optimization techniques without any changes to the main platform. Using the APIs the users can introduce the optimization technique that is most appropriate for solving the problem at hand. An example of an open framework dedicated to the design of metaheuristics is Paradiseo-MOEO (Liefooghe *et al.*, 2008).

12.3 List and Description of Software

Several software cover one or at most two of the previously discussed properties. There are several multiobjective optimization tools available; and each tool can solve a specific or more generic problem. Some tools are more appropriate for constrained optimization, others may be suitable for unconstrained continuous problems, or *tailored* for solving some specific problems. Unfortunately, real world applications often include one or more difficulties which make these tools inapplicable. Most of the time, objective functions are highly nonlinear or even may not be given in a closed form in terms of the design variables.

In this chapter, we describe only *general purpose* software tools that have been built from the ground up to solve multiobjective optimization problems. Therefore, we describe software that can efficiently handle several goals and constraints at the same time, allowing to choose the best solution from a set of solutions that represent the best trade-offs.

In this section we identify and select a collection of free and commercial general purpose software. An expanded list of other software and interesting libraries is also provided in Section 12.5.

12.3.1 modeFRONTIER

modeFRONTIER is a multiobjective optimization and design environment, written to allow easy coupling to almost any computer aided engineering (CAE) tool. modeFRONTIER provides an environment which allows product engineers and designers to integrate their various CAE tools, such as CAD, finite element structural analysis and computational fluid dynamics (CFD) software. There are also direct interfaces for Excel, Matlab and Simulink. Using a variety of state-of-the-art multiobjective optimization techniques, ranging from gradient-based methods to genetic algorithms, the process or design of interest can be optimized by specifying objectives and defining variables which affect factors such as geometric shape and operating conditions. modeFRONTIER (Fig. 12.6) in effect becomes a wrapper around the CAE tool, performing the optimization.

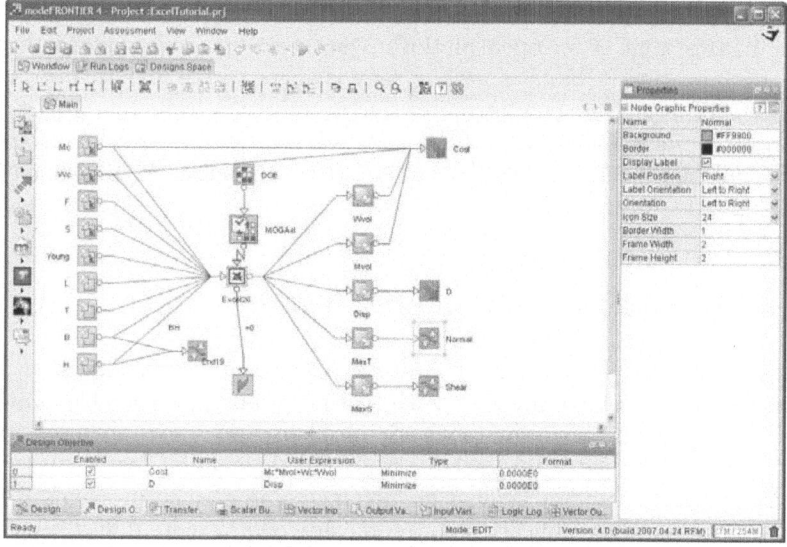

Fig. 12.6. A snapshot of modeFRONTER graphical users interface. In this panel the user can define the optimization problem.

modeFRONTIER includes a wide range of possible algorithms that can be selected for solving different problems. At present, the multiobjective methods available in modeFRONTIER are: Multiobjective Genetic Algorithm (MOGA), Adaptive Range MOGA, Multiobjective Simulated annealing (MOSA), Non-dominated Sorting Genetic Algorithm (NSGA-II) (Deb *et al.*, 2002), Multiobjective Game Theory, Evolutionary Strategies Methodologies and Normal Boundary Intersection (NBI) (Das and Dennis, 1998). Moreover, different algorithms can even be combined by the decision makers in order to obtain some hybrid approaches according to their preferences. A hybrid method can try to exploit the specific advantages of different approaches by combining more than one together. For example, it is possible to combine the robustness of a genetic algorithm together with the accuracy of a gradient-based method, using the former for initial screening and the latter for refinements. Whenever possible, modeFRONTIER's algorithms can be used in parallel, to run more than one evaluation at once and to take advantage of available queuing systems. modeFRONTIER is a commercial software developed by ESTECO; its website contains several examples of how to use the software for solving multiobjective optimization problems and decision making processes in engineering.

12.3.2 OPTIMUS

OPTIMUS is a world-leading process integration software, that bundles a collection of design exploration and numerical optimization methods. It allows users to build simulation workflows to automate their numerical simulation processes. These simulation workflows integrate one or more simulation codes and are executed by OPTIMUS - if possible - without user intervention. Once the simulation process is captured in an OPTIMUS workflow, users are able to explore their design space by modifying selected input parameters and hunt for new designs that are more reliable with better functional performance. All calculations are based on the integrated simulation tools that are part of the OPTIMUS workflows.

Methods available in OPTIMUS include:

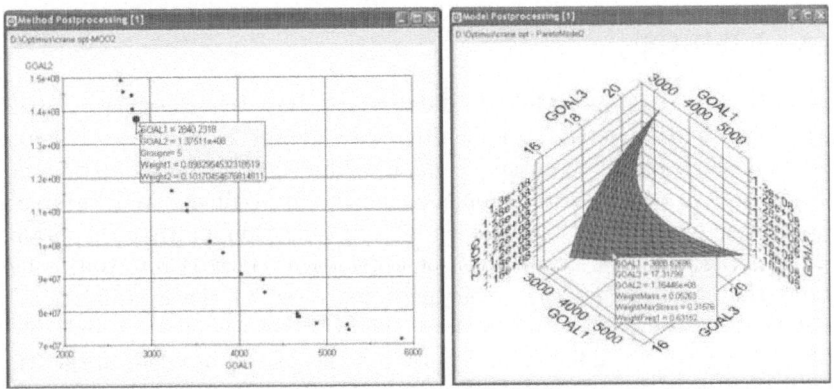

Fig. 12.7. Optimization Post-Processing with OPTIMUS. A scatter plot showing points on the Pareto optimal set for two objectives. When clicking on a point, the variable values are shown (left). A 3-dimensional plot of the Pareto optimal set (right).

- design space exploration, such as Design of Experiments (DOE) and Response Surface Modeling (RSM),
- numerical optimization, based on gradient-based local algorithms or genetic global algorithms, both for single or multiple objectives with continuous and/or discrete design variables and
- robustness and reliability engineering, including methods to assess and optimize the variability of design outputs based on variable design inputs.

The Multiobjective optimization methods include: non-dominated sorting evolutionary algorithm (NSEA and NSEA+, based on NSGA-II), normal-boundary intersection method as well as the weighting method, the weighted Chebyshev problem, the trade-off method, lexicographic ordering and the method of global criterion (see, for example, Chapters 1 and 2).

OPTIMUS is developed by Noesis Solutions, a subsidiary of LMS International, headquartered in Leuven, Belgium.

12.3.3 iSIGHT

Engineous' iSIGHT software integrates and manages the computer software required to execute simulation-based design processes, including commercial CAD/CAE software, internally developed programs, and Excel spreadsheets. iSIGHT drives toward optimal and reliable product designs using the library of advanced engineering tools. iSIGHT components include: optimization, design of experiments, Monte Carlo analysis, approximations. iSIGHT is continuously updated and contains state-of-the-art multiobjective genetic algorithm routines such as for example MOGA-NCGA and NSGA-II.

12.3.4 NIMBUS

NIMBUS (Miettinen and Mäkelä, 2006) is an interactive classification-based method for multiobjective optimization, see Chapter 2 (Miettinen, 1999). It is suitable for both differentiable and nondifferentiable multiobjective and single objective optimization problems subject to nonlinear and linear constraints with bounds for the variables. The classification information obtained from a decision maker is used to generate one to four Pareto optimal solutions that best reflect the preferences expressed in the classification. In practice, this means that one to four subproblems are created and solved with a solver appropriate for the characteristics of the problem in question. The subproblems include also reference point based subproblems from the reference point method, the STOM and the GUESS methods (see Chapter 2). WWW-NIMBUS is the implementation of the NIMBUS method operating via the Internet at http://nimbus.it.jyu.fi/. WWW-NIMBUS can be used free of charge for academic purposes. The implementation operating under Linux and MS-Windows operating systems is IND-NIMBUS (Miettinen, 2006). It is for sale and some information about it is available at http://ind-nimbus.it.jyu.fi/. An example of the user interface of IND-NIMBUS (classification window) is given in Figure 12.8.

In both the implementations, there are several underlying solvers available including a local proximal bundle method and a global genetic algorithm with different constraint-handling techniques. It is also possible to use a hybrid solver where the local solver is used after the global one. Both the implementations support the DM in comparing Pareto optimal solutions (s)he likes with graphical visualizations. Besides classification, the DM can also direct the search by asking for intermediate solutions between any two generated solutions.

12.3.5 PROMOIN

The interactive system PROMOIN (Caballero et al., 2002) has been designed as a decision aid tool for multiobjective problems, based on the use of interactive techniques. The current version of the software deals with linear problems, although a nonlinear version is currently under construction. The main idea underlying PROMOIN is the following. There are plenty of interactive techniques available in the literature. They differ in several aspects: the kind of problem handled, the type of final solution, the inner solution process and the information asked from the decision maker. The last issue is a key factor for the success of an interactive method. If the decision maker does not feel comfortable with the information (s)he has to provide, the method will hardly succeed in finding her/his most preferred solution. Therefore, the interactive technique should be chosen according to the decision maker's wishes

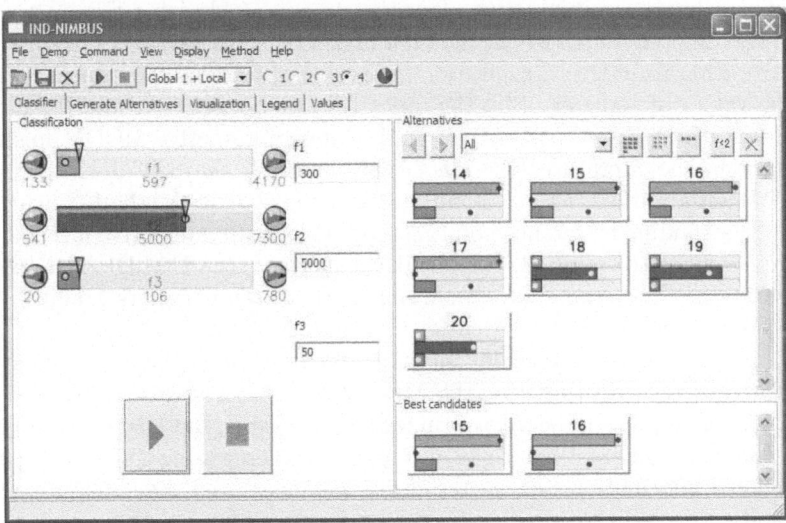

Fig. 12.8. IND-NIMBUS: graphical users interface

regarding how (s)he wants to give the information. Furthermore, the kind of information that the decision maker wishes to give may vary during the solution process, due to the fact that (s)he progressively learns about the problem, and gets a more accurate picture of the problem. The main interactive procedures have been incorporated into the system, including tradeoff based methods such as the Geoffrion-Dyer-Feinberg (GDF) method, the sequential proxy optimization (SPOT) method, the interactive surrogate worth trade-off method (ISWT) and other methods such as the Tchebycheff method and the Zionts-Wallenius method. Moreover, it contains reference point based methods such as STEM, STOM and the reference point method. (For further details of the methods, see, e.g., Chapter 2.) The method can be chosen according to the decision maker's wishes. On the other hand, the system offers the possibility to change between methods any time during the solution process. The program has been implemented under Windows environment, with the aim of providing the user with a friendly interface.

12.3.6 MKO-2

The interactive software system MKO-2 is a generalized multiobjective decision support system (Staykov, 2006; Vassilev *et al.*, 2006). It has been designed to support solving linear and linear integer multiobjective optimization problems. The system implements an innovative generalized classification-based interactive algorithm for multiobjective optimization with variable scalarizations and parameterizations, applicable for different types of problems (i.e., linear, nonlinear and mixed variables). The MKO-2 system will be extended to handle nonlinear multiobjective optimization problems as well. The incorporated generalized interactive algorithm is applicable for different ways of defining the preferences by the decision maker, such as weighting factors (priorities), aspiration levels, aspiration intervals, aspiration directions of change in the values of some or of all the objective functions, etc. Using the MKO-2 system, the decision makers can apply twelve interactive MCDM methods existing in the literature (the Chebyshev method, the STEM method, the STOM method, the reference point method, the GUESS Method (Buchanan, 1997), the modified reference point method, the visual interactive method, the reference direction method, the NIMBUS method, the DALDI method, the weighting method, and the ϵ-constraint method; see Chapters 1 and 2) and different strategies in the search for new Pareto optimal solutions, not only with the help of one particular method, but combining different interactive MCDM methods. In this way, the MKO-2 software system can be used not only for solving multiobjective optimization problems, but also for comparing and analyzing different solutions of a given problem, using different types of preference information, set by the decision maker, and different interactive methods. The MKO-2 software system operates under MS Windows operating system. The graphical user interface of the system enables decision makers with different degrees of qualification, referring to the methods and software tools, to operate easily with the system. MKO-2 decision support system can be used both for education and for solving real-life problems. It can be used free for academic purposes under a certain bilateral agreement.

12.3.7 Pareto Front Viewer

The software "Pareto Front Viewer" (PFV) provides interactive visualization of the Pareto frontier for multiobjective problems in the case of two to eight objective functions. It is assumed that an approximation of the Pareto frontier has already been constructed in the form of a finite list of objective vectors. The method for constructing the objective vectors plays no role. Thus, the PFV software can be combined with any Pareto frontier approximation technique. The method is based on the visualization of bi-objective slices of the Edgeworth-Pareto Hull (EPH) of the objective vectors, that is, the union of the domination cones with vertices located in the objective vectors of the approximation. The objective tradeoffs for any three objectives that are specified by a decision maker are visualized in the form of decision maps, which are collections of the overlaid bi-objective slices of the EPH. The influence of the other objectives can be studied by moving the sliders of the related scroll-bars. By this, the user is informed on the objective tradeoffs and is supported in the process of selecting the preferred Pareto optimal decision vector, which is based on the direct identification of the feasible goal at the computer screen. The software PFV was coded for the platforms MS Windows 98/NT/2000/XP. A demonstration version for up to five criteria and 500 criterion points can be downloaded for free from http://www.ccas.ru/mmes/mmeda/soft/third.htm.

12.3.8 Reasonable Goals Method for Databases

Reasonable Goals Method for Databases (RGDB) is the Web application server that supports selecting a small number of alternatives from large tables through Internet. The application server applies the Reasonable Goals Method, that is, visualization of the Pareto frontier of the envelope (convex hull) of the objective vectors related to the alternatives in the case of two to eight objectives. First, the user provides the table with alternatives to the server. Then, the server approximates the Edgeworth-Pareto Hull (EPH) of the convex hull of the objective vectors. Then, the user's computer receives the applet that supports interactive visualization of the Pareto frontier based on the visualization of bi-objective slices of the EPH. The user explores objective tradeoffs for any three selected objectives. The influence of the other objectives can be studied by moving the sliders of the related scroll-bars. Then, the reasonable goal is transmitted to the server, which selects a small number of the Pareto optimal decisions and transmits them back to the user. The Web application server coded in C++ can be used with the help of standard browsers. A demonstration version (5 objectives and up to 500 alternatives), can be found at http://www.ccas.ru/mmes/mmeda/rgdb/index.htm.

12.3.9 ParadisEO and GUIMOO

ParadisEO is a white-box object oriented generic framework dedicated to the flexible design of evolutionary multiobjective algorithms. This paradigm-free software aims to provide a set of classes allowing to ease and speed up the development of computationally efficient programs. It is based on a clear conceptual distinction between the solution methods and the multiobjective problems they are intended to solve. This separation confers a maximum design and code reuse. ParadisEO provides a broad range of archive-related features (such as elitism or performance metrics) and the most common Pareto-based fitness assignment strategies such as MOGA, NSGA, SPEA and Indicator-Based Evolutionary Algorithm (IBEA) (Zitzler and Künzli, 2004). Furthermore, parallel and distributed models as well as hybridization mechanisms can be applied to an algorithm designed within ParadisEO. This tool is developed by INRIA that, in addition, provides GUIMOO, a platform-independent free software dedicated to analysis of results of multiobjective problems. GUIMOO allows visualization of approximative Pareto frontiers and contains metrics for quantitative and qualitative performance evaluations.

12.4 Summary Table on Optimization Software

Table 12.1 summarizes the main characteristics of all the multiobjective optimization software tools described in the previous section. This table lists only tools that have been developed exclusively for multiobjective optimization. Many other software systems can be used for optimization or visualization of data but we limit our study to the tools that look for the Pareto frontier and have visualization capabilities dedicated to this type of results.

The table has 11 columns that can be read as follows:

1. **Software and Developers**: name of the software and information about companies or institutions taking care of the development. Whenever possible, web-pages or email contacts are reported.
2. **Platforms**: platforms where the software can run
3. **An easy-to-use GUI**
4. **EMO**: this column contains a brief description of the evolutionary multiobjective methods available in the software.
5. **MCDM**: This column contains a brief description of the MCDM methods available in the software.
6. **Rob.**: This column contains the symbol X if and only if the software contains at least a method to establish robustness of solutions.
7. **Meta**: This column reports whether or not the software contains one or more methods for meta-modeling.
8. **Vis**: visualization tool and statistical analysis. This column indicates whether or not the software contains one or more methods for visualizing the Pareto frontier and/or other results coming out from the optimization phase.

9. **Plug**: this column contains the symbol X if the software can be considered as an open platform where the users can add their own optimization methods as external plug-ins.
10. **//**: this symbol stands for **Parallelization**. Hence this column contains the symbol X if the software supports parallel computation and if the optimization algorithms can deal with queuing systems.
11. **License**: License type, commercial, free or academic.

Table 12.1. Summary of the main characteristics of the multiobjective optimization software described

Software and Developers	Platforms	GUI	EMO	MCDM	Rob.	Meta	Vis.	Plug	License	//
modeFRONTIER ESTECO www.esteco.com	All	×	MOGA-II, NSGA-II, ARMOGA, MOSA, Game Theory, MO evolutionary strategies	NBI, weighting method, reference point method	×	×	×	×	Commercial	×
OPTIMUS NOESIS Solutions www.noesissolutions.com	Windows and Linux	×	Non-Dominated Sorting Evolutionary Algorithms (NSEA and NSEA+)	NBI and 7 other methods	×	×	×	×	Commercial	×
iSIGHT Engineous Software www.engineous.com	All	×	NCGA, NSGA-II	LSGRG, MMFD, MOST, Stress Ratio, MIGA, Pointer	×	×	×	×	Commercial	×
WWW-NIMBUS, University of Jyväskylä, nimbus.it.jyu.fi	Web	×		NIMBUS, reference point method, GUESS, STOM			×		Free for Academic	
IND-NIMBUS, University of Jyväskylä, ind-nimbus.it.jyu.fi	Windows and Linux	×		NIMBUS, reference point method, GUESS, STOM			×		Commercial	
PROMOIN University of Malaga	Windows	×		SPOT, ISWT, STEM, STOM					Free	
MKO-2 Institute of Information Technologies - Bulgarian Academy of Sciences, Department of Decision Support Systems www.iit.bas.bg	Windows	×		Chebyshev, STEM and STOM methods, GUESS, reference point method, NIMBUS, DALDI, weighting method			×		Free for Academic	
Pareto Front Viewer www.ccas.ru/mmes/mmeda/soft	Windows	×					×	×	Commercial, Free up to 500 points	
Reasonable Goals Methods for Databases www.ccas.ru/mmes/mmeda/rgdb/	Web	×					×	×	Free	
ParadisEO Inria paradiseo.gforge.inria.fr	All	×	NSGA-II, IBEA				×	×	Commercial	×

12.5 List of Available Libraries

Much of the evolutionary multiobjective optimization studies use computer codes which are freely downloadable. Some of them are the NSGA-II code in C (http://www.iitk.ac.in/kangal/soft.htm), SPEA2 and other EMO codes in C++. An important platform containing a set of ready-to-go multiobjective optimization methods is *PISA* (Bleuler *et al.*, 2003). PISA consists of two parts: a set of optimization problems (variators) and a set of optimization algorithms (selectors). The selectors are state-of-the-art evolutionary multiobjective optimization methods (see Chapter 3). The user can write and submit a new module in the platform. All modules available can be used for academic purposes without a fee. Each module specifies its own licensing policy. PISA itself is a copyright of the Swiss Federal Institute of Technology, Computer Engineering and Networks Laboratory.

There are several other platforms that help the development process of evolutionary multiobjective optimizers, for example, *Open BEAGLE* which is an Object-Oriented software environment enabling the implementation of almost any kind of evolutionary algorithm, such as genetic algorithms and genetic programming, MOMHLib++, MOEA (Tan *et al.*, 2000).

Another important package is *DAKOTA*, a multilevel parallel object-oriented framework for design optimization, parameter estimation, uncertainty quantification, and sensitivity analysis developed by the Sandia National Laboratories.

For MCDM based multiobjective methods there are several items available in the Internet such as, for example, PROTASS developed by Rafal Cytrycki for linear multiobjective problems and available for download at *http://www.ekspert.szczecin.pl/protass/en*.

Even though designed for discrete problems, let us still mention one of the most famous decision support systems, the Analytic Hierarchy Process (AHP). Designed to reflect the way people actually think, AHP was developed in the 1970's (Saaty, 1996). AHP is now included into a commercial software called *Expert Choice*. This software is intuitive, graphically based and structured in a user-friendly fashion. With this tool, decision makers are able to drill down to their level of expertise, and apply judgments to the objectives to achieving their goals.

Finally, we can mention the Decisionarium project (http://www.decision-arium.net/) which focuses on the development of web based tools for interactive multicriteria decision support for individual decision making, for group collaboration and negotiation as well as for interaction and surveys over the web (mostly for discrete problems).

There are probably several other packages that should be listed in this paragraph. Unfortunately, most of the times, software products are implemented for academic testing purposes and are usually neither maintained nor advertised. Moreover, there exist several tools for solving single objective optimization problems that may contain some possibilities for multiobjective optimization. Those are deliberately excluded from this final list because this chapter wants to concentrate only on nonlinear multiobjective optimization software developed in the last decade.

12.6 Conclusions

The information collected and presented in this chapter is just a snapshot of the multiobjective optimization tools available. Setting up, installing and testing all these software packages on a number of different platforms has been a quite demanding job. It is obviously impossible to say which one is the best amongst all the listed software. A high number issues should be taken into account for evaluating a software such as ease of use, completeness, configurability, robustness, efficiency, user support and so on.

Software for multiobjective optimization and more complex integrated environments such as "Process Integration and Design Optimization" (PIDO) or "Problem Solving Environment" (PSE) have become popular in the last years and several new packages are probably coming on the market. The importance of multiobjective optimization for the commercial world can be readily seen by the fact that most of the industrial companies now support one or more of the available packages. There is clear evidence that both commercial and research/academic communities are becoming increasingly interested in multiobjective optimization software.

The data summarized in Table 12.1 let us conclude that commercial software are usually more complete and more close to an advisable multiobjective optimization software than a free or open sources tool. Anyhow, there are even some good libraries that are well-qualified starting points for people approaching multiobjective optimization.

References

Bleuler, S., Laumanns, M., Thiele, L., Zitzler, E.: PISA – A Platform and Programming Language Independent Interface for Search Algorithms. In: Fonseca, C.M., Fleming, P.J., Zitzler, E., Deb, K., Thiele, L. (eds.) EMO 2003. LNCS, vol. 2632, pp. 494–508. Springer, Heidelberg (2003)

Buchanan, J.T.: A Naiive Approach for Solving MCDM Problems: The GUESS Method. The Journal of the Operational Research Society 48(2), 202–206 (1997)

Caballero, R., Luque, M., Molina, J., Ruiz, F.: Promoin: An interactive system for multiobjective programming. Information Technologies and Decision Making 1, 635–656 (2002)

Das, I., Dennis, J.E.: Normal-boundary intersection: a new method for generating Pareto optimal points in multicriteria optimization problems. SIAM Journal on Optimization 8(3), 631–657 (1998)

Deb, K., Pratap, A., Agrawal, S., Meyarivan, T.: A fast and elitist multi-objective genetic algorithm: NSGA-II. IEEE Transactions on Evolutionary Computation 6(2), 181–197 (2002)

Gallopoulos, E., Houstis, E., Rice, J.R.: Future Research Directions in Problem Solving Environments for Computational Science (1991)

Houstis, E., Gallopoulos, E., Bramley, R., Rice, J.: Problem-Solving Environments for Computational Science. IEEE Computational Science and Engineering 4(3), 18–21 (1997)

Inselberg, A., Dimsdale, B.: Parallel Coordinates: a Tool for Visualizing Multi-Dimensional Geometry. In: VIS '90: Proceedings of the 1st conference on Visualization '90, San Francisco, California, pp. 361–378. IEEE Computer Society Press, Los Alamitos (1990)

Kaliszewski, I.: Out of the mist – Towards decision-maker-friendly multiple criteria decision making support. European Journal of Operational Research 158, 293–307 (2004)

Kohonen, T.: Self-organized formation of topologically correct feature maps. Biological Cybernetics 43, 59–69 (1982)

Liefooghe, A., Basseur, M., Jourdan, L., Talbi, E.-G.: ParadisEO-MOEO: A Framework for Evolutionary Multi-objective Optimization (2008)

Miettinen, K.: Nonlinear Multiobjective Optimization. Kluwer Academic Publishers, Boston (1999)

Miettinen, K.: IND-NIMBUS for demanding interactive multiobjective optimization. In: Trzaskalik, T. (ed.) Multiple Criteria Decision Making '05, pp. 137–150. Karol Adamiecki University of Economics, Katowice (2006)

Miettinen, K., Mäkelä, M.M.: Synchronous approach in interactive multiobjective optimization. European Journal of Operational Research 170(3), 909–922 (2006)

Obayashi, S., Sasaki, D.: Multi-objective optimization for aerodynamic designs by using armogas. In: HPCASIA '04: Proceedings of the High Performance Computing and Grid in Asia Pacific Region, Seventh International Conference on (HP-CAsia'04), Washington, DC, USA, pp. 396–403. IEEE Computer Society Press, Los Alamitos (2004)

Pryke, A., Mostaghim, S., Nazemi, A.: Heatmap Visualization of Population Based Multi Objective Algorithms. In: Obayashi, S., Deb, K., Poloni, C., Hiroyasu, T., Murata, T. (eds.) EMO 2007. LNCS, vol. 4403, pp. 361–375. Springer, Heidelberg (2007)

Saaty, T.L.: Multicriteria Decision Making: The Analytic Hierarchy Process; Planning, Priority Setting, Resource Allocation, 2nd edn. Analytic Hierarchy Process Series. RWS Publications, Pittsburgh (1996)

Staykov, B.: Multiobjective optimization software system. In: Problems of Engineering Cybernetics and Robotics, vol. 57, pp. 21–30 (2006)

Tan, K.C., Lee, T.H., Khoo, D., Khor, E.F.: MOEA Toolbox for Computer-Aided Multi-Objective Optimization. In: 2000 Congress on Evolutionary Computation, July 2000, vol. 1, pp. 38–45. IEEE Computer Society Press, Piscataway (2000)

Vassilev, V., Vassileva, M., Staykov, B., Miettinen, K.: Generalized multicriteria decision support systems. In: Proceedings of the International Workshop on Semantic Web and Knowledge Technologies Applications, 12th International Conference AIMSA, pp. 16–30 (2006)

Weistroffer, H.R., Smith, C.H., Narula, S.C.: Multiple criteria decision support software. In: Figueira, J., Greco, S., Ehrgott, M. (eds.) Multiple Criteria Decision Analysis: State of the Art Surveys, pp. 989–1018. Springer, Heidelberg (2005)

Zitzler, E., Künzli, S.: Indicator-based selection in multiobjective search. In: Yao, X., Burke, E.K., Lozano, J.A., Smith, J., Merelo-Guervós, J.J., Bullinaria, J.A., Rowe, J.E., Tiňo, P., Kabán, A., Schwefel, H.-P. (eds.) PPSN 2004. LNCS, vol. 3242, pp. 832–842. Springer, Heidelberg (2004)

13

Parallel Approaches for Multiobjective Optimization

El-Ghazali Talbi[1], Sanaz Mostaghim[2], Tatsuya Okabe[3], Hisao Ishibuchi[4], Günter Rudolph[5], and Carlos A. Coello Coello[6]

[1] Laboratoire d'Informatique Fondamentale de Lille
Université des Sciences et Technologies de Lille
59655 - Villeneuve d'Ascq cedex, France
talbi@lifl.fr
[2] Institute AIFB
University of Karlsruhe
76128 Karlsruhe, Germany
mostaghim@aifb.uni-karlsruhe.de
[3] Honda Research Institute Japan Co., Ltd.
8-1 Honcho, Wako-City, Saitama, 351-0188, Japan
okabe@jp.honda-ri.com
[4] Department of Computer Science and Intelligent Systems
Osaka Prefecture University
Osaka 599-8531, Japan
hisaoi@cs.osakafu-u.ac.jp
[5] Computational Intelligence Research Group
Chair of Algorithm Engineering (LS XI)
Department of Computer Science, University of Dortmund
44227 Dortmund, Germany Guenter.Rudolphuni-dortmund.de
[6] CINVESTAV-IPN (Evolutionary Computation Group)
Depto. de Computación, Av. IPN No 2508
Col. San Pedro Zacatenco, México, D.F., 07360 MEXICO
ccoello@cs.cinvestav.mx

Abstract. This chapter presents a general overview of parallel approaches for multiobjective optimization. For this purpose, we propose a taxonomy for parallel metaheuristics and exact methods. This chapter covers the design aspect of the algorithms as well as the implementation aspects on different parallel and distributed architectures.

Key words: Parallel algorithms, Parallel metaheuristics, Parallel multiobjective optimization, Parallel exact optimization

Reviewed by: Heinrich Braun, SAP AG, Walldorf, Germany
Jürgen Branke, University of Karlsruhe, Germany

J. Branke et al. (Eds.): Multiobjective Optimization, LNCS 5252, pp. 349–372, 2008.

13.1 Introduction

Multiobjective optimization problems are often NP-hard, complex and CPU time consuming. Exact methods can be used to find the exact Pareto front (or a subset of the front), but they are impractical to solve large problems as they are time and memory consuming. On the other hand, metaheuristics provide the approximated Pareto fronts in a reasonable time. However, they also remain time-consuming for solving large problems.

Parallel and distributed computing are used in the design and implementation of multiobjective optimization algorithms to speedup the search. Also, they are used to improve the precision of the used mathematical models, the quality of the obtained Pareto fronts, the robustness of the obtained solutions, and to solve large scale problems.

In this chapter, we present the main parallel models for metaheuristics and exact methods from the algorithmic design point of view. We consider continuous and combinatorial optimization problems as parallel models are suited either for combinatorial or continuous optimization problems. From the implementation point of view, we concentrate on the parallelization of multiobjective optimization algorithms on general-purpose parallel and distributed architectures as these architectures are the most widespread computation platforms. The rapid evolution of technology in terms of processors (multi-core), networks (Infiniband), and architectures (GRIDs, clusters) make the parallelization very popular.

Different architectural criteria which affect the efficiency of the implementation are shared memory / distributed memory, homogeneous / heterogeneous, dedicated / non dedicated, local network / large network. Indeed, these criteria have a strong impact on the deployment techniques such as load balancing and fault-tolerance. Depending on the type of the used architecture, different parallel and distributed programming environments such as message passing (PVM, MPI), shared memory (multi-threading, OpenMP), high throughput computing (Condor), and Grid computing (Globus) can be used.

This chapter is organized as follows. In the next section, we present the parallel models for designing metaheuristics for MOPs. In Section 3, we review the parallel models for exact algorithms. Section 4 deals with the implementation issues for metaheuristics and exact algorithms. Finally, we conclude the paper and discuss several lines for future research in Section 5.

13.2 Parallel Models for Metaheuristics

Different parallel models for metaheuristics have been proposed in the literature. They follow three major hierarchical models such as:

- Self-contained parallel cooperation (between different algorithms)

- Problem independent intra-algorithm parallelization
- Problem dependent intra-algorithm parallelization

where the last two models do not alter the behavior of the algorithms and therefore are generally used to speedup the search.

13.2.1 Level 1: Self-Contained Parallel Cooperation

Basic Concept

This group of parallel algorithms containing *the Island model* is used for parallel systems with very limited communication. In the island model, every processor runs an independent MOEA using a separate (sub)population. The processors might cooperate by regularly exchanging migrants which are good individuals in their subpopulations. These algorithms are also suitable for problems with large search spaces where a large population is being required. The large population is then being divided into several subpopulations.

In every processor, an optimization algorithm with selection and recombination operators is being carried out on a subpopulation. As written by Coello *et al.* (2002), there are several methods (also based on the island model) in the literature which we can categorize into two main groups. (1) Cooperating Subpopulations: These methods are based on partitioning the objective/search space. In this group, the population is divided into subpopulations. The number of subpopulations and the way the population is divided are the two key issues. (2) Multi-start Approach: Here, each processor independently runs an optimization algorithm.

Group 1: Cooperating Subpopulations

These algorithms attempt to distribute the task of finding the entire Pareto-optimal front among participating processors. By this way, each processor is destined to find a particular portion of the Pareto-optimal front. In fact, the population of a MOEA is divided into a number of independent and separate subpopulations resulting in several small separate MOEAs executing simultaneously which have the responsibility to find the (Pareto-)optimal solutions in their own search region. Each MOEA could have different operators, parameter values, as well as a different structure. In this model, some individuals within some particular subpopulations occasionally migrate to another one.

Generally, when distributing the task among the processors, the overlap between the solutions of two processors should be as small as possible. Also, the distribution algorithm must be scalable. Usually, the designer or a computational resource (master node) is responsible for distributing and dividing the population or the objective/search space.

In the literature, the very first approaches based on the island model do not directly divide the objective/search space into different regions, but implicitly result in the division as studied by Baita *et al.* (1995); Poloni (1995); Hiroyasu *et al.* (2000); Jozefowiez *et al.* (2002); Deb *et al.* (2003); Xiao and Armstrong (2003); de Toro Negro *et al.* (2004).

Baita *et al.* (1995) and Poloni (1995) use a local geographic selection scheme in which individuals are placed on a toroidal grid with one individual per grid intersection point. Hiroyasu *et al.* (2000) proposed the Divided Range Multi-Objective Genetic Algorithm (DRMOGA) in which the global population is sorted according to one of the objective functions (which is changed after a number of generations). Then, the population is divided into equally-sized sub-populations. Each of these sub-populations is allocated to a different processor in which a serial MOEA is applied. After a certain number of generations, the sub-populations are gathered and the process is repeated, but this time using some other objective function as the sorting criterion. The main goal of this approach is to focus the search effort of the population on different regions of the objective space. However, in this approach we cannot guarantee that the sub-populations remain in their assigned region. A similar approach is followed by de Toro Negro *et al.* (2004). Deb *et al.* (2003) use a modified domination criterion for assigning a specific region of the objective space to a processor.

Zhu and Leung (2002); Zhu (2002) proposed the Asynchronous Self-Adjustable Island Genetic Algorithm (aSAIGA) in which, rather than migrating a set of individuals, the islands exchange information related to their current explored region. Based on the information coming from other islands, a **self-adjusting** operation modifies the fitness of the individuals in the island to prevent two islands from exploring the same region. In a similar way to DRMOGA, this approach cannot guarantee that the sub-populations move tightly together throughout the search space, hence the information about the explored region may be meaningless.

Xiao and Armstrong (2003) use a generalized version of VEGA (Vector Evaluated Genetic Algorithm, Schaffer (1985)) to divide the population into subpopulations.

López-Jaimes and Coello (2005) proposed an approach called Multiple Resolution Multi-Objective Genetic Algorithm (MRMOGA), whose main idea is to encode the solutions using a different resolution in each island (heterogeneous nodes are assumed). Then, the variable decision space is divided into hierarchical levels with well-defined overlaps. Evidently, migration is only allowed in one direction (from low resolution to high resolution islands). A conversion scheme is required when migrating individuals, so that the resolution is properly adjusted. This approach uses an external population (or elitist archive) and the migration strategy considers such a population as well. The approach also uses a strategy to detect nominal convergence of the islands in order to increase their initial resolution. The rationale behind this approach is that the true Pareto front can be reached faster using this change of resolution in the islands, because the search space of the low resolution islands is proportionally smaller and, therefore, convergence is faster. This issue was originally identified by Parmee and Vekeria (1997) when they used an injection island strategy to solve a single-objective engineering optimization problem.

In the method proposed by Jozefowiez et al. (2005), each processor has its own population which is defined in the entire search space. The defined communication network between the processors is a ring where the processors send half of their populations to their two neighbors. The computations of a given processor do not begin until it has received the information from its two neighbors.

The first approach on dividing the objective space into several regions is introduced by Branke et al. (2004). This technique called Cone Separation divides the objective space into subspaces and assigns each subspace to one processor. They, however, do not divide the search space and therefore each processor explores the entire search space. The solutions outside the defined region in the objective space of each processor are considered as infeasible (although in reality they are feasible). Those infeasible solutions are migrated to other processors. This algorithm is scalable and there is no overlap between the solutions obtained by each processor. The so-called hypergraph has been used by Mehnen et al. (2004) to structure the populations in MOEAs and then applied it to parallel MOEAs. Streichert et al. (2005) refined the idea of the Cone Separation technique by using a clustering method for finding the right partitions in the objective space.

More recently, Bui et al. (2006) study an approach for dividing the search space. In their approach, they select a random (hyper-)sphere as the search space for every single processor. Then every processor runs a MOEA inside its defined region. The spheres are evaluated in terms of their solutions and their positions are being improved in the search space for the next iteration(s). This has been done beside other techniques like racing model using Multi-Objective Particle Swarm Optimization.

All of these methods work on processors which have similar properties in other words homogeneous systems. Mostaghim et al. (2007) study an approach which works asynchronously and is thus particularly suitable for heterogeneous computer clusters as occurring, e.g., in modern grid computing platforms.

Group 2: Multi-start Approach

This model introduced by Mezmaz et al. (2006) consists of several parallel local search algorithms which are independently run on several (also heterogeneous) processors. The basic idea of using such a model is that running several optimization algorithms with different initial seeds is more valuable than executing only one single run for a very long time. This is of particular importance for local search algorithms. Jozefowiez et al. (2007) use a parallel hybrid approach combining the multi-start model and the self contained parallel cooperation model. The Pareto front found by a parallel EA is partitioned and serves as a guide to multiple tabu search tasks.

Synchronous versus Asynchronous

Usually in MOEAs a set of non-dominated solution are found as the result of the optimization. In case of using the cooperating subpopulation model every single processor will cooperate to obtain one part of the non-dominated set. In elitist MOEAs like SPEA2 the non-dominated solutions are usually stored in an archive. In other algorithms like the NSGA-II, there is no archive as the main population contains the non-dominated solutions. In any of these cases, the set of non-dominated solutions must be updated as soon as a processor finishes its optimization task.

Apart from the way the subpopulations are created, we must ensure that the processors obtain good convergence and diversity of solutions. For this in some cases each processor can run the optimization several times as shown in Algorithm 3. Algorithm 3 is basically being used on a set of homogeneous

---

Algorithm 3 Synchronous cooperating subpopulations

Initiate subpopulations
repeat
  Wait for results of *all* processors
  Migration of individuals if any
  Update archive if any
until Termination condition met
Return archive

---

systems. The termination criterion could be a fixed number of runs on each processor (in many cases one iteration has been selected). In this algorithm "Initiate Subpopulation" deals with dividing the objective/search space in order to build the subpopulations. "Migration of individuals" refers to methods in which processors communicate with each other and exchange some of their individuals as migrants.

In reality, we typically deal with heterogeneous systems where this Algorithm is not suitable. In heterogeneous systems, there are different computing resources including very fast and very slow processors. According to Algorithm 3, all of the processors have to wait for the slowest one. In order to deal with these systems, Algorithm 4 is proposed. In this algorithm, whenever a processor returns its results, they can be immediately integrated into the archive. Based on the quality of the obtained archive a suitable new subpopulation can be selected for that processor. This makes the approach particularly suitable for heterogeneous computer clusters such as Grids, where very fast processors are used along with rather slow ones. It is not necessary to wait for the slowest processor to return its results. Here the processors can indirectly communicate through the archive. We must notice that migration is not straightforward as before.

Algorithm 4 Asynchronous cooperating subpopulations (Heterogeneous systems)

Initiate an empty archive
Initiate subpopulations
repeat
  if A processor returns results **then**
    Update archive
    Determine its new subpopulation
  end if
until Termination condition met
Return archive

Mostaghim *et al.* (2007) integrate a hypervolume based method into the optimization routine in every processor. For initializing a subpopulation, a guide is selected according to its marginal hypervolume. The hypervolume is the area dominated by all solutions stored in the archive (Chapter 14). The *marginal* hypervolume of a solution is the area dominated by the solution that is not dominated by any other solution. The guide is the solution from the archive which has not been selected before and which has the largest marginal hypervolume. After selecting the guide, a Multi-Objective Particle Swarm Optimization method is used to move its subpopulation toward the guide, hence searching the area around the guide.

13.2.2 Level 2: Problem Independent Parallel Intra-algorithm

Most of the metaheuristics are iterative methods. In this model, we will parallelize a single iteration of the algorithm. Our concern in this model are only search mechanisms which are problem independent such as the evaluation of the neighborhood in local search and the reproduction mechanism in evolutionary algorithms.

Basic Concept

During an optimization, we have to evaluate fitness values of candidates of solution (individuals). If we use benchmark problems/simple applications to evaluate fitness values, the calculation time is negligible. However, a real application sometimes needs huge computational time, e.g., using computational fluid dynamics (CFD), electro-magnetic field analysis, finite element method (FEM) etc. See Okabe *et al.* (2003); Okabe (2004). In this situation, total calculation time becomes too huge and it is generally impossible to obtain a certain result in a reasonable calculation time.

Let assume that the number of individuals, the maximum number of generations, the number of objectives and the calculation time of *i*th objective function are n, g, k and t_i, respectively. The total calculation time in evolutionary multiobjective optimization, denote T, can be easily calculated as follows:

$$T = gn \sum_{i=1}^{k} t_i + g\alpha = gnt + g\alpha, \qquad (13.1)$$

where, α is the time that genetic operator needs in one generation and $t = \sum_{i=1}^{k} t_i$. If $n = 100$, $g = 500$, $\alpha \approx 0$ and $t = 3$ ($days$) which is a certain real example using CFD solver, the total calculation time is about 411 years! Nevertheless, the problem should be optimized.

To tackle this problem, a parallel calculation is often used. The basic idea is shown in Fig. 13.1. This type of parallelization is called *master-slave model* or *global parallelization*, e.g., Branke *et al.* (2004); Cantu-Paz (1997a); Veldhuizen *et al.* (2003). The optimizer running on a master node carries out an overall calculation including initialization, crossover, mutation and selection except for evaluation of individuals. In evolutionary computation, several individuals exist in a population to be evaluated. However, the evaluation of each individual is completely independent from other evaluations. Therefore, in Fig. 13.1, each evaluation will be done on different slave nodes. The master node generates a population, e.g. car designs. Then, the master node distributes individuals to several independent slave nodes. In the slave nodes, the evaluations of individual, e.g. car design, are carried out simultaneously. Thereafter, the fitness values are gathered by the master node. Based on the fitness values, the master node selects promising individuals and generates new individuals by genetic operators. This flow is repeated until a given termination condition is met. Since several time-consuming evaluations are carried out at the same time, the total calculation time is dramatically reduced.

Calculation Time

Now, we will consider when we should parallelize a calculation using master-slave model. Assume that the total calculation time without/with parallelization and the number of nodes are T^{wo}, T^{w} and N, respectively. As an example, $n = N$ (the number of available nodes is the same as the number of individuals) is also assumed. Since a master node can be used not only for managing total calculation but also for fitness evaluation, a master node also contributes to fitness evaluation. One can easily obtain the following equations of T^{wo} and T^{w}:

$$T^{wo} = gnt + g\alpha = gNt + g\alpha, \qquad (13.2)$$

$$T^{w} = g\alpha + gt + g(N-1)T_{DT}, \qquad (13.3)$$

where, T_{DT} is the necessary time for data transfer from the master node to one slave node and from one slave node to the master node in one generation.

Now, the efficiency of parallelization, denoted as η, is calculated as:

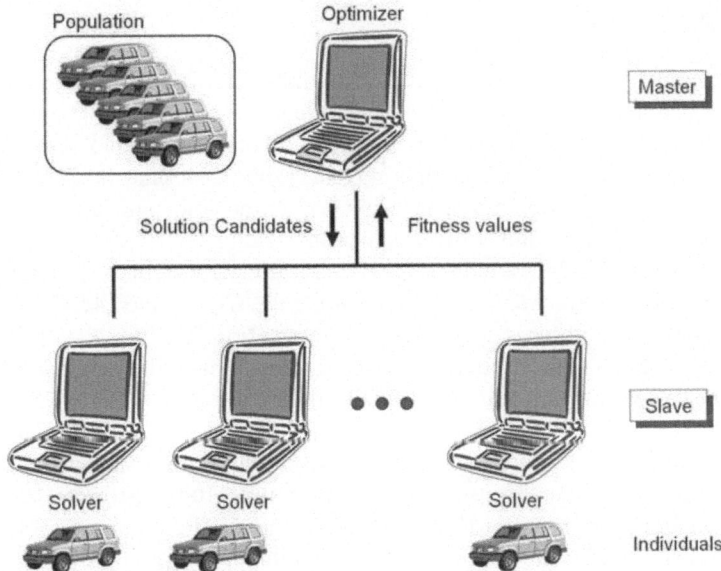

Fig. 13.1. Master-slave model for parallelization.

$$\eta = \frac{T^{wo}}{NT^w} \times 100(\%).\tag{13.4}$$

The numerator is the total resources for calculation when the parallelization is not used. The denominator is the total resources for calculation when parallelization is used. Since one master node and $(N-1)$ slave nodes are occupied for the time of T^w, the total resources are NT^w. If η becomes 100%, the parallelization is very useful. Oppositely, if η becomes 0%, the parallelization should not be done.

Using Eq. (13.2) and Eq. (13.3), Eq. (13.4) can be calculated as follows:

$$\eta = \frac{Nt + \alpha}{N(t + \alpha + (N-1)T_{DT})} \times 100.\tag{13.5}$$

Eq. (13.5) leads the following results:

- If $\alpha << t$ and $T_{DT} << t$, η is nearly 100%. This means that if the necessary calculation time for one fitness evaluation is sufficiently larger than α (for genetic operators) and T_{DT} (for data transfer), we should parallelize a calculation.
- If $\alpha \approx 0$, one can easily obtain the following relation:

$$\eta = \left(1 - \frac{(N-1)T_{DT}}{(N-1)T_{DT} + t}\right) \times 100.\tag{13.6}$$

This equation means that the smaller the value of t is, the worse is the efficiency of η.

Survey

From the beginning of the research for evolutionary algorithms on single objective optimization, parallelization technique has been paid attention due to population-based approach of evolutionary algorithms. There are many surveys in the literature, e.g., Schmeck *et al.* (2001); Bethke (1976); Adamidis (1994); Cantu-Paz (1997a,b). As a natural extension, parallelization is also used for evolutionary multiobjective optimization.

Stanley and Mudge (1995) propose the framework of parallel genetic algorithm called *Genetic Algorithm running on the INternet (GAIN)*. The usage of different architectures for parallel computation is often due to the fact that homogenous computers are not readily available. This situation leads to different computational time of fitness evaluations on slave nodes. If the computational time on a certain node is different from others, the efficiency of parallel computation decreases dramatically due to much idle time of faster slave nodes. To solve this problem, Stanley and Mudge propose the GAIN. Based on a given parameter that determines the maximum number of pending evaluations, the idle time is reduced. If the number of unevaluated individuals exceeds this number, the generation process sleeps. Otherwise, the generation process is carried out even if unevaluated individuals exist. The results of the GAIN show a robust and good performance.

Watanabe *et al.* (2002) extend an original master-slave model to maintain a higher diversity of the population. They call this extension as *Master-slave model with local cultivation (MSLC) model*. In this model, two randomly selected individuals are sent to a slave node. Using two individuals, most genetic operators are carried out in a slave node. However, in one generation, all individuals distributed to slave nodes are gathered and ranked again on the master node. Since most of calculation is done on slave node, the problem occurred on a master-slave node, i.e. higher computational cost of a master slave, is solved.

de Toro Negro *et al.* (2002) propose the parallel multiobjective evolutionary algorithm called *Parallel Single Front Genetic Algorithm (PSFGA)* as an extension of *Single Front Genetic Algorithm (SFGA)* based on master-slave model. The characteristic of the SFGA are as follows: Only the non-dominated individuals can join the recombination process, all non-dominated individuals are copied to the next population and the rest of individuals to complete the population are obtained by recombination and mutation of the non-dominated individuals. In PSFGA, the population is divided into several sub-populations based on fitness values. In the sub-population, the original SFGA is carried out. After execution of SFGA, all individuals are gathered by a master node. They conclude that parallelization is very helpful not only for the reduction of computational cost but also for the preservation of diversity.

Coello and Sierra (2004) study the parallelization of a coevolutionary multiobjective evolutionary algorithm. Based on the master-slave model, they parallelize their algorithm. The population is divided into several sub-population according to search region. In each generation, sup-populations cooperate or compete. In Coello and Sierra (2004), the parallel algorithm is compared with the serial (original) algorithm and shows better result from accuracy of solution and computational cost points of view.

Veldhuizen *et al.* (2003) discuss parallel evolutionary multiobjective optimization. In Veldhuizen *et al.* (2003), master-slave model, island model, diffusion model and hybrid model are discussed and the calculation time of them are also compared.

Dubreuil *et al.* (2006) analyze the master-slave model for distributed evolutionary computation theoretically. This paper builds a theoretical framework for the master-slave model and validates the framework empirically based on the Distributed BEAGLE C++ framework. They conclude that contrary to popular belief, the master-slave model can scale well.

Recently, many applications which need time-consuming fitness evaluations are successfully optimized using the master-slave model. Due to the page limitation, few of them are introduced, e.g., Jones *et al.* (1998); Sasaki *et al.* (2000); Okabe *et al.* (2003). Jones *et al.* (1998) parallelize a genetic algorithm on an aerodynamic and aeroacoustic optimization of airfoils. Despite the time-consuming multidisciplinary fitness evaluations, they successfully show good results by the usage of master-slave parallelization. Since their fitness evaluations need huge computational cost, their efficiency of parallelization achieves nearly 100%. Sasaki *et al.* (2000) optimize the design of a wing for supersonic transport using multiobjective genetic algorithm. To solve the huge computational cost, a simple master-slave model is used. They obtain the successful results with better performance. Okabe *et al.* (2003) optimize the shape of a micro heat exchanger problem using a commercial computational fluid dynamics software. To reduce huge computational cost, the algorithm is parallelized based on the master-slave model and successfully optimizes the shape. In Okabe *et al.* (2003), the necessary conditions of parallel optimization using a commercial solver are also discussed.

As introduced above, there are a lot of papers proposing new efficient method for the master-slave model and showing successful optimization results by master-slave model. Since real multiobjective optimization problems are more complicated, this type of parallelization will gather much attention in order to successfully obtain the optimal design of applications in reasonable time.

13.2.3 Level 3: Problem Dependent Parallelization

In this model, problem-dependent operations are parallelized. In general, the interest here is the parallelization of the evaluation of a single solution (different objectives and/or constraints). The parallel models may be based on the data partitioning or task partitioning. This model is useful in MOPs with time and/or memory intensive objectives and constraints. It may also be useful in MOPs with uncertainty which need in general an repeated evaluation of objective.

Basic Concept

In the last section, the evaluations in a generation are parallelized. However, even if the evaluations are parallelized, one fitness evaluation is sometimes still time-consuming. To solve this problem, we discuss the parallelization of each evaluation in this section. Possible parallelization of one fitness evaluations are listed as follows:

1. **Several solvers:**
   Consider a multiobjective optimization of a car design as an example. To design a car, several disciplines should be considered. Examples are to optimize the air flow around a car and the toughness of materials of a car. To optimize this problem, two independent solvers are necessary, i.e. CFD solver for the air flow and FEM for the toughness of materials. If we use one computer to evaluate two objectives, many users will firstly use the CFD to obtain the first objective function and secondly use the FEM to obtain the second objective function or vise versa. Some users will execute the CFD and the FEM at the same time. However, the total calculation time is nearly same with the above case because the computational resources are shared by two solvers. However, it is reasonable to execute the CFD and the FEM at the same time on different computers. Although the idle time, caused by the different calculation time of the CFD and the FEM, is not avoidable, the total calculation time becomes shorter.

2. **Decomposition of one fitness evaluation:**
   Consider the evaluation of a big product which consists of several parts. A simple idea to reduce the computational time is evaluation of each part and merging of them. Generally, CFD calculation for a big product is terribly time-consuming and has huge memory consumption. To tackle these problems, domain decomposition method (DDM) is often used in CFD research field, e.g., Elleighy and Tanaka (2001). Calculation domain is divided into several parts and assigned to different computers. Each computer calculates only the assigned part. To balance all calculation, the boundary condition is shared regularly. This division reduces the calculation time and memory consumption. However, since the boundary condition is shared regularly, rich connections among several computers are necessary. Furthermore, the user should take care of the division to reduce boundary and the balance of calculation cost on each computer.

3. **Multiple runs for one fitness evaluation:**
   Fitness evaluation sometimes needs several runs of a solver with different calculation conditions. An example is an optimization with uncertainty. Recently, robustness of fitness value against the variance of design parameters has gathered much attention, in particular, by researchers and

practitioners researching for a real application. In a real application, it is impossible to generate a product based on optimal design parameters because some variations are unavoidable. Therefore, it is very important to obtain a robust and (nearly) optimal design. To find robust and (nearly) optimal design, multiple runs of a solver are sometimes necessary. Assume that an optimizer obtains the design parameter x. To see the robustness against variance of the design parameter, the fitness values of $x + dx$ and $x - dx$ should also be evaluated. In this situation, it is reasonable to execute three solvers with different design parameters on different computers simultaneously. By simultaneous execution, calculation time will be reduced.

Calculation Time

For above situations, the total calculation time is considered here. Based on equations shown later, we will discuss when we should parallelize a calculation or not.

1. **Several solvers:**
   Assume a k-objective optimization problem where the time t_i is necessary to evaluate the ith objective function and N nodes are available for calculation. As an example, $N = k$ is assumed. The total calculation time without/with parallelization can be obtained as:

$$t^{wo} = \sum_{i=1}^{k} t_i = \sum_{i=1}^{N} t_i, \qquad (13.7)$$

$$t^{w} = \max(t_i) + T_{DT}, \qquad (13.8)$$

here, t^{wo}, t^{w}, and T_{DT} are the necessary time for k objective evaluations without/with parallelization, and the necessary time for data transfer, see Fig. 13.2 (a). The maximum time of all t_i is denoted by $\max(t_i)$. Using these equations, the efficiency of parallelization, denote η', can be obtained as:

$$\eta' = \frac{t^{wo}}{N t^{w}} \times 100 = \frac{\sum_{i=1}^{N} t_i}{N \max(t_i) + N T_{DT}} \times 100. \qquad (13.9)$$

This equation leads the following results when T_{DT} is negligible:
- If all t_i are the same, the efficiency of parallelization is 100%.
- Otherwise, the efficiency is reduced due to idle time of the computer with a shorter calculation.

If T_{DT} is not negligible, the efficiency will be reduced. In the worst case, the efficiency is nearly 0% when $T_{DT} >> t_i$. This means that the parallelization should not be used.

2. **Decomposition of one fitness evaluation:**
   Assume that N nodes are available for calculation and one problem will be decomposed into N sub-problems. The total calculation time of one problem and one sub-problem are assumed to be t_{all} and t_{sub}, respectively. Here, the decomposition is assumed as ideal, i.e., the time for all sub-problems is the same. The number of boundaries caused by decomposition and the time of internal data transfer per one boundary are assumed as B and T_{in}. By decomposition, each domain will be solved separately. However, to consider relations among neighbor domains, the boundary information should be adjusted regularly. One can obtain the following equations:

$$t^{wo} = t_{all} \tag{13.10}$$

$$t^w = t_{sub} + (N-1)T_{DT} + BT_{in}. \tag{13.11}$$

   Here, the number of boundaries of each decomposed domain is assume to be the same with others. The variable of T_{DT} is the time of data transfer for initial data. Since t_{all} is approximately Nt_{sub}, one can obtain the following efficiency (see Fig. 13.2 (b)):

$$\eta' = \frac{t^{wo}}{Nt^w} \times 100 = \left(1 - \frac{(N-1)T_{DT} + BT_{in}}{(N-1)T_{DT} + BT_{in} + t_{sub}}\right) \times 100 \tag{13.12}$$

   In domain decomposition method, T_{in} is generally very high. Therefore, by using rich connections among nodes, T_{in} should be reduced. Furthermore, the users should think of a way to reduce the number of boundaries, B.

3. **Multiple runs for one fitness evaluation:**
   Following the same way of master-slave model, it is easy to obtain the following equation:

$$\eta' = \left(1 - \frac{(N-1)T_{DT}}{(N-1)T_{DT} + t}\right) \times 100, \tag{13.13}$$

   here, N, T_{DT} and t are the number of necessary runs for one fitness value, the time for data transfer and the time for one fitness evaluation. As discussed in master-slave model, the calculation should be parallelized when $t \gg T_{DT}$.

The three models for parallel metaheuristics may be used in conjunction within a hierarchical structure. In Meunier *et al.* (2000); Talbi and Meunier (2006), this hierarchical architecture has been adopted to solve a complex multiobjective network design problem. At level 1, a parallel self contained cooperative model based on evolutionary algorithms (island model) and local search has been used. At level 2, a parallel evaluation model for a steady state evolutionary algorithm, in which the evaluation phase of the algorithm is done in parallel and in an asynchronous manner. Those two first parallel model are independent of the target MOP. Finally at level 3, a parallel synchronous decomposition model, in which the evaluation of a single solution is carried out in parallel by partitioning the geographical domain.

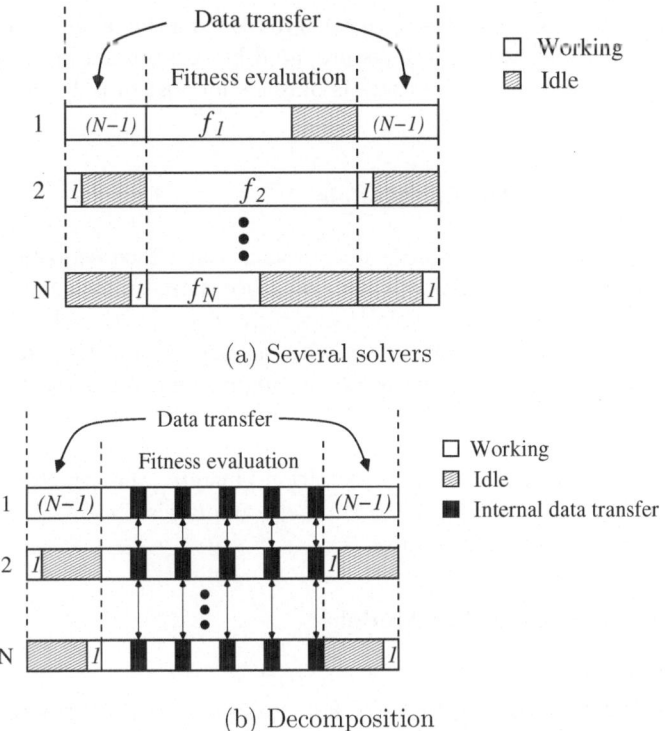

Fig. 13.2. Necessary calculation time of several solvers and decomposition.

13.3 Parallel Models for Non-heuristic Methods

Parallelization of exact optimization methods, particularly branch and bound ones, has been largely studied in the literature (refer to Talbi (2006)). However, to the best of our knowledge it is rarely tackled in the multiobjective context. For example, in the 11th MCDM conference (1994) Antunes and Tsoukiás (1997) survey new developments in computer science and they try to explore their specific relevance for the field of MCDA. The field of *distributed computing* is mentioned (p. 382f.) with the potential to integrate different MC models and methods in a single MCDA system. The benefit of *parallel computing* is seen in decomposing the problems. But no reference to any existing work is given[1]. Although there were presentations of parallel approaches at MCDM conferences, these papers have been rarely included in the official proceedings. For example, in the 15th MCDM conference (2000)

---

[1] The potential benefits of fuzzy sets and neural networks have been discussed where evolutionary algorithms and related metaheuristics were not an issue at that time.

there were at least two papers on parallel MCDM methods[2] but none of them appeared in the official proceedings prepared by Köksalan and Zionts (2001). This might be the reason why there is only a short list of publications that is presented next.

13.3.1 High Level Parallel Models

High level parallel models embrace approaches in which sequential MCDM methods are run independently with no or occasional information exchange in parallel. The simplest example is probably given by the idea to run several instances of the same algorithm with scalarized objective functions but different weights. This approach yields an approximation of the Pareto front and set. More flexibility is provided by the OpTiX-II software framework described by Grauer and Boden (1997). Here, the user may specify which MCDM methods are run in parallel on workstation clusters, and which information they are going to exchange. Unfortunately, numerical results are given for single-objective optimization only.

13.3.2 Low Level Parallel Models

Low level parallel models represent approaches in which parts of the sequential MCDM method are parallelized. For example, in 1992 Galperin Galperin (1992) proposed a new unscalarized method for MCDM based on his concept of balance numbers. In this procedure m subproblems can be solved in parallel independently (p. 81) in each iteration. Apparently, the parallelization was outlined only but not realized.

In case of interrelated multiobjective linear (MOLP) problems Volkovich (1997) parallelize the search for local solutions. Again, there are no results regarding speedup or efficiency. The situation changes for the parallel method for MOLPs presented by Wiecek and Zhang (1997). They achieve a speedup about 27 when using 32 processors for large problems. They deploy the technique of task partitioning by using an ADBASE solver on each processor.

Another reason for using parallel hardware is raised with interactive MCDM methods: If you like to foster interactivity during robustness analysis and that the decision maker does not give up too early (due to impatience and/or time schedule) then you must take care about fast response times. For this purpose, Costa and Climaco (1994) calculated solutions in parallel when using multiple reference points on a four processor system. They extended their work on parallelization to other interactive MCDM methods: in the course of the ELECTRE III method the creditability indices that define a fuzzy outranking relation can be calculated independently (and hence in parallel) for different pairs of alternatives. Moreover, they also parallelized the four subproblems of the distillation algorithm used in this method. A speedup

---

[2] See the program at http://mcdm2000.ie.metu.edu.tr/tentprog.htm

about 7 was achieved for 16 processors by Dias *et al.* (1997). Similarly, the preference indices of the interactive PROMETHEE method can be calculated independently and therefore in parallel. Dias *et al.* (1998) achieve a speedup about 15 for 16 processors in the best case.

Low level parallel models are also used for exact, combinatorial multi-objective problems. In particular, in case of biobjective flowshop problems Dhaenens *et al.* (2006) discuss parallelization of the weighted sum method with dichotomic search: after a new solution is found two new searches are launched. Consequently, many processors are idle in the early phase of the search resulting in poor speedup/efficiency measures. A similar approach can be deployed for the two phase method—with same disadvantages as in the previous method proposed by Lemesre *et al.* (2007a). Speedups do not exceed 1.7 for four processors. Lemesre *et al.* (2007b) achieved similar performance for the partitioning parallel method. In the first phase they use task partitioning and then space partitioning.

Needless to say, hybridizations of metaheuristics and exact methods do have some potential in case of parallelization. Basseur *et al.* (2004) propose a parallel hybrid model combining an exact approach (branch and bound) and a metaheuristic. The parallel metaheuristic is used to approximate the Pareto front. The parallel branch and bound is used to solve sub-problems and to improve the quality of the obtained Pareto front.

We are aware that this list of publications is not complete. But it reveals that there is considerably less work on the parallelization of non-heuristic and exact methods than for metaheuristics. Finally, we like to emphasize that the deployment of parallel hardware for interactive environments might be fruitful also for multiobjective metaheuristics.

13.4 Parallel Implementation and Deployment

Parallel and distributed architectures can have different memories (shared/ distributed), computation resources (homogeneous/heterogeneous), and networks (local/large). These different properties have a strong impact on the deployment technique such as load balancing and fault-tolerance. Depending on the type of architecture used, different parallel and distributed program ming environments such as message passing (PVM, MPI), shared memory (multi-threading, OpenMP), high throughput computing (Condor), and Grid computing (Globus) can be used. We briefly study some of this issues in the following.

13.4.1 Shared Memory versus Distributed Memory

The main advantage of parallel MOP algorithms implemented on shared memory architectures such as SMPs and multi-core processors is the simplicity.

For example, it is easier to share data such as upper bounds in exact algorithms and best found approximated non-dominated set of solutions in meta-heuristics. However, parallel distributed architectures offer a more flexible and fault-tolerant programming platform. Indeed, the memory access contention in shared memory architecture make the number of processors limited for this type of architectures.

13.4.2 Homogeneous and Dedicated versus Heterogeneous and Non-dedicated

Most massively parallel machines (MPP) and clusters of workstations (COW) such as IBM SP3 are composed of homogeneous processors and are generally dedicated to the application. The proliferation of powerful workstations and fast communication networks have shown the emergence of heterogeneous network of workstations (NOW) as platforms for high-performance computing. COWs and NOWs constitute a low-cost hardware alternative to run parallel algorithms. However, the efficient scheduling of tasks and fault tolerant mechanisms in NOWs is more complex to design and analyze due to the heterogeneity of those architectures (processors, networks, etc.) and a higher probability of faults.

Melab *et al.* (2006a) focus on solving large size problems using the Condor environment. It is an open source framework originally intended to the design and deployment of the three parallel models for meta-heuristics on dedicated clusters and networks of workstations. Relying on the Condor programming environment, it enables the execution of these applications on volatile non dedicated heterogeneous computational pools of resources. Efficient load balancing and fault-tolerance mechanisms have been designed for this purpose. Experimentations have been carried out on more than 100 PCs originally intended for education. The obtained results are convincing, both in terms of flexibility and easiness at implementation, and in terms of efficiency, quality and robustness of the provided solutions at run time.

13.4.3 Tightly Coupled (Local Networks) versus Loosely Coupled (Large Networks)

Massively parallel machines, clusters and local networks of workstations may be considered as tightly coupled architectures. Large network of workstations and Grid computing platforms are loosely coupled and are affected by higher cost of communication. The larger the granularity of a model, the better suited is the model for large networks.

Since the granularity of the self-contained parallel cooperation models (level 1) is very high, they can be easily deployed on large scale architectures which are in general loosely coupled and have high communication cost.

This model is also scalable in terms of the number of processors which is not the case for the other models (Problem Independent Parallel Intra-Algorithm and Problem Dependant Parallelization). These models are interesting when the evaluation of a single solution is CPU-time consuming and/or Input/Output intensive.

Melab *et al.* (2006b) report some results on parallel cooperative multiobjective meta-heuristics on computational grids. They particularly focus on the island model and the multi-start model and their cooperation. They propose a checkpointing-based approach to deal with the fault tolerance issue of the island model. Nowadays, existing DispatcherWorker grid middlewares are inadequate for the deployment of parallel cooperative applications. Indeed, these need to be extended with a software layer to support the cooperation. Therefore, they propose a Linda-like cooperation model and its implementation on top of XtremWeb. This middleware is then used to develop a parallel meta-heuristic applied to a bi-objective Flow-Shop problem using the two models. The work has been experimented on a multi-domain education network of 321 heterogeneous Linux PCs. The preliminary results, obtained after more than 10 days, demonstrate that the use of grid computing allows to fully exploit effectively different parallel models and their combination for solving large-size problem instances.

In terms of exact methods, the high level is more appropriate for large networks. The most popular parallelization approach of the branch and bound algorithm consists in building and exploring in parallel the search tree representing the problem being tackled. The deployment of such parallel model on a grid raises the crucial issue of dynamic load balancing. The major question is how to efficiently distribute the nodes of an irregular search tree among a large set of heterogeneous and volatile processors. Mezmaz *et al.* (2007) propose a new dynamic load balancing approach for the parallel branch and bound algorithm on the computational grid. The approach is based on a particular encoding of the tree nodes allowing a very simple description of the work units distributed during the exploration. Such description optimizes the communications involved by the huge amount of load balancing operations. The approach has been applied to one instance of the bi-objective flow-shop scheduling problem. The application has been experimented on a computational pool of more than 1000 processors belonging to seven Nation-wide clusters. The optimal Pareto front has been generated within almost 6 days with a parallel efficiency of 98%.

13.5 Conclusion and Future Trend

Parallel and distributed computing are powerful and necessary ways to reduce the computation time of multiobjective optimization algorithms and/or improve the quality of the obtained solutions. This chapter presents a general overview of parallel approaches for multiobjective optimization. For this purpose, we have proposed a taxonomy for parallel metaheuristics and exact methods. We have covered both the design aspect of algorithms and implementation on different parallel and distributed architectures. Different parallel models have been proposed in the design of multiobjective optimization algorithms. These models are largely experimented on a wide range of academic and real-life MOPs in different domains. The presented models may be used in conjunction within a hierarchical structure.

Multiobjective optimization algorithms have been implemented and deployed on different type of parallel and distributed architectures: clusters and networks of workstations and shared memory parallel architectures. An efficient implementation must consider the characteristics of the target parallel model (granularity, synchronous, etc.) and architecture (homogeneity, dedicated, etc.). For example, fine granularity models cannot easily be deployed on large scale distributed systems.

In the last decade, Grid computing and Peer-to-Peer (P2P) computing have become a real alternative to traditional high performance computing architectures for the development of large-scale distributed applications. This is a great challenge as Grid and P2P-enabled frameworks for multiobjective optimization algorithms are emerging.

Designing generic software frameworks to deal with the design and efficient transparent implementation of distributed multiobjective optimization algorithms is another important aspect. Software frameworks such as PARADISEO offer transparent implementation of different parallel models on different architectures using suitable programming environments as written by Cahon *et al.* (2004) and Liefooghe *et al.* (2007)[3].

In future, more and more applications will be concerned by parallel multiobjective optimization in different domains such as MDO (Multi-disciplinary Design Optimization), life sciences and industrial applications. Also, designing the interactive multiobjective optimization approaches which requires real-time parallel solving of MOPs is another important challenge.

References

Adamidis, P.: Review of Parallel Genetic Algorithms Bibliography. Technical Report, Aristotle University of Thessaloniki (1994)

Antunes, C., Tsoukiás, A.: Against fashion: A travel survival kit in "modern" MCDA. In: Multicriteria Analysis:International Conference on Multiple Criteria Decision Making, pp. 378–389. Springer, Berlin (1997)

Baita, F., Mason, F., Poloni, C., Ukovich, W.: Genetic Algorithm with Redundancies for the Vehicle Scheduling Problem. In: Biethahn, J., Nissen, V. (eds.) Evolutionary Algorithms in Management Applications, pp. 341–353. Springer, Berlin (1995)

---

[3] See the web site: `http://paradiseo.gforge.fr` for more details.

Basseur, M., Lemesre, J., Dhaenens, C., Talbi, E.-G.: Cooperation between branch and bound and evolutionary approaches to solve a bi-objective flow shop problem. In: Ribeiro, C.C., Martins, S.L. (eds.) WEA 2004. LNCS, vol. 3059, pp. 72–86. Springer, Heidelberg (2004)

Bethke, A.D.: Comparison of Genetic Algorithms and Gradient-based Optimizers on Parallel Processors: Efficiency of Use of Processing Capacity. Logic of Computers Group Technical Report 197, University of Michigan (1976)

Branke, J., Schmeck, H., Deb, K., Reddy, M.: Parallelizing Multi-Objective Evolutionary Algorithms: Cone Separation. In: IEEE Congress on Evolutionary Computation, pp. 1952–1957 (2004)

Bui, L.T., Abbass, H.A., Essam, D.: Local models - an approach to distributed multiobjective optimization. Technical Report TR-ALAR-200601002, The Artificial Life and Adaptive Robotics Laboratory, University of New South Wales, Australia (2006)

Cahon, S., Melab, N., Talbi, E.-G.: ParadisEO: A framework for the reusable design of parallel and distributed metaheuristics. Journal of Heuristics 10(3), 357–380 (2004)

Cantu-Paz, E.: A Survey of Parallel Genetic Algorithms. IlliGAL Report 97003, University of Illinois (1997a)

Cantu-Paz, E.: Designing Efficient Master-slave Parallel Genetic Algorithms. IlliGAL Report 97004, University of Illinois (1997b)

Coello Coello, C.A., Reyes Sierra, M.: A study of the parallelization of a coevolutionary multi-objective evolutionary algorithm. In: Monroy, R., Arroyo-Figueroa, G., Sucar, L.E., Sossa, H. (eds.) MICAI 2004. LNCS (LNAI), vol. 2972, pp. 688–697. Springer, Heidelberg (2004)

Coello Coello, C.A., Van Veldhuizen, D.A., Lamont, G.B.: Evolutionary Algorithms for Solving Multi-Objective Problems. Kluwer Academic Publishers, New York (2002)

Costa, J.P., Climaco, J.N.: A multiple reference point parallel approach in MCDM. In: International Conference on Multiple Criteria Decision Making, pp. 255–263. Springer, New York (1994)

de Toro Negro, F., Ortega, J., Fernandez, J., Diaz, A.: PSFGA: a parallel genetic algorithm for multiobjective optimization. In: Euromicro Workshop on Parallel, Distributed and Network-based Processing, pp. 384–391 (2002)

de Toro Negro, F., Ortega, J., Ros, E., Mota, S., Paechter, B., Martín, J.M.: PSFGA: Parallel Processing and Evolutionary Computation for Multiobjective Optimisation. Parallel Computing 30(5–6), 721–739 (2004)

Deb, K., Zope, P., Jain, S.: Distributed computing of Pareto-optimal solutions with evolutionary algorithms. In: Fonseca, C.M., Fleming, P.J., Zitzler, E., Deb, K., Thiele, L. (eds.) EMO 2003. LNCS, vol. 2632, pp. 534–549. Springer, Heidelberg (2003)

Dhaenens, C., Lemesre, J., Melab, N., Mezmaz, M., Talbi, E.-G.: Parallel exact methods for multi-objective combinatorial optimization. In: Parallel Combinatorial Optimization, John Wiley and Sons, Berlin (2006)

Dias, L.C., Costa, J.P., Climaco, J.N.: Conflicting criteria, cooperating processors—some experiments on implementing a multicriteria support method on a parallel computer. Computers and Operations Research 24(9), 805–817 (1997)

Dias, L.C., Costa, J.P., Climaco, J.N.: A parallel implementation of the PROMETHEE method. European Journal of Operational Research 104(3), 521–531 (1998)

Dubreuil, M., Gagne, C., Parizeau, M.: Analysis of a Master-slave Architecture for Distributed Evolutionary Computations. IEEE Transactions on Systems, Man, and Cybernetics 36(1), 229–235 (2006)

Elleighy, W.M., Tanaka, M.: Domain Decomposition Coupling of FEM and BEM. Transactions of the Japan Society for Computational Engineering and Science 4, 107–111 (2001)

Galperin, E.A.: Nonscalarized multiobjective global optimization. Journal of Optimization Theory and Applications 75(1), 69–85 (1992)

Grauer, M., Boden, H.: OpTiX-II: A software environment for MCDM based on distributed and parallel computing. In: Multicriteria Analysis: International Conference on Multiple Criteria Decision Making, pp. 199–208. Springer, Berlin (1997)

Hiroyasu, T., Miki, M., Watanabe, S.: The New Model of Parallel Genetic Algorithm in Multi-Objective Optimization Problems—Divided Range Multi-Objective Genetic Algorithm—. In: IEEE Congress on Evolutionary Computation, July 2000, vol. 1, pp. 333–340. IEEE Computer Society Press, Piscataway (2000)

Jones, B.R., Crossley, W.A., Lyrintzis, A.S.: Aerodynamic and Aeroacoustic Optimization of Airfoils via a Parallel Genetic Algorithm. In: AIAA 98-4811 (1998)

Jozefowiez, N., Semet, F., Talbi, E.-G.: Parallel and hybrid models for multiobjective optimization: Application to the vehicle routing problem. In: Guervós, J.J.M., Adamidis, P.A., Beyer, H.-G., Fernández-Villacañas, J.-L., Schwefel, H.-P. (eds.) PPSN 2002. LNCS, vol. 2439, pp. 271–280. Springer, Heidelberg (2002)

Jozefowiez, N., Semet, F., Talbi, E.-G.: Enhancements of nsga ii and its application to the vehicle routing problem with route balancing. In: Talbi, E.-G., Liardet, P., Collet, P., Lutton, E., Schoenauer, M. (eds.) EA 2005. LNCS, vol. 3871, pp. 131–142. Springer, Heidelberg (2006)

Jozefowiez, N., Semet, F., Talbi, E.-G.: Target aiming pareto search and its application to the vehicle routing problem with route balancing. Journal of Heuristics 13(5), 455–469 (2007)

Köksalan, M., Zionts, S.: International Conference on Multiple Criteria Decision Making. Springer, Berlin (2001)

Lemesre, J., Dhaenens, C., Talbi, E.-G.: An exact parallel method for a bi-objective permutation flowshop problem. European Journal of Operational Research 177(3), 1641–1655 (2007a)

Lemesre, J., Dhaenens, C., Talbi, E.-G.: Parallel partitioning method (PPM): A new exact method to solve bi-objective problems. Computers and Operations Research 34(8), 2450–2462 (2007b)

Liefooghe, A., Jourdan, L., Talbi, E.-G.: Paradiseo-MOEO: A framework for evolutionary multi-objective optimization. In: Evolutionary Multi-objective Optimization, Japan, pp. 457–471 (2007)

López-Jaimes, A., Coello Coello, C.A.: MRMOGA: Parallel Evolutionary Multiobjective Optimization using Multiple Resolutions. In: IEEE Congress on Evolutionary Computation, Edinburgh, Scotland, September 2005, vol. 3, pp. 2294–2301. IEEE Computer Society Press, Los Alamitos (2005)

Mehnen, J., Michelitsch, T., Schmitt, K., Kohlen, T.: pMOHypEA: Parallel evolutionary multiobjective optimization using hypergraphs. Technical Report of the SFB Project 531 Computational Intelligence CI–189/04, University of Dortmund (2004)

Melab, N., Cahon, S., Talbi, E.-G.: Grid computing for parallel bioinspired algorithms. Journal of Parallel and Distributed Computing (JPDC) 66(8), 1052–1061 (2006a)

Melab, N., Mezmaz, M., Talbi, E.-G.: Parallel cooperative metaheuristics on the computational grid: A case study - the biobjective flow-shop problem. Parallel computing 32(9), 643–659 (2006b)

Meunier, H., Talbi, E.-G., Reininger, P.: A multiobjective genetic algorithm for radio network design. In: IEEE Congress on Evolutionary Computation, Orlando, USA, pp. 317–324 (2000)

Mezmaz, M., Melab, N., Talbi, E.-G.: Using the multi-start and island models for parallel multi-objective optimization on the computational grid. In: IEEE International Conference on e-Science and Grid Computing (e-Science'06), pp. 112–120 (2006)

Mezmaz, M., Melab, N., Talbi, E.-G.: An efficient load balancing strategy for grid-based branch and bound. Parallel computing 33(4-5), 302–313 (2007)

Mostaghim, S., Branke, J., Schmeck, H.: Multi-objective particle swarm optimization on computer grids. In: The Genetic and Evolutionary Computation Conference, vol. 1, pp. 869–875 (2007)

Okabe, T.: Evolutionary Multi-objective Optimization -On the Distribution of Offspring in Parameter and Fitness Space-. Shaker Verlag, Aachen (2004)

Okabe, T., Foli, K., Olhofer, M., Jin, Y., Sendhoff, B.: Comparative Studies on Micro Heat Exchanger Optimization. In: IEEE Congress on Evolutionary Computation, pp. 647–654 (2003)

Parmee, I.C., Vekeria, H.D.: Co-operative Evolutionary Strategies for Single Component Design. In: Bäck, T. (ed.) International Conference on Genetic Algorithms, pp. 529–536. Morgan Kaufmann, San Francisco (1997)

Poloni, C.: Hybrid GA for Multi-Objective Aerodynamic Shape Optimization. In: Winter, G., Periaux, J., Galan, M., Cuesta, P. (eds.) Genetic Algorithms in Engineering and Computer Science, pp. 397–416. Wiley & Sons, Chichester (1995)

Sasaki, D., Obayashi, S., Sawada, K., Himeno, R.: Multiobjective Aerodynamic Optimization of Supersonic Wings Using Navier-Stokes Equations. In: European Congress on Computational Methods in Applied Sciences and Engineering (2000)

Schaffer, D.J.: Multiple objective optimization with vector evaluated genetic algorithms. In: International Conference on Genetic Algorithms and Their Applications, pp. 93–100 (1985)

Schmeck, H., Kohlmorgen, U., Branke, J.: Parallel Implementations of Evolutionary Algorithms. In: Solutions to Parallel and Distributed Computing Problems, pp. 47–68 (2001)

Stanley, T.J., Mudge, T.: A Parallel Genetic Algorithm for Multiobjective Microprocessor Design. In: The Sixth International Conference on Genetic Algorithms, pp. 597–604 (1995)

Streichert, F., Ulmer, H., Zell, A.: Parallelization of multi-objective evolutionary algorithms using clustering algorithms. In: Coello Coello, C.A., Hernández Aguirre, A., Zitzler, E. (eds.) EMO 2005. LNCS, vol. 3410, pp. 92–107. Springer, Heidelberg (2005)

Talbi, E.-G.: Parallel combinatorial optimization. Wiley, Chichester (2006)

Talbi, E.-G., Meunier, H.: Hierarchical parallel approach for gsm mobile network design. Journal of Parallel and Distributed Computing 66(2), 274–290 (2006)

Van Veldhuizen, D.A., Zydallis, J.B., Lamont, G.B.: Considerations in Engineering Parallel Multiobjective Evolutionary Algorithms. IEEE Transactions on Evolutionary Computation 7(2), 144–173 (2003)

Volkovich, V.L.: Distributed multiobjective optimization problems and methods for their solution. In: International Conference on Multiple Criteria Decision Making, pp. 222–232. Springer, Berlin (1997)

Watanabe, S., Hiroyasu, T., Miki, M.: Parallel Evolutionary Multi-Criterion Optimization for Mobile Telecommunication Networks Optimization. In: Evolutionary Methods for Design, Optimization and Control, pp. 162–172 (2002)

Wiecek, M.M., Zhang, H.: A parallel algorithm for multiple objective linear programs. Computational Optimization and Applications 8(1), 41–56 (1997)

Xiao, N., Armstrong, M.P.: A specialized island model and its application in multi-objective optimization. In: Cantú-Paz, E., Foster, J.A., Deb, K., Davis, L., Roy, R., O'Reilly, U.-M., Beyer, H.-G., Kendall, G., Wilson, S.W., Harman, M., Wegener, J., Dasgupta, D., Potter, M.A., Schultz, A., Dowsland, K.A., Jonoska, N., Miller, J., Standish, R.K. (eds.) GECCO 2003. LNCS, vol. 2724, pp. 1530–1540. Springer, Heidelberg (2003)

Zhu, Z.-Y.: An Evolutionary Approach to Multi-Objective Optimization Problems. Ph.D. thesis, The Chinese University of Hong Kong (2002)

Zhu, Z.-Y., Leung, K.-S.: Asynchronous Self-Adjustable Island Genetic Algorithm for Multi-Objective Optimization Problems. In: IEEE Congress on Evolutionary Computation, Piscataway, New Jersey, May 2002, vol. 1, pp. 837–842 (2002)

14

Quality Assessment of
Pareto Set Approximations

Eckart Zitzler[1], Joshua Knowles[2], and Lothar Thiele[1]

[1] ETH Zurich, Switzerland
   eckart.zitzler@tik.ee.ethz.ch, thiele@tik.ee.ethz.ch
[2] University of Manchester, UK
   j.knowles@manchester.ac.uk

Abstract. This chapter reviews methods for the assessment and comparison of Pareto set approximations. Existing set quality measures from the literature are critically evaluated based on a number of orthogonal criteria, including invariance to scaling, monotonicity and computational effort. Statistical aspects of quality assessment are also considered in the chapter. Three main methods for the statistical treatment of Pareto set approximations deriving from stochastic generating methods are reviewed. The *dominance ranking method* is a generalization to partially-ordered sets of a standard non-parametric statistical test, allowing collections of Pareto set approximations from two or more stochastic optimizers to be directly compared statistically. The *quality indicator method* — the dominant method in the literature — maps each Pareto set approximation to a number, and performs statistics on the resulting distribution(s) of numbers. The *attainment function method* estimates the probability of attaining each goal in the objective space, and looks for significant differences between these probability density functions for different optimizers. All three methods are valid approaches to quality assessment, but give different information. We explain the scope and drawbacks of each approach and also consider some more advanced topics, including multiple testing issues, and using combinations of indicators. The chapter should be of interest to anyone concerned with generating and analysing Pareto set approximations.

14.1 Introduction

In many application domains, it is useful to approximate the set of Pareto-optimal solutions, cf. (Ehrgott and Gandibleux, 2000; Deb, 2001; Coello Coello *et al.*, 2002). To this end, various approaches have been proposed ranging from exact methods to randomized search algorithms such as evolutionary algorithms, simulated annealing, and tabu search (see Chapters 2 and 3).

Reviewed by: Günter Rudolph, University of Dortmund, Germany
Serpil Sayin, Koç University, Turkey
Kalyanmoy Deb, Indian Institute of Technology Kanpur, India

J. Branke et al. (Eds.): Multiobjective Optimization, LNCS 5252, pp. 373–404, 2008.
© Springer-Verlag Berlin Heidelberg 2008

With the rapid increase of the number of available techniques, the issue of performance assessment has become more and more important and has developed into an independent research topic. As with single objective optimization, the notion of performance involves both the quality of the solution found and the time to generate such a solution. The difficulty is that in the case of stochastic optimizers the relationship between quality and time is not fixed, but may be described by a corresponding probability density function. Accordingly, every statement about the performance of a randomized search algorithm is probabilistic in nature. Another difficulty is particular to multiobjective optimizers that aim at approximating the set of Pareto-optimal solutions in a scenario with multiple criteria: the outcome of the optimization process is usually not a single solution but a set of trade-offs. This not only raises the question of how to define quality in this context, but also how to represent the outcomes of multiple runs in terms of a probability density function.

This chapter addresses both quality and stochasticity. Sections 2–5 are devoted to the issue of set quality measures; they define properties of such measures and discuss selected measures in the light of these properties. The question of how to statistically assess multiple sets generated by a stochastic multiobjective optimizer is dealt with in Sections 6–8. Both aspects are summarized in Section 9.

The chapter will be of interest to anyone concerned with generating methods of any type. Those who are interested in a preference based set of solutions should find this paper useful as well.

14.2 Quantifying Quality General Considerations

14.2.1 Pareto Set Approximations

Assume a general optimization problem $(X, Z, \mathbf{f}, rel)$ where X denotes the decision space, $Z = \mathbf{R}^k$ is the objective space, $\mathbf{f} = (f_1, f_2, \ldots, f_k)$ is the vector of objective functions, and rel represents a binary relation over Z that defines a partial order of the objective space, which in turn induces a preorder of the decision space.[1] In the presence of a single objective function ($k = 1$), the standard relation 'less than or equal' is generally used to define the corresponding minimization problem $(X, \mathbf{R}, (f_1), \leq)$. In the case of multiple objective functions, i.e., $k > 1$, usually the relation $\preceq$ with $\mathbf{z}^1 \preceq \mathbf{z}^2 \Leftrightarrow \forall i \in \{1, \ldots, k\} : \mathbf{z}_i^1 \leq \mathbf{z}_i^2$ is taken; it represents a natural extension of $\leq$ to $\mathbf{R}^k$ and is also known as *weak Pareto dominance*. The associated strict order $\prec$ with $\mathbf{z}^1 \prec \mathbf{z}^2 \Leftrightarrow \mathbf{z}^1 \preceq \mathbf{z}^2 \wedge \neg \mathbf{z}^2 \preceq \mathbf{z}^1$ is often denoted as *Pareto dominance*, and instead of $\mathbf{z}^1 \prec \mathbf{z}^2$ one also says $\mathbf{z}^1$ *dominates* $\mathbf{z}^2$. Using this terminology, the

---

[1] A binary relation is called a *preorder* iff it is reflexive and transitive. A preorder which is antisymmetric is denoted as *partial order*.

Pareto-optimal set comprises the set of decision vectors not dominated by any other element in the feasible set $S \subseteq X$.

The formal definition of an optimization problem given above assumes that only a single solution, any of those mapped to a minimal element, is sought. However, in a multiobjective setting one is often interested in the entire Pareto-optimal set rather than in a single, arbitrary Pareto-optimal solution. With many applications, e.g., engineering designs problems, knowledge about the Pareto-optimal set is helpful and provides valuable information about the underlying problem. This leads to a different optimization problem where the goal is to find a set of mutually incomparable solutions (for any two decision vectors $\mathbf{x}^1, \mathbf{x}^2$, neither weakly dominates the other one), which will be here denoted as *Pareto set approximations*; the symbol Ψ stands for the sets of all Pareto set approximations over X. Accordingly, sets of mutually incomparable objective vectors are here called *Pareto front approximations*, and the set of all Pareto front approximations over Z is represented by Ω.

Now, let $(X, Z, \mathbf{f}, rel)$ be the original optimization problem. It can be canonically transformed into a corresponding set problem $(\Psi, \Omega, \mathbf{f}', rel')$ by extending $\mathbf{f}$ and rel in the following manner:

- $\mathbf{f}'(E) = \{\mathbf{z} \in Z \mid \exists \mathbf{x} \in E : \mathbf{z} = \mathbf{f}(\mathbf{x})\}$
- $A \; rel' \; B \Leftrightarrow \forall \mathbf{z}^2 \in B \; \exists \mathbf{z}^1 \in A : \mathbf{z}^1 \; rel \; \mathbf{z}^2$

If rel is $\preceq$, then rel' represents the natural extension of weak Pareto dominance to Pareto front approximations. In the following, we will use the symbols $\preceq$ and $\prec$ as for dominance relations on objective vectors and decision vectors also for Pareto front approximations respectively Pareto set approximations—it will become clear from the context, which relation is referred to.

14.2.2 Outperformance and Quality Indicators

Suppose we would like to assess the performance of two multiobjective optimizers. The question of whether either outperforms the other one involves various aspects such as the quality of the outcome, the computation time required, the parameter settings, etc. Sections 2–5 of this chapter focus on the quality aspect and address the issue of how to compare two (or several) Pareto set approximations. For the time being, assume that we consider one optimization problem only and that the two algorithms to be compared are deterministic, i.e., with each optimizer exactly one Pareto set approximation is associated; the issue of stochasticity will be treated in later sections.

As discussed above, optimization is about searching in an ordered set. The partial order rel for an optimization problem $(X, Z, \mathbf{f}, rel)$ defines a preference structure on the decision space: a solution $\mathbf{x}^1$ is preferable to a solution $\mathbf{x}^2$ iff $\mathbf{f}(\mathbf{x}^1) \; rel \; \mathbf{f}(\mathbf{x}^2)$ and not $\mathbf{f}(\mathbf{x}^2) \; rel \; \mathbf{f}(\mathbf{x}^1)$. This preference structure is the basis on which the optimization process is performed. For the corresponding set problem $(\Psi, \Omega, \mathbf{f}', rel')$, this means that the most natural way to compare two Pareto set approximations A and B generated by two different multiobjective optimizers is to use the underlying preference structure rel'. In the context of weak Pareto dominance, there can be four situations: (i) A is better than B ($A \preceq B \wedge B \npreceq A$), (ii) B is better than A ($A \npreceq B \wedge B \preceq A$), (iii) A and B are incomparable ($A \npreceq B \wedge B \npreceq A$), or (iv) A and B are indifferent ($A \preceq B \wedge B \preceq A$), where 'better' means the first set weakly dominates the second, but the second does not weakly dominate the first. These are the types of statements one can make without any additional preference information. Often, though, we are interested in more precise statements that quantify the difference in quality on a continuous scale. For instance, in cases (i) and (ii) we may be interested in knowing how much better the preferable Pareto set approximation is, and in case (iii) one may ask whether either set is better than the other in certain aspects not captured by the preference structure— this is illustrated in Fig. 14.1. This is crucial for the search process itself, and almost all algorithms for approximating the Pareto set make use of additional preference information, e.g., in terms of diversity measures.

For this purpose, quantitative set quality measures have been introduced. We will use the term *quality indicator* in the following:

A (unary) quality indicator is a function $I : \Psi \to \mathbf{R}$ that assigns each Pareto set approximation a real number.

In combination with the $\leq$ or $\geq$ relation on $\mathbf{R}$, a quality indicator I defines a total order of Ω and thereby induces a corresponding preference structure: A is preferable to B iff $I(A) > I(B)$, assuming that the indicator values are to be maximized. That means we can compare the outcomes of two multi-objective optimizers, i.e., two Pareto set approximations, by comparing the corresponding indicator values.

Example 1. Let A be an arbitrary Pareto set approximation and consider the subspace Z' of the objective space $Z = \mathbf{R}^k$ that is, roughly speaking, weakly dominated by A. That means any objective vector in Z' is weakly dominated by at least one objective vector in $\mathbf{f}'(A)$.

The hypervolume indicator I_H (Zitzler and Thiele, 1999) gives the hypervolume of Z' (see Fig. 14.2). The greater the indicator value, the better the approximation set. Note that this indicator requires a reference point relatively to which the hypervolume is calculated.

Considering again Fig. 14.1, it can be seen that the hypervolume indicator reveals differences in quality that cannot be detected by the dominance relation. In the left scenario, $I_H(A) = 277$ and $I(B) = 231$, while for the scenario in the middle, $I_H(A) = 277$ and $I(B) = 76$; in the right scenario, the indicator values are $I_H(A) = 277$ and $I_H(B) = 174$.[2] This advantage, though, comes at the expense of generality, since every quality indicator represents certain assumptions about the decision maker's preferences. Whenever $I_H(A) > I_H(B)$,

---

[2] The objective vector $(20, 20)$ is the reference point.

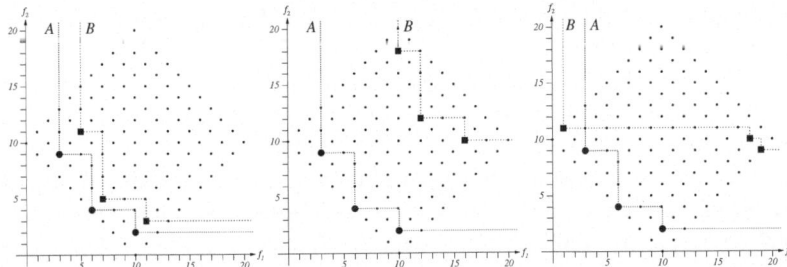

Fig. 14.1. Three examples to illustrate the limitations of statements purely based on weak Pareto dominance. In both the figures on the left, the Pareto set approximation A dominates the Pareto set approximation B, but in one case the two sets are much closer together than in the other case. On the right, A and B are incomparable, but in most situations A will be more useful to the decision maker than B. The background dots represent the image of the feasible set S in the objective space $\mathbf{R}^2$ for a discrete problem.

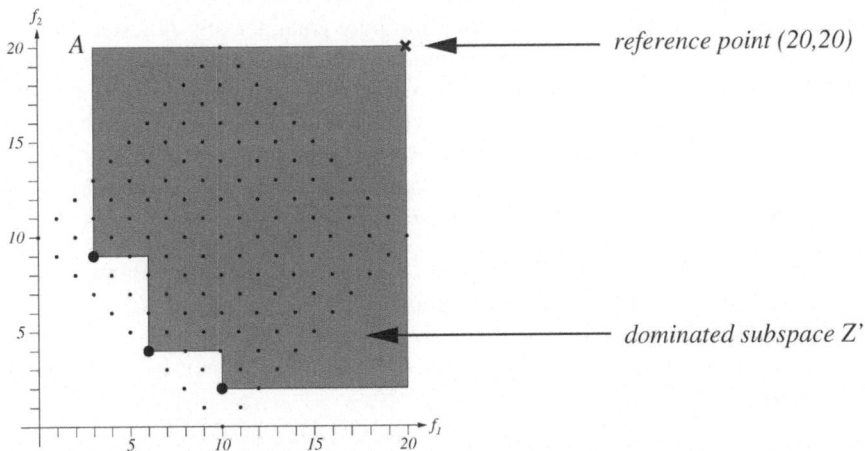

Fig. 14.2. Illustration of the hypervolume indicator. In this example, approximation set A is assigned the indicator value $I_H(A) = 277$; the objective vector $(20, 20)$ is taken as the reference point.

we can state that A is better than B with respect to the hypervolume indicator; however, the situation could be different for another quality indicator I' that assigns B a better indicator value than A. As a consequence, every comparison of multiobjective optimizers is not only restricted to the selected benchmark problems and parameter settings, but also to the quality indicator(s) under consideration. For instance, if we use the hypervolume indicator in a comparative study, any statement like "optimizer 1 outperforms optimizer 2 in terms of quality of the generated Pareto set approximation" needs to be qualified by adding "under the assumption that I_H reflects the decision maker's preferences".

Finally, note that the following discussion focuses on unary quality indicators, although an indicator can take in principle an arbitrary number of Pareto set approximations as arguments. Several quality indicators have been proposed that assign real numbers to pairs of Pareto set approximations, which are denoted as binary quality indicators (see Hansen and Jaszkiewicz, 1998; Knowles and Corne, 2002; Zitzler *et al.*, 2003, for an overview). For instance, the unary hypervolume indicator can be extended to a binary quality indicator by defining $I_H(A, B)$ as the hypervolume of the subspace of the objective space that is dominated by A but not by B.

14.3 Properties of Unary Quality Indicators

Quality indicators serve different goals: they may be used for comparing algorithms, but also during the optimization process as guidance for the search or as stopping criterion. In principle, one may consider any function from Ω to $\mathbf{R}$ as an indicator, but clearly there are certain properties that need to be fulfilled in order to make the indicator useful. These properties may vary depending on the purpose: for instance, when comparing several algorithms on a benchmark problem one may assume that the Pareto-optimal set is known, while such information is clearly not available in a real-world scenario. In the following, we will consider four main criteria:

Monotonicity: An indicator I is said to be *monotonic* iff for any Pareto set approximation that is compared to another Pareto set approximation holds: at least as good in terms of the dominance relation implies at least as good in terms of the indicator values. Formally, this can be expressed as follows:

$$\forall A, B \in \Psi : A \preceq B \Rightarrow I(A) \geq I(B)$$

where $\preceq$ stands for the underlying dominance relation, here weak Pareto dominance.

Monotonicity guarantees that an indicator does not contradict the partial order of Ω that is imposed by the weak Pareto dominance relation, i.e., consistency with the inherent preference structure of the optimization problem under consideration is maintained. However, it does not guarantee a unique optimum with respect to the indicator values; in other words, a Pareto set approximation that has the same indicator value as the Pareto-optimal set not necessarily contains only Pareto-optimal solutions. To this end, a stronger condition is needed which leads to the property of *strict monotonicity*:

$$\forall A, B \in \Psi : A \prec B \Rightarrow I(A) > I(B)$$

Currently, the hypervolume indicator is the only strictly monotonic unary indicator known, (see Zitzler *et al.*, 2007).

Scaling invariance: In practice, the objective functions are often subject to scaling, i.e., the objective function values undergo a strictly monotonic transformation. Most common are transformations of the form of $s(f(\mathbf{x})) = (f(\mathbf{x}) - f_l)/(f_u - f_l)$ where f_l and f_u are lower and upper bounds respectively for the objective function values such that each objective vector lies in $[0,1]^k$. In this context, it may be desirable that an indicator is not affected by any type of scaling which can be stated as follows: an indicator is denoted as *scaling invariant* iff for any strictly monotonic transformation $\mathbf{s} : \mathbf{R}^k \to \mathbf{R}^k$ the indicator values remain unaffected, i.e., for all $A \in \Psi$ the indicator value $I(A)$ is the same independently of whether we consider the problem $(\Psi, \Omega, \mathbf{f}', rel')$ or the scaled problem $(\Psi, \Omega, \mathbf{s} \circ \mathbf{f}', rel')$.[3] Scaling invariant indicators usually only exploit the dominance relation among solutions, but not their absolute objective function values.

Computation effort: A further property that is less easy to formalize addresses the computational resources needed to compute the indicator value for a given Pareto set approximation. We here consider the runtime complexity, depending on the number of solutions in the Pareto set approximation as well as the number of objectives, as a measure to compare indicators. This aspect becomes critical, if an indicator is to be used during the search process; however, even for pure performance assessment there may be limitations for certain indicators, e.g., if the running time is exponential in the number of objectives as with the hypervolume indicator (While, 2005).

Additional problem knowledge: Many indicators are parameterized and require additional information in order to be applied. Some assume the Pareto-optimal set to be known, while others rely on reference objective vectors or reference sets. In most cases, the indicator parameters are both user- and problem-dependent; therefore, it may be desirable to have as few parameters as possible.

There are many properties one may consider, and the interested reader is referred to (Knowles, 2002; Knowles and Corne, 2002) for a more detailed discussion.

---

[3] Alternatively, one may consider a weaker version of scaling invariance which is based on the order of the indicator values rather than on the absolute values. More precisely, the elements of Ψ would be sorted according to their indicator values; if the order remains the same for any type of scaling, then the indicator under consideration would be called scaling independent.

14.4 Discussion of Selected Unary Quality Indicators

The unary quality indicators that will be discussed in the following represent a selection of popular measures; however, the list of indicators is by no means complete. Furthermore, only deterministic indicators are considered. A summary of the indicators and properties we consider is given in Table 14.1.

14.4.1 Outer Diameter

The *outer diameter* measures the distance between the ideal objective vector and the nadir objective vector of a Pareto set approximation in terms of a specific distance metric. We here define the corresponding indicator I_{OD} as

$$I_{\mathrm{OD}}(A) = \max_{1 \leq i \leq n} w_i \left(\left(\max_{\mathbf{x} \in A} f_i(\mathbf{x}) \right) - \left(\min_{\mathbf{x} \in A} f_i(\mathbf{x}) \right) \right)$$

with weights $w_i \in \mathbf{R}^+$. If all weights are set to 1, then the outer diameter simply provides the maximum extent over all dimensions of the objective space.

The outer diameter is neither monotonic nor scaling invariant. However, it is cheap to compute (the runtime is linear in the cardinality of the Pareto set approximation A) and does not require any additional problem knowledge. The paramters w_i can be used to weight the different objectives, but they are as such not problem-specific.

14.4.2 Proportion of Pareto-Optimal Objective Vectors Found

Another measure to consider is the number of Pareto-optimal objective vectors that are weakly dominated by the image of a Pareto set approximation in objective space. The corresponding indicator I_{PF} has been introduced by Ulungu *et al.* (1999) as the *fraction of the Pareto-optimal front P weakly dominated* by a specific set $A \in \Psi$:

$$I_{\mathrm{PF}}(A) = \frac{\{\mathbf{z} \mid \exists \mathbf{x} \in A : \mathbf{f}(\mathbf{x}) \preceq \mathbf{z}\}}{|P|}$$

This measure assumes that the Pareto-optimal set resp. the Pareto-optimal front is known and that the number of optimal objective vectors is finite. The indicator value can be computed in $\mathcal{O}(|P| \cdot |A|)$ time, and they are invariant to scaling. The indicator is monotonic, but not strictly monotonic.

14.4.3 Cardinality

The *cardinality* $I_C(A)$ of a Pareto set approximation A can be considered both in decision space and objective space, (see, e.g. Van Veldhuizen, 1999). In either case, the indicator is not monotonic. However, it is cheap to compute, scaling invariant, and does not require any additional information.

Table 14.1. Summary of selected indicators and some of their properties. See accompanying text for full details

Indicator	Monotonicity	Scaling invariance	Computational effort	Additional problem knowledge needed
Outer Diameter	✗	✗	linear time	none
Proportion of Pareto Optimal Vectors Found	not strictly	invariant	quadratic	all Pareto optima
Cardinality	✗	invariant	linear time	none
Hypervolume	strictly monotonic	✗	exponential in k	needs upper bounding vector
Completeness	not strictly	invariant	anytime as it is based on sampling, but effort grows rapidly with decision space dimension	none
Epsilon Family	not strictly	✗	quadratic	reference set
D Family	not strictly	✗	quadratic	reference set
R Family	not strictly	✗	quadratic	reference set and point
Uniformity Measures	✗	✗	quadratic	varies

14.4.4 Hypervolume Indicator

The *hypervolume indicator* I_H, which was introduced in (Zitzler and Thiele, 1998), gives the volume of the portion of the objective space that is weakly dominated by a specific Pareto set approximation. It can be formally defined as

$$I_H^*(A) := \int_{\mathbf{z}_{lower}}^{\mathbf{z}_{upper}} \alpha_A(\mathbf{z}) d\mathbf{z}$$

where $\mathbf{z}_{lower}$ and $\mathbf{z}_{upper}$ are objective vectors representing lower resp. upper bounds for the portion of the objective space within which the hypervolume is calculated, and where the function α_A is the attainment function (Grunert da Fonseca *et al.*, 2001) for A

$$\alpha_A(\mathbf{z}) := \begin{cases} 1 & \text{if } \exists \mathbf{x} \in A : \mathbf{f}(\mathbf{x}) \preceq \mathbf{z} \\ 0 & \text{else} \end{cases}$$

that returns for an objective vector a 1 if and only if it is weakly dominated by A. In practice, the lower bound $\mathbf{z}_{lower}$ is not required to calculate the hypervolume for a set A. The hypervolume indicator is to be maximized.

The hypervolume indicator is currently the only unary indicator known to be strictly monotonic. This comes at the cost of high computational cost: the best known algorithms for computing the hypervolume have running times which are exponential in the number of objectives (see While, 2005; While et al., 2005, 2006; Fonseca et al., 2006; Beume and Rudolph, 2006). Furthermore, a reference point, an upper bound, needs to be specified; the indicator is sensitive to the choice of this upper bound, i.e. the ordering of Pareto set approximations induced by the indicator is affected by it, so the indicator is not scaling invariant by the above definition. Note: preference information can be incorporated into the hypervolume indicator, so that more emphasis can be placed on certain parts of the Pareto front than others (e.g. the middle, the extremes, etc.), whilst maintaining monotonicity (Zitzler et al., 2007).

14.4.5 Completeness Indicator

The *completeness indicator* I_{CP} was introduced in (Lotov et al., 2002, 2004) and goes back to the concept of completeness as defined by (Kamenev and Kondtrat'ev, 1992; Kamenev, 2001). The indicator gives the probability that a randomly chosen solution from the feasible set S is weakly dominated by a given Pareto set approximation A, i.e.,

$$I_{CP}(A) = \text{Prob}\left[A \preceq \{\mathcal{U}\}\right] \tag{14.1}$$

where $\mathcal{U}$ is a random variable representing the random choice from S. Provided that $\mathcal{U}$ follows a uniform probability density function, the indicator value $I_{CP}(A)$ can also be interpreted as the portion of the feasible set that is dominated by A. As such, the completeness indicator is strongly related to the hypervolume indicator; the difference is that the former takes the decision space into account, while the latter considers the objective space only.

Normally, one cannot compute the completeness directly. For this reason, the indicator values can be estimated by drawing samples from the feasible set and computing the completeness for these samples. As shown by Lotov et al. (Lotov et al., 2004), the confidence interval for the true value can be evaluated with any reliability, given sufficiently large samples. Furthermore, there is an extension of this indicator, namely $I_{CP}^{\epsilon}(A)$, where another dominance relation, e.g., the ϵ-dominance relation, $\preceq_{\epsilon}$, as defined above, is considered which reflects a specific ϵ-neighborhood of a Pareto set approximation in the objective space, see (Lotov et al., 2002, 2004) for details.

The completeness indicator is scaling-invariant as it does not rely on the absolute objective function values. Furthermore, the exact completeness indicator is as the hypervolume indicator strictly monotonic. However, as in practice sampling is necessary to estimate the exact indicator values, the indicator function based on estimates is monotonic (if always the same sample is used to compare two Pareto set approximation), while strict monotonicity cannot be ensured in general. Experimental studies have shown that the indicator estimates are effective only for relatively low-dimensional decision spaces (not more than a dozen decision variables) and for sufficiently slowly varying objective functions (Lotov et al., 2002, 2004). For a high-dimensional decision space, the Pareto-optimal set cannot be found via random point generation if it has an extremely small volume. For this reason, a generalized completeness estimate for the quality of approximation has been proposed for the case of a large number of variables and rapidly varying functions, see (Berezkin et al., 2006).

14.4.6 Epsilon Indicator Family

The *epsilon indicator family* has been introduced in (Zitzler et al., 2003) and comprises a multiplicative and an additive version—both exist in unary and binary form; the definition is closely related to the notion of epsilon efficiency (Helbig and Pateva, 1994). The binary multiplicative epsilon indicator, $I_\epsilon(A, B)$, gives the minimum factor ϵ by which objective vector associated with B can be multiplied such that the resulting transformed Pareto front approximation is weakly dominated by the Pareto front approximation represented by A:

$$I_\epsilon(A, B) = \inf_{\epsilon \in \mathbf{R}} \{\forall \mathbf{x}^2 \in B \; \exists \mathbf{x}^1 \in A : \mathbf{x}^1 \preceq_\epsilon \mathbf{x}^2\}. \tag{14.2}$$

This indicator relies on the ϵ-dominance relation, $\preceq_\epsilon$, defined as:

$$\mathbf{x}^1 \preceq_\epsilon \mathbf{x}^2 \iff \forall i \in 1..n : f_i(\mathbf{x}^1) \le \epsilon \cdot f_i(\mathbf{x}^2) \tag{14.3}$$

for a minimization problem, and assuming that all points are positive in all objectives. On this basis, the unary multiplicative epsilon indicator, $I_\epsilon^1(A)$ can then be defined as:

$$I_\epsilon^1(A) = I_\epsilon(A, R), \tag{14.4}$$

where R is any reference set of points. An equivalent unary additive epsilon indicator $I_{\epsilon+}^1$ is defined analogously, but is based on additive ϵ-dominance:

$$\mathbf{x}^1 \preceq_{\epsilon+} \mathbf{x}^2 \iff \forall i \in 1..n : f_i(\mathbf{x}^1) \le \epsilon + f_i(\mathbf{x}^2). \tag{14.5}$$

Both unary indicators are to be minimized. An indicator value less than or equal to 1 (I_ϵ^1) respectively 0 ($I_{\epsilon+}^1$) implies that A weakly dominates the reference set R.

The unary epsilon indicators are monotonic, but not strictly monotonic. They are sensitive to scaling and require a reference set relatively to which the epsilon value is calculated. For any finite Pareto set approximation A and any finite reference set R, the indicator values are cheap to compute; the runtime complexity is of order $\mathcal{O}(n \cdot |A| \cdot |R|)$.

14.4.7 The D Indicator Family

The *D indicators* are similar to the additive epsilon indicator and measure the average resp. worst case component-wise distance in objective space to the closest solution in a reference R, as suggested in (Czyzak and Jaskiewicz, 1998). Czyzak and Jaskiewicz (1998) introduced two versions, I_{D1} and I_{D2}; the first considers the average distance regarding the set R:

$$I_{D1}(A) = \frac{1}{|R|} \sum_{\mathbf{x}^2 \in R} \min_{\mathbf{x}^1 \in A} \max_{1 \leq i \leq k} \left(0, w_i(f_i(\mathbf{x}^1) - f_i(\mathbf{x}^2))\right)$$

where the w_i are weights associated with the specific objective functions. Alternatively, the worst case distance may be considered:

$$I_{D2}(A) = \max_{\mathbf{x}^2 \in R} \min_{\mathbf{x}^1 \in A} \max_{1 \leq i \leq k} \left(0, w_i(f_i(\mathbf{x}^1) - f_i(\mathbf{x}^2))\right)$$

As with the epsilon indicator family, the D indicators are monotonic, but not strictly monotonic, scaling dependent, and require a reference set. The running time complexity is of order $\mathcal{O}(n \cdot |A| \cdot |R|)$.

14.4.8 The R Indicator Family

The R *indicators* proposed in (Hansen and Jaszkiewicz, 1998) can be used to assess and compare Pareto set approximations on the basis of a set of utility functions. Here, a utility function u is defined as a mapping from the set $\mathbf{R}^k$ of k-dimensional objective vectors to the set of real numbers:

$$u : \mathbf{R}^k \mapsto \mathbf{R}.$$

Now, suppose that the decision maker's preferences are given in terms of a parameterized utility function u_λ and a corresponding set Λ of parameters. For instance, u_λ could represent a weighted sum of the objective values, where $\lambda = (\lambda_1, \ldots \lambda_n) \in \Lambda$ stands for a particular weight vector. Hansen and Jaszkiewicz (1998) propose several ways to transform such a family of utility functions into a quality indicator; in particular, the binary I_{R2} and I_{R3} indicators are defined as:[4]

$$I_{R2}(A, B) = \frac{\sum_{\lambda \in \Lambda} u^*(\lambda, A) - u^*(\lambda, B)}{|\Lambda|},$$

$$I_{R3}(A, B) = \frac{\sum_{\lambda \in \Lambda} [u^*(\lambda, B) - u^*(\lambda, A)]/u^*(\lambda, B)}{|\Lambda|}.$$

---

[4] The full formalism described in (Hansen and Jaszkiewicz, 1998) also considers arbitrary sets of utility functions in combination with a corresponding probability distribution over the utility functions. This is a way of enabling preference information regarding different parts of the Pareto front to be accounted for, e.g. more utility functions can be placed in the middle of the Pareto front in order to emphasise that region. The interested reader is referred to the original paper for further information.

where u^* is the maximum value reached by the utility function u_λ with weight vector λ on an Pareto set approximation A, i.e., $u^*(\lambda, A) = \max_{\mathbf{x} \in A} u_\lambda(\mathbf{f}(\mathbf{x}))$. Similarly to the epsilon indicators, the unary R indicators are defined on the basis of the binary versions by replacing one set by an arbitrary, but fixed reference set R: $I_{R2}^1(A) = I_{R2}(R, A)$ and $I_{R3}^1(A) = I_{R3}(A, R)$. The indicator values are to be minimized.

With respect to the choice of the parameterized utility function u_λ, there are various possibilities. A first utility function u that can be used in the above is a weighted linear function

$$u_\lambda(\mathbf{z}) = - \sum_{j \in 1..n} \lambda_j |z_j^* - z_j|, \tag{14.6}$$

where $\mathbf{z}^*$ is the ideal point, if known, or any point that weakly dominates all points in the corresponding Pareto front approximation. (When comparing approximation sets, the same $\mathbf{z}^*$ must be used each time).

A disadvantage of the use of a weighted linear function means that points not on the convex hull of the Pareto front approximation are not rewarded. Therefore, it is often preferable to use a nonlinear function such as the weighted Tchebycheff function,

$$u_\lambda(\mathbf{z}) = - \max_{j \in 1..n} \lambda_j |z_j^* - z_j|. \tag{14.7}$$

In this case, however, the utility of a point and one which weakly dominates it might be the same. To avoid this, it is possible to use the combination of linear and nonlinear functions: the augmented Tchebycheff function,

$$u_\lambda(\mathbf{z}) = - \left(\max_{j \in 1..n} \lambda_j |z_j^* - z_j| + \rho \sum_{j \in 1..n} |z_j^* - z_j| \right), \tag{14.8}$$

where ρ is a sufficiently small positive real number. In all cases, the set Λ of weight vectors should contain a sufficiently large number of uniformly dispersed normalized weight combinations λ with $\forall i \in 1..n : \lambda_n \geq 0 \wedge \sum_{j=1..n} \lambda_j = 1$.

The R indicators are monotonic, but not strictly, scaling dependent and require both a reference set as well as an ideal objective vector. The runtime complexity for computing the indicator values is of order $\mathcal{O}(n \cdot |\Lambda| \cdot |A| \cdot |R|)$.

14.4.9 Uniformity Measures

Various indicators have been proposed that measure how well the solutions of a Pareto set approximations are distributed in the objective space; often, the main focus is on a uniform distribution. To this end, one can consider the standard deviation of nearest neighbor distances, (see, e.g. Schott, 1995) and

(Deb *et al.*, 2002). Further examples can be found in (Knowles, 2002; Knowles and Corne, 2002).

In general, uniformity measures are not monotonic and not scaling invariant. The computation time required to compute the indicator values is usually quadratic in the cardinality of the Pareto set approximation under consideration, i.e., $\mathcal{O}(n \cdot |A|^2)$. Most measures of this class do not require additional information, but some involve certain problem-dependent parameters.

14.5 Indicator Combinations and Binary Indicators

The ideal quality indicator is strictly monotonic, scaling invariant, cheap to compute and does not require any additional information. However, it can be seen from the discussion above that such an ideal indicator does not exist. For instance, all monotonic unary quality indicators require a reference point and/or a reference set. The only strictly monotonic indicator currently known, the hypervolume indicator, is by far the most computationally expensive indicator. An obvious way to circumvent some of these problems is to combine multiple indicators. One has to define how exactly the resulting information is combined, for instance, one may consider a sequence of indicators. Suppose we would like to combine the epsilon indicator and the hypervolume indicator: one may say A is preferable to B if either the epsilon value for A is better or the epsilon values are identical and the hypervolume value for A is better. The resulting indicator combination would be strictly monotonic, but in average much less expensive than the hypervolume computation alone because in many cases the decision can be already made on the basis of the epsilon indicator. Another possibility is the use of binary quality indicators; see (Zitzler *et al.*, 2003) for a detailed discussion. Here, both scaling invariance and strict monotonicity can be achieved at the same time, e.g., with the coverage indicator (Zitzler and Thiele, 1998).

14.6 Stochasticity: General Considerations

So far, we have assumed that each algorithm under consideration always generates the same Pareto set approximation for a specific problem. However, many multiobjective optimizers are variants of randomized search algorithms and therefore stochastic in nature. If a stochastic multiobjective optimizer is applied several times to the same problem, each time a different Pareto set approximation may be returned. In this sense, with each randomized algorithm a random variable is associated whose possible values are Pareto Set approximations, i.e., elements of Ψ; the underlying probability density function is usually unknown.

One way to estimate this probability density function is by means of theoretical analysis. Since this approach is infeasible for many problems and algorithms used in practice, empirical studies are common in the context of the performance assessment of multiobjective optimizers. By running a specific algorithm several times on the same problem instance, one obtains a sample of Pareto set approximations. Now, comparing two stochastic optimizers basically means comparing the two corresponding samples of Pareto set approximations. This leads to the issue of statistical hypothesis testing. While in the deterministic case one can state, e.g., that "optimizer 1 achieves a higher hypervolume indicator value than optimizer 2", a corresponding statement in the stochastic case could be that "the expected hypervolume indicator value for algorithm 1 is greater than the expected hypervolume indicator value for algorithm 2 at a significance level of 5%".

In principle, there exist two basic approaches in the literature to analyze two or several samples of Pareto set approximations statistically. The more popular approach first transforms the samples of Pareto set approximations into samples of real values using quality indicators; then, the resulting samples of indicator values are compared based on standard statistical testing procedures.

Example 2. Consider two hypothetical stochastic multiobjective optimizers and assume that the outcomes of three independent optimization runs are as depicted in Fig. 14.3. If we use the hypervolume indicator with the reference point $(20, 20)$, we obtain two samples of indicator values: $(277, 171, 135)$ and $(277, 64, 25)$. These indicator value samples can then be compared and differences can be subjected to statistical testing procedures.

The alternative approach, the attainment function method, summarizes a sample of Pareto set approximations in terms of a so-called empirical attainment function. To explain the underlying idea, suppose that a certain stochastic multiobjective optimizer is run once on a specific problem. For each objective vector $\mathbf{z}$ in the objective space, there is a certain probability p that the resulting Pareto set approximation contains an element $\mathbf{x}$ such $\mathbf{f}(\mathbf{x}) \preceq \mathbf{z}$. We say p is the probability that $\mathbf{z}$ is *attained* by the optimizer. The *attainment function* gives for each objective vector $\mathbf{z}$ in the objective space the probability that $\mathbf{z}$ is attained in one optimization run of the considered algorithm. As before, the true attainment function is usually unknown, but it can be estimated on the basis of the approximation set samples: one simply counts the number of approximation sets by which each objective vector is attained and normalizes the resulting number with the overall sample size. The attainment function is a first order moment measure, meaning that it estimates the probability that $\mathbf{z}$ is attained in one optimization run of the considered algorithm *independently* of attaining any other $\mathbf{z}$. For the consideration of higher order attainment functions, Grunert da Fonseca *et al.* (2001) have developed corresponding statistical testing procedures.

Example 3. Consider Fig. 14.3. For the scenario on the right, the three Pareto front approximations cut the objective space into four regions: the upper right

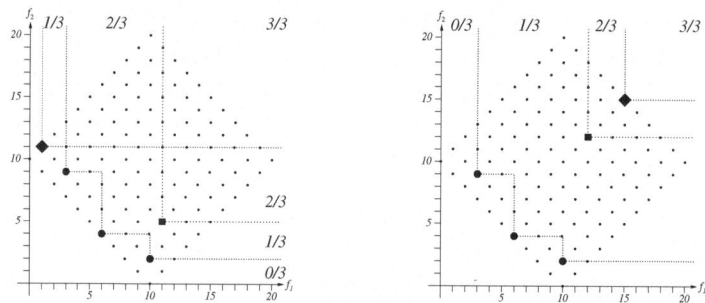

Fig. 14.3. Hypothetical outcomes of three runs for two different stochastic optimizers (left and right). The numbers in the figures give the relative frequencies according to which the distinct regions in the objective space were attained.

region is attained in all of the runs and therefore is assigned a relative frequency of 1, the lower left region is attained in none of the runs, and the remaining two regions are assigned relative frequencies of 1/3 and 2/3 because they are attained in one respectively two of the three runs. In the scenario on the left, the objective space is partitioned into six regions; the relative frequencies are determined analogously as shown in the figure.

A third approach to statistical analysis of approximation sets consists in ranking the obtained approximations by means of the dominance relation, in analogous fashion to the way dominance-based fitness assignment ranks objective vectors in evolutionary multiobjective optimization. Basically, for each Pareto set approximation generated by one optimizer it is computed by how many Pareto set approximations produced by another optimizer it is dominated. As a result, one obtains, for each algorithm, a set of ranks and can statistically verify whether the rank distributions for two algorithms differ significantly or not. We call this method, dominance ranking.

Example 4. To compare the outcomes of the two hypothetical optimizers depicted in Fig. 14.3, we check for each pair consisting of one Pareto set approximation of the first optimizer and one Pareto set approximation of the second optimizer whether either is better or not. For the Pareto front approximation represented by the diamond on the left hand side, none of the three Pareto front approximations on the right is better and therefore it is assigned the lowest rank 0. The Pareto front approximation associated with the diamond on the right hand side is worse than all three Pareto front approximations on the left and accordingly its rank is 3. Overall, the resulting rank distributions are $(0,0,1)$ for the algorithm on the left hand side and $(0,2,3)$ for the algorithm on the right hand side. A special statistical test can be used to determine whether the two rank distributions are significantly different.

14.7 Sample Transformations

The three comparison methodologies outlined in the previous section have in common that the sample of approximation sets associated with an algorithm is first transformed into another representation—specifically, a sample of indicator values, an empirical attainment function, or a sample of ranks—before the statistical testing methods are applied. In the following, we will review each of the different types of sample transformations in greater detail (but now considering the dominance ranking first); the issue of statistical testing will be covered in Section 14.8.

14.7.1 Dominance Ranking

Principles and Procedure

Suppose that we wish to compare the quality of Pareto set approximations generated by two stochastic multiobjective optimizers, where $A_1, A_2, \ldots, A_r$ represent the approximations generated by the first optimizer in r runs, while $B_1, B_2, \ldots, B_s$ denote the approximations generated by the second optimizer in s runs. Using the preference structure of the underlying set problem $(\Psi, \Omega, \mathbf{f}', rel')$, one can now compare all A_i with all B_j and thereby assign a figure of merit or a rank to each Pareto set approximation, similarly to the way that dominance-based fitness assignment works in multiobjective evolutionary algorithms. In principle, there are several ways to assign each Pareto set approximation a rank on the basis of a dominance relation, e.g., by counting the number of sets by which a specific approximation is dominated (Fonseca and Fleming, 1993) or by performing a nondominated sorting on the Pareto set approximations under considerations. Here, the former approach in combination with the extended weak Pareto dominance $\preceq$, cf. Section 14.2 on Page 374, is preferred as it produces a finer-grained ranking, with fewer ties, than nondominated sorting:

$$rank(A_i) = |\{B|B \in \{B_1, \ldots, B_s\} \ \wedge B \prec A_i\}|. \tag{14.9}$$

The ranks for $B_1, \ldots, B_s$ are determined analogously. The lower the rank, the better the corresponding Pareto set approximation with respect to the entire collection of sets associated with the other optimizer.

The result of this procedure is that each A_i and B_j is associated with a figure of merit. Accordingly, the samples of Pareto set approximations associated with each algorithm have been transformed into samples of ranks: $(rank(A_1), rank(A_2), \ldots, rank(A_r))$ and $(rank(B_1), rank(B_2), \ldots, rank(B_s))$.

An example performance comparison study using the dominance ranking procedure can be found in (Knowles *et al.*, 2006).

Discussion

The dominance ranking approach relies on the concept of Pareto dominance and some ranking procedure only, and thus yields quite general statements about the relative performance of the considered optimizers, fairly independently of any preference information. Thus, we recommend this approach to be the first step in any comparison: if one optimizer is found to be significantly better than the other by this procedure, then it is better in a sense consistent with the underlying preference structure. It may be interesting and worthwhile to use either quality indicators or the attainment function to characterize further the differences in the distributions of the Pareto set approximations, but these methods are not needed to conclude which of the stochastic optimizers generates the better sets, if a significant difference can be demonstrated using the ranking of approximation sets alone.

14.7.2 Quality Indicators

Principles and Procedures

As stated earlier, a *unary* quality indicator I is defined as a mapping from Ψ to the set of real numbers. The order that I establishes on Ω is supposed to represent the quality of the Pareto set approximations. Thus, given a pair of approximations, A and B, the difference between their corresponding indicator values $I(A)$ and $I(B)$ should reveal a difference in the quality of the two sets. This not only holds for the case that either set is better, but also when A and B are incomparable. Note that this type of information goes beyond pure Pareto dominance and represents additonal knowledge; we denote this knowledge as *preference information*.

Discussion

Using unary quality indicators in a comparative study is attractive as it transforms a sample of approximation sets into a sample of reals for which standard statistical testing procedures exist, cf. Section 14.8. In contrast to the dominance ranking approach, it is also possible to make quantitative statements about the differences in quality, even for incomparable Pareto set approximations. However, this comes at the cost of generality: every unary quality indicator represents specific preference information. Accordingly, any statement of the type 'algorithm A outperforms algorithm B' needs to be qualified in the sense of 'with respect to quality indicator I'—the situation may be different for another indicator.

14.7.3 Empirical Attainment Function

Principles and Procedures

The central concept in this approach is the notion of an *attainment function*. Since the multiobjective optimizers that we consider may be stochastic, the result of running the optimizer can be described by a distribution. Because the optimizer returns a Pareto set approximation in any given run, the distribution can be described in the objective space by a random *set* $\mathcal{Z}$ of random objective vectors $\breve{\mathbf{z}}^j$, with the cardinality of the set, m, also random, as follows:

$$\mathcal{Z} = \{\breve{\mathbf{z}}^j \in \mathbf{R}^k, j = 1, \ldots, m\}, \tag{14.10}$$

where k is the number of objectives of the problem. The attainment function is a description of this distribution based on the notion of goal-attainment: A goal, here meaning an objective vector, is attained whenever it is weakly dominated by the Pareto front approximation returned by the optimizer. It is defined by the function $\alpha_{\mathcal{Z}}(.) : \mathbf{R}^n \mapsto [0, 1]$ with

$$\alpha_{\mathcal{Z}}(z) = P(\breve{\mathbf{z}}^1 \preceq \mathbf{z} \vee \breve{\mathbf{z}}^2 \preceq \mathbf{z} \vee \ldots \vee \breve{\mathbf{z}}^m \preceq \mathbf{z}) \tag{14.11}$$

$$= P(\mathcal{Z} \preceq \{\mathbf{z}\}) \tag{14.12}$$

$$= P(\text{that the optimizer attains goal } \mathbf{z} \text{ in a single run}), \tag{14.13}$$

where $P(.)$ is the probability density function. The attainment function is a first order moment measure, and can be seen as a mean-measure for the set $\mathcal{Z}$. Thus, it describes the location of the Pareto set approximation distribution; higher order moments are needed if the variability across runs is to be assessed, and to assess dependencies between the probabilities of attaining two or more goals *in the same run* (see Fonseca *et al.*, 2005).

The attainment function can be estimated from a sample of r independent runs of an optimizer via the *empirical attainment function* (EAF) defined as

$$\alpha_r(\mathbf{z}) = \frac{1}{r} \sum_{i=1}^{r} I(\mathbf{f}'(A_i) \preceq \{\mathbf{z}\}), \tag{14.14}$$

where A_i is the ith Pareto set approximation (run) of the optimizer and $I(.)$ is the indicator function, which evaluates to one if its argument is true and zero if its argument is false. In other words, the EAF gives for each objective vector in the objective space the relative frequency that it was attained, i.e., weakly dominated by an Pareto front approximation, with respect to the r runs.

The outcomes of two optimizers can be compared by performing a corresponding statistical test on the resulting two EAFs, as will be explained in Section 14.8.4. In addition, EAFs can also be used for visualizing the outcomes of multiple runs of an optimizer. For instance, one may be interested in plotting all the goals that have been attained (independently) in 50% of the runs. This is defined in terms of a $k\%$-*attainment set*:

A Pareto set approximation A is called the $k\%$-*attainment set* of an EAF $\alpha_r(\mathbf{z})$, iff the corresponding Pareto front approximation weakly dominates exactly those objective vectors that have been attained in at least k percent of the r runs. Formally,

$$\forall \mathbf{z} \in Z : \alpha_r(\mathbf{z}) \geq k/100 \Leftrightarrow \mathbf{f}'(A) \preceq \{\mathbf{z}\} \qquad (14.15)$$

We can then plot the *attainment surface* of such an approximation set, defined as:

An attainment surface of a given Pareto set approximation A is the union of all tightest goals that are known to be attainable as a result of A. Formally, this is the set $\{\mathbf{z} \in \mathbf{R}^k \mid \mathbf{f}'(A) \preceq \mathbf{z} \wedge \not\exists \mathbf{z}^2 \in \mathbf{R}^k : \mathbf{f}'(A) \preceq \mathbf{z}^2 \prec \mathbf{z}\}$.

Roughly speaking, then, the $k\%$-attainment surface divides the objective space in two parts: the goals that have been attained and the goals that have not been attained with a frequency of at least k percent.

Example 5. Suppose a stochastic multiobjective optimizer returns the Pareto front approximations depicted in Fig. 14.4 for five different runs on a biobjective optimization problem. The corresponding attainment surfaces are shown in Fig. 14.5; they summarize the underlying empirical attainment function.

Discussion

The attainment function approach distinguishes itself from the dominance ranking and indicator approaches by the fact that the transformed samples are multidimensional, i.e., defined on Z and not on $\mathbf{R}$. Thereby, less information is lost by the transformation, and in combination with a corresponding statistical testing procedure detailed differences can be revealed between the EAFs of two algorithms (see Section 14.8). However, the approach is computationally

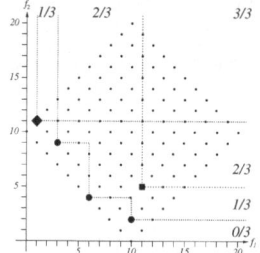 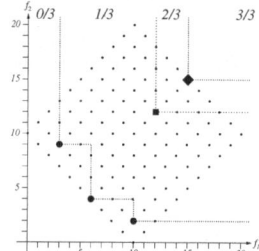

Fig. 14.4. A plot showing five Pareto front approximations. The visual evaluation is difficult, although there are only a few points per set, and few sets.

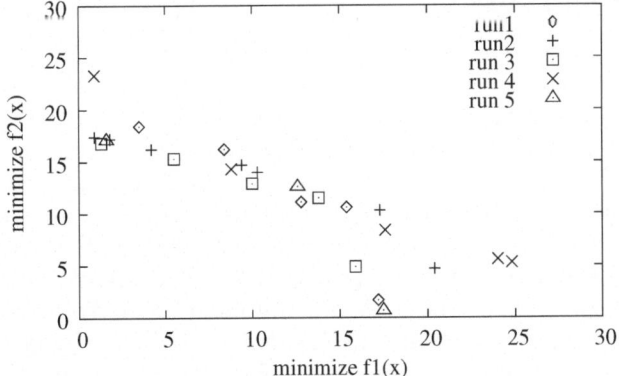

Fig. 14.5. Attainment surface plots for the Pareto fron approximations in Figure 14.4. The first (solid) line represents the 20%-attainment surface, the second line the 40%-attainment surface, and so forth; the fifth line stands for the 100%-attainment surface.

expensive and therefore only applicable in the case of a few objective functions. Concerning visualization of EAFs, recently, an approximate algorithm has been presented by Knowles (2005) that computes a given k%-attainment surface only at specified points on a grid and thereby achieves considerable speedups in comparison with the exact calculation of the attainment surface defined above.

14.8 Statistical Testing

14.8.1 Fundamentals

The previous section has described three different transformations that can be applied to a sample of Pareto set approximations generated from multiple runs of an optimizer. The ultimate purpose of generating the samples and applying the transformations is to allow us to (a) describe and (b) make inferences about the underlying random approximation set distributions of the (two or more) optimizers, thus enabling us to compare their performance.

It is often convenient to summarise a random sample from a distribution using *descriptive statistics* such as the mean and variance. The mean, median and mode are sometimes referred to as first order moments of a distribution, and they describe or summarise the *location* of the distribution on the real number line. The variance, standard deviation, and inter-quartile range are known as second-order moments and they describe the spread of the data. Using box-plots (Chambers *et al.*, 1983) or tabulating mean and standard deviation values are useful ways of presenting such data.

Statistical Inferences

Descriptive statistics are limited, however, and should usually be given only to supplement any statistical inferences that can be made from the data. The standard statistical inference we would like to make, if it is true, is that one optimizer's underlying Pareto set approximation *distribution* is better than another one's.[5] However, we cannot determine this fact definitively because we only have access to finite-sized *samples* of Pareto set approximations. Instead, it is standard practice to *assume* that the data is consistent with a simpler explanation known as the *null hypothesis*, H_0, and then to test how likely this is to be true, given the data. H_0 will often be of the form 'samples A and B are drawn from the same distribution' or 'samples A and B are drawn from distributions with the same mean value'. The probability of obtaining a finding at least as 'impressive' as that obtained, assuming the null hypothesis is true, is called the *p-value* and is computed using an inferential *statistical test*. The *significance level*, often denoted as α, defines the largest acceptable *p-value* and represents a threshold that is user-defined. A *p-value* lower than the chosen significance level α then signifies that the null hypothesis can be rejected in favour of an *alternative hypothesis*, H_A, at a significance level of α. The definition of the alternative hypothesis usually takes one of two forms. If H_A is of the form 'sample A comes from a better distribution than sample B' then the inferential test is a *one-tailed test*. If H_A does not specify a prediction about which distribution is better, and is of the form 'sample A and sample B are from different distributions' then it is a two-tailed test. A one-tailed test is more *powerful* than a two-tailed test, meaning that for a given alpha value, it rejects the null hypothesis more readily in cases where it is actually false.

Non-parametric Statistical Inference: Rank and Permutation Tests

Some inferential statistical tests are based on assuming the data is drawn from a distribution that closely approximates a known distribution, e.g. the normal distribution or Student's t distribution. Such known distributions are completely defined by their *parameters* (e.g. the mean and standard deviation), and tests based on these known distributions are thus termed parametric statistical tests. Parametric tests are powerful—that is, the null hypothesis is rejected in most cases where it is indeed false—because even quite small differences between the means of two normal distributions can be detected accurately. However, unfortunately, the assumption of normality cannot be theoretically justified for stochastic optimizer outputs, in general, and it is difficult to empirically test for normality with relatively small samples (less than 100 runs). Therefore, it is safer to rely on *nonparametric tests* (Conover, 1999), which make no assumptions about the distributions of the variables.

---

[5] Most statistical inferences are formulated in terms of precisely two samples, in this way.

Two main types of nonparametric tests exist: rank tests and permutation tests. Rank tests pool the values from several samples and convert them into ranks by sorting them, and then employ tables describing the limited number of ways in which ranks can be distributed (between two or more algorithms) to determine the probability that the samples come from the same source. Permutation tests use the original values without converting them to ranks but estimate the likelihood that samples come from the same source explicitly by Monte Carlo simulation. Rank tests are the less powerful but are also less sensitive to outliers and computationally cheap. Permutation tests are more powerful because information is not thrown away, and they are also better when there are many tied values in the samples, however they can be expensive to compute for large samples.

In the following, we describe selected methods for nonparametric inference testing for each of the different transformations. We follow this with a discussion of issues relating to matched samples, multiple inference testing, and assessing worst- and best-case performance.

14.8.2 Comparing Samples of Dominance Ranks

Dominance ranking converts the samples of approximation sets from two or more optimizers into a sample of dominance ranks. A test statistic is computed from these ranks by summing over the ranks in each of the two samples and taking the difference of these sums. In order to determine whether the value of the test statistic is significant, a permutation test must be used. The standard Mann-Whitney rank sum test and tables (Conover, 1999) cannot be used here because the rank distributions are affected by the fact that the sets are partially ordered (rather than totally ordered numbers). Thus, to compute the null distribution, the assignment of the Pareto set approximations to the optimizers must be permuted. Basically, the set $\{A_1, A_2, \ldots, A_r, B_1, B_2, \ldots, B_s\}$ is partitioned into one set of r approximations and another set of s approximations; for each partitioning the difference between the rank sums can be determined, finally yielding a distribution of rank sum differences. Details for this statistical testing procedure are given in (Knowles et al., 2006).

14.8.3 Comparing Sample Indicators Values

The use of a quality indicator reduces the dimension of a Pareto set approximation to a single figure of merit. One of the main advantages, and underlying motivations, for using indicators is that this reduction to one dimension allows statistical testing to be carried out in a relatively straightforward manner using standard univariate statistical tests, i.e. as is done when comparing best-of-population fitness values (or equivalents) in single-objective algorithm comparisons. Here, the Mann-Whitney rank sum test or Fisher's permutation test can be used (Conover, 1999); the Kruskal-Wallis test may be more appropriate if multiple (more than two) algorithms are to be compared.

In the case that a combination of multiple quality indicators is considered (see Page 386), slightly different preferences are assessed by each of the indicators and this may help to build up a better picture of the overall quality of the Pareto set approximations. On the other hand, using several indicators does bring into play multiple testing issues if the distributions from different indicators are being tested independently, cf. Section 14.8.5.

14.8.4 Comparing Empirical Attainment Functions

The EAF of an optimizer is a generalization of a univariate empirical cumulative distribution function (ECDF) (Grunert da Fonseca *et al.*, 2001). In order to test if two ECDFs are different, the Kolmogorov-Smirnov (KS) test can be applied. This test measures the maximum difference between the ECDFs and assesses the statistical significance of this difference. An algorithm that computes a KS-like test for two EAFs is described in (Shaw *et al.*, 1999). The test only determines if there is a significant *difference* between the two EAFs, based on the maximum difference. It does not determine whether one algorithm's entire EAF is 'above' the other one:

$$\forall z \in Z, \alpha_r^A(z) \geq \alpha_r^B(z),$$

or not. In order to probe such specific differences, one must use methods for visualizing the EAFs.

For two-objective problems, plotting significant differences in the empirical attainment functions of two optimizers, using a pair of plots, can be done by colour-coding either (i) levels of difference in the sample probability, or (ii) levels of statistical significance of a difference in sample probability, of attaining a goal, for all goals. Option (ii) is more informative and can be computed from the fact that there is a correspondence between the statistical significance level α of the KS-like test and the maximum distance between the EAFs that needs to be exceeded. Thus the KS-like test can be run for different selected α values to compute these different distances. Then, the actual measured distances between the EAFs at every z can be converted to a significance level.

An example of such a pair of plots is shown in Figure 14.6. This kind of plot has been used to good effect in (López-Ibáñez *et al.*, 2006). Note also that Fonseca *et al.* (2005) have devised plots that can indicate second-order information, i.e. the probability of an optimizer attaining pairs of goals simultaneously.

14.8.5 Advanced Topics

Matched Samples

When comparing a pair of stochastic optimizers, two slightly different scenarios are possible. In one case, each run of each optimizer is a completely

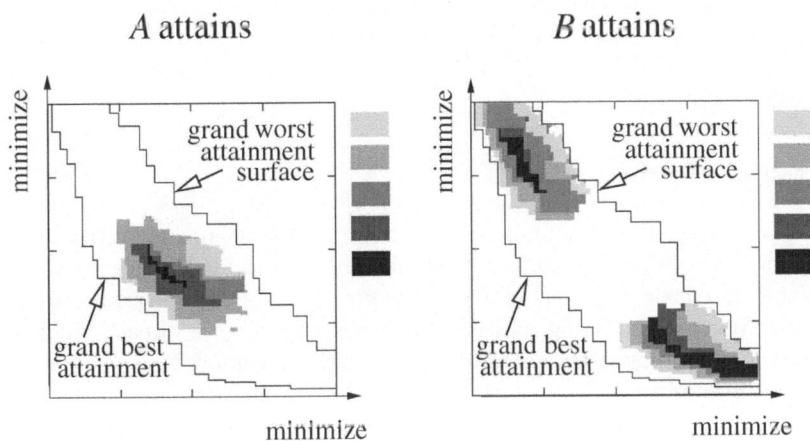

Fig. 14.6. Individual differences between the probabilities of attaining different goals on a two-objective minimization problem with optimizer O_1 and optimizer O_2, shown using a greyscale plot. The grand best and worst attainment surfaces (the same in both plots) indicate the borders beyond which the goals are never attained or always attained, computed from the *combined* collection of Pareto set approximations. Differences in the frequency with which certain goals are met by the respective algorithms O_1 and O_2 are then represented in the region between these two surfaces. In the left plot, darker regions indicate goals that are attained more frequently by O_1 than by O_2. In the right plot, the reverse is shown. The intensity of the shading can correspond to either the magnitude of a difference in the sample probabilities, or to the level of statistical significance of a difference in these probabilities.

independent random sample; that is, the initial population (if appropriate), the random seed, and all other random variables are drawn independently and at random on each run. In the other case, the influence of one or more random variables is partially removed from consideration; e.g. the initial population used by the two algorithms may be matched in corresponding runs, so that the runs (and hence the final quality indicator values) should be taken as pairs. In the former scenario, the statistical testing will reveal, in quite general terms, whether there is a difference in the distributions of indicator values resulting from the two stochastic optimizers, from which a general performance difference can be inferred. In the latter scenario—taking the particular case where initial populations are matched—the statistical testing reveals whether there is a difference in the indicator value distributions *given the same initial population*, and the inference in this case relates to the optimizer's ability to *improve* the initial population. While the former scenario is more general, the latter may give more statistically significant results.

If matched samples have been collected, then the Wilcoxon signed rank test (Conover, 1999) or Fisher's matched samples test (Conover, 1999) can be used instead of the Mann-Whitney rank sum test respectively Fisher's permutation test.

Multiple Testing

Multiple testing (Benjamini and Hochberg, 1995; Bland and Altman, 1995; Miller, 1981; Perneger, 1998; Westfall and Young, 1993) occurs when one wishes to consider several statistical hypotheses (or comparisons) simultaneously. When considering multiple tests, the significance of each single result needs to be adjusted to account for the fact that, as more tests are considered, it becomes more and more likely that some (unspecified) result will give an extreme value, resulting in a rejection of the null hypothesis for that test.

For example, imagine we carry out a study consisting of twenty different hypothesis tests, and assume that we reject the null hypothesis of each test if the p-value is 0.05 or less. Now, the chance that at least one of the inferences will be a type-1 error (i.e. the null hypothesis is wrongly rejected) is $1 - (0.95^{20}) \simeq 64\%$, when assuming that the null hypothesis was true in every case. In other words, more often than not, we wrongly claim a significant result (on at least one test). This situation is made even worse if we only report the cases where the null hypothesis was rejected, and do not report that the other tests were performed: in that case, results can be utterly misleading to a reader.

Multiple testing issues in the case of assessing stochastic multiobjective optimizers can arise for at least two different reasons:

- There are more than two algorithms and we wish to make inferences about performance differences between all or a subset of them.
- There are just two algorithms, but we wish to make multiple statistical tests of their performance, e.g., considering more than one indicator.

Clearly, this is a complicated issue and we can only touch on the correct procedures here. The important thing to know is that the issue exists, and to do something to minimize the problem. We briefly consider five possible approaches:

i). Do all tests as normal (with uncorrected p-values) but report all tests done openly and notify the reader that the significance levels are not, therefore, reliable.

ii). In the special case where we have multiple algorithms but just one statistic (e.g. one indicator), use a statistical test that is designed explicitly for assessing several independent samples. The Kruskal-Wallis test (Conover, 1999), is an extension of the two-sample Mann-Whitney test that works for multiple samples. Similarly, the Friedman test (Conover, 1999) extends the paired Wilcoxon signed rank test to any number of related samples.

iii). In the special case where we want to use multiple statistics (e.g. multiple different indicators) for just two algorithms, and we are interested *only* in an inference derived *per-sample* from all statistics, (e.g. we want to test the significance of a difference in hypervolume between those pairs A_i and B_i *where* the diversity difference between them is positive), then the permutation test can be used to derive the null distribution, as usual.

iv). Minimize the number of different tests carried out on a pair of algorithms by carefully choosing which tests to apply before collecting the data. Collect independent data for each test to be carried out.

v). Apply the tests on the same data but use methods for correcting the *p*-values for the reduction in confidence associated with data re-use.

Approach (i) does not allow powerful conclusions to be drawn, but it at least avoids mis-representation of results. The second approach is quite restrictive as it only applies to a single test being applied to multiple algorithms—and uses rank tests, which might not be appropriate in all circumstances. Similarly, (iii) only applies in the special case noted. A more general approach is (iv), which is just the conservative option; the underlying strategy is to perform a test only if there is some realistic chance that the null hypothesis can be rejected (and the result would be interesting). This careful conservatism can then be accommodated. However, while following (iv) might be possible much of the time, sometimes it is essential to do several tests on limited data and to be as confident as possible about any positive results. In this case, one should then use approach (v).

The simplest and most conservative, i.e., weakest approach for correcting the *p*-values is the Bonferroni correction (Bland and Altman, 1995). Suppose we would like to consider an overall significance level of α and that altogether n comparisons, i.e., distinct statistical tests, are performed per sample. Then, the significance level α_s for each distinct test is set to

$$\alpha_s = \frac{\alpha}{n} \tag{14.16}$$

Explicitly, given n tests T_i for hypotheses $H_i (1 \leq i \leq n)$ under the assumption H_0 that all hypotheses H_i are false, and if the individual test critical values are $\leq \alpha/n$, then the experiment-wide critical value is $\leq \alpha$. In equation form, if

$$P(T_i \text{ passes } | H_0) \leq \frac{\alpha}{n} \text{ for } 1 \leq i \leq n, \tag{14.17}$$

then

$$P(\text{some } T_i \text{ passes } | H_0) \leq \alpha. \tag{14.18}$$

In most cases, the Bonferroni approach is too weak to be useful and other methods are preferrable (Perneger, 1998), e.g., resampling based methods (Westfall and Young, 1993).

Assessing Worst-Case or Best-Case Performance

In certain circumstances, it may be important to compare the worst-case or best-case performance of two optimizers. Obtaining statistically significant inferences for these is more computationally demanding than when assessing differences in mean or typical performance, however, it can be done using permutation methods, such as bootstrapping or variants of Fisher's permutation test (Efron and Tibshirani, 1993, chap. 15).

For example, let us say that we wish to estimate whether there is a difference in the expected worst indicator value of two algorithms, when each is run ten times. To assess this, one can run each algorithm for 30 batches of 10 runs, and find the mean of the worst-in-a-batch value, for each algorithm. Then, to compute the null distribution, the labels of all 600 samples can be randomly permuted, and the worst indicator value from those with a label in $1, \ldots, 10$ are determined. By sampling this statistic many times, the desired p-value that the mean of the worst-in-a-batch statistics are significantly different, can be computed. Quite obviously, such a testing procedure is quite general and it can be tailored to answer many questions related to worst-case or best-case performance.

14.9 Summary

This chapter deals with the issue of assessing and comparing the quality of Pareto set approximations. Two current principal approaches, the quality indicator method and the attainment function method, are discussed, and, in addition, a third approach, the dominance-ranking technique, is presented.[6]

As discussed, there is no 'best' quality assessment technique with respect to both quality measures and statistical analysis. Instead, it appears to be reasonable to use the complementary strengths of the three general approaches. As a first step in a comparison, it can be checked whether the considered optimizers exhibit significant differences using the dominance-ranking approach, because such an analysis allows the strongest type of statements. Quality indicators can then be applied in order to quantify the potential differences in quality and to detect differences that could not be revealed by dominance ranking. The corresponding statements are always restricted as they only hold for the preferences that are represented by the considered indicators. The computation and the statistical comparison of the empirical attainment functions are especially useful in terms of visualization and to add another level of detail; for instance, plotting the regions of significant difference gives hints on *where* the outcomes of two algorithms differ.

---

[6] Implementations for selected quality indicators as well as statistical testing procedures can be downloaded at *http://www.tik.ee.ethz.ch/sop/pisa/* under the heading 'performance assessment'.

We noted when discussing quality indicators that, as well as their traditional use to assess optimization outcomes, they can also be used within optimizers, to guide the generating process (Beume *et al.*, 2007; Fleischer, 2003; Smith *et al.*, 2008; Wagner *et al.*, 2007; Zitzler and Künzli, 2004). Optimizers that seek to maximize a quality indicator directly are effectively conducting the search in the space of approximation sets, rather than in the space of solutions or points. This seems a logical and attractive approach when attempting to generate a Pareto front approximation, because ultimately the outcome will be assessed using a quality indicator (usually). However, although such approaches are improving, some of them still rely on approximation of the set-based indicator function, or they do not rely solely on the indicator, but make use of heuristics concerning individuals (point/solutions) (e.g., an individual's nondominated rank) as well. A recent study even compared set-based selection with individual-based selection, and found the latter to be generally preferable.

Quality indicators for assessing Pareto front approximations are sometimes used without explicitly stating what the DM preferences are. Really, the indicator(s) used should reflect any information one has about the DM preferences, so that approximation sets are assessed appropriately. The work of Hansen and Jaszkiewicz (1998) defined some quality indicators in terms of sets of utility functions, a framework that easily allows for DM preferences to be incorporated into assessment. A similar approach was recently proposed by Zitzler *et al.* (2007) for the hypervolume. Both of these indicator families can be used to incoporate preferences within generating methods (potentially in an interactive fashion).

Finally, note that there are several further issues that have not been treated in this chapter, e.g., binary quality indicators; indicators taking the decision vectors into account; computation of indicators on parallel or distributed architectures. Many of these issues represent current research directions which will probably lead to modified or additional performance assessment methods in the near future.

Acknowledgements

Sections 14.1 to 14.5 summarize the results of the discussion of the working group on set quality measures during the Dagstuhl seminar on evolutionary multiobjective optimization 2006. The working group consisted of the following persons: Jörg Fliege, Carlos M. Fonseca, Christian Igel, Andrzej Jaszkiewicz, Joshua D. Knowles, Alexander Lotov, Serpil Sayin, Lothar Thiele, Andrzej Wierzbicki, and Eckart Zitzler. The authors would also like to thank Carlos M. Fonseca for valuable discussion and for providing the EAF tools.

References

Benjamini, Y., Hochberg, Y.: Controlling the false discovery rate: a practical and powerful approach to multiple testing. Journal of the Royal Statistical Society, Series B (Methodological) 57, 125–133 (1995)

Berezkin, V.E., Kamenev, G.K., Lotov, A.V.: Hybrid adaptive methods for approximating a nonconvex multidimensional pareto frontier. Computational Mathematics and Mathematical Physics 46(11), 1918–1931 (2006)

Beume, N., Rudolph, G.: Faster S-Metric Calculation by Considering Dominated Hypervolume as Klee's Measure Problem. In: Proceedings of the Second IASTED Conference on Computational Intelligence, pp. 231–236. ACTA Press, Anaheim (2006)

Beume, N., Naujoks, B., Emmerich, M.: SMS-EMOA: Multiobjective selection based on dominated hypervolume. European Journal on Operational Research 181, 1653–1669 (2007)

Bland, J.M., Altman, D.G.: Multiple significance tests: the bonferroni method. British Medical Journal 310, 170 (1995)

Chambers, J., Cleveland, W., Kleiner, B., Tukey, P.: Graphical Methods for Data Analysis. Wadsworth, Belmont (1983)

Coello Coello, C.A., Van Veldhuizen, D.A., Lamont, G.B.: Evolutionary Algorithms for Solving Multi-Objective Problems. Kluwer Academic Publishers, New York (2002)

Conover, W.J.: Practical Nonparametric Statistics, 3rd edn. John Wiley and Sons, New York (1999)

Czyzak, P., Jaskiewicz, A.: Pareto simulated annealing—a metaheuristic for multi-objective combinatorial optimization. Multi-Criteria Decision Analysis 7, 34–47 (1998)

Deb, K.: Multi-objective optimization using evolutionary algorithms. Wiley, Chichester (2001)

Deb, K., Pratap, A., Agrawal, S., Meyarivan, T.: A fast and elitist multi-objective genetic algorithm: NSGA-II. IEEE Transactions on Evolutionary Computation 6(2), 181–197 (2002)

Efron, B., Tibshirani, R.: An introduction to the bootstrap. Chapman and Hall, London (1993)

Ehrgott, M., Gandibleux, X.: A Survey and Annotated Bibliography of Multiobjective Combinatorial Optimization. OR Spektrum 22, 425–460 (2000)

Fleischer, M.: The Measure of Pareto Optima. In: Fonseca, C.M., Fleming, P.J., Zitzler, E., Deb, K., Thiele, L. (eds.) EMO 2003. LNCS, vol. 2632, pp. 519–533. Springer, Heidelberg (2003)

Fonseca, C.M., Fleming, P.J.: Genetic Algorithms for Multiobjective Optimization: Formulation, Discussion and Generalization. In: Forrest, S. (ed.) Proceedings of the Fifth International Conference on Genetic Algorithms, pp. 416–423. Morgan Kaufmann, San Mateo (1993)

Fonseca, C.M., Grunert da Fonseca, V., Paquete, L.: Exploring the performance of stochastic multiobjective optimisers with the second-order attainment function. In: Coello Coello, C.A., Hernández Aguirre, A., Zitzler, E. (eds.) EMO 2005. LNCS, vol. 3410, pp. 250–264. Springer, Heidelberg (2005)

Fonseca, C.M., Paquete, L., López-Ibáñez, M.: An Improved Dimension-Sweep Algorithm for the Hypervolume Indicator. In: Congress on Evolutionary Computation (CEC 2006), Sheraton Vancouver Wall Centre Hotel, Vancouver, BC Canada, pp. 1157–1163. IEEE Computer Society Press, Los Alamitos (2006)

Grunert da Fonseca, V., Fonseca, C.M., Hall, A.O.: Inferential Performance Assessment of Stochastic Optimisers and the Attainment Function. In: Zitzler, E., Deb, K., Thiele, L., Coello Coello, C.A., Corne, D.W. (eds.) EMO 2001. LNCS, vol. 1993, pp. 213–225. Springer, Heidelberg (2001)

Hansen, M.P., Jaszkiewicz, A.: Evaluating the quality of approximations of the non-dominated set. Technical report, Institute of Mathematical Modeling, Technical University of Denmark. IMM Technical Report IMM-REP-1998-7 (1998)

Helbig, S., Pateva, D.: On several concepts for ϵ-efficiency. OR Spektrum 16(3), 179–186 (1994)

Kamenev, G., Kondtratíev, D.: Method for the exploration of non-closed nonlinear models (in Russian). Matematicheskoe Modelirovanie 4(3), 105–118 (1992)

Kamenev, G.K.: Approximation of completely bounded sets by the deep holes method. Computational Mathematics And Mathematical Physics 41, 1667–1676 (2001)

Knowles, J.: A summary-attainment-surface plotting method for visualizing the performance of stochastic multiobjective optimizers. In: Computational Intelligence and Applications, Proceedings of the Fifth International Workshop on Intelligent Systems Design and Applications: ISDA'05 (2005)

Knowles, J., Corne, D.: On Metrics for Comparing Non-Dominated Sets. In: Congress on Evolutionary Computation (CEC 2002), pp. 711–716. IEEE Press, Piscataway (2002)

Knowles, J., Thiele, L., Zitzler, E.: A Tutorial on the Performance Assessment of Stochastic Multiobjective Optimizers. TIK Report 214, Computer Engineering and Networks Laboratory (TIK), ETH Zurich (2006)

Knowles, J.D.: Local-Search and Hybrid Evolutionary Algorithms for Pareto Optimization. Ph.D. thesis, University of Reading (2002)

López-Ibáñez, M., Paquete, L., Stützle, T.: Hybrid population-based algorithms for the bi-objective quadratic assignment problem. Journal of Mathematical Modelling and Algorithms 5(1), 111–137 (2006)

Lotov, A.V., Kamenev, G.K., Berezkin, V.E.: Approximation and Visualization of Pareto-Efficient Frontier for Nonconvex Multiobjective Problems. Doklady Mathematics 66(2), 260–262 (2002)

Lotov, A.V., Bushenkov, V.A., Kamenev, G.K.: Interactive Decision Maps. Approximation and Visualization of Pareto Frontier. Kluwer Academic Publishers, Boston (2004)

Miller, R.G.: Simultaneous Statistical Inference, 2nd edn. Springer, New York (1981)

Perneger, T.V.: What's wrong with Bonferroni adjustments. British Medical Journal 316, 1236–1238 (1998)

Schott, J.: Fault Tolerant Design Using Single and Multicriteria Genetic Algorithm Optimization. Master's thesis, Department of Aeronautics and Astronautics, Massachusetts Institute of Technology (1995)

Shaw, K.J., Nortcliffe, A.L., Thompson, M., Love, J., Fonseca, C.M., Fleming, P.J.: Assessing the Performance of Multiobjective Genetic Algorithms for Optimization of a Batch Process Scheduling Problem. In: 1999 Congress on Evolutionary Computation, Washington, D.C., pp. 37–45. IEEE Computer Society Press, Los Alamitos (1999)

Smith, K.I., Everson, R.M., Fieldsend, J.E., Murphy, C., Misra, R.: Dominance-based multiobjective simulated annealing. IEEE Transactions on Evolutionary Computation. In press (2008)

Ulungu, E.L., Teghem, J., Fortemps, P.H., Tuyttens, D.: Mosa method: A tool for solving multiobjective combinatorial optimization problems. Journal of Multi-Criteria Decision Analysis 8(4), 221–236 (1999)

Van Veldhuizen, D.A.: Multiobjective Evolutionary Algorithms: Classifications, Analyses, and New Innovations. Ph.D. thesis, Graduate School of Engineering, Air Force Institute of Technology, Air University (1999)

Wagner, T., Beume, N., Naujoks, B.: Pareto-, Aggregation-, and Indicator-Based Methods in Many-Objective Optimization. In: Obayashi, S., Deb, K., Poloni, C., Hiroyasu, T., Murata, T. (eds.) EMO 2007. LNCS, vol. 4403, pp. 742–756. Springer, Heidelberg (2007), extended version published as internal report of Sonderforschungsbereich 531 Computational Intelligence CI-217/06, Universität Dortmund (September 2006).

Westfall, P.H., Young, S.S.: Resampling-based multiple testing. Wiley, New York (1993)

While, L.: A New Analysis of the LebMeasure Algorithm for Calculating Hyper-volume. In: Coello Coello, C.A., Hernández Aguirre, A., Zitzler, E. (eds.) EMO 2005. LNCS, vol. 3410, pp. 326–340. Springer, Heidelberg (2005)

While, L., Bradstreet, L., Barone, L., Hingston, P.: Heuristics for Optimising the Calculation of Hypervolume for Multi-objective Optimisation Problems. In: Congress on Evolutionary Computation (CEC 2005), IEEE Service Center, Edinburgh, Scotland, pp. 2225–2232. IEEE Computer Society Press, Los Alamitos (2005)

While, L., Hingston, P., Barone, L., Huband, S.: A Faster Algorithm for Calculating Hypervolume. IEEE Transactions on Evolutionary Computation 10(1), 29–38 (2006)

Zitzler, E., Künzli, S.: Indicator-Based Selection in Multiobjective Search. In: Yao, X., Burke, E.K., Lozano, J.A., Smith, J., Merelo-Guervós, J.J., Bullinaria, J.A., Rowe, J.E., Tiňo, P., Kabán, A., Schwefel, H.-P. (eds.) PPSN 2004. LNCS, vol. 3242, pp. 832–842. Springer, Heidelberg (2004)

Zitzler, E., Thiele, L.: Multiobjective Optimization Using Evolutionary Algorithms - A Comparative Case Study. In: Eiben, A.E., Bäck, T., Schoenauer, M., Schwefel, H.-P. (eds.) PPSN 1998. LNCS, vol. 1498, pp. 292–301. Springer, Heidelberg (1998)

Zitzler, E., Thiele, L.: Multiobjective Evolutionary Algorithms: A Comparative Case Study and the Strength Pareto Approach. IEEE Transactions on Evolutionary Computation 3(4), 257–271 (1999)

Zitzler, E., Thiele, L., Laumanns, M., Fonseca, C.M., Grunert da Fonseca, V.: Performance Assessment of Multiobjective Optimizers: An Analysis and Review. IEEE Transactions on Evolutionary Computation 7(2), 117–132 (2003)

Zitzler, E., Brockhoff, D., Thiele, L.: The Hypervolume Indicator Revisited: On the Design of Pareto-compliant Indicators Via Weighted Integration. In: Obayashi, S., Deb, K., Poloni, C., Hiroyasu, T., Murata, T. (eds.) EMO 2007. LNCS, vol. 4403, pp. 862–876. Springer, Heidelberg (2007)

15

Interactive Multiobjective Optimization from a Learning Perspective

Valerie Belton[1], Jürgen Branke[2], Petri Eskelinen[3], Salvatore Greco[4], Julián Molina[5], Francisco Ruiz[5], and Roman Słowiński[6]

[1] Department of Management Science, University of Strathclyde, 40 George Street, Glasgow, UK, G1 1QE,< `val.belton@strath.ac.uk`
[2] Institute AIFB, University of Karlsruhe, 76128 Karlsruhe, Germany, `branke@aifb.uni-karlsruhe.de`
[3] Helsinki School of Economics P.O. Box 1210, FI-00101 Helsinki, Finland, `Petri.Eskelinen@hse.fi`
[4] Faculty of Economics, University of Catania, Corso Italia 55, 95129 Catania, Italy, `salgreco@unict.it`
[5] Department of Applied Economics (Mathematics), University of Málaga, Calle Ejido 6, E-29071 Málaga, Spain, `julian.molina@uma.es, rua@uma.es`
[6] Institute of Computing Science, Poznań University of Technology, 60-965 Poznań, Poland, and Systems Research Institute, Polish Academy of Sciences, 01-447 Warsaw, Poland, `roman.slowinski@cs.put.poznan.pl`

Abstract. Learning is inherently connected with Interactive Multiobjective Optimization (IMO), therefore, a systematic analysis of IMO from the learning perspective is worthwhile. After an introduction to the nature and the interest of learning within IMO, we consider two complementary aspects of learning: individual learning, i.e., what the decision maker can learn, and model or machine learning, i.e., what the formal model can learn in the course of an IMO procedure. Finally, we discuss how one might investigate learning experimentally, in order to understand how to better support decision makers. Experiments involving a human decision maker or a virtual decision maker are considered.

15.1 Introduction

The aim of this chapter is to explore the notion of learning in the context of Interactive Multiobjective Optimization (IMO) where Classical Multiobjective Optimization (CMO) (see Chapter 2) or Evolutionary Multiobjective Optimization (EMO) (see Chapter 7) are used. This is an important subject because, on one hand, IMO enables the Decision Maker (DM) to learn about the optimization problem, and, on the other hand, it allows the formal model

---

Reviewed by: Kalyanmoy Deb, Indian Institute of Technology Kanpur, India
Kaisa Miettinen, University of Jyväskylä, Finland

J. Branke et al. (Eds.): Multiobjective Optimization, LNCS 5252, pp. 405–433, 2008.
© Springer-Verlag Berlin Heidelberg 2008

to evolve in response to additional information about preferences of the DM, which can be viewed as learning of the formal model. In consequence, the quality of an IMO procedure is related to what and how the DM and the model can learn in the course of the search for the most preferred solution. A characterization of IMO procedures from this perspective requires answers to several questions. What are characteristic features of individual learning? How can individual learning be supported? How do different models learn about preferences of the DM? What relations exist between individual learning and model learning? We try to answer these questions in order to assess potential benefits of incorporating interactive procedures within the evolutionary process, taking into account both behavioural and technical aspects. In our investigation, we have sought to take account of past and current research in this area; however, we are aware that the analysis presented is not to the depth that the importance of and potential interest in the topic would merit. The subject is simply worth another book. Our modest aim in this chapter is to point out interesting issues and to present some preliminary conclusions supported by a selective review of literature. We would wish to encourage future research in this area, and to recommend particular attention to learning aspects when developing and implementing IMO procedures.

The chapter begins with a brief broad overview of learning and why it is of interest to us. The following two sections go on to explore in greater detail individual learning and model learning in the context of IMO. The fourth section focuses on methodologies and procedures to investigate learning within IMO, and how we can characterize IMO procedures from the learning point of view, in order to transmit the best practice from CMO to EMO.

15.2 What Is Learning?

Literature on learning is extensive and cuts across a number of domains, most traditionally Psychology and Education which are both concerned with individual learning (see (Merriam and Caffarella, 1991) for a well regarded overview), but more recently (since the mid 20th century) including: Artificial Intelligence and Machine Learning (see (Russell and Norvig, 2003) for a comprehensive review), Organisational Studies, with the influential work of Argyris and Schon (1978) on single and double loop learning; and the work of Lave and Wenger (1991) on social and situated learning in communities of practice.

At the general level of seeking to support complex decision making in organisations, all of these perspectives on learning are relevant. However, in this chapter we focus on an individual participating in an interactive decision making process, with the dual objectives of directly facilitating the DM's learning (individual learning) and of progressive modelling of DM's preferences (model learning).

15.2.1 Individual Learning

Individual learning is a concept which is intuitively meaningful to us all, but one which conjures up many different interpretations - for example, you might learn that the world's population is currently about 6.5 billion (CIA, 2008), how to ride a bike, that you prefer apples to oranges but your friend prefers bananas to both, that drinking more than 2 cups of coffee a day leaves you a nervous wreck, or that you now believe it is unethical to eat meat and intend to become a vegetarian. These interpretations incorporate a range of *learning outcomes*. Some of the outcomes are objective in the sense of being subject to external validation, for example, facts and conceptual relationships, demonstration of skills or changes in behaviour. Others, such as beliefs, values, and self-awareness, are internal and subjective, in the sense of pertaining to the individual; others relate to your understanding of others values and are inter-subjective. These outcomes are the consequence of *learning processes* and associated stimuli. Possible learning processes range from the planned and explicit, such as formal education or conscious reflection, to the unconscious learning that leads to the building of tacit knowledge, or the learning may be triggered by an unexpected event. Stimuli for learning may be: exposure to new ideas or knowledge; a good or bad experience; the need to respond to questions about one's own ideas; the motivation to challenge others' ideas; the need to take action; or the process of conscious reflection. It is important also to recognise different "levels" of learning; from the acquisition of new knowledge or facts driven by external stimuli, through the internal processes of assimilating and making sense of what we know or have experienced, to a transformation of perspective as a consequence of critical reflection.

Thus, learning is both process and product, so we need to pay attention to how individuals learn and what they learn in course of an IMO procedure.

15.2.2 Model Learning

Model learning is a concept which underlines an evolution of the model in view of facts observed through sensors in an external world. Instead of evolution, one can also speak about adaptation to a new situation. A model is usually implemented as a computer program on a machine, hence the term machine learning is often used instead of model learning. Thus, machine, or model, learning is broadly defined as the ability of a computer program to improve its performance with regard to a defined task by learning from data or examples. Machine, or model, learning is an important subfield of Artificial Intelligence (Michalski *et al.*, 1998). Of course, a model learns (evolves, adapts) in a way and with regard to a task defined by a human. The model relates an output (response or dependent variable, decision attribute, conclusion) with an input (explanatory or independent variables, condition attributes, premise), either analytically, using a *function*, or logically, using *decision rules* or *trees*. Traditionally, the learning of functions falls into a broad term of *regression* (ordinal regression, statistical regression, neural network training), while learning of decision rules or trees is called *induction* (learning by example, data mining, knowledge discovery). Several model learning techniques are surveyed in Chapter 10.

In case of IMO, the input information for model learning is a reflection of subjective matters, such as an individual's preferences. It is exhibited by pieces of information such as: classification, rating or pairwise comparisons of selected solutions; specification of a reference or a reservation point; or indication of acceptable trade-offs. Note that the information provided by the DM may be holistic, when it concerns overall evaluation of particular solutions expressed in terms of classification, rating or some pairwise comparisons, and/or decomposed, when it concerns directly some parameters of the decision model (weights, tradeoffs, marginal value functions, preference thresholds, etc.). The output of model learning is a more or less explicit preference model of the DM who provided the input information. It may be used to guide the search for a "preferred" solution, and evolve from iteration to iteration with regard to new input information. In Chapters 4 and 5 of this book, two methods in which the model learning component is of primary importance are presented. In the GRIP method (Chapter 4) involving ordinal regression, the model learns a set of additive value functions compatible with pairwise comparisons of some reference solutions, including some indications of intensity of preference. In the Dominance-based Rough Set Approach (Chapter 5), the model learns association rules characterizing the Pareto optimal set, and decision rules characterizing good solutions in a given population. Model learning also plays a role in interactive evolutionary multiobjective optimization discussed in Chapters 6 and 7, at least for those approaches that attempt to infer a higher-level description of DM's preferences from the interaction. Model learning and individual learning are strongly inter-related, since the model learns from reactions of the DM and the DM learns from explanations provided by the model, as discussed in the next section.

15.2.3 The Learning Cycle

The overall cycle of IMO is depicted in Figure 15.1. Using preference information articulated by the DM, the inference engine constructs or updates the computer's model of the DM's preferences. The inference engine depends on the one hand on the nature of the preference information, and on the other hand on the nature of the considered model of DM's preferences. Thus, for example, the inference engine can determine:

- a preference model in terms of a linear value function built using preference information expressed in terms of subjective tradeoffs (Geoffrion *et al.*, 1972),
- a preference model in terms of an achievement scalarizing function built using preference information expressed in terms of a reference point consisting of aspiration levels assigned to considered criteria (Wierzbicki, 1980),

- a preference model in terms of one or a set of additive value functions built using preference information expressed in terms of a partial or complete preorder of some reference solutions (Jacquet-Lagrèze *et al.* (1987), Chapter 4).
- a preference model in terms of a set of *"if..., then... "* decision rules induced from rough approximation of an ordinal classification of some representative solutions in the Pareto optimal set (see Chapter 5).

Based on this model, the computer attempts to find better solutions using an optimizer. These solutions are then presented to the DM, possibly with some additional information and/or specific requests for a new input. The outputs of the inference and optimization phase help the DM to learn about the problem and may influence her preferences. This general outline shows the complex interaction between individual learning (which is influenced by the optimizer's output) and the model learning (which is influenced by the preference information provided by the DM). Thus, in the context of IMO, one cannot be considered independently of the other. The interaction with the DM should be designed such that the DM's response allows the best update of the preference model, but also to help the DM learn as much as possible about the problem. It is important to realize that these goals may be in conflict. One implication of this is that it may be a good idea to give the DM more information than the minimum necessary to provide the required preference information. For example, instead of proposing only one solution to the DM as the best in the current state of interaction (which is the minimum information that one could expect from a calculation stage of an IMO procedure), the DM can learn much more from a visualization of the Pareto optimal set (see Chapter 9), or from its description in terms of *"if..., then... "* association rules explaining that if certain levels are attained on some objectives, it is not reasonable to expect more than some other levels on correlated objectives (see Chapter 5).

As indicated above, this general framework can be instantiated in many different ways according to:

- How the decision maker's preferences are articulated: for example, via pairwise comparisons or classification of a set of solutions, specification of a reference point, bounds or maximal/minimal trade-offs.
- The nature of the inference engine: for example, an ordinal regression, a rule induction mechanism, an artificial neural network or evolutionary algorithms.
- The adopted preference model: for example, a value function, achievement scalarizing function, outranking relation or set of decision rules.
- The output of the inference and optimization tool: for example, new solution candidates, ranking or classification of solutions, decision rules or association rules.

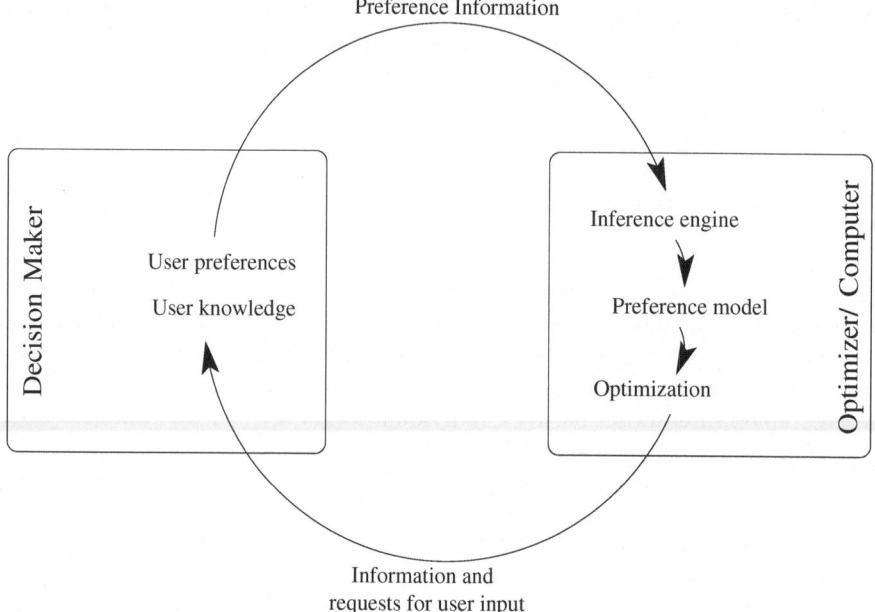

Fig. 15.1. Learning cycle of IMO

15.2.4 Motivation for the Interest in Learning

Consideration of learning is important in the context of interactive decision processes because it is one of a number of factors we might wish to consider in evaluating the performance of different approaches (Vanderpooten, 1990; Olson, 1992; Miettinen, 1999). The broad aim of both "classical" and "evolutionary" approaches to multiobjective optimisation is to guide the DM, in an efficient and effective manner, to a preferred solution that is Pareto optimal. The DM does not know in advance what are the "good" solutions from the technical, objective perspective (Pareto optimality) and thus she has to learn what is possible. Analogously, the model does not know which of these solutions the DM will prefer from the subjective perspective. The extent to which the DM's preferences are pre-formed is an open question. The answer to this depends in part on one's philosophical perspective, but there is much support for the constructivist view of modelling for decision support (Morecroft and Sterman, 1992; Roy, 1993; Belton and Elder, 1994) which argues that the whole IMO process is instrumental in helping the DM learn about her preferences. The expectation is that an effective learning process would lead to increased satisfaction with and confidence in a decision, as well as a better understanding of the underlying rationale. However, there are many unanswered questions. Are these benefits achievable? Are some interactive processes more effective than others? Does it depend on characteristics of the DM or the problem? To what extent is the technical performance of an approach in conflict with the learning effectiveness and if so, how can an appropriate balance be achieved? An important meta-level question is "How can we research these issues?" We return to this discussion in Section 15.5. In Sections 15.3 and 15.4 we expand on the discussions of individual learning and model, or machine, learning.

15.3 Aspects of Individual Learning

15.3.1 What Does a Decision Maker Learn?

The aim of any decision aiding process is that the DM should discover a way forward which is both feasible and desirable - we might say that she learns her "preferred" solution. We position the word preferred in quotation marks, because, as Korhonen and Wallenius (1996) highlight, the notion of "preferred" (or "most preferred", or "satisfactory") is problematical; they propose a definition based on a DMs conviction that, given a realistic understanding of the problem, it is not possible to find a solution she likes better. Thus there are two important, interdependent components to the learning necessary to arrive at a "preferred" solution - an understanding of the problem and an understanding of what the DM's values are.

Understanding the problem first calls for knowledge both of what is technically relevant and possible, alongside what matters to the DM, i.e., what her values are. It is important that appropriate attention is paid to the process of problem structuring, as advocated by Keeney (1992)'s value focused thinking and Belton and Stewart (2002)'s integrated approach to Multiple Criteria Decision Making (MCDM), in order to properly learn what are the key objectives and constraints. This is particularly important if an analysis incorporates only a few key objectives. The interactive process may lead to the surfacing or emergence of new criteria, which may lead to the analyst and DM deciding to re-analyse the problem.

Having specified what defines a potential solution then the foregoing description of a preferred solution requires the DM to have a realistic understanding of what solutions are possible. This could be interpreted in many ways - the set of all possible solutions, the set of Pareto optimal solutions, an understanding of how objectives are causally related (i.e., what trade-offs are necessary), etc. The IMO process will enable the DM to learn about these issues to a greater or lesser degree.

A good interactive process should enable the DM to learn about her preferences and there is a strong belief on the part of analysts that learning does take place (Miettinen, 1999; Vanderpooten, 1990; Olson, 1992), and in our own experiences this is echoed by DMs asked to reflect on their involvement in case studies. However, the notion is also problematic. The field of MCDM has an extensive language, together with associated parameters and structures, to reflect the notion of preference within its models, for example: importance weights, global weights, swing weights, preference thresholds, indifference thresholds, veto thresholds, aspiration levels, goals, acceptance levels, value functions, preference functions, ... to mention a few. The extent to which these parameters are psychologically meaningful to DMs and the questions used to elicit them are behaviourally realistic is often questioned and contributes to the cognitive complexity of the task. Larichev's (Larichev, 1992) consideration of these issues is one of the most significant contributions in the MCDM community to date, and there have been several subsequent calls (for example, (Dyer *et al.*, 1992; Korhonen and Wallenius, 1996)) for the need to pay attention to behavioural issues and the work of the Behavioural Decision Theory community.

The doubt about psychological meaningfulness of specific preference model parameters to DMs, has also led several researchers to develop approaches which eliminate these parameters from the dialogue between the DM and the model. Instead, the DM is asked to provide a holistic preference information, e.g., in form of pairwise comparisons of selected solutions, which serve as examples for model learning. The preference model learned from such information is used, in turn, on the complete set of solutions in order to obtain a preference relation on this set. A proper exploitation of this relation leads to "preferred" solutions. This approach emphasizes the discovery of intentions as an interpretation of actions rather than as a priori position. It is thus concordant with the principle of *posterior rationality* proposed by March (1978) and with *aggregation-disaggregation logic* of Jacquet-Lagrèze (1981). The best-known implementations in the field of MCDM are: the UTA method (Jacquet-Lagrèze and Siskos, 1982), with its extensions UTAGMS (Greco *et al.*, 2008) and the previously mentioned GRIP (Figueira *et al.*, 2008) - in which the model learns a set of compatible value functions; and the Dominance-based Rough Set Approach (DRSA) (Greco *et al.*, 2001, 2005, 2008; Słowinński *et al.*, 2005) in which the model learns a set of "if..., then..." decision rules.

15.3.2 How Do We Know if a Decision Maker Has Learned?

A DM's confidence in the decision and understanding of the related issues (such as the range of potential solutions, the necessary trade-offs, etc) and the ability to explain her choice would seem to be good indicators that she has learned. However, we should be careful not to look only for the DM learning how to speak the technical language of MCDM rather than truly about preferences. It may also be possible to detect learning as an interactive process progresses. For example, if the DM provides preference information inconsistent with previously provided information, this could be due to learning. Learning about the process itself and about her preferences would lead a DM to being able to anticipate the next step and to become more comfortable with making judgements at a later stage in the process. Furthermore, a willingness to use the process again, or to recommend it to someone else, is an indicator of satisfaction which may be associated with learning.

15.3.3 How Does a Decision Maker Learn?

The Chinese proverb "Tell me and I will forget, show me and I may remember, involve me and I will understand", which is often quoted in an educational context, is equally applicable here and highlights the potential power of interactive methods to facilitate learning. However, the way in which someone learns depends on many factors, including what is being learned, the circumstances in which this happens, the characteristics of the individual and her motivation to do so. The circumstances of the DM may also influence her willingness to seriously engage in the interactive process, for example, her degree of ownership of the problem, her motivation or the pressure to find a good solution, the time available, her general mental state (whether she is relaxed, stressed, alert, etc.), as well as the presence or an absence of a facilitator (analyst).

The characteristics of the approach, in conjunction with the DM characteristics, will determine whether the overall chemistry is successful, for example: How is information presented? What questions does the DM have to respond to and how easy or difficult is it, firstly to understand the question and, secondly, to answer? How flexible is the approach? How transparent is the approach? What opportunities are there to explore and experiment? How demanding is the approach, for example, how many interactions are required? What level of time commitment is required? How long is the time span between interactions? For example, Korhonen and Wallenius (1996) indicate that DMs often appear to tire of interactive procedures, and may conclude the process prematurely.

Much learning relies on feedback systems of the kind characterised as single and double loop learning by Argyris and Schon (1978). Single loop learning, which can be described as learning for improvement, achieves that improvement without questioning underlying assumptions or values. Action is initiated with a view to achieving a specified goal; the outcome is compared with the objective and if this is not met corrective action is taken. A thermostat is an oft-cited example of single loop learning. Double loop learning, which may be described as learning for understanding or transformation, results from a challenge to and rethinking of assumptions or values. An interactive process such as IMO is a feedback system of exactly this nature. However, it is important to consider when and how it is appropriate to encourage single or double loop learning. In some circumstances the DM may be strongly focused on quickly achieving specific goals, a form of single loop learning. Another DM may be more interested in exploration and experimentation. An environment which permits experimentation or play (as advocated by March (1988) in "The technology of foolishness") encourages trial and error and is more likely to lead to double loop learning. Learning for understanding was an important objective of early visual interactive systems to support MCDM such as VIG (Korhonen, 1987) and V.I.S.A. (Belton and Vickers, 1989). The ability to see a situation through a different lens, facilitated in one sense by powerful tools for visualisation (see Chapters 8 and 9) can also be a catalyst for learning. The involvement of other stakeholders in the decision can also highlight different perspectives on an issue. If their view is significantly different, this is one way of challenging the DM's thinking; another potential catalyst for learning. Other challenges might arise if the DM finds it very difficult to express a preference between options, if she does not see any progress in the interactive search process, perceives inconsistencies in where the search takes her, or is surprised or disappointed by results. In much of the literature on learning a "disorienting dilemma" is cited as being necessary for learning, which leads to a significant change in understanding the problem permitting a more inclusive, discriminating and integrating perspective (Mezirow, 1991). Perhaps it might be interesting to build such challenges into a method if it is felt that the DM would benefit from the challenge? It is also important to allow space for reflection, to give the DM time to mull over her thought processes and to ensure that no new insights arise, or if they do, to provide the opportunity to retrace one's steps.

Many models of individual "learning style" have been proposed, two of the most common being the Learning Styles Inventory (LSI) (Kolb, 1984) based on Kolb's learning cycle (Kolb and Fry, 1975; Kolb, 1984) and Visual /Auditory /Kinaesthetic (VAK) models (Dunn *et al.*, 1984) together with extended versions of this associated with Gardner (1993)'s theory of multiple intelligences. Although the validity of these models is questioned by some and they are primarily situated in the educational domain, they may offer insight into factors which may influence how a DM responds to interactive methods in general and to a particular approach. For example, the VAK classification provides an indication of an individual's preferred mode(s) of receiving information. We might expect auditory learners to respond poorly to the more usual, visual (text or graphics) way of presenting information in interactive methods. Kolb's LSI categorises learners on two dimensions which define four learning styles. The dimensions are: the way in which they approach a situation (through reflective observation - watching what happens and reflecting on it, or active experimentation - just getting on and doing it); and the way in which they make sense of the experience (through abstract conceptualisation - i.e., thinking about it, or through concrete experience - or "feeling" it). Active experimenters are perhaps more likely to respond positively to interactive methods as usually implemented.

An important issue related to the question how a DM learns, is the *transparency* and *traceability* of the decision support process. If information provided by the DM is processed in a way that is not clear, the consequence may be that she cannot see how a final recommendation is derived from her inputs. Such a decision model is essentially a *black box*, the output of which has to be accepted on the basis of trust in the analyst's expertise and authority. The DM has not learned about the problem, about her preferences or why she should decide in a particular way. Ideally, the decision model should be a *glass box* which fulfils both representation and recommendation tasks in a transparent way (see (Roy, 1993; Słowinński *et al.*, 2005)). However, it is likely that different representations are seen as more or less transparent by different DMs. Some may be comfortable with a simple functional representation augmented by a natural language explanation (for example, see (Papamichail and French, 2003)). Others may prefer a rule-based approach in which the model can be expressed in quasi-natural language, such as that described by Greco *et al.* (2005, 2008) and by Słowinński *et al.* (2005). In this approach each decision rule can be clearly identified with those parts of the preference information (decision examples) which support the rule, and the preference model is decomposed into simple scenarios which inform the DM in a quasi-natural language about the local trade-offs. In this way, the rules permit traceability of the decision support process and give understandable justifications for the decision to be made.

It is also important to pay attention to the extensive work from the field of Behavioural Decision Theory (BDT) on the exploration of factors which may influence how a DM perceives a problem situation and how the way in which she responds to questions designed to elicit preferences can be influenced by the framing of the problem and the phrasing of the question. There is a substantial literature on this topic (for example, see (Kahneman *et al.*, 1982; von Winterfeldt and Edwards, 1982)) and it is not possible to review this in detail here. One phenomenon which is particularly relevant in the context of IMO is the notion of anchoring, one form of which is the status-quo bias; this demonstrates that DM's are reluctant to move away from an initial position, or solution, even if that position is only recently or even hypothetically assigned (see, (Tversky and Kahneman, 1991), or (Keeney *et al.*, 2006), for examples). The effects of anchoring in IMO were investigated by Buchanan and Corner (1997). Other potentially relevant phenomena are loss aversion (decision makers are more significantly influenced by differences framed as losses rather than gains - see Tversky and Kahneman 1991) and the effect of decoy options (whereby the presence of an option C can influence a DM's choice between A and B - see (Huber *et al.*, 1982)). The reader is also referred to Korhonen and Wallenius (1996) who review behavioural issues in MCDM and a recent paper by Morton and Fasolo (2008), which reviews work in BDT of particular relevance to the use of methods based on multiattribute value theory.

Having considered what and how a DM might learn from engaging in an interactive process to support her decision making, together with the factors that might influence the extent to which the experience is a positive and productive one, we have tried to summarise these in Figure 15.2. The key drivers of learning are shown in the hexagonal boxes, with the solid arrows indicating the learning cycle of experimentation and reflection; the rectangular boxes depict "what" is learned, linked to "how" by the dashed arrows; and influential factors are grouped and displayed in the ovals. In the next section we go on to consider factors which define the ability of different models to learn about the DM's preferences.

15.4 Aspects of Model (or Machine) Learning

In the decision making phase of an interactive procedure, the DM provides some information about her preferences and this information is used to construct a preference model, which is usually a value function, even if it is interpreted as a distance (metric) between a reference point and an attainable point or, more generally, as an achievement scalarizing function. This model is used in the next calculation phase to find a compromise solution that fits the presumed DM's preferences. In the following decision making phase, the DM can accept this solution or give some new preference information that updates the preference model. By adapting the model, the computer *learns* about the DM's preferences. It is not our aim in this chapter to discuss technical details of the processes by which this might happen, but to reflect on the generic characteristics of these and to introduce a model of the quality of the learning process.

A fundamental feature of this learning process is the *flexibility* of the interaction procedure, meaning the capacity to incorporate any preference information coming from the DM. The flexibility is related to the *generality* of the preference model and to the *reversibility* of the procedure, understood as a possibility for the DM to return to a previous iteration in order to change the preference information provided at that stage.

From the point of view of generality of the preference model, an important characteristic is consideration of a *plurality* of value functions which are compatible with the preference information provided by the DM. This is an important distinctive feature of some interactive methods, contrasting with methods which consider only one instance of the value function within a given class (for example the linear value functions).

To illustrate by an example: consider the case in which the decision maker wants to minimize two objective functions g_1 and g_2, and the decision maker says that she prefers solution x, for which $g_1(x) = 4$ and $g_2(x) = 8$, to solution y, for which $g_1(y) = 6$ and $g_2(y) = 4$. If we consider the class of linear value functions $U(a) = \lambda_1 g_1(a) + \lambda_2 g_2(a)$, then two cases are possible:

- one can look for just one value function compatible with the preferences of the decision maker and, for example, consider the instance of the form $U(a) = 2g_1(a) + 3g_2(a)$, which verifies the representation condition:

$$U(x) = 2g_1(x) + 3g_2(x) = 32 > U(y) = 2g_1(y) + 3g_2(y) = 24$$

The twenty-four methods of estimating additive utilities described by Fishburn (1967) are classical examples of such an approach.

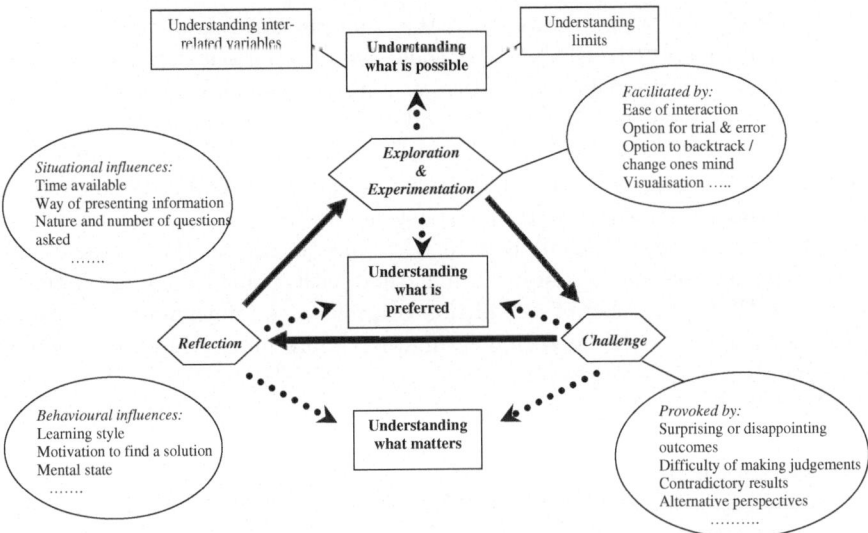

Fig. 15.2. What and how a DM learns and the factors that influence this.

- one can look for the whole set of value functions compatible with the preferences of the decision maker, i.e. all instances of value functions $U(a) = \lambda_1 g_1(a) + \lambda_2 g_2(a)$, such that

$$\lambda_1 g_1(x) + \lambda_2 g_2(x) > \lambda_1 g_1(y) + \lambda_2 g_2(y)$$

Another interesting feature related to the learning process is *universality*, i.e. the non specificity of the form of value functions: the less specific the form, the greater the chance that the model learns in a sequence of iterations. For example, the preference model admitting the form of a linear value function is less universal than the preference model admitting the form of an additive value function $(U(a) = U_1[g_1(a)] + U_2[g_2(a)]$, with U_1 and U_2 being non-decreasing functions, in the above example), which, in turn, is less universal than the model admitting the form of an increasing monotonic value function $(U(a) = F(g_1(a), g_2(a))$, with F being non-decreasing in both its arguments, in the above example).

However, we note that there are two sides of universality. On the one hand, the more universal the preference model, the higher the chance of being able to properly represent the DM's true preferences. On the other hand, a more complex underlying model requires more information, and thus more tiresome interaction, to properly tune the many parameters. In particular, if the DM gives a noisy response, a complex underlying preference model may overfit, while a simple function will resist to the noise. Finally, even a less general model may be sufficient to appropriately reflect the DM's preferences in the local vicinity of the current point of interest.

Another point to be considered is the *zooming capacity* of the preference model. With respect to this point of view, we have on one side methods using the preference model only locally, i.e., representing DM's preferences in a limited region of the objective space, and on the other side, globally, i.e., representing the DM's preferences in the whole objective space. For example, the GDF method is mainly local, and the reference point method is mainly global. Between these two extremes, there are methods which use a global model for the DM's preferences, but allow it to be refined locally in the course of the interactive process. One example for a method with such a "zooming" capacity would be UTA^{GMS}, where the addition of preference information relative to solutions from a particular region results in a more precise representation of preferences in this local region.

An overall representation of the features characterizing the learning capacity of the preference model in an interactive procedure is presented in a tree form in Figure 15.3.

In the table displayed in Figure 15.4, we show how the above features are extant in few representative interactive procedures, outlined briefly below. We have selected a set of sufficiently diversified procedures, from the viewpoint of the model learning - this diversification is connected with the type of preference information provided by the DM, type of preference model being learned,

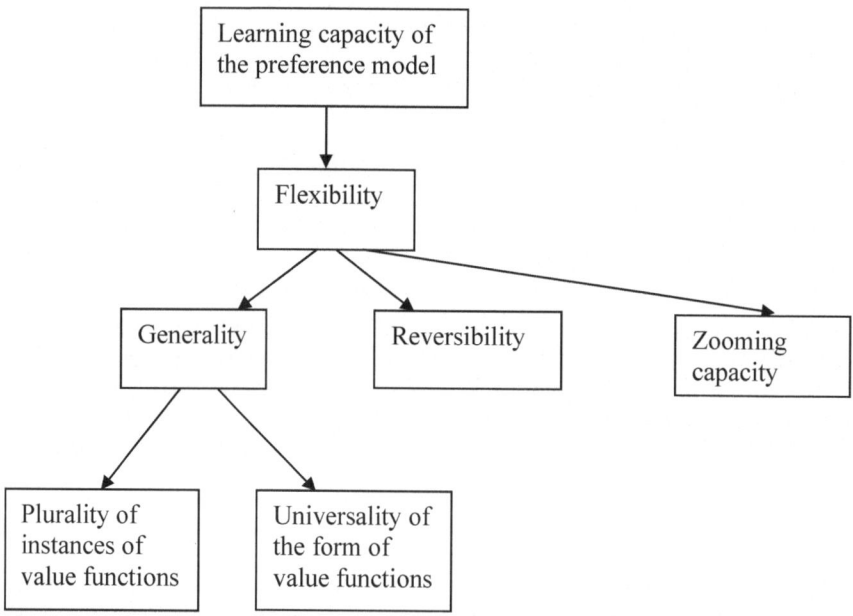

Fig. 15.3. Tree of features characterizing the learning capacity of the preference model in an interactive procedure.

and type of information given by the method to the DM. For an exhaustive overview of interactive methods see Chapter 2.

STEM method (Benayoun *et al.*, 1971). The procedure progressively reduces the set of admissible Pareto optimal solutions by adding constraints expressing the concessions that the DM is willing to concede with respect to considered criteria. The compromise solutions proposed to the DM are obtained by minimizing the distance (weighted Tchebychev metric) from the utopia point to the set of admissible solutions. Let us remark that STEM is devoted to linear problems, but nonlinear variants of STEM have been proposed, for example, in (Vanderpooten and Vincke, 1989; Eschenauer *et al.*, 2000). Observe, moreover, that STEM asks the DM to classify the objective functions into those whose values are satisfactory in the current solution and those whose values are unsatisfactory. This means that it is a classification-based method (see Chapter 2).

Geoffrion, Dyer and Feinberg method ((Geoffrion *et al.*, 1972), Chapter 2). The procedure maximizes a value function which is not known explicitly, but is assumed to be differentiable and concave. In each iteration the DM provides preference information specifying a subjective local trade-off between different criteria. The subjective trade-off information defines a search direction for the next calculation phase. A solution that maximizes the unknown value function in the established search direction is proposed to the DM, until the DM is satisfied.

Reference Point method ((Wierzbicki, 1980), Chapter 2) finds in each iteration a Pareto optimal point, called the current point, which optimizes an achievement scalarizing function referring to a reference point (attainable or not) specified by the DM (as well as several other solutions by shifting the reference point). The DM is free to modify the reference points as she is supposed to use this freedom to learn about the shape of the Pareto optimal set and to explore its interesting parts.

Light Beam Search (LBS) method (Jaszkiewicz and Słowiński, 1999) organizes the search on the Pareto set through a sampling of a neighborhood of the current point which moves as a result of either a change of decision maker's reference point, or a shift of the current point within its neighborhood. An outranking relation is used as a local preference model in the neighborhood of the current point. The neighborhood is composed of Pareto optimal points that outrank the current point, so the neighborhood includes points that are comparable with the current point; the Pareto optimal points from outside the neighborhood are either incomparable with or are outranked by the current point. The procedure requires threshold information and is called LBS due to analogy of the search with a projection of a focused beam of light from a spotlight at the reference point onto the Pareto optimal set.

Multiobjective Optimization using UTA^{GMS} or GRIP Methods (Chapter 4). This method is based on the ordinal regression methodology for elicitation of preference models (Greco *et al.*, 2008; Figueira *et al.*, 2008). In each iteration, the DM gives preference information in terms of preference

	STEM method	GDF method	Reference Point method	Light Beam Search method	UTA^{GMS} or GRIP methods	Dominance-based Rough Set Approach
Plurality	the value function is the Tchebychev metric, thus only one value function is considered in each iteration	the value function is an unknown value function representing a local trade-off between different criteria specified by the DM; only one value function is considered in each iteration, even if many value functions may represent the same trade-off at a certain point	the value function is an achievement scalarizing function for a particular reference point, thus only one value function is considered in each iteration	the value function is an achievement scalarizing function for a particular reference point, however, each point in the neighbourhood of the current point is attainable by an achievement scalarizing function with a different direction, thus many value functions are considered in each iteration	the method considers the whole set of additive value functions compatible with the preference information given by the DM	implicitly, the method considers a whole set of value functions making the same classification of solutions from the Pareto optimal sample into "good" and "others" as the selected decision rule; thus, the whole set of value functions concordant with the decision rules is considered in each iteration
Universality	the considered value function has only one form of a weighted Tchebychev metric measuring the distance form the utopia point to the set of admissible Pareto optimal solutions	the considered value function is very general: it is supposed to be differentiable and concave	the achievement scalarizing function is composed of an additive and a max-min component; it enjoys some good theoretical properties, like controllability and acceptance of both attainable and non-attainable reference points	the achievement scalarizing function is the same as in the Reference Point method, however, it is controlled not only by reference points but also by the relative weights of objectives within the neighbourhood of the current point defined by the outranking relation	this method considers additive value functions with monotone marginal value functions; such an additive function is a very general value function; consequently, it reaches a very good level of universality	it has been proved that a set of decision rules is equivalent to the most general value function (monotonic with respect to the value of the considered objectives); thus, this method considers the most general formulation of a value function and, consequently, it reaches the maximum of universality
Generality	the considered value function has only one form of a weighted Tchebychev metric measuring the distance form the utopia point to the set of admissible solutions	for the features of plurality of instances and universality of value functions, there is some generality	for the features of plurality of instances and universality of value functions, there is a fair generality	for the features of plurality of instances and universality of value functions, and for consideration of a neighbourhood of the current point, there is a good credit for generality.	for the features of plurality of instances and universality of value functions, there is a large credit for generality	for the features of plurality of instances and universality of value functions, there is a large credit for generality
Reversibility	each concession given on a criterion is definite and no more concessions can be considered on the same criterion, thus the procedure is not reversible	the method could be used as an explanatory search process and in this case it would be reversible	the method is fully reversible	the method is fully reversible	the method could be used as an explanatory search process and in this case it would be reversible	the method could be used as an explanatory search process and in this case it would be reversible

	STEM method	GDF method	Reference Point method	Light Beam Search method	UTA^{GMS} or GRIP methods	Dominance-based Rough Set Approach
Flexibility	for the lack of generality and reversibility, there is no space for flexibility of the preference model	for the moderate generality and reversibility, there is some credit for flexibility of the preference model	for the fair generality and the reversibility, there is a fair credit for flexibility of the preference model	for the good generality and the reversibility, there is a good credit for flexibility of the preference model	for the good generality and the reversibility, there is enough space for flexibility of the preference model	for the good generality and the reversibility, there is enough space for flexibility of the preference model
Learning capacity	the model can learn very little from the interaction with the DM	the model can learn the direction of improvement in each iteration	the model can learn a single point on the Pareto front in each iteration	the model can learn the interesting area of comparable points on the Pareto front in each iteration	due to the universality, the model can learn a lot from the interaction with the DM	due to the universality, the model can learn a lot from the interaction with the DM
Zooming capability	has zooming capacity	local use of the preference model	global use of the preference model	global with respect to the reference point, local with respect to the outranking relation on the Pareto set	has zooming capacity	has zooming capacity

Fig. 15.4. Features influencing the learning capacity of selected interactive methods.

comparisons of selected Pareto optimal solutions. Then, the whole set of additive value functions compatible with the preference information given by the DM is used to build a possible and a necessary order of all solutions considered in the Pareto optimal set. In the possible order, solution a is weakly preferred to solution b if there is at least one value function giving to solution a a value not smaller than to solution b. In the necessary order, solution a is weakly preferred to solution b if all value functions give to solution a a value not smaller than to solution b. The procedure ends when the possible and necessary orders give to the DM enough information for choosing the satisfactory solution.

Dominance-based Rough Set Approach (DRSA) to Multiobjective Optimization (Chapter 5). This method is based on MCDA methodology for data mining on ordinal data (Greco *et al.*, 2001). In each iteration, the procedure presents a set of representative Pareto optimal solutions to the DM who is asked to indicate the solutions that are relatively good. A set of *"if..., then..."* decision rules explaining this classification is induced using DRSA. The DM chooses from this set one decision rule characterizing good solutions. This decision rule imposes some additional constraints on the acceptable set of solutions to the MOO problem. Then, a new Pareto optimal set is calculated and a representative sample of this set is presented to the DM. The procedure stops when the DM finds a satisfactory solution in the presented sample. In each decision phase of this procedure, the DM is informed about the shape of the current Pareto optimal set by association rules showing relationships between values of particular objectives.

In conclusion, if we take into account the features presented in Figure 15.3, we can see that the interactive procedures which consider the whole set of value functions compatible with preference information have a greater capacity of learning the preference of the DM than the procedures considering only one such value function. The procedure based on the UTA^{GMS} and GRIP methods has all favourable features; in particular, the form of the considered value functions is general, just additive. The interactive procedure based on DRSA has a similar capacity; it takes into account the whole set of compatible value functions, i.e., those which make the same classification of solutions from the Pareto optimal sample (into good and others) as the selected decision rule, however, it is even more general because value functions corresponding to rules used in DRSA are just non-decreasing for gain-type objectives.

15.5 How do We Investigate Learning?

Having defined what we mean by effective learning from the perspective of both the decision maker and the model which defines the supporting interactive process, in this final section of the chapter we go on to discuss how we might investigate learning in order to understand how to better support decision makers. Is it possible to determine, a priori or interactively, the most appropriate way of helping a particular decision maker learn and to change or adapt the model in use accordingly? This is a very challenging task. Not only will different decision makers like or dislike different methods, the behaviour of an individual may differ from one situation to another, depending on factors such as the importance and complexity of the decision, time available, and knowledge of the problem. Furthermore, as Korhonen and Wallenius (1996) state, "learning one's preferences is a gradual process" and may lead to a change in the decision maker's needs and behaviour during the resolution of a problem. Thus our aim would be to understand the characteristics and needs of a decision maker and how these evolve during an interactive process, in order to chose and, if possible, adapt the interactive method such that it matches both the decision maker and the problem. This leads us to the question: is it possible to determine and/or classify different "learning patterns" for both parties? In this section we discuss open questions with the aim of giving a series of guidelines for further research in this field, which should combine behavioural and psychological aspects related to the decision maker, with technical features and performances of the interactive techniques.

Following on from earlier discussion of how a decision maker might learn from an interactive process and what could influence that learning, it would seem reasonable to utilise these factors in defining decision maker profiles and, potentially, a taxonomy of decision makers (or, more precisely, of decision making behaviours). The next step would be to seek to classify interactive techniques according to their performance with respect to the relevant aspects. As a result of these two classifications, a correspondence between methods and behavioural patterns could be established. Easier said than done! We will try to suggest some ideas that could be useful in order to produce such classifications.

Whilst a case study or action research approach (see (Easterby-Smith et al., 2002) for an overview of qualitative research methods, (Reason and Bradbury, 2001), for an in depth coverage of action research, and (Montibeller, 2007), for discussion of the use of action research in MCDM) might be used to develop a taxonomy of DMs and understand how they learn in "real" situations, we feel that more could be learned and more quickly in a realistic experimental context. These experiments should confront real DMs with real decision problems, and with the use of different interactive techniques. A number of such experiments are described in the literature and although none were explicitly concerned with assessing learning, we can learn from them about how DMs respond to different methods. Wallenius (1975) compares the performances of the GDF method (Geoffrion et al., 1972), the STEP method (Benayoun et al., 1971) and a simple trial and error procedure (somewhat similar to an iterative goal programming scheme, see (Dyer, 1973)). A hypothetical problem of planning production, inventory and labour was presented to 36 participants (18 students and 18 managers), who solved the problem using all the three methods. The following criteria were used to evaluate the performance of each method: DM's confidence in the final solution; ease of use of the method; ease of understanding the logic of the method; usefulness of the information provided; rapidity of convergence (number of cycles and total time) and CPU time. In addition, the DMs were asked to perform an overall preference ranking of the methods.

Buchanan and Daellenbach (1987) carried out a similar experiment (using the same hypothetical problem and the same evaluation criteria) to compare the performance of the Zionts-Wallenius method (Zionts and Wallenius, 1976, 1983), the Surrogate Worth Tradeoff Method (Chankong and Haimes, 1987, 1983, pp. 371–379), Steuer's Chebyshev method (Steuer and Choo, 1983, Steuer, 1985, pp. 419–450) and a naive approach. Miettinen and Mäkelä (1999) and Miettinen et al. (2006) report two studies about reference point (and classification) based interactive procedure NIMBUS (Miettinen, 1999; Miettinen and Mäkelä, 2006) and the reference direction method (Narula et al., 1994), taking into account the computational efficiency (number of objective function evaluations), duration of the procedure (overall time, number of iterations, time per iteration) and the opinion of the DMs on the controllability of the method and the final solution. Buchanan and Gardiner (2003) also compare the performance of two reference point approaches, seeking the DM's evaluation of the quality of the final solution obtained.

In all the studies outlined above, two common issues emerge. First, although the studies reveal that a majority of participants share opinions on many aspects, there are still DMs who have a different view. This implies that the performance of a given interactive method could perhaps be described in general for a "typical" DM, but not for all individuals. Second, the criteria used to evaluate the methods have in all cases been determined by the authors of the studies, and no indication is given of how important or relevant the DMs consider each criterion to be. This suggests an initial unstructured approach which allows DMs to define factors consider relevant in evaluating different interactive processes. Taking these together with those factors derived from the literature and already outlined in this paper this will provide a comprehensive basis for the characterisation of DM behaviours, to be further validated through a series of realistic experiments.

The next step would be to categorize the different interactive procedures with respect to their match with the DM characteristics. One way to do this is to test them with human DMs. Such experiments can highlight desirable properties of each procedure, but we must always be aware of the potential for experimental bias. For example, if the same problem is solved several times by a DM using different methods, it is impossible to make her "forget" what she has learnt about the problem before starting each resolution. If, on the other hand, the problem is changed, then it might be difficult to explain whether a change in the DM's behaviour is due to the change of method or to the change of problem. These potential drawbacks may be overcome if a "virtual" decision maker is used instead. In many such experiments (see, e.g., (Miettinen, 1999)), this "virtual" decision maker has always taken the form of a utility or value function. But in most of the cases, a value function may be too limited to actually model the real behaviours of a DM. So the most challenging question of this section is: is it possible to model the behavioural characteristics previously identified? Is it possible to substitute the classical value functions by more sophisticated modelling tools?

15.5.1 Designing Experiments Involving a Human Decision Maker

As already indicated, a key aim of the proposed experiments is to better understand how DMs learn from the use of interactive methods in order to better match and adapt methods to a particular DM in a particular situation. Previous experimental work has indicated that DMs often act in an apparently "irrational" way, for example when the cognitive load imposed by a process becomes heavy (Larichev *et al.*, 1987). However, it is important to recognise that although such behaviour might indicate limitations on the part of the DM, it may equally well reveal serious gaps in current understanding and theories related to learning and rational decision making (see (Olson, 1992), and references therein). In particular, the lack of a theory to explain the process of learning means that a DM changing her mind, because she has better understood the problem or her preferences, may be mis-interpreted as irrationality. These factors highlight the importance of designing experiments which engage real DMs with real problems, supported by Hobbs' assertions that experiments rather than reasoning are necessary in assessing subjective factors such as DM perceptions (Hobbs, 1986).

However, the nature of experiments is wide-ranging and it is important to consider the range of possibilities. On the one hand, a so called true experiment, which randomly allocates participants in the experiment to a control or experimental group and seeks to control for extraneous factors, is consistent with a positivist/deductive research methodology, focuses on establishing causal relationships, has high specificity (concentrates on a few variables) and high internal validity, but suffers from limited realism and limited generalisability. On the other hand, a quasi experiment (which lacks the random allocation to controlled conditions) in a field setting (which gives less control over extraneous factors) (Robson, 1993) may have greater realism but at the expense of internal validity. A quasi-experiment comes closer to the case study or naturalistic paradigm and there tends to be more emphasis on emergent design and inductive reasoning as well as greater reliance on qualitative data. Thus, the latter approach may be appropriate to help us better understand and develop models of learning, whilst the former may be better suited to test these.

Whilst there is a substantial general literature on the use of experiments as a research method and the experimental method is widely used in the field of behavioural decision making, there are relatively few published papers on its use in MCDM in general and in evaluating interactive methods in particular. Key overview papers are those by Hobbs (1986), Olson (1992) and Aksoy *et al.* (1996). The paper by Olson (1992), which is strongly focused on interactive multiobjective methods summarises 11 studies conducted up to 1990 and Aksoy *et al.* (1996) extend this to 1994, incorporating 6 studies involving human DMs and 8 in which the DM's responses are simulated by a model of some kind. These studies highlight a number of important considerations in designing experiments to evaluate or compare MCDM methods, as well as a range of practices. For Hobbs (1986) the ideal comparative experiment should be: well controlled and use a sufficiently large sample of DMs to be able to discern significant differences and generalize results; compare widely used methods that represent divergent philosophies of decision making, or claimed methodological improvements; compare methods across a variety of problem types; be a realistic representation of real world decision making. He also points out the need to be aware of order bias and possible DM fatigue. Olson (1992)'s view of an ideal study is one which involves substantive problems with real DMs, a situation which often limits the number of DMs that can be involved (thus the small sample size makes it difficult to discern significant differences or generalize results) but permits more in depth enquiry - circumstances which equate more to the quasi experiment in a field setting than the true experiment outlined above. It is clear that in designing an experiment it is necessary to balance multiple factors which may be in conflict, a multicriteria problem in itself. We outline briefly some of the key issues:

i). Defining realistic problems which are meaningful to the DMs, in the sense that they understand the issue and have responsibility for implementing a solution. This is very difficult to achieve in practice and often constructed problems are utilized with students as DMs. Although this latter point may be seen as a limitation, it is reported that "expert" DMs may favour a trial and error approach (which have performed surprisingly well in comparison to more sophisticated methods in a couple of studies, especially in the case of simple problems), whereas more inexperienced DMs tend to welcome more guidance. Olson (1992) suggests that the use of business school students, who are potential DMs but with little experience, can be justified and may in fact represent an appropriate sample set in which to discern learning effects.

ii). The level of randomness to incorporate in an experiment, ranging from a tightly controlled approach which exposes all DMs to the same methods and problems in the same order, etc, to one which seeks to vary all factors in an attempt to eliminate potential effects. A particular consideration might be the effect of the order in which a DM is exposed to methods; if the DM feels that the first method gives a good understanding of the problem and generates a solution in which she has confidence, then she may simply try to find the same solution with subsequent methods.

iii). The issue of DM fatigue as a consequence of the cognitive burden of the experimental process, due simply to the time elapsed, and also to subsequent questioning.

iv). The need to consider not only the methods, but the way in which they are implemented and the extent to which this is comparable across methods. Miettinen (1999) points out that a poor user interface might spoil a good method; similarly a poor method can be given credibility through an effective interface.

To summarize some limitations and disadvantages of experiments involving human DMs, we refer to the list of Zanakis *et al.* (1998), which draws mostly on the same sources as Olson (1992) and Aksoy *et al.* (1996): sample size (number of DMs involved) and range of problems studied is very limited; DMs are typically students rather than real DMs; the way information is elicited may influence the results more than the model used; the learning effect biases outcomes, especially when the DM employs various methods sequentially; inherent human differences might affect decisions, and thus, different performances can be due to the methods used or to which DM applies it; and finally, it is impossible or difficult to answer questions such as: Which method is more appropriate for what type of problem? What are the advantages or disadvantages of using one method over another? Does a decision change when using different methods and if so, why and to what extent?

However, although we can learn from the experiments on which the above discussions are based, the proposed research will go beyond comparative evaluation of methods in terms of DM satisfaction to try to understand the nature of DM learning and the behavioural, situational and method factors or characteristics which impact on this. Thus we may need to look more broadly at ways of enhancing the experimental process in order to stimulate learning and to capture subjective accounts of it.

15.5.2 Designing Experiments Involving a Virtual Decision Maker

Experimentation involving human DMs is essential to capture behavioural and subjective issues such as how a DM is reacting to the method and her confidence in the outcome. However, the use of a simulated DM has certain advantages and may provide a convenient means of examining more theoretical properties of methods. These advantages include, for example: the ability to fully control the experimental situation; the possibility of repeating experiments; the ability to expose the simulated DM to sequential trials without the problems of order bias and learning transfer; and considerations such as availability and lack of fatigue. Last but not least, experiments with a simulated DM are much cheaper and less time consuming than experiments with a human DM. Hence we return now to the question of whether is it possible to model the behavioural patterns which we would hope to be revealed by experimentation with human DMs. There have been some attempts to model DM behaviour and use it to test software (for example, Gibson *et al.*, 1987; Mote *et al.*, 1988; Reeves and Gonzalez, 1989; Aksoy *et al.*, 1996; Shackelford and Corne, 2004). However, these approaches are limited in the extent to which they reflect actual DM's behaviour and hence we suggest it is appropriate to consider if we can develop more sophisticated modelling tools. As a first step we need to consider what behavioural characteristics we would like to imitate. On the basis of the foregoing discussions we suggest that in addition to an underlying preference structure and stopping rule, the virtual DM model should capture:

- Cognitive complexity of the task, which should reflect the type of judgements required, the nature and amount of information a DM must process.
- DM fatigue, which might be a function of the number of iterations, the number of required judgements and cognitive complexity. This might be countered by level of motivation.
- DM learning, which might be reflected by a change in preference structure, possibly dependent on the previous iterations.
- DM inconsistency, which may be consequence of the tendency of a DM to "experiment", a reflection of the cognitive complexity of the required judgements, or simple "noise" in preference judgements.

The second step is to consider how we might model these characteristics and their consequences. Making the assumption of a base preference model (expressed in a functional or logical form) some factors may lead to a gradual or stepwise increase in associated noise level (for example, noise level might increase as a function of the number of iterations beyond a specified threshold), others to a change of form of the preference function.

A virtual DM could be exposed to different problems and interactive approaches allowing comparisons to be made without problems of bias or fatigue. Performance may be measured in terms of time to reach a solution, effort expended, level of satisfaction achieved (which might be expressed as the distance from the achieved solution to the theoretically preferred solution).

A long term aim of such experimentation could be to determine a match between different identified behavioural patterns and the techniques which are most suitable to support that particular way of working. However, we would like to insist on the fact that the behaviours of the DM can always be more complex and dynamic than those suggested by the previously mentioned approaches. Thus, it would be an interesting challenge to seek to develop flexible problem solving environments which can detect behavioural changes and react by offering different methods during the resolution process. In this line, Caballero et al. (2002), proposed an integrated interactive system in which several interactive methods of different kinds have been implemented, and where the DM can change between them at any time during the process. In this system, the decision to change the style of interaction rests with the analyst, but this idea could be complemented with those ideas explored above in order to assist the analyst to better choose the most adequate technique in each case.

15.6 Summary

The aim of this chapter was to explore the notion of learning in the context of interactive multiobjective optimization. There are two fundamental and interdependent sides to learning: the individual learning of the DM, and the model learning of the inference and optimization engine. Our discussion has been wide ranging, covering considerations of how individual learning can be characterised and facilitated, and the ways in which different types of model learn, through interaction with individuals, about preference structure. We have also looked at the interdependence of the two sides of learning, and the potential research challenges. It is clear that there is scope for both positive and negative feedback between these two forms of learning in any interactive decision process and it is essential that those who develop and implement interactive approaches need to pay attention to these issues. So far, there has been rather little focused research in this area and thus we hope that this chapter will provide both a stimulus and foundation for more in depth research which pays explicit attention to the issues discussed here.

Acknowledgements

This chapter is based on working group discussions during the Dagstuhl seminar on "Practical Approaches to Multiobjective Optimization" in December 2006. Besides the authors, the following people participated in the working group and contributed to the discussion: Jerzy Błaszczyński, José Figueira, Pablo Funes, Vincent Mousseau. We gratefully acknowlege their input during the seminar.
The second, the forth and the seventh author wish to acknowledge the support of COST Action IC0602 "Algorithmic Decision Theory".

References

Aksoy, Y., Butler, T.W., Minor, E.D.: Comparative studies in interactive multiple objective mathematical programming. European Journal of Operational Research 89(2), 408–422 (1996)

Argyris, C., Schon, D.A.: Organisational Learning: A Theory of Action and Perspective. Addison-Wesley, Reading (1978)

Belton, V., Elder, M.D.: Decision support systems: Learning from visual interactive modelling. Decision support systems 12, 355–364 (1994)

Belton, V., Stewart, T.J.: Muliple Criteria Decision Analysis: An Integrated Approach. Kluwer, Boston (2002)

Belton, V., Vickers, S.P.: V.I.S.A - VIM for MCDA. In: Lockett, G., Islei, G. (eds.) Improving Decision Making in Organizations, pp. 287–304. Springer, Berlin (1989)

Benayoun, R., de Montgolfier, J., Tergny, J., Larichev, O.: Linear programming with multiple objective functions: STEP method (STEM). Mathematical Programming 1, 366–375 (1971)

Buchanan, J.T., Corner, J.L.: The effects of anchoring in interactive MCDM solution methods. Computers and Operations Research 24(10), 907–918 (1997)

Buchanan, J.T., Daellenbach, H.G.: Comparative evaluation of interactive solution methods for multiple objective decision models. European Journal of Operational Research 29, 353–359 (1987)

Buchanan, J.T., Gardiner, L.: A comparison of two reference point methods in multiple objective mathematical programming. European Journal of Operational Research 149, 17–34 (2003)

Caballero, R., Luque, M., Molina, J., Ruiz, F.: PROMOIN: an interactive system for multiobjective programming. Information Technologies and Decision Making 1, 635–656 (2002)

Chankong, V., Haimes, Y.Y.: Multiobjective Decision Making Theory and Methodology. Elsevier Science Publishing Co., New York (1983)

Chankong, V., Haimes, Y.Y.: The interactive surrogate worth trade-off (ISWT) method for multiobjective decision-making. In: Zionts, S. (ed.) Multiple Criteria Problem Solving, pp. 42–67. Springer, Berlin (1987)

CIA: The 2008 World Factbook. Central Intelligence Agency (2008), https://www.cia.gov/library/publications/the-world-factbook/index.html

Dunn, R., Dunn, K., Price, G.E.: Learning Style Inventory. Price Systems, Lawrence, KS (1984)

Dyer, J.S.: An empirical investigation of a man-machine interactive approach to the solution of the multiple criteria problem. In: Cochrane, J., Zeleny, M. (eds.) Multiple Criteria Decision Making, pp. 202–216. University of South Carolina Press, Columbia (1973)

Dyer, J.S., Fishburn, P.C., Steuer, R.E., Wallenius, J., Zionts, S.: Multiple criteria decision making, multiattribute utility theory: The next 10 years. Management Science 38, 645–654 (1992)

Easterby-Smith, M., Thorpe, R., Lowe, A.: Management Research: an Introduction. Sage, London (2002)

Eschenauer, H.E., Osyczka, A., Schäfer, E.: Interactive multicriteria optimisation in design process. In: Eschenauer, H., Koski, J., Osyczka, A. (eds.) Multicriteria Design Optimization Procedures and Applications, pp. 71–114. Springer, Berlin (2000)

Figueira, J., Greco, S., Słowiński, R.: Building a set of additive value functions representing a reference preorder and intensities of preference: GRIP method. European Journal of Operational Research. doi:10.1016/j.ejor.2008.02.006 (2008)

Fishburn, P.C.: Methods of estimating additive utilities. Management Science 13(7), 435–453 (1967)

Gardner, H.: Frames of Mind: The Theory of Multiple Intelligences. Basic Books, New York (1993)

Geoffrion, A., Dyer, J., Feinberg, A.: An interactive approach for multi-criterion optimization, with an application to the operation of an academic department. Management Science 19(4), 357–368 (1972)

Gibson, M., Bernardo, J.J., Cheng, C., Badinelli, R.: A comparison of interactive multiple-objective decision making procedures. Computers and Operations Research 50(1), 97–105 (1987)

Greco, S., Matarazzo, B., Słowiński, R.: Rough sets theory for multicriteria decision analysis. European Journal of Operational Research 129(1), 1–47 (2001)

Greco, S., Matarazzo, B., Słowiński, R.: Decision rule approach. In: Figueira, J., Greco, S., Ehrgott, M. (eds.) Multiple Criteria Decision Analysis: State of the Art Surveys, pp. 507–563. Springer, Berlin (2005)

Greco, S., Mousseau, V., Słowiński, R.: Ordinal regression revisited: Multiple criteria ranking using a set of additive value functions. European Journal of Operational Research 191(2), 415–435 (2008)

Hobbs, B.F.: What can we learn from experiments in multiobjective decision analysis? IEEE Transactions on Systems, Man, and Cybernetics 3, 384–394 (1986)

Huber, J., Payne, J.W., Puto, C.: Adding asymmetrically dominated alternatives:violations of regularity and the similarity hypothesis. Journal of Consumer Research 9, 90–98 (1982)

Jacquet-Lagrèze, E.: Systèmes de décision et acteurs multipies: Contribution à une théorie de l'action pour les sciences des organisations. Thèse d'Etat, Université de Paris-Dauphine, Paris (1981)

Jacquet-Lagrèze, E., Siskos, Y.: Assessing a set of additive utility functions for multicriteria decision making: the UTA method. European Journal of Operational Research 10(2), 151–164 (1982)

Jacquet-Lagrèze, E., Meziani, R., Słowiński, R.: MOLP with an interactive assessment of a piecewise-linear utility function. European Journal of Operational Research 31(3), 350–357 (1987)

Jaszkiewicz, A., Słowiński, R.: The 'Light Beam Search' approach - an overview of methodology and applications. European Journal of Operational Research 113(2), 300–314 (1999)

Kahneman, D., Slovic, P., Tversky, A. (eds.): Judgment under Uncertainty. Cambridge University Press, Cambridge (1982)

Keeney, R.: Value Focussed Thinking: A Path to Creative Decision Making. Harvard University Press, Cambridge (1992)

Keeney, R.L., Raiffa, H., Hammond, J.S.: The hidden traps in decision making. In: Harvard Business Review Online (2006)

Kolb, D.: Experiential Learning: Experience as the Source of Learning. Prentice-Hall, Englewood Cliffs (1984)

Kolb, D., Fry, R.: Toward an applied theory of experiential learning. In: Cooper, C.L. (ed.) Theories of Group Processes, pp. 27–56. John Wiley, London (1975)

Korhonen, P.: VIG - a visual interactive support system for multiple criteria decision making. JORBEL 27(1), 4–15 (1987)

Korhonen, P., Wallenius, J.: Behavioural issues in MCDM: Neglected research questions. Journal of Multi-Criteria Decision Analysis 5, 178–182 (1996)

Larichev, O.I.: Cognitive validity in design of decision-aiding techniques. Journal of Multi-Criteria Decision Analysis 1, 127–138 (1992)

Larichev, O.I., Polyakov, O.A., Nikiforov, A.D.: Multicriterion linear programming problems - analytical survey. Journal of Economic Psychology 8, 389–407 (1987)

Lave, J., Wenger, E.: Situated Learning. Legitimate peripheral participation. University of Cambridge Press, Cambridge (1991)

March, J.G.: Bounded rationality, ambiguity, and the engineering of choice. The Bell Journal of Economics 9(2), 587–608 (1978)

March, J.G.: The technology of foolishness. In: March, J.G. (ed.) Decisions and Organizations, pp. 253–265. Basil Blackwell, New York (1988)

Merriam, S.B., Caffarella, R.S.: Learning in Adulthood. A comprehensive guide. Jossey-Bass, San Francisco (1991)

Mezirow, J.: Transformative Dimensions of Adult Learning. Jossey-Bass, San Francisco (1991)

Michalski, R.S., Bratko, I., Kubat, M.: Machine Learning and Data Mining - Methods and Applications. Wiley, New York (1998)

Miettinen, K.: Nonlinear Multiobjective Optimization. Kluwer, Boston (1999)

Miettinen, K., Mäkelä, M.M.: Comparative evaluation of some interactive reference point-based methods for multi-objective optimisation. Journal of the Operational Research Society 50, 949–959 (1999)

Miettinen, K., Mäkelä, M.M.: Synchronous approach in interactive multiobjective optimization. European Journal of Operational Research 170(3), 900–922 (2006)

Miettinen, K., Mäkelä, M.M., Kaario, K.: Experiments with classification-based scalarizing functions in interactive multiobjective optimization. European Journal of Operational Research 29, 931–947 (2006)

Montibeller, G.: Action researching multiple criteria decision analysis interventions. In: Shaw, D. (ed.) 49th Operational Research Society Conference Keynote Papers (2007), http://personal.lse.ac.uk/MONTIBEL/OR49_Action_researching_MCDA.pdf

Morecroft, J.D.W., Sterman, J.D.: Modelling for Learning. North Holland, Amsterdam (1992)

Morton, A., Fasolo, B.: Behavioural decision theory for multi-criteria decision analysis: a guided tour. In: Journal of the Operational Research Society. To appear (2008), doi:10.1057/palgrave.jors.2602550

Mote, J., Olson, D.L., Venkataramanan, M.A.: A comparative multiobjective programming study. Mathematical and Computer Modelling 10(10), 719–729 (1988)

Narula, S.C., Kirilov, L., Vassilev, V.: Reference direction approach for solving multiple objective nonlinear programming problems. IEEE Transactions on Systems, Man, and Cybernetics 24, 804–806 (1994)

Olson, D.L.: Review of empirical studies in multiobjective mathematical programming: Subject reflection of nonlinear utility and learning. Decision Sciences 23, 1–20 (1992)

Papamichail, K.N., French, S.: Explaining and justifying the advice of a decision support system: a natural language generation approach. Expert Systems with Applications 24(1), 35–48 (2003)

Reason, P., Bradbury, H.: Handbook of Action Research. Sage, London (2001)

Reeves, G.R., Gonzalez, J.J.: A comparison of two interactive MCDM procedures. European Journal of Operational Research 41(2), 203–209 (1989)

Robson, C.: Real World Research: a Resource for Social Scientists and Practitioner Researchers, 2nd edn. Blackwell, Oxford (1993)

Roy, B.: Decision science or decision-aid science. European Journal of Operational Research 66(2), 184–203 (1993)

Russell, S.J., Norvig, P.: Artificial Intelligence: a Modern Approach. Prentice Hall, Upper Saddle River (2003)

Shackelford, M.R.N., Corne, D.W.: A technique for evaluation of interactive evolutionary systems. In: Parmee, I.C. (ed.) Proceedings of the 6th International Conference on Adaptive Computing in Design and Manufacture (ACDM 2004), pp. 197–208. Springer, Heidelberg (2004)

Słowinński, R., Greco, S., Matarazzo, B.: Rough set based decision support. In: Burke, E.K., Kendall, G. (eds.) Search Methodologies: Introductory Tutorials in Optimization and Decision Support Techniques, pp. 475–527. Springer, New York (2005)

Steuer, R.E.: Multiple Criteria Optimization: Theory, Computation, and Applications. John Wiley & Sons, New York (1985)

Steuer, R.E., Choo, E.U.: An interactive weighted tchebycheff procedure for multiple objective programming. Mathematical Programming 26, 326–344 (1983)

Tversky, A., Kahneman, D.: Loss aversion in riskless choice: a reference-dependent model. The Quarterly Journal of Economics 106, 1039–1061 (1991)

Vanderpooten, D.: The interactive approach in MCDA a conceptual framework and some basic conceptions. Mathematical and Computer Modelling 12, 1213–1220 (1990)

Vanderpooten, D., Vincke, P.: Description and analysis of some representative interactive multicriteria procedures. Mathematical and Computer Modelling 12, 1221–1238 (1989)

von Winterfeldt, D., Edwards, W.: Decision Analysis and Behavioral Research. Cambridge University Press, Cambridge (1982)

Wallenius, J.: Comparative evaluation of some interactive approaches to multicriterion optimization. Management Science 21(2), 1387–1396 (1975)

Wierzbicki, A.P.: The use of reference objectives in multiobjective optimization. In: Fandel, G., Gal, T. (eds.) Multiple Criteria Decision Making, Theory and Applications, pp. 468–486. Springer, Berlin (1980)

Zanakis, S.H., Solomon, A., Wishart, N., Dublish, S.: Multi-attribute decision making: A simulation comparison of selected methods. European Journal of Operational Research 107(3), 507–529 (1998)

Zionts, S., Wallenius, J.: An interactive programming method for solving the multiple criteria problem. Management Science 22, 652–663 (1976)

Zionts, S., Wallenius, J.: An interactive multiobjective linear programming method for a class of underlying nonlinear utility functions. Management Science 29, 519–529 (1983)

16

Future Challenges

Kaisa Miettinen[1,*], Kalyanmoy Deb[2,*], Johannes Jahn[3],
Wlodzimierz Ogryczak[4], Koji Shimoyama[5], and Rudolf Vetschera[6]

[1] Department of Mathematical Information Technology, P.O. Box 35 (Agora),
   FI-40014 University of Jyväskylä, Finland, kaisa.miettinen@jyu.fi
[2] Department of Mechanical Engineering, Indian Institute of Technology Kanpur,
   PIN 208 016, India, deb@iitk.ac.in
[3] Department of Mathematics, University of Erlangen-Nürnberg, Martensstrasse
   3, 91058 Erlangen, Germany, jahn@am.uni-erlangen.de
[4] Institute of Control & Computation Engineering, Faculty of Electronics &
   Information Technology, Warsaw University of Technology, ul. Nowowiejska
   15/19, 00-665 Warsaw, Poland, w.ogryczak@ia.pw.edu.pl
[5] Institute of Fluid Science, Tohoku University, 2-1-1 Katahira, Aoba-ku, Sendai,
   980-8577, Japan, shimoyama@edge.ifs.tohoku.ac.jp
[6] Department of Business Administration, University of Vienna, Brünnerstrasse
   72, 1210 Wien, Austria, rudolf.vetschera@univie.ac.at

Abstract. Many important topics in multiobjective optimization and decision making have been studied in this book so far. In this chapter, we wish to discuss some new trends and challenges which the field is facing. For brevity, we here concentrate on three main issues: new problem areas in which multiobjective optimization can be of use, new procedures and algorithms to make efficient and useful applications of multiobjective optimization tools and, finally, new interesting and practically usable optimality concepts. Some research has already been started and some such topics are also mentioned here to encourage further research. Some other topics are just ideas and deserve further attention in the near future.

16.1 Introduction

Handling problems with multiple conflicting objectives has been studied for decades (as discussed, e.g., in Chapters 1 to 3); yet there still exist many interesting topics for future research. There are both theoretical questions as well as challenges set by real applications to be tackled. Some of the questions

---

* In 2007 also Helsinki School of Economics, P.O. Box 1210, FI-00101 Helsinki,
  Finland
  Reviewed by: Jörg Fliege, University of Southampton, UK
  Joshua Knowles, University of Manchester, UK
  Jürgen Branke, University of Karlsruhe, Germany

J. Branke et al. (Eds.): Multiobjective Optimization, LNCS 5252, pp. 435–461, 2008.
© Springer-Verlag Berlin Heidelberg 2008

can be answered, for example, by hybridizing or integrating ideas from the MCDM and EMO literature.

Here we do not even pursue covering all relevant future challenges but concentrate on three major topics, mainly due to space limitation. First, we discuss new and challenging problem domains in which multiobjective optimization and decision making (or decision support) techniques can be applied. Second, we discuss some new methodologies for multiobjective optimization which allow a synergetic application of optimization and decision making or provide a more global approach to optimization. Third, we describe new and innovative definitions of optimality in multiobjective optimization, which allows one to find a subjective or preferred set of optimal solutions.

Many topics discussed in this chapter are currently under study in various research groups. We still discuss such topics here mainly from the point of view of propagating such ideas to more people. We would like to encourage readers to pursue research along these directions, but our compilation will be successful if future researchers make due acknowledgment of the cited references and this compilation. In our view, the ideas presented are important and have a long-term implication to the field of multiobjective optimization. Collaborative and focused research efforts to implement some of such ideas will be the next step towards making the field more applicable, sustainable and enjoyable.

16.2 Challenging Multiobjective Optimization Problems

Besides solving typical optimization problems having multiple objectives, multiobjective optimization methodologies can also be used in other kinds of problem solving tasks. In this section, we briefly mention some of such research directions.

16.2.1 New Problem Domains

Multiobjective optimization problems arise in many applied fields of research. Although many of these problem types are already investigated, there are also some important new problem classes which deserve to be examined in detail. In the following, we discuss a few selected problems.

Multiobjective Bilevel Optimization

In multiobjective bilevel optimization (Dempe, 2002) one considers the optimization problem

$$\text{minimize } \mathbf{f}(\mathbf{x}, \mathbf{y})$$
$$\text{subject to } \mathbf{y} \in Y \text{ and}$$
$$\mathbf{x} \text{ solves } \begin{cases} \text{minimize } \mathbf{g}(\mathbf{x}, \mathbf{y}) \\ \text{subject to } \mathbf{x} \in X. \end{cases}$$

Here $Y \subset \mathbf{R}^n$ and $X \subset \mathbf{R}^m$ are given feasible sets possibly defined by inequalities and equalities and $\mathbf{f} : \mathbf{R}^m \times \mathbf{R}^n \to \mathbf{R}^k$ and $\mathbf{g} : \mathbf{R}^m \times \mathbf{R}^n \to \mathbf{R}^l$ are vector-valued functions. So, on the "lower level" one has to solve a multiobjective optimization problem for an arbitrary parameter $\mathbf{y} \in Y$. The problem on the "upper level" is a multiobjective optimization problem where the feasible set is defined by Y and the whole Pareto optimal set of the lower problem. Actually, we have two coupled multiobjective problems on two levels. This so-called multiobjective bilevel problem is difficult to solve because we need the complete Pareto optimal set of the problem on the lower level for every parameter $\mathbf{y} \in Y$. The use of interactive methods on the lower level is not helpful in this case. An overview on these complicated problem types in the single objective case can be found in (Dempe, 2002, 2003).

There are interesting applications for this problem class. The bilevel problem in its original form goes back to von Stackelberg (1934), who has introduced a special case of these problems. The so-called Stackelberg games are special bilevel problems. In our case the leader and the follower (in the context of Stackelberg games) have multiple objectives.

In addition to games and economical applications, there are also various applications in engineering (Bard, 1998; Dempe, 2002, 2003). For instance, certain equilibrium problems in chemical engineering can be formulated as bilevel problems (Dempe, 2002).

Semidefinite Optimization

Semidefinite Optimization is a field in optimization which has rapidly grown since the beginning of the 1990's. A multiobjective semidefinite optimization problem can be formulated as

$$\text{minimize } \mathbf{f}(\mathbf{x})$$
$$\text{subject to } G(\mathbf{x}) \text{ is positive semidefinite and}$$
$$\mathbf{x} \in \mathbb{R}^m.$$

Here we assume that $\mathbf{f} : \mathbb{R}^m \to \mathbb{R}^k$ is a vector-valued function and $G : \mathbb{R}^m \to \mathcal{S}^n$ is a matrix-valued function, where $\mathcal{S}^n$ denotes the Hilbert space of symmetric (n, n)-matrices with real coefficients. Although the case that the objective function is real-valued has been studied in detail and numerical methods are available for linear and nonlinear semidefinite optimization problems, investigations of the multiobjective case and the development of numerical methods are still expected.

Many applications lead to semidefinite optimization problems (Jahn, 2007). Among others, we only mention the design of a rib in the front of the wing of the new Airbus A380. This complicated problem of material optimization has been solved by semidefinite optimization where one minimizes the weight of the structure and the compliance is treated as a constraint. Thus, the design of one rib is a solution of an ε-constraint problem (see Section 1.3.2 in Chapter 1), where ε has not been varied. In this sense, this rib is a result of a special scalarization technique known from multiobjective optimization.

Set Optimization

Since the end of the 1980's, multiobjective optimization has been extended to set optimization. These are problems of the type

$$\text{minimize } F(\mathbf{x})$$
$$\text{subject to } \mathbf{x} \in S,$$

where $S \subset \mathbb{R}^n$ is a feasible set being defined by inequalities and equalities and $F : \mathbb{R}^n \rightrightarrows \mathbb{R}^k$ is set-valued. So, for a given feasible point $\mathbf{x}$ the image $F(\mathbf{x})$ is a set of vectors in $\mathbb{R}^k$. Although there are investigations on these set problems as an extension of multiobjective optimization, we need new approaches taking into account that we have to work with partial orderings for sets (and not only for points). First steps have been taken with the KNY (Kuroiwa-Nishnianidze-Young) partial ordering (Jahn, 2004) but many significant theoretical questions are still open.

Problems of this type may occur if the objective is not clearly defined but only specified in a vague set-oriented way. If one cannot define a function value of the objective but only the range for this value, one has to solve a set problem.

An example of an industrial application is the navigation of autonomous transportation robots. Here one uses ultrasonic sensors determining the smallest distance to an obstacle in the emission cone. Since the direction of the object cannot be identified in the cone, the location of the object is set-valued. Therefore, questions of navigation may lead to problems of set optimization.

Further Problem Types

There are many other multiobjective problem types to be explored more intensively. Among others, we need more investigations in *multiobjective dynamic optimization*. Dynamic optimization is a significant field of optimization with important applications and it has been used for decades. It is essential to extend these studies to the multiobjective case and only a few studies exist to date (Bingul, 2007; Deb *et al.*, 2007a; Farina *et al.*, 2000; Palaniappan *et al.*, 2001).

Another problem class can be called *multiobjective clustering* (Delattre and Hansen, 1980). If one applies cluster analysis to a set of points in order to find out appropriate clusters, in some cases standard methods do not give the desired result. Recent investigations on multiobjective clustering show that the use of multiple objectives may result in better clusters (e.g., formulating a biobjective optimization of minimizing intra-cluster distance and maximizing inter-cluster distance and finding a set of trade-off solutions will provide solutions not accessible by methods that optimize only one of the criteria) (Handl and Knowles, 2007). This topic is certainly a future challenge as well.

16.2.2 Large-Scale Problems

By large-scale problems we understand problems with many variables, constraints or objectives. In general, an exact lower bound for the number of variables, constraints or objectives is not specified. These problems arise very naturally in concrete applications. For instance, if one discretizes a system of partial differential equations defining the constraints one gets immediately many constraints and many variables. If the considered variable of our problem is a function, the discretization of this function leads to many variables. In order to get a good approximation one has to work with many variables in this case.

Problems with many objectives may occur, for example, in engineering. For instance, the design of suspension bridges may lead to several hundred objectives being difficult to handle. Other problems in material optimization also belong to this class of large-scale problems. Standard methods of multiobjective optimization cannot be applied to these large-scale problem types without simplifying the original problem. Therefore, we need new methods being able to treat problems with many variables, constraints or objectives.

It seems to be difficult to design interactive methods for large-scale problems. There are various reasons. For the decision maker (DM) it is difficult to handle a lot of objectives (or variables). The auxiliary problems which have to be solved during every calculation phase may be so time-consuming that an interaction with the DM does not make sense. Here we have to find new concepts for interaction. Let us add that in some problems function evaluations may be very costly even though the dimensions are small. From the computational point of view, such problems can also be regarded as large-scale ones and interaction may suffer as discussed above.

Like in the case of single objective optimization, one important step for the reduction of computation time in multiobjective optimization is the parallelization of algorithms and their implementation with a distributed computing system. Some earlier studies have demonstrated the use of a distributed computing paradigm for the parallel computation of automatically allocated non-overlapping regions of the Pareto optimal set (Deb et al., 2003; Branke et al., 2004b). A successful treatment of large-scale problems can be reached by using computers in parallel. New approaches such as grid computing allow to use entire networks of computers as one huge parallel computer. New methods have to be designed for parallel architectures. The change from sequential structures to parallel structures will accelerate in the future. More discussions on possibilities of parallel multiobjective optimization can be found in Chapter 13.

16.2.3 Using Multiobjective Optimization to Aid in Other Problem Solving Tasks

Besides solving multiobjective optimization problems, multiobjective concepts and approaches can also be exploited to solve other optimization problems:

- *Constraint handling:* In single objective optimization problems, an additional objective of minimizing the overall constraint violation can be employed. Furthermore, in problems where the constraints form an empty feasible region, the constraints can each be converted as objective functions. This enables solving the problem by taking constraint violations as objectives to be minimized (Miettinen *et al.*, 1998).

- *Optimization with an additional requirement:* In many problems, although the goal is to minimize or maximize a single or multiple objectives, the solution should also exhibit other desirable properties. For example, in the context of evolving computer programs for performing a task using the genetic programming approach, the goal is often to execute the task as accurately as possible but with a hidden agenda of developing a strategy (program) which is as simple as possible. Bleuler *et al.* (2001) minimized the *size* of a genetic program in addition to the supplied objective functions. Since the minimization of the program size is also an important objective, the genetic programming attempts to find the optimal program without making the program unnecessarily large.

- *Improving the search landscape:* Furthermore, some recent studies (Knowles *et al.*, 2001; Neumann and Wegener, 2005) have shown that decomposing the original single objective function carefully into multiple functionally different objectives and treating the problem as a multiobjective optimization problem makes the problem easier to solve than the usual single objective optimization procedure.

- *Revisit traditional problems with multiple objectives for a better and more informative solution strategy:* Sometimes, adding extra objectives as so-called helper (or proxy) objectives allows a better handling of single objective optimization problems (Jensen, 2004). Certain problems are traditionally solved using a particular procedure. A reconsideration of such problems using a multiobjective optimization strategy can be useful in many problem solving tasks, like the multiobjective clustering problem discussed earlier.

- *Knowledge discovery from multiobjective optimization results:* A recent concept of *innovization*, innovation through optimization, makes a post-optimality analysis of obtained trade-off solutions for deciphering principles which are commonly appearing in most obtained trade-off solutions (Deb and Srinivasan, 2006). Since the solutions obtained by an EMO or an a posteriori MCDM method are close to being (or are) Pareto optimal, they are expected to have certain features which remain common to qualify these solutions to be close to the Pareto optimal set and certain features which allow them to have a trade-off among objectives. An effort to try to decipher such valuable information from a set of trade-off near-optimal solutions is a unique way of discovering salient information about "how to solve a problem in a near-optimal manner?". In many engineering design problems and game playing problems, interesting and new design principles and strategies can be unearthed by such a procedure.

The possibility of adding additional objectives to make the search more flexible or even deleting one or more objectives to restrict the search in certain directions provides flexible ways of performing various search tasks (Fliege, 2007). These possibilities certainly open up new avenues and new ways of solving problems and should be exploited more in the near future. The above and a number of other possibilities of aiding different problem solving tasks through multiobjective optimization are discussed in (Knowles *et al.*, 2008).

16.3 Challenging Methods of Finding Optimal Solutions

Having discussed new problem domains for multiobjective optimization, we now discuss new and challenging methodologies of arriving at optimal solutions to multiobjective optimization problems.

16.3.1 Hybrid Methods

As mentioned earlier in this book, in the MCDM literature, solving multiobjective optimization problems has typically been understood as a task of helping a DM in finding the most preferred solution in the presence of conflicting objectives. In this kind of a problem setting, DM's preference information plays an important role. However, until recently, EMO approaches have mostly concentrated on approximating the whole set of Pareto optimal solutions. This brings about a natural question of how MCDM and EMO approaches can complement each other. For example, EMO methodologies can be used to include preference information (Fonseca and Fleming, 1998; Parmee *et al.*, 2000) (see Chapter 6 for more studies). As an example of hybridizing ideas and methods of MCDM and EMO fields, we can mention that some reference point (see Section 2.3 in Chapter 2) based EMO methods have already been introduced (Deb *et al.*, 2006; Thiele *et al.*, 2007), but there is much more potential in preparing new hybrid methods. Other examples of augmenting interactive MCDM methods with EMO ideas include (Deb and Kumar, 2007a,b), where the reference direction approach (Korhonen, 1988) and the "light beam search" (Jaszkiewicz and Słowiński, 1999) are utilized, respectively. Overall, the goal is to analyze the strengths of different approaches and utilize and combine them.

A very simple hybridizing idea is to use continuous local search methods (with scalarizing functions used in MCDM, see Chapter 1) together with EMO. This can be useful, for example, in order to improve (or even guarantee Pareto optimality of) different solutions produced by an EMO algorithm.

Hybridizations of approximation algorithms (approximating the Pareto optimal set) and interactive MCDM methods have, for example, been given in (Klamroth and Miettinen, 2008; Miettinen *et al.*, 2003). Similar ideas can be applied with EMO and interactive MCDM methods. By first using an approximation algorithm, the DM gets a general understanding about the problem as a whole, its possibilities and limitations and it is easier for him/her to specify preference information for the interactive method used. It is, for example, easier to specify the starting point for the interactive method or to specify a reference point.

One possibility of creating hybrid methods is to apply MCDA methods developed for dealing with a discrete set of solution alternatives (Olson, 1996) to the set of solutions generated by an EMO algorithm or a subset thereof. In this way, decision support tools of MCDA could help the DM in analyzing multidimensional objective vectors and finding the most preferred solution. For an example, see (Thiele *et al.*, 2007), where using the reference direction based VIMDA method (Korhonen, 1988) is discussed. A simple implementation of an EMO-MCDA hybrid procedure is also suggested in (Deb and Chaudhuri, 2007).

Sometimes, it is difficult for DMs to move from one Pareto optimal solution to another because this necessitates giving up in some objective function values. If, for example, an EMO algorithm is used to generate an approximation of the Pareto optimal set, the solutions in the population produced are not yet necessarily Pareto optimal. This leads to an idea of a method with a natural win-win situation. Namely, populations generated during the EMO search are shown to the DM, (s)he can direct the search and get better and better solutions.

16.3.2 Global Solvers

Many multiobjective optimization problems arising in engineering are problems defined by nonconvex, nondifferentiable and multi-modal functions. These functions are highly nonlinear. In this case, we must be able to employ a global solver to find the globally optimal solutions. Often, the auxiliary single objective problems which have to be solved as subproblems in multiobjective methods do not have the necessary mathematical structure, like generalized convexity, ensuring that computed points are global solutions. In many algorithmic investigations the question whether a global solution of the auxiliary problems can be determined, is very often not discussed. But in practice this is a significant point. In single objective optimization, locally optimal solutions can still be of some use, as economists or engineers already accept a computed point if a drastic reduction of costs can be obtained by the obtained (local or global) solution. But in multiobjective optimization finding global solutions is crucial, as often the optimization task is followed by a decision making task. If the solvers used do not compute global solutions, one obtains an approximation of the set of Pareto optimal points which may be completely awkward to make decisions with (see also discussion in Chapter 1). Such an ill-functioning appears, for example, if the set of Pareto optimal points is not connected, that is, it consists of several disconnected parts. Then the gaps between these parts are difficult to identify from a numerical point of view. Evolutionary or stochastic optimization methods are better equipped in dealing with such problems and must be investigated more rigorously.

Let us point out that instead of using solvers that can guarantee only the local optimality of solutions generated, it is possible to use some global single objective solver for solving the auxiliary problems produced by MCDM methods, for example, evolutionary algorithms or a hybrid solver where a local solver is used after an evolutionary algorithm, both suggested by Miettinen and Mäkelä (2006).

In the context of single objective optimization, a recent study (Eremeev and Reeves, 2003) has suggested that after a solution is found by using an approximate solver (such as an evolutionary algorithm), a validation procedure must be used to support the result. The study suggested a sampling procedure to estimate the frequency of falling into local optima. Extensions of such studies can be made in the context of multiobjective optimization. However, there still is much to do in this field in order to find the most appropriate solvers to be used in each problem considered.

In general, because evolutionary algorithms or stochastic methods are potential global solvers, the question of interest is how to improve their algorithmic behavior with techniques using derivatives. The above-mentioned way is the simplest possibility. For instance, if one applies an evolutionary algorithm to a complicated problem with smooth functions, it certainly makes sense to combine the evolutionary algorithm with a local solver which uses information on derivatives. Such a combination may improve the evolutionary algorithm. These *memetic* algorithms are difficult to design because one has to determine when to switch from the evolutionary algorithm to the local solver and back. For example, memetic methods combining an evolutionary algorithm with the well-known sequential quadratic programming (SQP) method produce promising results. Here we need comprehensive investigations on the interface of these evolutionary methods and derivative-based methods being qualified for such a combination. These investigations should lead to modern metaheuristic approaches resulting in new global solvers. For instance, hybrid solvers involving simulated annealing and the (local) proximal bundle method are introduced in (Miettinen *et al.*, 2006).

Based on the remarks listed, there is a need for efficient global solvers (as also concluded by Aittokoski and Miettinen (2008)). We should develop hybrid methods combining standard methods with global strategies. These global hybrid solvers are very desirable and they would bring a breakthrough in finding guaranteed global Pareto optimal solutions in multiobjective optimization.

16.4 New Trends in Optimality

Finally, let us devote some thoughts to a few new trends in defining optimality in the context to multiobjective optimization: subjective preferences, different optimality concepts, and robust solutions.

16.4.1 Subjective Preferences

The MCDM literature typically places the DM at the center of the solution process of a multiobjective decision problem (as, e.g., Belton and Stewart (2001)). The DM's preferences determine which objective functions are more or less important, and how different objective values are to be rated. Consequently, aggregation across objectives has to be performed in a way which is consistent with the DM's preferences. This subjectivity is often seen as one of the characteristic features of multiobjective optimization problems, which distinguish this area from single objective optimization, where an objectively optimal solution can be found.

This subjective view is not entirely shared in the EMO literature, or more generally in multiobjective optimization. When solving multiobjective optimization problems, the aggregation across the individual objectives is often specified by model developers, with little involvement of the actual DMs. From a subjective perspective, this might seem a grave omission: an analyst who selects an aggregation mechanism across objective functions (like an additive function), and specifies weights of individual objectives to be used in this aggregation, takes away decision authority from the actual DM. Even seemingly objective concepts like dominance or Pareto optimality contain subjective elements because dominance requires at least information about the direction of preference within each objective function. But not all multiobjective optimization problems exhibit this level of subjectivity. Sometimes, even an aggregation across objective functions can be performed quite objectively, and a model developer might be even in a better position to perform such an "objective" optimization than the actual DM. However, if subjective information is not available, EMO can be used to get an idea of the Pareto optimal set, at least in the case of optimization problems with two to four objectives.

Rather than establishing a strict dichotomy between "objective" single objective optimization problems and "subjective" multiobjective problems, we can propose a taxonomy of different levels of subjectivity in multiobjective optimization problems as four cases. In this taxonomy, the classical view of multiobjective problems does not even form an endpoint, but an intermediate stage.

The proposed taxonomy consists of four cases:

 i) Multiobjective optimization problems as a technical solution device.
 ii) Multiobjective optimization problems as an approximation of a higher level objective.
iii) Subjective multiobjective optimization problems.
iv) Problems involving meta-criteria.

Multiobjective Optimization Problems as a Technical Solution Device

In some cases, heuristics work better on multiobjective optimization problems than on problems with a single objective function. In these cases, it might make sense to perform a "multi-objectivization" of the problem (Knowles *et al.*, 2001): to split an explicitly given criterion into several functions and solve the resulting multiobjective optimization problem, as discussed in Section 16.2.3. Of course, the aggregation procedure in this case is fully determined and has to reconstruct the original objective function.

Multiobjective Optimization Problems as an Approximation of a Higher Level Objective

In many applications of multiobjective optimization methods, the DM actually wants to maximize some higher level criterion, but this criterion can either not be directly measured, or the relationship of the decision variables to that higher level criterion is not clear. Therefore, one uses several substitute criteria and solves a multiobjective optimization problem, instead. For more details, refer to (Miettinen, 1999). A study on EMO (Handl *et al.*, 2007) called these substitute criteria 'proxy objectives'. In many MCDM applications in business, the long run profit of the firm is the ultimate goal. But the impact of many decisions on a long run profit can hardly be quantified. For example, when hiring a new executive, one cannot predict how much a particular person will contribute to profit, so substitute criteria like education and experience are used to approximate that person's productivity. Another example demonstrating the benefits of using multiobjective optimization is discussed in (Hakanen *et al.*, 2005), where estimating amortization time and interest rate for capital is avoided when balancing between investment and running costs in the case of designing a heat recovery system of a paper mill.

In these cases, neither the choice of a preference model nor the selection of parameters (e.g., weights) to be used in that model is purely subjective, but both should approximate the likely relationship of substitute criteria to the higher level criterion. A higher weight in this case does not indicate that a substitute (lower level) criterion is considered more important, but that it is considered to have a stronger influence on the higher level criterion.

Subjective Multiobjective Optimization Problems

Subjective problems are typically considered in MCDM, where the aggregation of objective functions depends solely on the subjective preferences of the DM. This type of decision problems are often illustrated by referring to personal decisions like the purchase of a car, where attributes like comfort, speed or costs need to be compared.

In this case, no objective aggregation model exists, which would be valid for all DMs. Of course, modeling can still be performed by an analyst, but only in close contact with the actual DM who has to provide the relevant preference information.

Problems Involving Meta-criteria

Many multiobjective optimization problems are related to decisions in which the interests of multiple stakeholders have to be taken into account. Such decisions occur, for example, in public policy. Even when the decision is ultimately made by one individual, for example, a politician, that individual has to consider the interests of different parties. While in the decision problems discussed so far, an improvement in any objective function could be considered to improve the overall evaluation of a decision alternative (this assumption underlies the whole concept of Pareto optimality), this is no longer true when aspects like fairness need to be taken into account. Here, further improvements of the situation of stakeholders who are already better off than the others might be considered as unfair and, thus, make a solution less preferable.

Such "meta-criteria", which evaluate the distribution of results across several criteria, occur not only in multi-person decisions. For example, when time streams of income are evaluated, income in each period could be considered as an objective. Apart from maximizing income in each period (which would correspond to the standard multiobjective formulation), DMs might prefer a constant income stream over a stream which exhibits large variations over time. This preference for particular patterns should not be confused with risk aversion; it can occur even if all payments are known in advance with certainty. To handle this type of problems, Kostreva *et al.* (2004) developed the concept of *equitable* multiobjective decision making and showed how several multiobjective optimization methods, in particular reference point methods, can be extended to handle such problems.

Further Comments on the Taxonomy

Our taxonomy of multiobjective optimization problems has several consequences for the way in which "preferences" are elicited, modelled and aggregated. The first difference concerns the person, or group of persons, from whom preference information can be obtained. In highly subjective problems, only the DM him/herself can provide information about preferences. But in problems where multiple criteria are used to approximate a higher level objective, it might be reasonable to obtain input from several experts in order to get a clearer picture of how substitute criteria will actually influence the higher level objective. In the remaining two cases no real "preference elicitation" can take place. When multiple criteria are introduced for technical reasons, their aggregation is also a technical problem. In the case of meta-criteria, the way in which individual criteria are aggregated is based on the meta-criteria involved, which can be considered as axioms an aggregation method must fulfill.

This distinction has also consequences for the likely stability of preference information. While there is some empirical evidence that individual preferences towards multiple criteria remain stable over time (Blackmond and Fischer, 1987; San Miguel *et al.*, 2002), they are still subject to more external influences than causal relationships between substitute criteria and higher level objective. Consequently, "preference" information obtained for the latter type of problems, as well as for the other two classes, needs to be elicited less often in repeated decisions than for subjective problems.

One might also view properties of solution concepts, like efficiency or independence of irrelevant alternatives, differently in the four cases of our taxonomy. In the first case, such axioms are more or less irrelevant. Aggregation has to reconstruct the original objective, regardless of whether it fulfills common axioms of decision analysis or not. In the second case, rationality (in the form of axioms) becomes more important, since in most problems, it can be expected that the true relationship between substitute criteria and the actual higher level objective also follows these principles. In the third case, acceptance of axioms is entirely up to the DM. Empirical research on bias phenomena in decision making has provided considerable evidence that subjects consciously choose to violate axioms of decision analysis, even when this violation is pointed out to them (von Winterfeldt and Edwards, 1986). Finally, in the last case, meta-criteria are themselves axioms, and their acceptance by all stakeholders is a prerequisite for acceptable solutions.

By formulating the above taxonomy, we have just started to explore the impact of different levels of subjectivity on the solution process, as well as the underlying theory of multiobjective optimization, both with MCDM and EMO methods. This could become an interesting area of future research.

16.4.2 Generalized Dominance and Redefining Optimality

Most multiobjective optimization studies use the concept of Pareto optimality for driving their search. However, there exist a number of other trends of redefining the usual Pareto optimality. Such considerations usually reduce the size of the optimal set and in some occasions make it easier for the search algorithms to handle the complexity associated with multiobjective optimization. Here we discuss a number of such trends of redefining optimality in multiobjective optimization.

In this book, the basic concept of optimality has been that of Pareto optimality, but a closely related, relaxed, concept of weak Pareto optimality is sometimes used because the latter is computationally simpler and many straightforward approaches to multiobjective optimization generate weakly Pareto optimal solutions (see, e.g., Preface and Chapter 1). However, weak Pareto optimality is not satisfactory for applications because it ignores clear possibilities of solution improvement with respect to some objectives. Actually, even the concept of Pareto optimality may be too weak for many applications. As discussed in Chapter 1, the notion of proper Pareto optimality (Geoffrion, 1968) assumes that all the trade-offs are bounded (see also Chapter 2). Sometimes, more useful for applications are solutions that are properly Pareto optimal with an a priori given bound on trade-offs.

Several dominance (and thereby efficiency or Pareto optimality) concepts can be introduced as the so-called dominance cone (Yu, 1974) as also briefly discussed in Chapter 1. The partial order of the dominance relation is implied by a convex cone D in such a sense that $\mathbf{y}'$ dominates $\mathbf{y}''$ if and only if $\mathbf{y}' - \mathbf{y}'' \in D \setminus \{\mathbf{0}\}$. The standard Pareto optimality or Pareto dominance is defined by using an orthant cone (negative orthant for minimization). A narrower cone restricts the dominance relation thus expanding the corresponding efficient set. On the other hand, a wider cone enforces more dominated outcome vectors, thus narrowing down the efficient set.

A corresponding dominance cone can be constructed by combining the orthant with the half-space (Kaliszewski, 1994; Wierzbicki, 1986). Actually, the reference point method and many other scalarizing function model such dominance by taking the sum of objective values (the half space) with a small weight to regularize the basic term of the max-min aggregation (the orthant). See also Chapters 1 and 2 as well as (Miettinen, 1999).

Most traditional MCDM approaches to multiobjective optimization seek for the best solution according to the DM's preferences while treating the dominance relation as a common principle of all rational preference models. Thus the concept of Pareto optimality is rather used as a necessary condition to establish the boundary of acceptable choices. Therefore, strengthening the dominance concept is not so crucial for the implementation of interactive MCDM procedures, although still important. On the other hand, EMO procedures use three different features: emphasis on nondominated solutions in the current population, emphasis on previously-found nondominated solutions, and emphasis on less crowded solutions in the objective space (see Chapter 3). Many studies related to different dominance relations and approaches utilizing them have been published during the years in the MCDM field. Lately, they have also attracted attention in the EMO field. For example, wider dominance cones can be used to focus an EMO search on a part of the Pareto optimal set (Branke *et al.*, 2001; Laumanns *et al.*, 2002), instead on the complete set. In particular, the cone dominance enables to formalize concepts of narrowing the Pareto optimal set related to limitations on trade-offs.

Note that the dominance cone can be changed during the solution process. Such a dominance structure appears, for instance, in the case of a given value (or utility) function maximization. The dominance structure corresponding to the comparison of the value function values is represented by the tangent cone to the isoline contours of the value function at any objective vector. For poorly characterized preferences in multiobjective problems, it is often desirable to seek (approximate) optimal solutions for a large class of value functions. Solutions corresponding to the optimal value of a large variety of linear value functions can be emphasized within the EMO procedure, thereby aiding to find *knee* objective vectors in certain problems (Branke *et al.*, 2004a). An approximate majorization relation enables the search for solutions maximizing all symmetric concave value functions (Goel and Meyerson, 2006).

There are many applications leading to problems with a large number of uniform criteria considered impartially which makes the distribution of outcomes more important than the assignment of several outcomes to the specific criteria. Such models are generally related to the evaluation and optimization of various systems which serve many users where quality of service for every individual user defines the criteria. This applies to various technical and social systems. An example arises in locating public facilities where the decisions often concern the placement of a service center or another facility in a position so that the users are allocated in an impartial way. Thus, we are interested in comparing distributions of values within the objective vectors rather than componentwise comparison of objective vectors (Ogryczak, 1999). Note that having two possible location patterns generating objective vectors $(5, 0, 5)$ and $(0, 1, 0)$, we would recognize both the location patterns as Pareto optimal in terms of (distance) minimization. However, the first location pattern generates two objectives (distances) equal to 5 and one objective equal to 0, whereas the second pattern generates one objective equal to 1 and two objectives equal to 0. Thus, in terms of the distribution of objective values, the second location pattern is clearly better.

The need to search for some optimal distribution of objective values is commonly recognized in problems which may be viewed as resource allocation models. While allocating limited resources to maximize the system efficiency they also attempt to provide a fair treatment of all the competing activities. For instance, in networking, a central issue is how to allocate bandwidth to flows efficiently and fairly (Denda *et al.*, 2000; Pióro and Medhi, 2004). Furthermore, uniform individual criteria may be associated with some events rather than physical users, like in many dynamic optimization problems where uniform individual criteria represent a similar event in various periods or in decision problems under uncertainty where uniform individual criteria represent the outcome realizations under various scenarios. Another type of model is that of approximation of discrete data by a functional form. The residuals may be viewed as objectives to be minimized, and there is no reason to treat them in any way but impartially.

In many models fair consideration of all criteria requires more than only impartiality. In order to ensure fairness in a system, all system entities have to be equally well provided with the system's services. This means that more equal objective vectors are preferred to unequal ones or, more formally, a transfer of any small amount from an objective function to any other relatively worse-off objective results in a more preferred objective vector. For instance, a solution generating all three objective values equal to 2 is considered better than any solution generating individual values 4, 2 and 0. This leads to concepts of fairness expressed by the equitable efficiency as a specific refinement of Pareto optimality taking into account impartiality and inequality minimization (Kostreva et al., 2004). Thus, seeking for the optimal distribution of objective values is actually a new multiobjective problem type. However, the dominance structure for objective vectors does not represent any cone (Kostreva and Ogryczak, 1999).

Currently, some specific solution concepts are used for various application areas. Biobjective aggregations to the mean and some dispersion measure are used in the areas of decisions under risk and location analysis as well. The max-min approach additionally regularized with the lexicographic order (the so-called max-min fairness) is commonly used in resource allocation problems (Luss, 1999). Approaches exploiting the multiobjective nature of distribution optimization problems are rather rare (Ogryczak et al., 2008). Actually, such problems are hard for preference modeling and identification within the interactive MCDM methods as well as for the EMO approaches. Nevertheless, they deserve to be investigated more intensively.

16.4.3 Robust Solutions

A conventional optimization approach that considers only the optimality of a decision or a design, that is, performance at decision or design condition, should work fine in a controlled environment. Real-world applications, on the other hand, inevitably involve errors and uncertainties (be it, e.g., in the design process, manufacturing process, and/or operating conditions); so that the resulting performance may be lower than expected. For instance, the aerodynamic performance of an airplane wing design is very sensitive to the wing shape and flight conditions and, thus, it may deteriorate drastically when subject to wing manufacturing errors and wind variations even if the wing design is optimized.

Several approaches have been developed to deal with uncertain or imprecise data. The approaches focused on the quality or on the variation (stability) of the solution for some data domains are considered robust. The notion of robustness applied to decision problems was first introduced almost 50 years ago by Gupta and Rosenhead (1968). Practical importance of the performance sensitivity against data uncertainty and errors has later attracted considerable attention to the search for robust solutions. Actually, as suggested by Roy (1998), the concept of robustness should be applied not only to solutions but, more generally to various assertions and recommendations generated within a decision support process. A brief comparison between conventional optimization and robust optimization is illustrated in Fig. 16.1 a). Solution A obtained by a conventional optimization is the best in terms of optimality, but disperses widely in terms of the objective function against the dispersion of design variable or environmental variable, and this dispersion may extend to an infeasible range. On the other hand, solution B obtained by a robust optimization is moderately good in terms of optimality and also good in terms of robustness, that is, dispersion of objective function is narrow against dispersion of design variable.

On the other hand, the optimal solution despite generating objective values dispersed quite widely may be clearly better than a solution not dispersed at all. As depicted in Fig. 16.1 b), solution A though characterized by dispersed results remains under all conditions better than the stable solution B. Hence, solution B is obviously dominated and it cannot be considered a robust optimal solution.

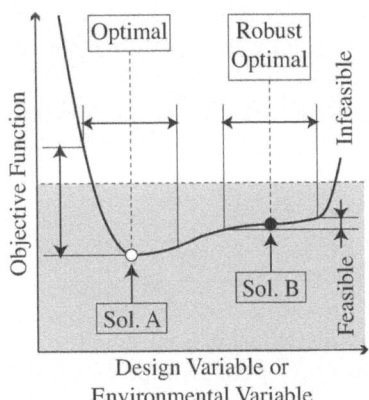 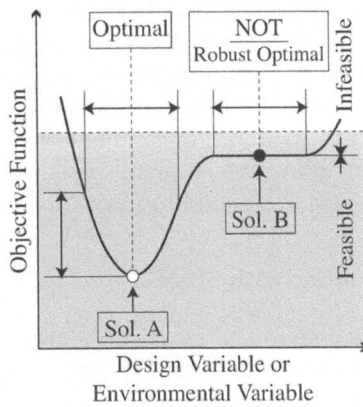

Fig. 16.1. Comparison between conventional optimization and robust optimization (for a minimization problem): a) conventional optimal solution A vs. robust optimal solution B; b) stable but not robust optimal solution B.

The precise concept of robustness depends on the way the uncertain data domains and the quality or stability characteristics are introduced. Typically, in robust analysis one does not attribute any probability distribution to represent uncertainties. Data uncertainty is rather represented by non-attributed scenarios, which means there is no specific rule to determine the data uncertainty characteristics. Since one wishes to optimize results under each scenario, robust optimization might be in some sense viewed as a multiobjective optimization problem where objectives correspond to the scenarios. However, despite of many similarities of such robust optimization concepts to multiobjective models, there are also some significant differences (Hites *et al.*, 2006). Actually, robust optimization is a problem of optimal distribution of objective values under several scenarios (c.f. Section 16.4.2) rather than a standard multiobjective optimization model.

A conservative notion of robustness focusing on worst case scenario results is widely accepted and the min-max optimization is commonly used to seek robust solutions. The worst case scenario analysis can be applied either to the absolute values of objectives (the absolute robustness) or to the regret values (the deviational robustness) (Kouvelis and Yu, 1997). The latter, when considered from the multiobjective perspective, represents a simplified reference point approach with the utopian (ideal) objective values for all the scenario used as aspiration levels. Recently, a more advanced concept of ordered weighted averaging was introduced into robust optimization (Perny et al., 2006), thus, allowing to optimize combined performances under the worst case scenario together with the performances under the second worst scenario, the third worst and so on. Such an approach exploits better the entire distribution of objective vectors in search for robust solutions and, more importantly, it introduces some tools for modeling robust preferences. Actually, while more sophisticated concepts of robust optimization are considered within the area of discrete programming models, only the absolute robustness is usually applied to the majority of decision and design problems.

Taking into account the current computational capabilities of both EMO and MCDM techniques, one may expect development of new robust optimization approaches in many areas. Here, we do not make any attempt to discuss all such existing implementations.

Dealing with Risk

When an (objective or subjective) probability distribution is specified to characterize the data uncertainty, robust optimization becomes a problem of decision under risk. In this context, robustness is represented by the notion of risk aversion, and typically by a "strong" risk aversion. There exists a well-developed methodology for decisions under risk and it can be directly applied to robust optimization. In particular, the mean-risk (MR) approach (Markowitz model) quantifies the problem in a lucid form of only two objectives: the mean (expected) outcome μ and the risk ϱ, a scalar measure of the variability (dispersion) of outcomes. The latter may be equally interpreted as a robustness measure of solutions, thus allowing the MR model to be read as mean-robustness in an appropriate setting.

The MR approach allows to formalize robust optimization with two separate criteria: optimality (μ) and robustness (ϱ). Indeed, in many real-life problems improvements in optimality and robustness are competing while the MR model allows to formalize it and to analyze the trade-off between these two criteria. The classical Markowitz model uses the standard deviation σ (or variance σ^2) as the risk measure. Similarly, the biobjective model $\min\{\mu, \sigma\}$ is applied for robust optimization, although frequently in the scalarized form $\min\{\mu + \alpha\sigma\}$ with the trade-off parameter $\alpha < 0$. Unfortunately, while the mean-variance model is well suited for normal distributions, it may lead to inferior conclusions in general. Referring to the case depicted in Fig. 16.1 b), one may notice that obviously a worse solution B is characterized by $\sigma = 0$, thus, in terms of the biobjective MR model it is not dominated by solution A with a positive measure of dispersion, despite the fact that the latter is clearly better under all scenarios. This flaw of MR models may be overcome by the use of asymmetric dispersion measures focused only on disturbances negative to the optimization and combining them with the mean values.

For instance, the biobjective model $\min\{\mu, \mu + \bar{\sigma}\}$ with $\bar{\sigma}$ representing the upper side standard deviation will generate only solutions with nondominated distributions of results (Ogryczak and Ruszczyński, 1999), namely, solutions which cannot be improved under all scenarios simultaneously. One may notice that, while considering the maximum upper deviation (from the mean) Δ as a probability independent dispersion measure, one gets the criterion $\mu + \Delta$ expressing the worst case scenario result, that is, the classical conservative notion of robustness. Multiobjective approaches to decisions under risk (Ogryczak, 2002) allow to model various robust solution concepts.

Shimoyama *et al.* (2005) have proposed a multiobjective robust optimization approach called design for multiobjective six sigma (DFMOSS). The DFMOSS builds on the ideas of design for six sigma (DFSS) (Enginous Software, Inc., 2002), coupled with an EMO algorithm (Deb, 2001), for an enhanced capability to reveal trade-off information considering both optimality and robustness of design. Jin and Sendhoff (2003) have also discussed the trade-off between optimality and robustness in the context of multiobjective optimization. The DFSS is based on the "six sigma" concept, which was originally established as a measure of excellence for business processes. The aim is to achieve a process with such a small dispersion that the range of $\pm 6\sigma$ (where σ is standard deviation) around the mean value μ is included in an acceptable range for the performance parameter. The level of dispersion can be defined as "sigma level n" satisfying the following constraints:

$$\mu - n\sigma \geq \text{LSL} \quad \text{and} \quad \mu + n\sigma \leq \text{USL}, \tag{16.1}$$

where LSL and USL are lower and upper specification limits, respectively. A larger sigma level indicates smaller dispersion. In the context of robust design optimization, smaller dispersion translates to a more robust characteristic.

For a general single objective optimization problem where an objective function $f(\mathbf{x})$ of design variable $\mathbf{x}$ must be minimized, the DFMOSS deals with the biobjective optimization problem where the mean value (μ_f) and the standard deviation (σ_f) of $f(\mathbf{x})$ must be minimized when $\mathbf{x}$ disperses around the design condition due to errors and uncertainties. During the optimization process itself, multiple solutions (individuals) are dealt with simultaneously using EMO. For each individual, μ_f and σ_f are evaluated as two separate objective functions from $f(\mathbf{x})$ at the sample points around $\mathbf{x}$. From them, better solutions are selected based on the Pareto optimality concept between μ_f and σ_f. New solutions for the next step are reproduced by crossover and mutation from the selected solutions. This optimization process is iterated until the trade-off relation between μ_f and σ_f has converged, and multiple robust optimal solutions have been obtained. After the optimization, the sigma level n satisfying (16.1) is post-evaluated for the obtained optimal solutions. This allow one to select a robust solution with the highest sigma level (preferably with the level 6).

Note that some optimization problems do not have robust solutions that satisfy six sigma. In such cases, it is preferable to find a solution with a sigma level n as high as possible, even if it is less than six sigma. In addition, (16.1) can still be considered during the μ_f–σ_f optimization; it is better to do this when n to be satisfied is strongly determined by a certain design requirement. Let us also mention that Deb and Gupta (2005) have suggested two types of robustness in the context of multiobjective optimization. Certainly, many other variations are possible.

Uncertainty in Presence of Constraints

Uncertainty in problem parameters may affect not only the objective functions but also the feasible set, thus, threatening the feasibility of solutions. Solving such problems is frequently referred to as reliability-based optimization, where one seeks the best solution among those remaining feasible for various data perturbations. Again, the precise concept of solution depends on the way the uncertain data domains are introduced. When uncertainty is represented by non-attributed scenarios, the worst case approach can be applied. When probability distribution is specified (either objective or subjective) to characterize the data uncertainty, one gets a typical stochastic programming problem. Fig. 16.2 shows a hypothetical problem with two inequality constraints. Typically, the optimal solution lies on a constraint boundary or at the intersection of more than one constraints, as shown in the figure. In the event of uncertainties in design variables (as shown in the figure with a probability distribution around the optimal solution) in many instances such a solution will be infeasible. In order to find a solution which is more reliable (meaning that there is a very small probability of instances producing an infeasible solution), the true optimal solution must be sacrificed and a solution interior to the feasible

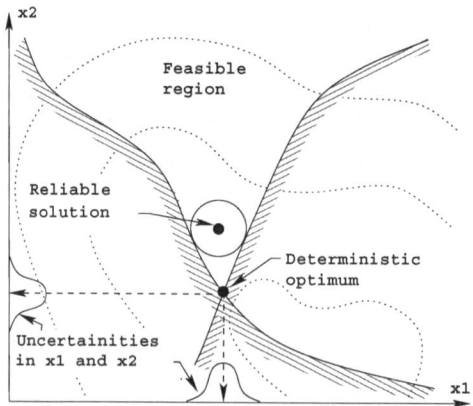

Fig. 16.2. The concept of reliability-based optimization.

region may be chosen. For a desired reliability measure R, it is then desired to find that feasible solution which will ensure that the probability of having an infeasible solution instance created through uncertainties from this solution is at most $(1 - R)$. To arrive at such a solution, a stochastic optimization problem can be converted to its deterministic equivalent (Birge and Louveaux, 1997; Romeijn et al., 2006).

To handle such cases with a large reliability requirement, probabilistic methodologies involving a double loop, single loop and decoupled methods are used. For example, one can incorporate a decoupled method with an EMO procedure (Deb et al., 2007b) to find reliable sets, instead of a sensitive Pareto optimal set, corresponding to a specified reliability value. More such studies are needed to make the approach computationally viable and applicable to practical multiobjective problem solving tasks.

Another issue involving uncertainty in solution evaluation comes from dealing with noisy environments, in which objective and constraint function evaluations introduce inherent noise. Although this issue has received a lot of attention in the context of single objective evolutionary algorithms (see a survey by Jin and Branke (2005)), some attempts have recently been made in EMO as well (Bui et al., 2005; Hughes, 2001; Teich, 2001). Clearly, more studies are needed to fully understand the effect of noise in solution evaluation procedures in multiobjective optimization.

16.5 Conclusions

In this chapter, we have discussed some ideas for future research in the context of multiobjective optimization and decision making. However, plenty of research is still needed in many aspects of decision making. In this respect, some of the future directions mentioned by Miettinen (1999) still stand as relevant challenges. Let us hope that the future years will bring new light in them and many other fruitful and rewarding topics.

An important challenge is to increase awareness of the possibilities and potential of multiobjective optimization because there still are many application fields where multiobjective optimization is not used at all or is used in a very simplistic way even though the problems solved clearly involve multiple conflicting objectives. Often, the existence of decision support tools is simply not known to many researchers. Here, the importance of strong and encouraging case studies cannot be emphasized enough. For people dealing with applications, case studies give a possibility to see the benefits obtainable in a concrete and understandable way. A necessary and natural step of bringing multiobjective optimization tools closer to real DMs is the important challenge of designing user-friendly software for decision support. We need software that is easily accessible (like the WWW-NIMBUS® system (Miettinen and Mäkelä, 2000, 2006) operating via the Internet) and this certainly is a field needing more attention.

In many applications, multiple objectives are hidden and simplified in the modeling phase in order to produce a problem that seems to be solvable. Increasing awareness of the existence of multiobjective optimization methods and tools also encourages questioning the existing models in order to avoid simplifications that blur the possibility of studying the interdependencies between the conflicting objectives in real problems.

The possibilities and importance of interactive methods have been emphasized a lot in this book because interactive methods give the DM a possibility to learn about the problem considered. If the problem is complex and function evaluations take a lot of time, the interactive nature of the solution process may suffer because the DM has to wait for new and improved solutions. This sets requirements and challenges on the computational efficiency of the methods used. Besides using meta-modeling (see Chapter 10) and optimization techniques with increased accuracy (as the solution process proceeds), new approaches and ideas are needed. For example, ideas related to learning are discussed in Chapter 15.

One aspect has clearly emerged from the chapters of this book: multiobjective optimization using evolutionary algorithms or otherwise and decision making aids must be put together synergistically, computationally efficiently, and, above all, interactively for a DM to examine possible candidate solutions and choose one particular preferred solution at the end. Such a task requires one to first know both optimization and decision making literature well. This book has shown a number of possibilities of such mergers from various points of view. This chapter has also suggested a number of avenues for moving forward in this direction. With collaborative efforts from various research groups involving multiobjective optimization and decision making, we should witness more holistic approaches, interactive algorithms and software systems to be developed for practical use in the coming years.

Acknowledgements

The work of K. Miettinen was partly supported by the Foundation of the Helsinki School of Economics. The work of K. Deb was partly supported by the Academy of Finland (grant # 118319). The work of W. Ogryczak was partially supported by the Ministry of Science and Information Society Technologies under grant 3T11C 005 27.

References

Aittokoski, T., Miettinen, K.: Cost effective simulation-based multiobjective optimization in performance of internal combustion engine. Engineering Optimization 40(7), 593–612 (2008)

Bard, J.F.: Practical Bilevel Optimization: Algorithms and Applications. Kluwer Academic Publishers, Dordrecht (1998)

Belton, V., Stewart, T.J.: Multiple Criteria Decision Analysis. Kluwer Academic Publishers, Dordrecht (2001)

Bingul, Z.: Adaptive genetic algorithms applied to dynamic multiobjective problems. Applied Soft Computing 7(3), 791–799 (2007), doi:10.1016/j.asoc.2006.03.001.

Birge, J.R., Louveaux, F.: Introduction to Stochastic Programming. Springer, Heidelberg (1997)

Blackmond, L.K., Fischer, G.W.: Estimating utility functions in the presence of response error. Management Science 33, 965–980 (1987)

Bleuler, S., Brack, M., Zitzler, E.: Multiobjective genetic programming: Reducing bloat using SPEA2. In: Proceedings of the 2001 Congress on Evolutionary Computation, pp. 536–543. IEEE Computer Society Press, Piscataway (2001)

Branke, J., Kaußler, T., Schmeck, H.: Guidance in evolutionary multi-objective optimization. Advances in Engineering Software 32, 499–507 (2001)

Branke, J., Deb, K., Dierolf, H., Osswald, M.: Finding knees in multi-objective optimization. In: Yao, X., Burke, E.K., Lozano, J.A., Smith, J., Merelo-Guervós, J.J., Bullinaria, J.A., Rowe, J.E., Tiño, P., Kabán, A., Schwefel, H.-P. (eds.) PPSN 2004. LNCS, vol. 3242, pp. 722–731. Springer, Heidelberg (2004a)

Branke, J., Schmeck, H., Deb, K., Reddy, M.: Parallelizing multi-objective evolutionary algorithms: Cone separation. In: Proceedings of the Congress on Evolutionary Computation (CEC-2004), pp. 1952–1957. IEEE Press, Piscataway (2004b)

Bui, L.T., Abbass, H.A., Essam, D.: Fitness inheritance for noisy evolutionary multi-objective optimization. In: Proceedings of the International Conference on Genetic and evolutionary computation (GECCO-2005), pp. 779–785. ACM Press, New York (2005)

Deb, K.: Multi-Objective Optimization using Evolutionary Algorithms. Wiley, Chichester (2001)

Deb, K., Chaudhuri, S.: I-MODE: An interactive multi-objective optimization and decision-making using evolutionary methods. In: Obayashi, S., Deb, K., Poloni, C., Hiroyasu, T., Murata, T. (eds.) EMO 2007. LNCS, vol. 4403, pp. 788–802. Springer, Heidelberg (2007)

Deb, K., Gupta, H.: Searching for robust Pareto-optimal solutions in multi-objective optimization. In: Coello Coello, C.A., Hernández Aguirre, A., Zitzler, E. (eds.) EMO 2005. LNCS, vol. 3410, pp. 150–164. Springer, Heidelberg (2005)

Deb, K., Kumar, A.: Interactive evolutionary multi-objective optimization and decision-making using reference direction method. In: Proceedings of the Genetic and Evolutionary Computation Conference (GECCO-2007), pp. 781–788. ACM Press, New York (2007a)

Deb, K., Kumar, A.: 'Light beam search' based multi-objective optimization using evolutionary algorithms. In: Proceedings of the Congress on Evolutionary Computation (CEC-2007), pp. 2125–2132. IEEE Computer Society Press, Piscataway (2007b)

Deb, K., Srinivasan, A.: Innovization: Innovating design principles through optimization. In: Proceedings of the Genetic and Evolutionary Computation Conference (GECCO-2006), pp. 1629–1636. ACM Press, New York (2006)

Deb, K., Zope, P., Jain, S.: Distributed computing of Pareto-optimal solutions with evolutionary algorithms. In: Fonseca, C.M., Fleming, P.J., Zitzler, E., Deb, K., Thiele, L. (eds.) EMO 2003. LNCS, vol. 2632, pp. 534–549. Springer, Heidelberg (2003)

Deb, K., Sundar, J., Reddy, U., Chaudhuri, S.: Reference point based multi-objective optimization using evolutionary algorithms. International Journal of Computational Intelligence Research 2(6), 273–286 (2006)

Deb, K., Rao N., U.B., Karthik, S.: Dynamic multi-objective optimization and decision-making using modified NSGA-II: A case study on hydro-thermal power scheduling. In: Obayashi, S., Deb, K., Poloni, C., Hiroyasu, T., Murata, T. (eds.) EMO 2007. LNCS, vol. 4403, pp. 803–817. Springer, Heidelberg (2007a)

Deb, K., Padmanabhan, D., Gupta, S., Mall, A.K.: Reliability-based multi-objective optimization using evolutionary algorithms. In: Obayashi, S., Deb, K., Poloni, C., Hiroyasu, T., Murata, T. (eds.) EMO 2007. LNCS, vol. 4403, pp. 66–80. Springer, Heidelberg (2007b)

Delattre, M., Hansen, P.: Bicriterion cluster analysis. IEEE Transaction Pattern Analysis and Machine Intelligence 2(4), 277–291 (1980)

Dempe, S.: Foundations of Bilevel Programming. Kluwer Academic Publishers, Dordrecht (2002)

Dempe, S.: Annotated bibliography on bilevel programming and mathematical programs with equilibrium constraints. Optimization 52, 333–359 (2003)

Denda, R., Banchs, A., Effelsberg, W.: The fairness challenge in computer networks. In: Crowcroft, J., Roberts, J., Smirnov, M.I. (eds.) Quality of Future Internet Services, pp. 208–220. Springer, Heidelberg (2000)

Engineous Software, Inc.: iSIGHT Reference Guide Version 7.1, pp. 220–233. Engineous Software, Inc. (2002)

Eremeev, A.V., Reeves, C.R.: On confidence intervals for the number of local optima. In: Raidl, G.R., Cagnoni, S., Cardalda, J.J.R., Corne, D.W., Gottlieb, J., Guillot, A., Hart, E., Johnson, C.G., Marchiori, E., Meyer, J.-A., Middendorf, M. (eds.) EvoIASP 2003, EvoWorkshops 2003, EvoSTIM 2003, EvoROB/EvoRobot 2003, EvoCOP 2003, EvoBIO 2003, and EvoMUSART 2003. LNCS, vol. 2611, pp. 224–235. Springer, Heidelberg (2003)

Farina, M., Deb, K., Amato, P.: Dynamic multiobjective optimization problems: Test cases, approximations, and applications. IEEE Transactions on Evolutionary Computation 8(5), 425–442 (2000)

Fliege, J.: The effects of adding objectives to an optimisation problem on the solution set. Operations Research Letters 35(6), 782–790 (2007)

Fonseca, C.M., Fleming, P.J.: Multiobjective optimization and multiple constraint handling with evolutionary algorithms–Part I: A unified formulation. IEEE Transactions on Systems, Man and Cybernetics 28(1), 26–37 (1998)

Geoffrion, A.M.: Proper efficiency and the theory of vector maximization. Journal of Mathematical Analysis and Applications 22(3), 618–630 (1968)

Goel, A., Meyerson, A.: Simultaneous optimization via approximate majorization for concave profits or convex costs. Algorithmica 44, 301–323 (2006)

Gupta, S., Rosenhead, J.: Robustness in sequential investment decisions. Management Science 15, 18–29 (1968)

Hakanen, J., Miettinen, K., Mäkelä, M., Manninen, J.: On interactive multiobjective optimization with NIMBUS in chemical process design. Journal of Multi-Criteria Decision Analysis 13(2–3), 125–134 (2005)

Handl, J., Knowles, J.: An evolutionary approach to multiobjective clustering. IEEE Transactions on Evolutionary Computation 11(1), 56–76 (2007)

Handl, J., Kell, D.B., Knowles, J.: Multiobjective optimization in bioinformatics and computational biology. ACM/IEEE Transactions on Computational Biology and Bioinformatics 4(2), 279–292 (2007)

Hites, R., De Smet, Y., Risse, N., Salazar-Neumann, M., Vincke, P.: About the applicability of MCDA to some robustness problems. European Journal of Operational Research 174, 322–332 (2006)

Hughes, E.J.: Evolutionary multi-objective ranking with uncertainty and noise. In: Zitzler, E., Deb, K., Thiele, L., Coello Coello, C.A., Corne, D.W. (eds.) EMO 2001. LNCS, vol. 1993, pp. 329–343. Springer, Heidelberg (2001)

Jahn, J.: Vector Optimization – Theory, Applications, and Extensions. Springer, Heidelberg (2004)

Jahn, J.: Introduction to the Theory on Nonlinear Optimization. Springer, Heidelberg (2007)

Jaszkiewicz, A., Słowiński, R.: The 'light beam search' approach – an overview of methodology and applications. European Journal of Operational Research 113, 300–314 (1999)

Jensen, M.T.: Helper-objectives: Using multi-objective evolutionary algorithms for single-objective optimisation. Journal of Mathematical Modelling and Algorithms 3(4), 323–347 (2004)

Jin, Y., Branke, J.: Evolutionary optimization in uncertain environments. IEEE Transactions on Evolutionary Computation 9(3), 303–317 (2005)

Jin, Y., Sendhoff, B.: Trade-off between performance and robustness: An evolutionary multiobjective approach. In: Fonseca, C.M., Fleming, P.J., Zitzler, E., Deb, K., Thiele, L. (eds.) EMO 2003. LNCS, vol. 2632, pp. 237–251. Springer, Heidelberg (2003)

Kaliszewski, I.: Quantitative Pareto Analysis by Cone Separation Technique. Kluwer Academic Publishers, Dodrecht (1994)

Klamroth, K., Miettinen, K.: Integrating approximation and interactive decision making in multicriteria optimization. Operations Research 56(1), 222–234 (2008)

Knowles, J., Corne, D., Deb, K. (eds.): Multiobjective Problem Solving from Nature. Springer, Heidelberg (2008)

Knowles, J.D., Watson, R.A., Corne, D.W.: Reducing local optima in single-objective problems by multi-objectivization. In: Zitzler, E., Deb, K., Thiele, L., Coello Coello, C.A., Corne, D.W. (eds.) EMO 2001. LNCS, vol. 1993, pp. 269–283. Springer, Heidelberg (2001)

Korhonen, P.: A visual reference direction approach to solving discrete multiple criteria problems. European Journal of Operational Research 34, 152–159 (1988)

Kostreva, M., Ogryczak, W., Wierzbicki, A.: Equitable aggregations and multiple criteria analysis. European Journal of Operational Research 158(2), 362–377 (2004)

Kostreva, M.M., Ogryczak, W.: Linear optimization with multiple equitable criteria. RAIRO Operations Research 33, 275–297 (1999)

Kouvelis, P., Yu, G.: Robust Discrete Optimization and Its Applications. Kluwer Academic Publishers, Dodrecht (1997)

Laumanns, M., Thiele, L., Deb, K., Zitzler, E.: Combining convergence and diversity in evolutionary multi-objective optimization. Evolutionary Computation 10(3), 263–282 (2002)

Luss, H.: On equitable resource allocation problems: A lexicographic minimax approach. Operations Research 47, 361–378 (1999)

Miettinen, K.: Nonlinear Multiobjective Optimization. Kluwer Academic Publishers, Boston (1999)

Miettinen, K., Mäkelä, M.M.: Interactive multiobjective optimization system WWW-NIMBUS on the Internet. Computers & Operations Research 27(7–8), 709–723 (2000)

Miettinen, K., Mäkelä, M.M.: Synchronous approach in interactive multiobjective optimization. European Journal of Operational Research 170, 909–922 (2006)

Miettinen, K., Mäkelä, M.M., Männikkö, T.: Optimal control of continuous casting by nondifferentiable multiobjective optimization. Computational Optimization and Applications 11(2), 177–194 (1998)

Miettinen, K., Lotov, A.V., Kamenev, G.K., Berezkin, V.E.: Integration of two multiobjective optimization methods for nonlinear problems. Optimization Methods and Software 18, 63–80 (2003)

Miettinen, K., Mäkelä, M.M., Maaranen, H.: Efficient hybrid methods for global continuous optimization based on simulated annealing. Computers & Operations Research 33(4), 1102–1116 (2006)

Neumann, F., Wegener, I.: Minimum spanning trees made easier via multi-objective optimization. In: Proceedings of the Genetic and Evolutionary Computation Conference (GECCO-2005), pp. 763–769. ACM Press, New York (2005)

Ogryczak, W.: On the distribution approach to location problems. Computers & Industrial Engineering 37, 595–612 (1999)

Ogryczak, W.: Multiple criteria optimization and decisions under risk. Control and Cybernetics 31, 975–1003 (2002)

Ogryczak, W., Ruszczyński, A.: From stochastic dominance to mean-risk models: Semideviations as risk measures. European Journal of Operational Research 116, 33–50 (1999)

Ogryczak, W., Wierzbicki, A., Milewski, M.: A multi-criteria approach to fair and efficient bandwidth allocation. Omega 36, 451–463 (2008)

Olson, D.: Decision Aids for Selection Problems. Springer, New York (1996)

Palaniappan, S., Zein-Sabatto, S., Sekmen, A.: Dynamic multiobjective optimization of war resource allocation using adaptive genetic algorithms. In: Proceedings of the IEEE Southeast Conference, pp. 160–165. Clemson University, Clemson, SC (2001)

Parmee, I.C., Cevtković, D., Watson, A.W., Bonham, C.R.: Multiobjective satisfaction within an interactive evolutionary design enviornment. Evolutionary Computation Journal 8(2), 197–222 (2000)

Perny, P., Spanjaard, O., Storme, L.-X.: A decision-theoretic approach to robust optimization in multivalued graphs. Annals of Operations Research 147, 317–341 (2006)

Pióro, M., Medhi, D.: Routing, Flow and Capacity Design in Communication and Computer Networks. Morgan Kaufmann, San Francisco (2004)

Romeijn, H.E., Ahuja, R.K., Dempsey, J.F., Kumar, A.: A new linear programming approach to radiation therapy treatment planning problems. Operations Research 54, 201–216 (2006)

Roy, B.: A missing link in OR-DA: Robustness analysis. Foundations of Computing and Decision Sciences 23, 141–160 (1998)

San Miguel, F., Ryan, M., Scott, A.: Are preferences stable? The case of health care. Journal of Economic Behavior and Organization 48, 1–14 (2002)

Shimoyama, K., Oyama, A., Fujii, K.: A new efficient and useful robust optimization approach – design for multi-objective six sigma. In: Proceedings of the IEEE Congress on Evolutionary Computation, vol. 1, pp. 950–957. IEEE Computer Society Press, Piscataway (2005)

Teich, J.: Pareto-front exploration with uncertain objectives. In: Zitzler, E., Deb, K., Thiele, L., Coello Coello, C.A., Corne, D.W. (eds.) EMO 2001. LNCS, vol. 1993, pp. 314–328. Springer, Heidelberg (2001)

Thiele, L., Miettinen, K., Korhonen, P.J., Molina, J.: A preference-based interactive evolutionary algorithm for multiobjective optimization. Working Papers W-412, Helsinki School of Economics, Helsinki (2007)

von Stackelberg, H.: Marktform und Gleichgewicht. Springer, Berlin (1934)

von Winterfeldt, D., Edwards, W.: Decision Analysis and Behavioral Research. Cambridge University Press, Cambridge (1986)

Wierzbicki, A.: On completeness and constructiveness of parametric characterizations to vector optimization problems. OR Spectrum 8, 73–87 (1986)

Yu, P.: Cone convexity, cone extreme points, and nondominated solutions in decision problems with multiple objectives. Journal of Optimization Theory and Applications 14, 319–377 (1974)

Index